De Gruyter Studies in Mathematics 49

Mikhail I. Ostrovskii

Metric Embeddings

Bilipschitz and Coarse Embeddings into Banach Spaces

De Gruyter

Mathematical Subject Classification 2010: Primary: 46B85; Secondary: 05C12, 30L05, 46B20, 54E35.

Author
Prof. Dr. Mikhail I. Ostrovskii
St. John's University
St. John's College of Liberal Arts and Sciences
Department of Mathematics and Computer Science
8000 Utopia Parkway
Queens NY 11439
United States
ostrovsm@stjohns.edu

ISBN 978-3-11-026340-4
e-ISBN 978-3-11-026401-2
Set-ISBN 978-3-11-916622-5
ISSN 0179-0986

Library of Congress Cataloging-in-Publication Data

A CIP catalog record for this book has been applied for at the Library of Congress.

Bibliographic information published by the Deutsche Nationalbibliothek

The Deutsche Nationalbibliothek lists this publication in the Deutsche Nationalbibliografie; detailed bibliographic data are available in the internet at http://dnb.dnb.de.

Typesetting: PTP-Berlin Protago-TEX-Production GmbH, www.ptp-berlin.de
Printing and binding: Hubert & Co. GmbH & Co. KG, Göttingen
∞ Printed on acid-free paper

Printed in Germany

www.degruyter.com

Preface

Embeddings of discrete metric spaces (such as graphs with graph distances, finitely generated groups with word distances) into Banach spaces (the most important are the Hilbert space and the space L_1) recently became an important tool in computer science and topology. The purpose of the book is to present some of the most important techniques and results, mostly on bilipschitz and coarse embeddings. The topics include: (1) Embeddability of locally finite metric spaces into Banach spaces is finitely determined; (2) Constructions of embeddings; (3) Distortion in terms of Poincaré inequalities; (4) Constructions of families of expanders and of families of graphs with unbounded girth and lower bounds on average degrees; (5) Banach spaces which do not admit coarse embeddings of expanders; (6) Structure of metric spaces which are not coarsely embeddable into a Hilbert space; (7) Applications of Markov chains to embeddability problems; (8) Metric characterizations of properties of Banach spaces; (9) Lipschitz free spaces.

A substantial part of the book is devoted to a detailed presentation of relevant results of Banach space theory and graph theory. The final chapter contains a list of open problems. An extensive bibliography is also included. Each chapter, except the open problems chapter, contains exercises and a Notes and Remarks section containing references, discussion of related results, and suggestions for further reading.

The book will help readers to enter into and to work in a very rapidly developing area having many important connections with different parts of mathematics and computer science. I also hope that the readers will become interested in the open problems listed in this book. I plan to maintain a web site containing information on progress on problems listed in Chapter 11 (at `http://facpub.stjohns.edu/ostrovsm/`). I would be very thankful for information on errors in this book and on progress on the open problems presented here, my e-mail: `ostrovsm@stjohns.edu`.

It should be mentioned that the area currently develops in numerous different directions and employs very diverse techniques. For this reason it is not possible to present even the most important results and techniques in a book of moderate size. I suggest readers consult [333], [334], and [345] for some of the directions which are underrepresented in this book, more references are given in the Notes and Remarks sections. The selection of topics for this book is strongly influenced by my interests and expertise.

I am very grateful to all of my colleagues who helped me with useful comments, helpful discussions, and constructive criticism. These include Florent Baudier, Florin Catrina, Genady Grabarnik, Hamed Hatami, William B. Johnson, the late Nigel J. Kalton, Jerome Kaminker, Gilles Lancien, Nathan Linial, Calvin Mittman, Assaf Naor,

Piotr W. Nowak, Mikhail M. Popov, Doron Puder, Matias Raja, Beata Randrianantoanina, David Rosenthal, Mark Sapir, Gideon Schechtman, and Romain Tessera.

Also I would like to thank my home institution, St. John's University, for supporting me during my work on this book by providing a research leave in Fall 2012. I was also partially supported by NSF Grant DMS-1201269.

Queens, February 2013 Mikhail I. Ostrovskii

Contents

Chapter 1

Introduction: examples of metrics, embeddings, and applications

1.1 Metric spaces: definitions and main examples

The notions of being "close" or "far" are used in many different contexts and are understood in many different ways. In this book we consider one of the most standard ways for mathematics to look at these notions, the notion of a metric space.

Definition 1.1. A *metric space* is a set X endowed with a function $d : X \times X \to \mathbb{R}_+$ (where $\mathbb{R}_+$ is the set of all nonnegative real numbers) satisfying the conditions:

(a) *Triangle inequality:* $\forall x, y, z \in X \quad d(x,z) \le d(x,y) + d(y,z)$.

(b) *Symmetry:* $\forall x, y \in X \quad d(x,y) = d(y,x)$.

(c) *Separation axiom:* $\forall x, y \in X \quad x \ne y \Rightarrow d(x,y) \ne 0$.

(d) $\forall x \in X \quad d(x,x) = 0$.

The function d is called a *metric* on X.

Example 1.2 (Normed spaces). Any normed and any Banach space $(X, \|\cdot\|)$ is a metric space with respect to the metric $d(x,y) = \|x - y\|$. Usually we shall consider real Banach spaces (that is, the case where the field of scalars is $\mathbb{R}$). In some results and proofs it will be natural to consider complex Banach spaces (the case where the field of scalars is $\mathbb{C}$).

Example 1.3 (Banach spaces ℓ_p, ℓ_p^n, $1 \le p \le \infty, n \in \mathbb{N}$). Their definitions (in all of these spaces addition and multiplication by scalars are defined componentwise):

$$\ell_p^n = \left\{\{x_i\}_{i=1}^n : x_i \in \mathbb{R},\ \|\{x_i\}_{i=1}^n\| = \left(\sum_{i=1}^n |x_i|^p\right)^{1/p}\right\}, \qquad 1 \le p < \infty.$$

$$\ell_\infty^n = \left\{\{x_i\}_{i=1}^n : x_i \in \mathbb{R},\ \|\{x_i\}_{i=1}^n\| = \max_{1 \le i \le n} |x_i|\right\}.$$

$$\ell_p = \left\{\{x_i\}_{i=1}^\infty : x_i \in \mathbb{R},\ \|\{x_i\}_{i=1}^\infty\| = \left(\sum_{i=1}^\infty |x_i|^p\right)^{1/p} < \infty\right\}, \quad 1 \le p < \infty.$$

$$\ell_\infty = \left\{\{x_i\}_{i=1}^\infty : x_i \in \mathbb{R},\ \|\{x_i\}_{i=1}^\infty\| = \sup_{i \in \mathbb{N}} |x_i| < \infty\right\}.$$

We also introduce the Banach space c_0 which is defined as a subset of ℓ_∞ consisting of those sequences $\{x_i\}_{i=1}^\infty$ for which $\lim_{i\to\infty} x_i = 0$, with the same norm as in ℓ_∞, which can now be written as

$$\|\{x_i\}_{i=1}^\infty\| = \max_{i\in\mathbb{N}} |x_i|.$$

Observe that for $p = 1$ the formula for the ℓ_p-norm becomes simpler:

$$\ell_1 = \left\{\{x_i\}_{i=1}^\infty : x_i \in \mathbb{R},\ \|\{x_i\}_{i=1}^\infty\| = \sum_{i=1}^\infty |x_i| < \infty\right\}.$$

Also, if p is even, we can omit the absolute value in the formula for the norm: $\|\{x_i\}_{i=1}^n\| = \left(\sum_{i=1}^n x_i^p\right)^{1/p}$. This makes it more convenient to work with.

Example 1.4 (Banach function spaces). $L_p(0,1)$, $1 \le p < \infty$ is the space of those measurable functions on $[0,1]$ for which the Lebesgue integral $\int_0^1 |f(x)|^p\,dx$ is finite. Addition and scalar multiplication are defined pointwise, the norm is given by $\|f\| = (\int_0^1 |f(x)|^p\,dx)^{1/p}$. Usually we omit $(0,1)$ from the notation of this space and denote it just L_p.

The space $L_p(\Omega, \Sigma, \mu)$, $1 \le p < \infty$, where (Ω, Σ, μ) is a measure space, is defined similarly. In the case where Ω has a natural topology and Σ is the algebra of all Borel sets of Ω, we write $L_p(\Omega, \mu)$.

Remark 1.5. Readers without knowledge of measure theory should not worry too much: in most cases an intuitive understanding of $L_p(0,1)$ as some kind of completion of continuous functions or step functions with respect to the norm described above (where the integral is understood as the Riemann integral) is sufficient.

Example 1.6 (Hilbert spaces). A very important role in functional analysis and metric geometry is played by Banach spaces whose norm is induced by an inner product. Such Banach spaces are called *Hilbert spaces*.

Definition 1.7. An *inner product* (sometimes also called *scalar product*) on a (complex or real) linear space X is a function $\langle\cdot,\cdot\rangle : X \times X \to \mathbb{K}$ (where $\mathbb{K}$ is $\mathbb{C}$ or $\mathbb{R}$, respectively) satisfying the following conditions:

(a) $\langle x, x\rangle > 0$ for all $x \in X$ except $x = 0$.

(b) $\langle 0, 0\rangle = 0$.

(c) $\forall \alpha_1, \alpha_2 \in \mathbb{K}\ \forall x_1, x_2, x \in X\ \langle \alpha_1 x_1 + \alpha_2 x_2, x\rangle = \alpha_1\langle x_1, x\rangle + \alpha_2\langle x_2, x\rangle$.

(d) $\forall x, y \in X\ \langle x, y\rangle = \overline{\langle y, x\rangle}$, where the bar denotes the complex conjugation (the bar can be omitted if $\mathbb{K} = \mathbb{R}$).

We say that a norm $\|\cdot\|$ on a linear space X is *induced by an inner product* if $\forall x \in X\ \|x\|^2 = \langle x, x\rangle$ for some inner product $\langle\cdot,\cdot\rangle$ on X.

Among the Banach spaces mentioned above the following spaces are Hilbert spaces: ℓ_2, ℓ_2^n and $L_2(\Omega, \Sigma, \mu)$. The corresponding inner products are:

$$\sum_{i=1}^{\infty} x_i \overline{y_i}, \quad \sum_{i=1}^{n} x_i \overline{y_i}, \quad \int_\Omega x(t)\overline{y(t)}d\mu(t).$$

Example 1.8 (Graphs with graph distances). Let $G = (V(G), E(G))$ be a graph, so $V(G)$ is a set of objects called *vertices* and $E(G)$ is some set of unordered pairs of vertices called *edges*. We denote an unordered pair consisting of vertices u and v by uv or by $\{u, v\}$ and say that u and v are *ends* or *end vertices* of uv. A *walk* in G is a finite sequence of the form $W = v_0, e_1, v_1, e_2, \dots, e_k, v_k$ whose terms are alternately vertices and edges such that, for $1 \le i \le k$, the edge e_i has ends v_{i-1} and v_i. We say that W *starts* at v_0 and ends at v_k, and that W is a v_0v_k*-walk*. The number k is called the *length* of the walk. A graph G is called *connected* if for each $u, v \in V(G)$ there is a uv-walk in G. If G is connected, we endow $V(G)$ with the metric $d_G(u, v) =$ the length of the shortest uv-walk in G. The metric d_G is called the *graph distance* or the *shortest path metric*. When we say "graph G with its graph distance (shortest path metric)" we mean the metric space $(V(G), d_G)$.

Example 1.9 (Groups with word metrics). Let G be a group and S be a finite subset of G. We say that G is *generated* by the set S, or that S is a *generating set* of G, if each $g \in G$ can be written as a finite product of elements from S (elements from S can be repeated in this product arbitrarily many times). A group is called *finitely generated* if it has a finite generating set.

Now let G be a finitely generated group and S be a generating set in G. We assume that S does not contain the identity and is symmetric, that is, contains g if and only if it contains g^{-1}. The *Cayley graph* of G corresponding to the generating set S is the graph whose vertices are elements of G, elements $g_1 \in G$ and $g_2 \in G$ are connected by an edge if and only if $g_1^{-1}g_2 \in S$. The graph distance of this graph is called the *word metric* because the distance between group elements g and h is the shortest representation of $g^{-1}h$ in terms of elements of S, such representations are called *words* in the *alphabet* S.

Definition 1.10. A *semimetric* is like a metric except that the separation axiom is not required.

Example 1.11 (Cut semimetrics (sometimes called *elementary cut metrics*)). Let S be a subset of a set A, $\bar{S}$ be the complement of S. The pair $(S, \bar{S})$ is called a *cut* in A and $S, \bar{S}$ are called *parts* of this cut. The *cut semimetric* on A corresponding to the cut

$(S, \bar{S})$ is defined by

$$d_S(u,v) = \begin{cases} 0 & \text{if } u \text{ and } v \text{ are in the same part} \\ 1 & \text{if } u \text{ and } v \text{ are in different parts} \end{cases}$$

Example 1.12 (Weighted graphs). A *weighted graph* is a graph $G = (V, E)$ in which each edge e has been assigned a nonnegative real number $w(e)$. This number is called the *weight* of e. The *weight of a walk* $W = v_0, e_1, v_1, e_2, \dots, e_k, v_k$ is defined as $w(W) = \sum_{j=1}^{k} w(e_j)$. We endow the vertex set V with the (semi)metric

$$d_{G,w}(u,v) = \inf\{w(W) : \ W \text{ is a } uv\text{-walk in } G\}.$$

This metric is called a *weighted graph distance* or *weighted shortest path semimetric* (*semi* is omitted if we know that we get a metric). Sometimes we use the term *unweighted graph* to emphasize that we consider a graph with no weights, or, what is the same, with all weights equal to 1.

Observe that for infinite graphs the weighed graph distance can fail the separation axiom even in the case where all weights are strictly positive.

Remark 1.13. Each finite semimetric space (X, d_X) can be obtained as a weighted graph with its weighted graph distance. In fact, consider the complete graph with the vertex set X. (Recall that a graph is called *complete* if each pair of vertices in it is joined by one edge and there are no *loops*, that is, edges joining a vertex to itself.) We introduce weights of edges by $w(xy) = d_X(x, y)$. It is clear that the weighted graph distance of the obtained weighted graph coincides with d_X.

Example 1.14 (Semimetrics in combinatorial optimization). Semimetrics arise naturally in construction of approximation algorithms in computer science. We are going to present one of the most important and well-known examples of this type.

The *sparsest cut problem* is the following problem of combinatorial optimization. We are given a connected graph $G = (V, E)$, with a positive weight (called a *capacity*) $c(e)$ associated to each edge $e \in E$, and a nonnegative number (called a *demand*) $D(u, v)$ associated to each (unordered) pair of vertices $u, v \in V$. By a *cut* of G we mean a partition of the vertex set V into two disjoint sets: S and its complement $\bar{S}$. The *sparsity* of the cut $(S, \bar{S})$ is defined as

$$\frac{\sum_{u\in S, v\in\bar{S}, uv\in E} c(uv)}{\sum_{u\in S, v\in\bar{S}} D(u,v)}, \tag{1.1}$$

that is, the sparsity is the ratio between the capacities and the demands which "cross" the cut. The *sparsest cut problem* is: find a cut of minimum sparsity among cuts with nonzero $\sum_{u\in S, v\in\bar{S}} D(u, v)$. This problem is known to be computationally hard (see the "Notes and Remarks" section), for this reason the following version of the sparsest cut problem is also of interest: to approximate the minimum sparsity.

One of the approaches to this approximate version of the sparsest cut problem starts with writing the quantity (1.1) in terms of a cut semimetric d_S:

$$\frac{\sum_{uv\in E} c(uv)d_S(u,v)}{\sum_{u,v\in V} D(u,v)d_S(u,v)}. \tag{1.2}$$

(The quantities in (1.1) and (1.2) are not equal, their quotient is equal to 2, but this does not change the problem.)

Then, as is easy to see, the minimum decreases if instead of the minimum over cut metrics d_S we consider the minimum over all nontrivial semimetrics d on V (by a nontrivial semimetric here we mean a semimetric for which $\sum_{u,v\in V} D(u,v)d(u,v) \neq 0$). The point is that the problem of minimization of

$$\frac{\sum_{uv\in E} c(uv)d(u,v)}{\sum_{u,v\in V} D(u,v)d(u,v)}. \tag{1.3}$$

over the set of all nontrivial semimetrics d on V belongs to the class of so-called *linear programming* (LP) problems for which reasonably fast algorithms are known (see Notes and Remarks section for references). In more detail, by homogeneity, we may restrict our attention to the metrics satisfying $\sum_{u,v\in V} D(u,v)d(u,v) = 1$. Then we can write the problem as: *Minimize the sum* $\sum_{uv\in E} c(uv)d(u,v)$ *over the set of all collections* $\{d(u,v)\}_{u,v\in V}$ *satisfying the conditions*

$$\begin{aligned} & \sum_{u,v\in V} D(u,v)d(u,v) = 1 \\ \forall u,v,w \in V \quad & d(u,w) \le d(u,v) + d(v,w) \\ \forall u,v \in V \quad & d(u,v) = d(v,u) \\ \forall u,v \in V \quad & d(u,v) \ge 0 \\ \forall u \in V \quad & d(u,u) = 0. \end{aligned} \tag{1.4}$$

We get a problem of minimization of a linear form subject to finitely many linear inequalities, such a problem is called a *linear programming problem.*

Of course, it is not immediately clear how to estimate the sparsest cut from above in terms of the minimum computed for this LP problem (and the corresponding semimetric $d_{\min}$). Later we shall see (Section 1.4 in this chapter) that this can be done in terms of the possible quality of embeddings of the semimetric space $(V, d_{\min})$ into the Banach space ℓ_1.

1.2 Types of embeddings: isometric, bilipschitz, coarse, and uniform

By an *embedding* of a set X into Y we mean any (not necessarily injective or surjective) map of X into Y. We are interested in the case where X and Y are metric (or at least semimetric) spaces and we are interested in embeddings which do not "distort too much" the metric structure. We start with embeddings which preserve distances.

1.2.1 Isometric embeddings

Definition 1.15. A map $f : X \to Y$ between two metric spaces is called an *isometric embedding* if it preserves distances, that is $d_Y(f(u), f(v)) = d_X(u, v)$ for all $u, v \in X$. If there exists an isometric embedding of X into Y we say that X is *isometric* to a subset (subspace) of Y. If an isometric embedding of X into Y is a bijection of X and Y, we say that X and Y are *isometric*.

Example 1.16. The space ℓ_p^n is isometric to a linear subspace of ℓ_p^m if $n < m$, and to a linear subspace of ℓ_p.

In fact, the map which extends a sequence $\{x_i\}_{i=1}^n \in \ell_p^n$ with the corresponding amount of zeros is an isometric embedding of ℓ_p^n onto a linear subspace. (In the described situations there are many other isometric embeddings.)

The theory of isometric embeddings is a very rich theory which was developed from several different perspectives (we provide some references in the Notes and Remarks section). In many cases, in particular, in many important applications of embeddings there is no hope for existence of isometric embeddings (see e.g. Example 1.32 below), for this reason many weaker types of embeddings have been introduced and studied. We are going to consider three of them: bilipschitz, coarse, and uniform. Our main interest in this book is the theory of these wider classes of embeddings.

We do not intend to present much of the theory of isometric embeddings. The purpose of this section is to present some simple but useful facts about isometric embeddings (much more general results are known and can be found in references provided in the Notes and Remarks section).

Proposition 1.17 (Fréchet embedding). *Each countable metric space embeds isometrically into ℓ_∞. Each metric space with n elements embeds isometrically into ℓ_∞^n.*

Proof. Let $X = \{u_i\}_{i=0}^\infty$ be a countable metric space. We introduce a map $f : X \to \ell_\infty$ by

$$f(v) = \{d(v, u_i) - d(u_i, u_0)\}_{i=1}^\infty.$$

Observe that

$$\|f(v) - f(w)\| = \sup_{i \in \mathbb{N}} |d(v, u_i) - d(w, u_i)|.$$

The triangle inequality implies

$$\sup_{i\in\mathbb{N}} |d(v,u_i) - d(w,u_i)| \le d(v,w).$$

On the other hand, if $v \ne w$, then at least one of v, w is among $\{u_i\}_{i=1}^{\infty}$. Suppose that $v \in \{u_i\}_{i=1}^{\infty}$. We get

$$\sup_{i\in\mathbb{N}} |d(v,u_i) - d(w,u_i)| \ge |d(v,v) - d(w,v)| = d(v,w).$$

This proves the first statement. The second statement can be proved similarly. We can make its proof simpler if we observe that for a *bounded metric space* $X = \{u_i\}$ (that is, for a space X for which $\sup_{u,v\in X} d(u,v)$ is finite) the definition of $f(v)$ can be simplified to $f(v) = \{d(v,u_i)\}_i$. In particular, if $X = \{u_i\}_{i=1}^{n}$, then $f(v) = \{d(v,u_i)\}_{i=1}^{n}$ defines an isometric embedding into ℓ_∞^n. □

Remark 1.18. One can isometrically embed any n-element metric space into ℓ_∞^{n-1} (see Exercise 1.56). For large n further lowering of the dimension is possible, see the Notes and Remarks section and Exercise 1.60.

Proposition 1.19. *ℓ_p $(1 \le p < \infty)$ is isometric to a subspace of $L_p(0,1)$.*

Proof. Consider a sequence $\{\sigma_i\}_{i=1}^{\infty}$ of disjoint measurable subsets of [0, 1] with non-zero Lebesgue measures. Let $m(\sigma_i)$ denote the Lebesgue measure of σ_i. Straightforward computation of the norm shows that the embedding $T : \ell_p \to L_p(0,1)$ defined by

$$(T(\{x_i\}_{i=1}^{\infty}))(t) = \begin{cases} \dfrac{x_i}{(m(\sigma_i))^{1/p}} & \text{if } t \in \sigma_i \\ 0 & \text{if } t \notin \cup_{i=1}^{\infty}\sigma_i \end{cases}$$

is an isometric embedding of ℓ_p into $L_p(0,1)$. □

We are going to use also the following result (which we do not prove, see the Notes and Remarks section for references).

Fact 1.20. Any separable subspace of $L_p(\Omega,\Sigma,\mu)$ $(1 \le p < \infty)$ is isometric to a subspace of $L_p(0,1)$.

This fact shows that the space $L_p(0,1)$ is the "largest" $L_p(\Omega,\Sigma,\mu)$ which we need to consider if we restrict our attention to separable spaces. We denote $L_p(0,1)$ simply by L_p.

Proposition 1.21. *If a finite metric space is isometric to a subset of ℓ_2, then it is isometric to a subset of ℓ_1.*

Proof. The result follows from Lemmas 1.22 and 1.24 stated below. If M is an n-element metric space isometric to a subset of ℓ_2, then its image is contained in an n-dimensional subspace of ℓ_2. As is well known, an n-dimensional subspace of a Hilbert space is isometric to ℓ_2^n. By Lemma 1.22, M is isometric to a subset of $L_1(S^{n-1},\mu)$, where S^{n-1} is the unit sphere in ℓ_2^n and μ is a Lebesgue measure on S^{n-1} (these notions are well-defined because ℓ_2^n can be identified with the n-dimensional Euclidean space $\mathbb{R}^n$). By Lemma 1.24, a finite subset of $L_1(S^{n-1},\mu)$ is isometric to a subset of ℓ_1. □

Lemma 1.22. *The space ℓ_2^n is isometric to a subspace of $L_1(S^{n-1},\mu)$.*

Proof. Since μ is rotationally invariant, the value of the integral

$$\int_{S^{n-1}} |\langle e,\theta\rangle|\, d\mu(\theta)$$

is the same for all $e \in S^{n-1}$. Let α be the common value of all such integrals.

Define $T : \ell_2^n \to L_1(S^{n-1},\mu)$ by

$$(Tx)(\theta) = \frac{\langle x,\theta\rangle}{\alpha}, \quad x \in \ell_2^n, \quad \theta \in S^{n-1}.$$

Linearity of T is obvious. We check that T is distance-preserving by writing

$$\|Tx\|_{L_1} = \int_{S^{n-1}} \left|\frac{\langle x,\theta\rangle}{\alpha}\right| d\mu(\theta) = \int_{S^{n-1}} \left|\frac{\left\langle \frac{x}{\|x\|_{\ell_2^n}},\theta\right\rangle}{\alpha}\right| \|x\|_{\ell_2^n} d\mu(\theta) = \|x\|_{\ell_2^n}. \quad \square$$

Observation 1.23. Similar argument shows that ℓ_2^n embeds isometrically into the space $L_p(S^{n-1},\mu)$ $(1 \le p < \infty)$.

Lemma 1.24. *Each finite subset of $L_1(\Omega,\Sigma,\mu)$ is isometric to a subset of ℓ_1.*

Proof. Let $M = \{m_1,\dots,m_n\}$ be a finite subset of $L_1(\Omega,\Sigma,\mu)$. Consider all possible permutations σ_α $(\alpha = 1,\dots,n!)$ of the set $\{1,\dots,n\}$. We are going to construct an isometric embedding of M into $\ell_1^{n!}$.

Define the measurable sets (not necessarily disjoint, and possibly empty)

$$\Omega_\alpha = \{x \in \Omega : m_{\sigma_\alpha(1)}(x) \ge m_{\sigma_\alpha(2)}(x) \ge \cdots \ge m_{\sigma_\alpha(n)}(x)\}.$$

Note that $\cup_{\alpha=1}^{n!}\Omega_\alpha = \Omega$. Let

$$\Lambda_1 = \Omega_1,\ \Lambda_2 = \Omega_2 \setminus \Omega_1,\ \Lambda_\alpha = \Omega_\alpha \setminus \left(\cup_{i=1}^{\alpha-1}\Omega_i\right), 3 \le \alpha \le n!.$$

We have $\cup_{\alpha=1}^{n!}\Lambda_\alpha = \Omega$, and $\Lambda_\alpha \cap \Lambda_\beta = \emptyset$ for $\alpha \ne \beta$. Let $T : M \to \ell_1^{n!}$ be given by

$$T(m_i) = \left\{\int_{\Lambda_\alpha} m_i\, d\mu\right\}_{\alpha=1}^{n!}.$$

The following computation shows that T is an isometry:

$$\begin{aligned}\|m_i - m_j\|_{L_1(\Omega,\Sigma,\mu)} &= \int_\Omega |m_i - m_j|\ d\mu = \sum_{\alpha=1}^{n!} \int_{\Lambda_\alpha} |m_i - m_j|\ d\mu \\ &= \sum_{\alpha=1}^{n!} \left| \int_{\Lambda_\alpha} (m_i - m_j)\ d\mu \right| = \|Tm_i - Tm_j\|_{\ell_1^{n!}}.\end{aligned}$$

The first equality in the second line is justified by the fact that $m_i - m_j$ does not change sign on Λ_α. □

Proposition 1.25. *There exist* 4*-element metric spaces which are isometric to subsets in* ℓ_1*, but cannot be embedded isometrically into* ℓ_2*.*

Proof. The *complete bipartite graph* $K_{m,n}$ is a graph with $m+n$ vertices in which the vertex set consists of two classes: one class containing m vertices and the other class containing n vertices. Two vertices in $K_{m,n}$ are joined by an edge if and only if they belong to different classes.

We consider the graph $K_{1,3}$ with its graph distance. Let w be the vertex in the one-element class, and v_1, v_2, v_3 be vertices in the 3-element class.

Embedding into ℓ_1. We denote by $\{e_i\}_{i=1}^\infty$ the *unit vectors* in ℓ_1, that is, e_i is the vector whose i-th coordinate is 1 and all other coordinates are 0. (We are going to use the same definition of *unit vectors* in the spaces ℓ_p and ℓ_p^n for all values of $p \in [1,\infty]$, as well as for c_0.) With this notation the map $v_i \mapsto e_i$; $w \mapsto 0$ is an isometric embedding of $K_{1,3}$ into ℓ_1.

On the other hand, the space ℓ_2 has the following property: if $a, b, c \in \ell_2$ are such that $\|a - b\| + \|b - c\| = \|a - c\|$, then a, b, c lie on the same line and b is between a and c. Now, assume that there is an isometric embedding of $K_{1,3}$ into ℓ_2. By the previous observation the image of w should be the midpoint of three line segments: the line segment joining the image of v_i and v_j for $i \neq j$, $i, j \in \{1,2,3\}$. This is clearly impossible. □

1.2.2 Bilipschitz embeddings

Definition 1.26. Let $C < \infty$. A map $f : (X, d_X) \to (Y, d_Y)$ between two metric spaces is called *C-Lipschitz* if

$$\forall u, v \in X \quad d_Y(f(u), f(v)) \le C d_X(u, v).$$

A map f is called *Lipschitz* if it is C-Lipschitz for some $C < \infty$. For a Lipschitz map f we define its *Lipschitz constant* by

$$\operatorname{Lip}(f) := \sup_{d_X(u,v)\neq 0} \frac{d_Y(f(u), f(v))}{d_X(u, v)}.$$

A map $f : X \to Y$ is called a *C-bilipschitz embedding* if there exists $r > 0$ such that

$$\forall u, v \in X \quad r d_X(u, v) \leq d_Y(f(u), f(v)) \leq r C d_X(u, v). \tag{1.5}$$

A *bilipschitz embedding* is an embedding which is C-bilipschitz for some $C < \infty$. The smallest constant C for which there exist $r > 0$ such that (1.5) is satisfied is called the *distortion* of f. (It is easy to see that such smallest constant exists.)

Remark 1.27. For metric spaces *bilipschitz embeddings* can be defined as injective maps f for which both f and f^{-1}, defined as a function on $f(X)$, are Lipschitz maps. In such a case the distortion of f is $\mathrm{Lip} f \cdot \mathrm{Lip}(f^{-1})$.

It is clear that each finite metric space X admits a bilipschitz embedding into each at least 1-dimensional normed space and into each metric space Y which has more elements than X. In the study of families of finite metric spaces the following definition is important.

Definition 1.28. A family $\{f_\alpha\}$ of embeddings is called *uniformly bilipschitz* if they have uniformly bounded distortions.

Remark 1.29. If we consider linear mappings of normed spaces, one into another, a different terminology is used:

Definition 1.30. A linear bilipschitz embedding $T : X \to Y$ of one normed space into another is called an *isomorphic embedding*. This definition can be restated as: T is an isomorphic embedding if T is injective, $\|T\| < \infty$ and $\|T^{-1}\| < \infty$, where T^{-1} is considered as a linear map $T^{-1} : T(X) \to X$.

The notion of a bilipschitz embedding is important because the notion of an isometric embedding is too restrictive for many applications. We illustrate this statement with the following three examples.

Definition 1.31 (Some graph theory definitions). A *complete graph* with n vertices in which any two distinct vertices are joined by exactly one edge is denoted K_n. A *path* with n vertices is a graph whose vertices form a sequence $\{v_i\}_{i=1}^n$ and edges are determined by the following: v_k, $k = 2, \ldots, n-1$ is joined by exactly one edge with v_{k-1} and v_{k+1}. The vertex v_1 is joined with v_2 only and the vertex v_n is joined with v_{n-1} only. The path with n vertices is denoted P_n. It is somewhat confusing, but it is customary to use the term *path* also for a walk in which no vertex is repeated. (See page 3 for definition of a walk.) If we add to P_n an edge joining v_1 and v_n, we get a graph called a *cycle* of length n and denoted by C_n. A graph is called *simple* if any two vertices in it are joined by at most one edge and there are no loops. The *degree* of a vertex is the number of edges incident to it. Vertices joined by an edge are called *neighbors*.

Example 1.32. A finite simple connected graph G admits an isometric embedding into ℓ_2 if and only if it is either a complete graph K_n or a path P_n for some n.

Proof. It is easy to find isometric embeddings of K_n and P_n into ℓ_2. Let $\{e_k\}_{k=1}^{\infty}$ be the unit vector basis in ℓ_2. For K_n we map $v_k \mapsto \frac{e_k}{\sqrt{2}}$. For P_n we map $v_k \mapsto ke_1$. It is easy to see that both maps are isometric embeddings.

To prove the "only if" part of the statement we assume that G is a finite simple connected graph, which is not a path, but is such that $(V(G), d_G)$ is isometric to a subset of ℓ_2, and denote the isometric embedding by f. Our goal is to show that these conditions imply that G is a complete graph.

The fact that G is not a path immediately implies that G is either a cycle or has a vertex of degree at least 3. In the case where G is a cycle we observe that the cycle C_3 is simultaneously a complete graph K_3, and we are done in this case. As for longer cycles, we prove that they do not admit isometric embeddings into ℓ_2 in the following way. Since vertices v_{k-1}, v_k, v_{k+1} in a cycle with at least 4 vertices satisfy $d_G(v_{i-1}, v_{i+1}) = d_G(v_{i-1}, v_i) + d_G(v_i, v_{i+1})$, we get that $f(v_{k-1})$, $f(v_k)$, and $f(v_{k+1})$ should be on the same line, with $f(v_k)$ being a midpoint of the line segment joining $f(v_{k-1})$ and $f(v_{k+1})$. Since this observation is applicable also to v_n, v_1, and v_2, we get a contradiction.

Now let $v \in V(G)$ be a vertex of degree ≥ 3, and let u_1, u_2, u_3 be its neighbors. We show that u_i are pairwise adjacent. If two pairs of them (say u_1, u_2 and u_2, u_3) are not adjacent, we get a contradiction because $f(v)$ should be simultaneously a midpoint of the line segment joining $f(u_1)$ and $f(u_2)$ and a midpoint of the line segment joining $f(u_2)$ and $f(u_3)$.

If only one edge, say u_1u_3, is missing then both $f(u_2)$ and $f(v)$ should be midpoints of the line segment joining $f(u_1)$ and $f(u_3)$.

Therefore v and all of its neighbors should form a complete subgraph in G. Since the same should hold for each of the neighbors of v, we get that G should be a complete graph. □

Example 1.33. There are no isometric embeddings of the unit circle S in the Euclidean plane, with the metric inherited from the plane, into the Banach space c_0. On the other hand, for each $1 < C < \infty$ there is $n \in \mathbb{N}$ such that S admits a C-bilipschitz embedding into ℓ_∞^n.

Proof. Assume that there is an isometric embedding $f : S \to c_0$. Since the norm in c_0 is $\max |x_i|$, the definition of an isometric embedding implies that the set $S \times S$ is covered by the union $\bigcup_{i=1}^{\infty} T_i$, where T_i is the set of pairs which are 'taken care of' by the i-th coordinate in the following sense. Denote by $e_i^* : c_0 \to \mathbb{R}$ the functional whose value is the i-th coordinate (or i-th term of the sequence). We say that a pair $(u, v) \in S \times S$ is *taken care of* by e_i^* if $|e_i^*(f(u)) - e_i^*(f(v))| = d_S(u, v)$. By continuity, each set T_i is closed in the standard product topology on $S \times S$.

Out next purpose is to show that T_i does not contain interior points. Assume the contrary. Then there is a pair $(u, v) \in S$ such that for some neighborhoods N_u and N_v and any $x \in N_u$, $y \in N_v$ we have $d_S(x, y) = |e_i^*(f(x)) - e_i^*(f(y))|$. This assumption can be translated into geometric language in the following way: we can find a map P of N_u and N_v onto the line L passing through u and v, such that $P(u) = u$, $P(v) = v$ and $d_L(P(x), P(y)) = d_S(x, y)$ for all $x \in N_u$ and $y \in N_v$. Here d_L is the standard Euclidean distance on the line L. Consider $x_0 \in N_u$ and $y_0 \in N_v$ on the shorter circular arc of S joining u and v (any arc if the arcs are of the same length) such that $d_S(u, x_0) = d_S(v, y_0)$. Since $d_L(P(x_0), v) = d_S(x_0, v)$, the point $P(x_0)$ is closer to u than the image $P_L(x_0)$ of the orthogonal projection onto L. For the same reason $P(y_0)$ is closer to v than $P_L(y_0)$. But then $d_L(P(x_0), P(y_0)) > d_L(P_L(x_0), P_L(y_0)) = d_S(x_0, y_0)$ (the last equality follows from the assumption $d_S(u, x_0) = d_S(v, y_0)$), and we get a contradiction. Hence T_i does not contain interior points.

Therefore our assumption that f is an isometric embedding contradicts the *Baire category theorem* (its statement: *A complete metric space cannot be covered by a sequence of closed subsets which do not have interior points.*) Hence S does not admit isometric embeddings into c_0.

As for C-bilipschitz embeddings into ℓ_∞^n, we can find them not only for S, but for the whole space ℓ_2^2, (the construction below is a standard Banach-space-theoretical construction which can be easily extended to the case where ℓ_2^2 is replaced with an arbitrary finite-dimensional normed space). We find a finite ε-net $\{v_i\}_{i=1}^n$ on S (if needed, see the definition of an ε-net on page 34). Let $\{v_i^*\}_{i=1}^n$ be linear functionals on ℓ_2^2 satisfying $\|v_i^*\| = v_i^*(v_i) = 1$, $i = 1, \ldots, n$. We introduce an embedding $f : \ell_2^2 \to \ell_\infty^n$ by $f(x) = \{v_i^*(x)\}_{i=1}^n$. The condition $\|v_i^*\| = 1$ implies that f is 1-Lipschitz.

To estimate the Lipschitz constant of the inverse map consider any $x, y \in S$. There exists i such that $\left\| \frac{x-y}{\|x-y\|} - v_i \right\| \le \varepsilon$. Therefore $|v_i^*(x - y)| \ge (1 - \varepsilon)\|x - y\|$ and $\|f(x) - f(y)\| \ge (1 - \varepsilon)\|x - y\|$. □

Example 1.34. There is no isometric embedding of a metric space consisting of n equidistant points into ℓ_2^m with $m < n - 1$.

Proof. We may restrict our attention to the case where distances are equal to 1 and to maps which map one of the equidistant points to $0 \in \ell_2^m$. With these assumptions it suffices to prove that any vectors $\{x_i\}_{i=1}^k$ in ℓ_2 satisfying $\|x_i\| = 1$ and $\|x_i - x_j\| = 1$ for $i \ne j$ are linearly independent.

This can be done by observing that under these assumptions we have $\langle x_i, x_j \rangle = \frac{1}{2}$ if $i \ne j$, and therefore the matrix $(\langle x_i, x_j \rangle)_{i,j=1}^k$ (this matrix is called the *Gram matrix* of vectors $x_1, \dots, x_k$) is of the form

$$\begin{pmatrix} 1 & \frac{1}{2} & \frac{1}{2} & \cdots & \frac{1}{2} \\ \frac{1}{2} & 1 & \frac{1}{2} & \cdots & \frac{1}{2} \\ \frac{1}{2} & \frac{1}{2} & 1 & \cdots & \frac{1}{2} \\ \vdots & \vdots & \vdots & \ddots & \vdots \\ \frac{1}{2} & \frac{1}{2} & \frac{1}{2} & \cdots & 1 \end{pmatrix}.$$

It is easy to verify that this matrix is invertible. Therefore any equality of the form

$$\sum_{i=1}^{k} a_i x_i = 0$$

implies $a_1 = a_2 = \cdots = a_k = 0$. (Instead of proving the last statement one can also apply the well-known fact that a Gram matrix of a set of vectors is invertible if and only if the vectors are linearly independent.) □

On the other hand, our next result (Theorem 1.35) shows that we can find bilipschitz embeddings of the metric space of Example 1.34 (observe that this space can be identified with the complete graph K_n with its graph distance) into ℓ_2^k with k much smaller than n and distortion very close to 1. The value of Theorem 1.35 goes far beyond proving this. The Johnson–Lindenstrauss lemma is one of the most frequently used in applications results of the theory of embeddings.

Theorem 1.35 (Johnson–Lindenstrauss lemma). *For each $\varepsilon \in (0,1)$ there is a constant $K = K(\varepsilon)$, $0 < K < \infty$, such that for each $n \in \mathbb{N}$ ($n \ge 2$), each n-element subset $V = \{v_1, \dots, v_n\} \subset \ell_2^n$, and each $k \ge K \ln n$ there is a mapping $f : \ell_2^n \to \ell_2^k$ satisfying*

$$(1-\varepsilon)\|v_i - v_j\| \le \|f(v_i) - f(v_j)\| \le (1+\varepsilon)\|v_i - v_j\| \tag{1.6}$$

for all $i, j = 1, 2, \dots, n$.

Proof. We are going to proceed according to the following scheme: For given n, ε, and an appropriate k we find a probability space $(\Omega, \Sigma, \mathbb{P})$ (see Section 1.3 for probability theory terminology and notation which we are going to use) and a map $\omega \mapsto T_\omega$ from Ω to the set of all linear operators $\ell_2^n \to \ell_2^k$ satisfying for each $x \in \ell_2^n$ the following condition

$$\mathbb{P}\{\omega : (1-\varepsilon)\|x\| \le \|T_\omega(x)\| \le (1+\varepsilon)\|x\|\} \ge 1 - \frac{1}{n^2}. \tag{1.7}$$

If we find such a map, we can finish the proof in the following way. There are $\binom{n}{2}$ different pairs of distinct points in V. By (1.7), the probability that a random linear map does not satisfy the condition (1.6) for a given pair (v_i, v_j) is at most $\frac{1}{n^2}$. Therefore the probability that a random linear map does not satisfy condition (1.6) for at least one of the pairs (v_i, v_j) does not exceed $\binom{n}{2} \cdot \frac{1}{n^2} < \frac{1}{2}$. Therefore there exist linear maps satisfying the condition (1.6) for all pairs (v_i, v_j).

We construct T_ω using real-valued random variables of the following type. Let X be a symmetric random variable satisfying $\mathrm{Var}(X) = 1$ and

$$\mathbb{E}e^{tX} \leq e^{ct^2} \tag{1.8}$$

for some $0 < c < \infty$ and all $t \geq 0$. Recall that $\mathrm{Var}(X) = \mathbb{E}(X - \mathbb{E}X)^2$. In our case $\mathbb{E}X = 0$ since X is assumed to be symmetric, so the condition $\mathrm{Var}(X) = 1$ can be written as $\mathbb{E}(X^2) = 1$.

As an example of such a variable one can consider a Bernoulli random variable which has values ± 1 with the probability $\frac{1}{2}$ each. In fact, in this case $\mathrm{Var}(X) = 1$ and $\mathbb{E}e^{tX} = \frac{1}{2}(e^t + e^{-t})$, and an easy analysis with power series implies that $\frac{1}{2}(e^t + e^{-t}) \leq e^{t^2}$.

Let $\{X_{ij}\}_{j=1,i=1}^{n\ \ k}$ be independent random variables having the same distribution as X. We introduce T_ω as the operator $\ell_2^n \mapsto \ell_2^k$ whose matrix $A(\omega)$ in the unit vector bases of ℓ_2^n and ℓ_2^k has entries $a_{ij}(\omega) = \frac{1}{\sqrt{k}} X_{ij}(\omega)$.

To get (1.7) we need to show that for fixed $x \in \ell_2^n$, $\|x\| = 1$, and a suitably chosen $k \in \mathbb{N}$ we have

$$\begin{aligned} \mathbb{P}\{\omega : \quad \|T_\omega(x)\| \geq 1 + \varepsilon\} &\leq \frac{1}{2n^2}, \\ \mathbb{P}\{\omega : \quad \|T_\omega(x)\| \leq 1 - \varepsilon\} &\leq \frac{1}{2n^2}. \end{aligned} \tag{1.9}$$

The vector $T_\omega(x)$ has components $\frac{1}{\sqrt{k}} \sum_{j=1}^n X_{ij} x_j$. Since X_{ij} are independent random variables with the same distribution, all of the components have the same distribution. To simplify the notation we introduce $\{X_j\}_{j=1}^n$ as independent random variables with the same distribution as X. All of the components of $T_\omega(x)$ have the same distribution as $\frac{1}{\sqrt{k}} \sum_{j=1}^n x_j X_j$.

Let $x \in \ell_2^n$ be such that $\|x\| = 1$. Our next purpose is to compute $\mathbb{E}_\omega \|T_\omega(x)\|^2$. We have

$$\mathbb{E}\|T_\omega(x)\|^2 = \mathbb{E}\Big(\sum_{i=1}^k \Big(\frac{1}{\sqrt{k}} \sum_{j=1}^n X_{ij} x_j\Big)^2\Big).$$

We have

$$\mathbb{E}\Big(\frac{1}{\sqrt{k}} \sum_{j=1}^n X_{ij} x_j\Big)^2 = \frac{1}{k} \mathbb{E}\Big(\sum_{j=1}^n \sum_{m=1}^n X_{ij} X_{im} x_j x_m\Big).$$

Symmetry and independence of $\{X_{ij}\}$ imply that $\mathbb{E}(X_{ij}X_{im}) = 0$ if $j \neq m$. The condition $\mathbb{E}(X^2) = 1$ implies $\mathbb{E}(X_{ij}X_{ij}) = 1$. Therefore

$$\mathbb{E}\Big(\frac{1}{\sqrt{k}}\sum_{j=1}^{n} X_{ij}x_j\Big)^2 = \frac{1}{k}\sum_{j=1}^{n} x_j^2 = \frac{1}{k}$$

and $\mathbb{E}\|T_\omega(x)\|^2 = 1$. The purpose of the rest of the proof is to show that the random variable $\|T_\omega(x)\|^2$ is concentrated near its expectation.

Let $U_i = \sum_{j=1}^{n} X_{ij}x_j$. We need to estimate from above $\mathbb{P}_1 = \mathbb{P}\{\omega : \|T_\omega(x)\| \geq 1+\varepsilon\}$ and $\mathbb{P}_2 = \mathbb{P}\{\omega : \|T_\omega(x)\| \leq 1-\varepsilon\}$. We have

$$\begin{aligned}\mathbb{P}_1 &= \mathbb{P}\Big\{\omega : \frac{1}{k}\sum_{i=1}^{k} U_i^2(\omega) \geq (1+\varepsilon)^2\Big\}\\ &\leq \mathbb{P}\Big\{\omega : \sum_{i=1}^{k} U_i^2(\omega) \geq k(1+\varepsilon)\Big\}.\end{aligned}$$

We consider $s \in \big[0, \frac{1}{8c}\big]$, where c is the constant in (1.8). We have

$$\mathbb{E}e^{s\sum_{i=1}^{k} U_i^2} \geq \mathbb{P}_1 \cdot e^{sk(1+\varepsilon)}.$$

Therefore

$$\mathbb{P}_1 \leq e^{-sk(1+\varepsilon)}\mathbb{E}e^{s\sum_{i=1}^{k} U_i^2} \overset{\text{(by independence)}}{=} e^{-sk(1+\varepsilon)}\prod_{i=1}^{k}\mathbb{E}e^{sU_i^2}. \tag{1.10}$$

We introduce a random variable $U = \sum_{j=1}^{n} x_jX_j$. It is clear that each of the random variables U_i has the same distribution as U. Our next step is to show that there is a constant $0 < D < \infty$ depending on c only such that for $s \in \big[-\frac{1}{8c}, \frac{1}{8c}\big]$ we have

$$\mathbb{E}e^{sU^2} \leq e^{s+Ds^2}. \tag{1.11}$$

As a preliminary step we show that for a (standard) normal random variable $g(\beta)$ (also known as a Gaussian random variable) and a real number u we have $\mathbb{E}_g e^{ug} = e^{u^2/2}$. (Here we did not specify the probability space, writing $\mathbb{E}_g$ we mean the expectation with respect to the probability space on which g is a normal random variable.) In fact,

$$\begin{aligned}\mathbb{E}_g e^{ug} &= \frac{1}{\sqrt{2\pi}}\int_{-\infty}^{\infty} e^{ut}e^{-\frac{t^2}{2}}\,dt = \frac{1}{\sqrt{2\pi}}\int_{-\infty}^{\infty} e^{-\left(\frac{t^2}{2}-ut+\frac{u^2}{2}\right)+\frac{u^2}{2}}\,dt\\ &= \frac{1}{\sqrt{2\pi}}e^{\frac{u^2}{2}}\int_{-\infty}^{\infty} e^{-\frac{(t-u)^2}{2}}\,dt = e^{\frac{u^2}{2}}.\end{aligned}$$

Therefore the random variable U (which we assume to be defined on a probability space different from the one on which g is defined) satisfies for each real number $t \geq 0$ (below $\mathbb{E}$ denotes the expectation with respect to the probability measure on which U is defined and $\mathbb{E}_g$ denotes the expectation with respect to the probability measure on which g is defined)

$$\mathbb{E}e^{tU^2} = \mathbb{E}\mathbb{E}_g e^{\sqrt{2t}\,gU} = \mathbb{E}_g\mathbb{E}e^{\sqrt{2t}g\sum_{j=1}^n x_j X_j} \overset{\text{(by independence)}}{=} \mathbb{E}_g \prod_{j=1}^n \mathbb{E}e^{\sqrt{2t}\,g x_j X_j}$$

$$\overset{(1.8)}{\leq} \mathbb{E}_g \prod_{j=1}^n e^{c\cdot 2tg^2x_j^2} \overset{(\text{since } \|x\|=1)}{=} \mathbb{E}_g e^{2tcg^2}.$$

Now we observe that $\mathbb{E}_g e^{2tcg^2} \leq \sqrt{2}$ if $t \leq \frac{1}{8c}$. In fact, in such a case

$$\mathbb{E}_g e^{2tcg^2} \leq \mathbb{E}e^{g^2/4} = \frac{1}{\sqrt{2\pi}}\int_{-\infty}^{\infty} e^{t^2/4}e^{-t^2/2}dt$$

$$= \frac{1}{\sqrt{2\pi}}\int_{-\infty}^{\infty} e^{-t^2/4}dt \overset{(t=\sqrt{2}s)}{=} \frac{1}{\sqrt{2\pi}}\int_{-\infty}^{\infty} e^{-2s^2/4}\sqrt{2}\,ds = \sqrt{2}.$$

Now we continue our proof of (1.11). Since $\mathbb{E}(U^2) = 1$, using the power series representation of e^x we get

$$\mathbb{E}e^{sU^2} = 1 + s + \mathbb{E}\sum_{m\geq 2}\frac{s^mU^{2m}}{m!} = 1 + s + \mathbb{E}\sum_{m\geq 2}(8cs)^m\frac{\left(\frac{1}{8c}\right)^m U^{2m}}{m!}$$

$$= 1 + s + (8cs)^2\mathbb{E}\sum_{m\geq 2}(8cs)^{m-2}\frac{\left(\frac{1}{8c}\right)^m U^{2m}}{m!} \leq 1 + s + Ds^2$$

where $D = (8c)^2\sqrt{2}$. To get the last inequality we use $|8cs| \leq 1$ and

$$\mathbb{E}\sum_{m\geq 2}\frac{\left(\frac{1}{8c}\right)^m U^{2m}}{m!} \leq \mathbb{E}e^{\frac{1}{8c}U^2} \leq \sqrt{2},$$

the latter inequality has been proved above.

We use the well-known inequality $1 + x \leq e^x$ (which follows immediately from the convexity of e^x) and get

$$\mathbb{E}e^{sU^2} \leq e^{s+Ds^2} \text{ for } s \in \left[-\frac{1}{8c}, \frac{1}{8c}\right].$$

Now we can continue the estimate started in (1.10). We have

$$\mathbb{P}_1 \leq e^{-sk(1+\varepsilon)}e^{k(s+Ds^2)} = e^{-k(s\varepsilon - Ds^2)}.$$

We maximize $s\varepsilon - Ds^2$ over s. The maximum is attained when $s = \frac{\varepsilon}{2D}$, and is equal to $\frac{\varepsilon^2}{4D}$. We recall that we consider $0 \leq s \leq \frac{1}{8c}$ only, so the choice $s = \frac{\varepsilon}{2D}$ is suitable only if $\frac{\varepsilon}{2D} \leq \frac{1}{8c}$, but one can see that this is not a real obstacle because we may assume that $\varepsilon < 1$ and $2D \geq 8c$. Thus we have

$$\mathbb{P}_1 \leq e^{-k\cdot\varepsilon^2/4D}.$$

We need to find the condition under which the right-hand side is $\leq \frac{1}{2n^2}$. This happens if $-k \cdot \varepsilon^2/4D \leq -2\ln n - \ln 2$ or

$$k \geq \frac{4D(2\ln n + \ln 2)}{\varepsilon^2}.$$

Obviously the right-hand side of the last inequality is $\leq K(\varepsilon)\ln n$ for a suitably chosen $K(\varepsilon)$ and $n \geq 2$.

To estimate

$$\mathbb{P}_2 = \mathbb{P}\{\omega : \|T_\omega(x)\| \leq 1-\varepsilon\} \leq \mathbb{P}\left\{\omega : \sum_{i=1}^{k} U_i^2(\omega) \leq k(1-\varepsilon)\right\}$$

we consider $s \in [-\frac{1}{8c}, 0]$ and use a similar argument. We have

$$\mathbb{E}e^{s\sum_{i=1}^k U_i^2} \geq \mathbb{P}_2 e^{sk(1-\varepsilon)}.$$

Using the same computation as above we get

$$\mathbb{P}_2 \leq e^{-sk(1-\varepsilon)}e^{k(s+Ds^2)} = e^{-k(-s\varepsilon - Ds^2)}.$$

We maximize $-s\varepsilon - Ds^2$ over s. The maximum is attained when $s = -\frac{\varepsilon}{2D}$ and is equal to $\frac{\varepsilon^2}{4D}$. The rest of the argument is the same as for $\mathbb{P}_1$. □

1.2.3 Coarse and uniform embeddings

Definition 1.36. Let $f : (X, d_X) \to (Y, d_Y)$ be a map between two (semi)metric spaces. We introduce two real-valued functions of a real variable related to such a map:

$$\omega_f(t) = \sup\{d_Y(f(u), f(v)) : u, v \in X,\ d_X(u,v) \leq t\}, \quad t \geq 0,$$

and

$$\varphi_f(t) = \inf\{d_Y(f(u), f(v)) : u, v \in X,\ d_X(u,v) \geq t\}, \quad t \geq 0,$$

so that

$$\forall u, v \in X \quad \varphi_f(d_X(u,v)) \leq d_Y(f(u), f(v)) \leq \omega_f(d_X(u,v)). \tag{1.12}$$

We say that f is a *coarse embedding* and X *coarsely embeds* into Y if $\omega_f(t) < \infty$ for all t and $\lim_{t\to\infty}\varphi_f(t) = \infty$. We say that f is a *uniform embedding* and X *uniformly embeds* into Y if $\varphi_f(t) > 0$ for all $t > 0$ and $\lim_{t\to 0}\omega_f(t) = 0$.

Remark 1.37. To make $\varphi_f(t)$ well-defined in the case where the metric space X is bounded, we agree that the infimum over an empty set is ∞. It should be mentioned that the definitions of $\varphi_f(t)$ and of a coarse embedding are useful only for unbounded metric spaces.

The definitions above are closely related to the following classical definitions:

Definition 1.38. The function $\omega_f(t)$ is called the *modulus of continuity* of f. The map f is called *uniformly continuous* if $\lim_{t\to 0}\omega_f(t) = 0$.

Observe that for metric spaces the definition of a uniform embedding can be restated as: An injective map $f : X \to Y$ between two metric spaces is called a *uniform embedding* if it is uniformly continuous and its inverse f^{-1}, defined as a map from $f(X)$ to X, is also uniformly continuous. This definition makes sense and is used also for bounded metric spaces.

Observation 1.39. If X is a Banach space and Y is any metric space, then for any map $f : X \to Y$, the modulus ω_f is a subadditive function in the sense that $\omega_f(s+t) \le \omega_f(s)+\omega_f(t)$. If s and t are integers, the same inequality holds if X is an unweighted graph with its graph distance.

Proof. In fact, if X is a Banach space, then for any pair $u, v \in X$ satisfying $\|u-v\| \le s+t$ there is a vector w satisfying $\|u-w\| \le s$ and $\|w-v\| \le t$. By the triangle inequality we have $d_Y(f(u), f(v)) \le d_Y(f(u), f(w)) + d_Y(f(w), f(v)) \le \omega_f(s) + \omega_f(t)$. Therefore $\omega_f(s+t) \le \omega_f(s) + \omega_f(t)$. The same proof works in the second case. □

Example 1.40. For $1 \le p \le q < \infty$ there exists an embedding $T : L_p(\mathbb{R}) \to L_q(\mathbb{R}\times\mathbb{R})$ which is simultaneously coarse and uniform.

Proof. Our proof is simpler for $p = 1$, so we start with this case. We define a map $T : L_1(\mathbb{R}) \to L_\infty(\mathbb{R}\times\mathbb{R})$ by:

$$T(f)(t,s) \stackrel{\text{def}}{=} \begin{cases} 1 & \text{if } 0 < s \le f(t), \\ -1 & \text{if } f(t) < s < 0, \\ 0 & \text{otherwise.} \end{cases}$$

For all $f, g \in L_1(\mathbb{R})$ we have:

$$|T(f)(t,s) - T(g)(t,s)| = \begin{cases} 1 & \text{if } g(t) < s \le f(t) \text{ or } f(t) < s \le g(t), \\ 0 & \text{otherwise.} \end{cases}$$

Thus, for all $q \ge 1$ we have

$$\begin{aligned} \|T(f) - T(g)\|^q_{L_q(\mathbb{R}\times\mathbb{R})} &= \int_{\mathbb{R}} \left(\int_{(g(t),f(t)]\cup(f(t),g(t)]} ds \right) dt \\ &= \int_{\mathbb{R}} |f(t) - g(t)|\, dt = \|f - g\|_{L_1(\mathbb{R})}. \end{aligned} \tag{1.13}$$

It is clear that this equality implies that the embedding $T : L_1(\mathbb{R}) \to L_q(\mathbb{R} \times \mathbb{R})$ is both coarse and uniform.

Our proof in the case $p > 1$ is more complicated. First we consider the complex-valued space $L_q(\mathbb{R} \times \mathbb{R})$. We comment on the real-space-case later.

Let $0 < \alpha < 2\beta$. It is easy to see that the integral

$$\int_{-\infty}^{\infty} \frac{(1 - \cos(tx))^\beta}{|t|^{\alpha+1}} dt$$

converges for each $x \in \mathbb{R}$. Observe that if we replace x with sx, $s > 0$, in the integral, we get

$$\begin{aligned}\int_{-\infty}^{\infty} \frac{(1 - \cos(tsx))^\beta}{|t|^{\alpha+1}} dt &\overset{(u=st)}{=} \int_{-\infty}^{\infty} \frac{(1 - \cos(ux))^\beta}{|u|^{\alpha+1}/s^{\alpha+1}} \frac{du}{s} \\ &= s^\alpha \int_{-\infty}^{\infty} \frac{(1 - \cos(ux))^\beta}{|u|^{\alpha+1}} du.\end{aligned}$$

Also the value of the integral does not change if we replace x with $-x$. Therefore there exists a constant $c_{\alpha,\beta} > 0$ such that

$$\int_{-\infty}^{\infty} \frac{(1 - \cos tx)^\beta}{|t|^{\alpha+1}} dt = c_{\alpha,\beta} |x|^\alpha.$$

Define a map $T : L_p(\mathbb{R}) \to L_q(\mathbb{R} \times \mathbb{R})$ by

$$T(f)(s,t) = \frac{1 - e^{itf(s)}}{|t|^{(p+1)/q}}.$$

We need to show that this function is in $L_q(\mathbb{R} \times \mathbb{R})$. We use the following elementary trigonometric formula

$$((1 - \cos w)^2 + (\sin w)^2)^{1/2} = \sqrt{2}\,(1 - \cos w)^{1/2}. \tag{1.14}$$

Using this formula we get

$$\begin{aligned}\iint |T(f)(s,t)|^q \, dt\, ds &= \iint \frac{|1 - e^{itf(s)}|^q}{|t|^{p+1}} dt\, ds \\ &= \iint \frac{2^{q/2}(1 - \cos(tf(s)))^{q/2}}{|t|^{p+1}} dt\, ds \\ &= 2^{q/2} \int c_{p,q/2} |f(s)|^p \, ds = 2^{q/2} c_{p,q/2} \|f\|_{L_p(\mathbb{R})}^p.\end{aligned}$$

If $f, g \in L_p(\mathbb{R})$, using a similar computation we get:

$$\begin{aligned}\|T(f) - T(g)\|^q_{L_q(\mathbb{R}\times\mathbb{R})} &= \left\| \frac{1 - e^{itf(s)} - (1 - e^{itg(s)})}{|t|^{(p+1)/q}} \right\|^q_{L_q(\mathbb{R}\times\mathbb{R})} \\ &= \iint \frac{|1 - e^{it(f(s)-g(s))}|^q}{|t|^{p+1}}\, dt\, ds \\ &= 2^{q/2} c_{p,q/2} \|f - g\|^p_{L_p(\mathbb{R})},\end{aligned}$$

so T is the required embedding.

Returning to the real case: considering real and imaginary parts of a complex-valued function we may identify a complex-valued space $L_q(\Omega)$ with the real-valued space $L_q(\Omega \cup \Omega)$. The norms of a complex-valued function on Ω and the corresponding real-valued function on $\Omega \cup \Omega$ are not the same, but the map is a bilipschitz map. Using Fact 1.20 we complete the proof in the real case. □

Remark 1.41. If $q = \infty$, even an isometric embedding $f : L_p(\mathbb{R}) \to L_q(0,1)$ exists. In fact, let $\{(a_i, b_i)\}_{i=1}^{\infty}$ be a sequence of disjoint subintervals of $(0,1)$. It is easy to see that the subspace of $L_\infty(0,1)$ consisting of functions f which are constant on each of (a_i, b_i) (that is, $x, y \in (a_i, b_i) \Rightarrow f(x) = f(y)$) is isometric to ℓ_∞. Since $L_p(\mathbb{R})$, $1 \le p < \infty$ is separable, the conclusion follows from Exercise 1.54.

Remark 1.42. The proof in the case $p = 1$ presented above can be used to find a simultaneously coarse and uniform embedding of $L_1(\Omega)$ into $L_q(\Omega \times \mathbb{R})$ for an arbitrary measure space Ω. In the case $q = 2$ we get

Corollary 1.43. *The metric space* $(L_1(\Omega,\mu), \|f - g\|^{1/2}_{L_1(\Omega,\mu)})$ *admits an isometric embedding into a Hilbert space.*

Definition 1.44. A metric space X is called *uniformly discrete* if there exists a constant $\delta > 0$ such that

$$\forall u, v \in X\ (u \neq v) \Rightarrow (d_X(u,v) \ge \delta).$$

For uniformly discrete metric spaces the notion of a uniform embedding is not useful because any bijection between a uniformly discrete metric space and a uniformly discrete subset of another metric space is a uniform embedding.

The definition of a coarse embedding is usually given in the form:

Definition 1.45. A map $f : (X, d_X) \to (Y, d_Y)$ between two metric spaces is called a *coarse embedding* if there exist non-decreasing functions $\rho_1, \rho_2 : [0,\infty) \to [0,\infty)$ (observe that this condition implies that ρ_2 has finite values) such that $\lim_{t\to\infty} \rho_1(t) = \infty$ and

$$\forall u, v \in X\ \rho_1(d_X(u,v)) \le d_Y(f(u), f(v)) \le \rho_2(d_X(u,v)).$$

As in the case of bilipschitz embeddings we need the following definition for families.

Definition 1.46. Let $\{(X_\alpha, d_\alpha)\}_{\alpha\in A}$ and $\{(Y_\alpha, D_\alpha)\}_{\alpha\in A}$ be two families of metric spaces. A family $f_\alpha : X_\alpha \to Y_\alpha$ of embeddings is called *uniformly coarse* if there exist two non-decreasing functions $\rho_1, \rho_2 : [0,\infty) \to [0,\infty)$ such that

- $\forall \alpha \in A\ \forall u, v \in X_\alpha \quad \rho_1(d_\alpha(u,v)) \le D_\alpha(f_\alpha(u), f_\alpha(v)) \le \rho_2(d_\alpha(u,v))$.
- ρ_2 takes finite values only.
- $\lim_{t\to\infty} \rho_1(t) = \infty$.

Remark 1.47 (Warning). M. Gromov [175, p. 211] used the term *uniform embedding* for what we call a *coarse embedding* and defined it a bit differently, but for all problems which we are going to consider his definition is equivalent to Definition 1.36. We prefer to use the term *coarse embedding* because in nonlinear functional analysis the term *uniform embedding* is used in the sense of Definition 1.36.

The main purpose of this remark is to alert the reader that in some publications the term *uniform embedding* is used for the class of embeddings which we call *coarse*.

Example 1.48. Embeddings of an infinite binary tree.

We define the *infinite binary tree* T_∞ as the graph whose vertices are in bijective correspondence with all finite sequences of 0 and 1, including an empty sequence, two vertices are adjacent if the lengths of the corresponding sequences differ by 1, and one of them is an extension of the other. (In more detail: a sequence $\{a_i\}_{i=1}^m$ is an *extension* of $\{b_i\}_{i=1}^n$ if $m > n$ and $a_1 = b_1, a_2 = b_2, \dots, a_n = b_n$.)

The infinite binary tree with its graph distance admits an isometric embedding into ℓ_1. To get such an embedding we consider a bijection between edges of T_∞ and the unit vectors of ℓ_1. We map the vertex v_0 corresponding to the empty sequence to 0. We map any other vertex v of T_∞ to the sum of unit vectors corresponding to edges of the unique path from v_0 to v in T_∞. The fact that this defines an isometric embedding can be verified in a straightforward way. Using Example 1.40 we get the existence of a coarse embedding of T_∞ into ℓ_2. It is worth mentioning that one can get a coarse embedding of T_∞ into ℓ_2 much easier, just by using the same construction as above with the unit vectors of ℓ_2 instead of the unit vectors of ℓ_1.

Remark 1.49. It is known (see Section 9.5.1) that there are no bilipschitz embeddings of T_∞ into ℓ_2.

Remark 1.50. If the metric space M is an unweighted graph with its graph distance, then a coarse embedding $f : M \to X$ is a Lipschitz map with Lipschitz constant $\le \rho_2(1)$.

Proof. Let $x, y \in M$. There exists a sequence $x_0, \dots, x_n$ such that $n = d_M(x, y)$, $x_0 = x, x_n = y$ and $d_M(x_i, x_{i+1}) = 1$. Then $d_X(f(x_i), f(x_{i+1})) \le \rho_2(1)$; therefore $d_X(f(x), f(y)) \le n\rho_2(1)$. □

1.3 Probability theory terminology and notation

By a *probability space* we mean a triple $(\Omega, \Sigma, \mathbb{P})$, where Ω is a set and Σ is a collection of subsets of Ω satisfying the conditions

(i) If $A \in \Sigma$, then so is its complement: $A' \in \Sigma$

(ii) If $\{A_n\}$ is any finite or countable collection of sets in Σ, then $(\cup_n A_n) \in \Sigma$ and $(\cap_n A_n) \in \Sigma$.

(iii) $\emptyset \in \Sigma$ and $\Omega \in \Sigma$.

A collection Σ of subsets satisfying the conditions (i)–(iii) is called a σ*-algebra*. Elements of Σ are called *events*.

The last element of the triple $\mathbb{P}$ is a real-valued function on Σ satisfying the conditions:

- $\mathbb{P}(A) \ge 0$ for each $A \in \Sigma$.
- $\mathbb{P}(\Omega) = 1$.
- If $\{A_n\}$ is any finite or countable collection of non-overlapping sets in Σ, then $\mathbb{P}(\cup_n A_n) = \sum_n \mathbb{P}(A_n)$. (This condition implies $\mathbb{P}(\emptyset) = 0$.)

A function $\mathbb{P}$ satisfying these conditions is called a *probability measure* on Σ, the number $\mathbb{P}(A)$ is called the *probability* of the event A.

The set of all subsets of any set is always a σ-algebra.

If X is a metric space, the smallest σ-algebra of subsets of X containing all closed sets of X is called the *Borel algebra* of X and the sets in it are called *Borel sets*.

Let $(\Omega, \Sigma, \mathbb{P})$ be a probability space. A function f on Ω with values in a metric space X is called *measurable* if the set $f^{-1}(I)$ is in Σ for each Borel set I in X (it is easy to check that in the case of real-valued functions it suffices to require that $f^{-1}(I)$ is in Σ for each interval I of the real line). A *random variable* on a probability space $(\Omega, \Sigma, \mathbb{P})$ is a measurable real-valued, complex-valued, or Banach space valued function on Ω. An X-valued random variable is called *symmetric* if for each Borel set $I \subset X$ the equality $\mathbb{P}\{\omega : X(\omega) \in I\} = \mathbb{P}\{\omega : -X(\omega) \in I\}$ holds.

For some of the random variables we define the *expectation*. We introduce the *indicator function* of a set S by

$$\mathbf{1}_S(x) = \begin{cases} 1 & \text{if } x \in S \\ 0 & \text{if } x \notin S. \end{cases} \tag{1.15}$$

First we define expectations for functions with values in a Banach space X (which can be $\mathbb{R}$ or $\mathbb{C}$) presented as finite sums of the form

$$\sum_i a_i \mathbf{1}_{A_i}, \tag{1.16}$$

where $a_i \in X$ and $A_i \in \Sigma$. Such random variables are called *simple*.

We define the expectation $\mathbb{E}(F)$ of a simple random variable $F = \sum_i a_i \mathbf{1}_{A_i}$ by

$$\mathbb{E}(F) = \sum_i a_i \mathbb{P}(A_i).$$

One can check that this sum does not depend on the choice of representation.

To introduce expectations of other random variables we observe that $\mathbb{E}(\|F\|)$ is a norm on the linear space of all simple X-valued (X is a fixed Banach space) random variables. If a random variable F can be represented as a limit of a sequence of simple random variables $\{F_i\}$ with respect to this norm, we define $\mathbb{E}(F) = \lim_{i\to\infty} \mathbb{E}(F_i)$. It can be checked that this definition does not depend on the choice of the sequence $\{F_i\}_{i=1}^{\infty}$.

It is easy to verify that for a random variable F the expectation $\mathbb{E}(F)$ is defined if and only if the expectation $\mathbb{E}(\|F\|)$ is defined, and

$$\|\mathbb{E}(F)\| \leq \mathbb{E}(\|F\|).$$

The space of all X-valued random variables F for which $\mathbb{E}(F)$ and $\mathbb{E}(\|F\|)$ are defined is denoted $L_1(\Omega, \Sigma, \mathbb{P}, X)$. We use the notation $L_1(X)$ in the case where Ω is the interval $[0, 1]$, Σ is the σ-algebra of Borel subsets of $[0, 1]$, and $\mathbb{P}$ is the usual (Lebesgue) measure restricted to $[0, 1]$.

We also introduce the Banach space $L_p(\Omega, \Sigma, \mathbb{P}, X)$ $(1 < p < \infty)$ as linear subspaces of $L_1(\Omega, \Sigma, \mathbb{P}, X)$ for which $\mathbb{E}(\|X\|^p) < \infty$ is defined. We use the notation $L_p(X)$ in the case of $[0, 1]$.

The *variance* $\mathrm{Var}(F)$ of a real-valued random variable F is defined by

$$\mathrm{Var}(F) = \mathbb{E}\big((F - \mathbb{E}F)^2\big).$$

Let $A, B \in \Sigma$ be two events. The *conditional probability of A with respect to B* is defined as $\mathbb{P}(A|B) = \frac{\mathbb{P}(A\cap B)}{\mathbb{P}(B)}$ (defined only if $\mathbb{P}(B) \neq 0$). Events $A, B \in \Sigma$ are called *independent* if $\mathbb{P}(A \cap B) = \mathbb{P}(A) \cdot \mathbb{P}(B)$. If A and B are independent and $\mathbb{P}(B) \neq 0$, then $\mathbb{P}(A|B) = \mathbb{P}(A)$.

Random variables $F_1 : \Omega \to X_1$ and $F_2 : \Omega \to X_2$ are called *independent* if for any two Borel sets $I_1 \subset X_1$ and $I_2 \subset X_2$ the events $F_1^{-1}(I_1)$ and $F_2^{-1}(I_2)$ are independent.

Random variables $\{F_i\}_{i=1}^n$ with values in $\{X_i\}_{i=1}^n$, respectively, are called *independent* if

$$\mathbb{P}\Big(\bigcap_{i=1}^n F_i^{-1}(I_i)\Big) = \prod_{i=1}^n \mathbb{P}\big(F_i^{-1}(I_i)\big)$$

for any collection $\{I_i\}_{i=1}^n$ of Borel sets satisfying $I_i \subset X_i$. An infinite sequence $\{F_i\}_{i=1}^\infty$ of random variables is called *independent* if each finite subset of it is independent in the sense of the previous definition.

It is not difficult to prove that if F_1 is a real-valued random variable on Ω, F_2 is a Banach space valued random variable on Ω, and F_1 and F_2 are independent, then $\mathbb{E}(F_1 F_2) = \mathbb{E}(F_1)\mathbb{E}(F_2)$.

Now we introduce some basic definitions and facts about martingales. Let $(\Omega, \Sigma, \mathbb{P})$ be a probability space, and $\mathcal{F}$ be a σ-subalgebra of Σ (that is, a σ-algebra of sets of Ω satisfying $\mathcal{F} \subset \Sigma$). Let X be a Banach space and $f : \Omega \to X$ be a map. The map is called *measurable with respect to* $\mathcal{F}$ (or $\mathcal{F}$*-measurable*) if for each Borel set $A \subset X$ the pre-image $f^{-1}(A)$ is an element of $\mathcal{F}$.

For $F \in L_1(\Omega, \Sigma, \mathbb{P}, X)$ we define the *conditional expectation* $\mathbb{E}(F|\mathcal{F})$ with respect to the σ-subalgebra $\mathcal{F}$ as an $\mathcal{F}$-measurable X-valued random variable $\widetilde{F}$ for which

$$\int_A F\,d\mathbb{P} = \int_A \widetilde{F}\,d\mathbb{P} \tag{1.17}$$

for each $A \in \mathcal{F}$, where $\int_A F\,d\mathbb{P}$ is defined as $\mathbb{E}(F \cdot \mathbf{1}_A)$. We emphasize that (1.17) is required only for $A \in \mathcal{F}$.

The proof of the existence of the conditional expectation for general real-valued random variables requires the Radon–Nikodým theorem on integral representation of absolutely continuous measures. As for Banach space valued random variables, the existence of conditional expectation in this case is proved in the following way: it is easy to see that for a variable of the form $f \cdot x$, where $f \in L_1(\Omega, \Sigma, \mathbb{P})$ and $x \in X$, we have $\mathbb{E}(f \cdot x|\mathcal{F}) = \mathbb{E}(f|\mathcal{F}) \cdot x$. On the other hand, it can be shown that linear combinations of functions of the form $f \cdot x$, where $f \in L_1(\Omega, \Sigma, \mathbb{P})$ and $x \in X$, are dense in $L_1(\Omega, \Sigma, \mathbb{P}, X)$.

For applications in this book we are going to use the conditional expectations with respect to finite σ-subalgebras $\mathcal{F}$ only. In this case the existence is obvious: It is easy to check that if $\mathcal{F}$ is finite, it is generated by its minimal subsets, which form a partition of Ω. Let $\mathcal{F}$ be generated by disjoint sets $\{A_1, \dots, A_n\}$ partitioning Ω, and F be an X-valued random variable. Then the random variable $\tilde{F}$ which takes the value

$$\frac{\int_{A_i} F\,d\mathbb{P}}{\mathbb{P}(A_i)}$$

at each point of A_i is the conditional expectation of F with respect to $\mathcal{F}$.

Let $(\Omega, \Sigma, \mathbb{P})$ be a probability space. A (finite or infinite) sequence of random variables $\{M_0, M_1, \dots, M_n, \dots\}$ in $L_1(\Omega, \Sigma, \mathbb{P}, X)$ is called a *martingale* if there exists an increasing sequence $\mathcal{F}_0 \subseteq \mathcal{F}_1 \subseteq \cdots \subseteq \mathcal{F}_n \subseteq \dots$ of σ-subalgebras of Σ (such sequence of σ-subalgebras is called a *filtration*) such that for each $n \geq 0$ the random variable M_n is $\mathcal{F}_n$-measurable and satisfies $M_n = \mathbb{E}(M_{n+1}|\mathcal{F}_n)$.

1.4 Applications to the sparsest cut problem

We return to the discussion of the sparsest cut problem (see page 4). Let $d_{\min}$ be the semimetric which solves the minimization problem (1.4).

Observation 1.51. If the semimetric space $(V, d_{\min})$ is isometric to a subset of L_1, then the quantity $\sum_{uv \in E} c(uv) d_{\min}(u, v)$ coincides with the sparsity of the sparsest cut.

To prove this observation we need the following result which is useful in many other contexts.

Proposition 1.52. *Let (X, d_X) be a finite semimetric space. The space X admits an isometric embedding into L_1 if and only if its metric is a positive linear combination of cut semimetrics, in the sense that there are nonnegative numbers $\alpha_1, \dots, \alpha_m$ and cuts $(S_1, \bar{S}_1), \dots, (S_m, \bar{S}_m)$ such that*

$$\forall u, v \in X \quad d_X(u, v) = \alpha_1 d_{S_1}(u, v) + \dots + \alpha_m d_{S_m}(u, v).$$

Proof. If a metric d_X is given by a linear combination of the described type we define an embedding $f : X \to \ell_1^m$ whose coordinates are given by

$$(f(x))_i = \begin{cases} 0 & \text{if } x \in S_i \\ \alpha_i & \text{if } x \notin S_i. \end{cases}$$

It is easy to check that

$$\forall u, v \in X \quad \|f(u) - f(v)\| = \alpha_1 d_{S_1}(u, v) + \dots + \alpha_m d_{S_m}(u, v),$$

so that f is an isometric embedding.

On the other hand, if X admits an isometric embedding into L_1 we may assume (see Lemma 1.24) that X is a finite subset of some ℓ_1^k.

Let $\{\beta_{i,j}\}_{j=1}^{p(i)}$ be the set of all possible values of the i-th coordinate of elements of X, $i = 1, \dots, k$, arranged in an increasing order, so $\beta_{i,j+1} > \beta_{i,j}$. Let $S_{i,j}$ be the set of elements whose i-th coordinate is $\le \beta_{i,j}$, and $d_{i,j}$ be the corresponding cut semimetric. Since the coordinates of each element $u \in X$ have the form $\beta_{1,j(1)}, \beta_{2,j(2)}, \dots, \beta_{k,j(k)}$, one can easily check that

$$d_X(u, v) = \sum_{i=1}^{k} \sum_{j=1}^{p(i)-1} (\beta_{i,j+1} - \beta_{i,j}) d_{i,j}(u, v). \qquad \square$$

Now we return to the proof of Observation 1.51. We start by proving, for any nonnegative numbers $\gamma_1, \dots, \gamma_m$ and $z_1, \dots, z_m$, the inequality

$$\frac{\gamma_1 + \dots + \gamma_m}{z_1 + \dots + z_m} \ge \min\left\{\frac{\gamma_1}{z_1}, \dots, \frac{\gamma_m}{z_m}\right\}. \tag{1.18}$$

This can be done by induction starting with the inequality

$$\frac{\gamma_1 + \gamma_2}{z_1 + z_2} \geq \min \left\{ \frac{\gamma_1}{z_1}, \frac{\gamma_2}{z_2} \right\}.$$

To get this inequality we may assume that $\frac{\gamma_1}{z_1} \leq \frac{\gamma_2}{z_2}$ or

$$\gamma_2 z_1 \geq \gamma_1 z_2 \tag{1.19}$$

In such a case we need to show $\frac{\gamma_1+\gamma_2}{z_1+z_2} \geq \frac{\gamma_1}{z_1}$ or

$$\gamma_1 z_1 + \gamma_2 z_1 \geq \gamma_1 z_1 + \gamma_1 z_2. \tag{1.20}$$

We see that (1.20) follows immediately from (1.19).

By Proposition 1.52, if $(V, d_{\min})$ is isometric to a subset of L_1, then $d_{\min} = \alpha_1 d_{S_1} + \cdots + \alpha_m d_{S_m}$. By (1.18) we get

$$\frac{\sum_{uv \in E} c(uv) d_{\min}(u, v)}{\sum_{u,v \in V} D(u, v) d_{\min}(u, v)} \geq \min_{1 \leq i \leq m} \frac{\sum_{uv \in E} c(uv) d_{S_i}(u, v)}{\sum_{u,v \in V} D(u, v) d_{S_i}(u, v)}.$$

The conclusion of the observation follows.

An immediate consequence of the proof of Observation 1.51 is that if the metric space $(V, d_{\min})$ is not isometric to a subset of L_1, but there is a bilipschitz embedding of this space into L_1 with distortion $\leq C$, then the sparsity of the sparsest cut does not exceed $C \times$ (the sparsity of $d_{\min}$).

1.5 Exercises

Recall that a metric space is called *separable* if it has a countable dense set.

Exercise 1.53. Prove that any separable metric space admits an isometric embedding into ℓ_∞.

By a *linear isometric embedding* of one normed linear space into another we mean an isometric embedding which is simultaneously a linear map.

Exercise 1.54. Prove that any separable normed linear space admits a linear isometric embedding into ℓ_∞.

If needed, see a hint in the Hints to exercises section below.

Exercise 1.55. We say that vectors $a = \{a_i\}_{i=1}^m$ and $b = \{b_i\}_{i=1}^m$ in ℓ_p^m are *disjoint* if the sets $\{i : a_i \neq 0\}$ and $\{i : b_i \neq 0\}$ are disjoint.

Let $1 < p < \infty$, $p \neq 2$, and $n \leq m$. Show that each linear isometric embedding $T : \ell_p^n \to \ell_p^m$ is given by $T(e_i) = d_i$, where $\{e_i\}_{i=1}^n \subset \ell_p^n$ are the unit vectors and $\{d_i\}_{i=1}^n$ are disjoint vectors of norm 1 in ℓ_p^m.

If needed, see a hint in the Hints to exercises section below.

Exercise 1.56. Decrease the dimension in Proposition 1.17 to $n-1$.

The definition of a uniformly convex normed linear space is given on page 31.

Exercise 1.57. Show that ℓ_2 is uniformly convex.

The definition of bounded geometry is given on page 31.

Exercise 1.58. Show that each finitely generated group with the word distance has bounded geometry.

Exercise 1.59. Find a sharp estimate from below of the distortion of embeddings of $K_{1,3}$ into ℓ_2 (see Proposition 1.25).

If needed, see a hint in the Hints to exercises section below.

Exercise 1.60. Show that for $n \geq 4$ any metric space with n elements admits an isometric embedding into ℓ_∞^{n-2}.

In the Hints to exercises section we suggest an approach to Exercise 1.60.

Exercise 1.61. Let $1 < p < q < \infty$. Show that there exists an embedding $T : \ell_p \to \ell_q$ which is simultaneously coarse and uniform.

In the Hints to exercises section we suggest an approach to Exercise 1.61.

Exercise 1.62. Consider two different finite sets of generators, S_1 and S_2, in a finitely generated group G. Let d_1 and d_2 be the corresponding word metrics on G. Show that the map $(G, d_1) \to (G, d_2)$ which maps each element to itself is a (bijective) bilipschitz embedding.

1.6 Notes and remarks

1.6.1 To Section 1.1

Metric geometry first appeared in the first half of the 20th century in the works of Fréchet, Menger, Blumenthal, and Schoenberg. Expositions of results obtained in this period can be found in [65], [66]. In recent years metric geometry has experienced explosive growth, as demonstrated by expositions in the following books, surveys, and lecture notes: [55], [82], [87], [177], [196], [197], [198], [237], [289], [333], [334], [345], [394], [443].

Each of the examples of the metric space examples presented in Section 1.1 corresponds to a well-developed area of mathematics. We shall use some results of the corresponding areas without proofs, here we mention the corresponding sources. Banach spaces are among the most important objects in functional analysis. As a general introduction to functional analysis we recommend [130]. Recommended sources for basic Banach space theory are [9], [50], and [141]. Extremely useful advanced monographs on Banach space theory are [55], [120], [287], and [288]. Another highly recommended source is the two-volume handbook [221], [222].

We are going to follow the graph theory terminology and notation of a very comprehensive textbook [71], another relevant text is [122]. An extremely useful source in graph theory is the collection of problems [297] containing hints and complete solutions for all of its problems.

A famous essay on geometric group theory is [175]. The books [195] and [323] can be used to enter the subject.

Cut semimetrics and their applications in combinatorial optimization are the subject of the book [118].

The sparsest cut problem is one of the most important problems in the field of approximation algorithms. The problem of computing the sparsest cut is known to be NP hard [409]. The sparsest cut problem is used as a subroutine in many approximation algorithms for NP-hard problems; see the survey article [411], as well as the references in [280],[25],[26],[98],[431],[438].

See [408] for information on algorithms available for linear programming problems.

1.6.2 To Section 1.2

Different aspects of the theory of isometric embeddings are presented in [65], [118], [151], [152], [181], [220], [252], [262], [436].

Note on Fréchet. In 1910 Fréchet [155, p. 161–162] defined ℓ_∞ as a metric space (he denoted it D) and proved that each separable metric space can be isometrically embedded into ℓ_∞. Some authors use the term *Kuratowski embedding* or *Hausdorff–Kuratowski embedding* for what we call the *Fréchet embedding*. This is because in 1935 Kuratowski [258, p. 543] proved that a bounded metric space X can be isometrically embedded into the space $C(X)$ of bounded continuous functions on X with the supremum norm using the map: $p \mapsto f_p(x) := d_X(x, p)$. In [259, p. 224] the unbounded case is also treated using the embedding similar to the one in our proof of Proposition 1.17; at this point Kuratowski refers to the paper by Kunugui [257], where the proof of the mentioned above Fréchet result is reproduced. Hausdorff is mentioned in this context in [257] where it is observed that by embedding an arbitrary metric space into some Banach space we get a new proof of the Hausdorff theorem on existence of a completion of a metric space. After this investigation the present author decided that the term *Fréchet embedding* is the most suitable for embeddings of the type considered in Proposition 1.17.

The mentioned Fact 1.20 can be derived from the following results:

Lemma 1.63 ([130, Lemma III.8.5]). *For each separable subset G of $L_p(\Omega, \Sigma, \mu)$, $1 \leq p < \infty$, there is a set $S_1 \in \Sigma$ and a σ-subalgebra of Σ-restricted-to-S_1, such that the restriction μ_1 of μ to Σ_1 has the following properties*

- $L_p(S_1, \Sigma_1, \mu_1)$ *is separable.*
- $G \subset L_p(S_1, \Sigma_1, \mu)$.

The pair (Σ_1, μ_1) is called a *separable measure algebra* if $L_p(S_1, \Sigma_1, \mu_1)$, $1 \leq p < \infty$, is separable. (The standard definition, see [396, p. 321] is different, but is equivalent.)

The Carathéodory theorem (see [396, p. 321]) states that each separable measure algebra is isomorphic to some algebra of measurable subsets of $[0, 1]$ endowed with the Lebesgue measure. This isomorphism induces an isometric embedding of $L_p(S_1, \Sigma_1, \mu_1)$ into $L_p(0, 1)$.

Proposition 1.21 and Lemmas 1.22 and 1.24 apparently should be considered as folklore. The exact estimates for the dimension of ℓ_1^k containing an n-element subset of L_1 were found in [440] and [35].

Examples 1.32 and 1.33 were suggested by the present author, although with high probability they were known before. Example 1.32 can be easily generalized to an arbitrary strictly convex Banach space.

The *Baire category theorem* can be found in many sources, see e.g. [130, p. 20] or [396, p. 139].

Example 1.34 is well known. Theorem 1.35 is from [219]. As a well-known and very valuable tool in computer science, the Johnson–Lindenstrauss lemma has been proven in various different ways, each method characterized by different algorithmic features, see [4],[6],[109],[154],[209],[250],[316] and references therein. The sharpness of the Johnson–Lindenstrauss lemma was studied in [12],[13]. Our proof of the Johnson–Lindenstrauss lemma is based on the lecture of Assaf Naor given at Workshop in Analysis and Probability at Texas A & M University, July 2006, see also [316]. A significant amount of effort was put into finding analogues of the Johnson–Lindenstrauss lemma for other spaces, see [54],[83],[93],[172],[223],[226],[276], see also k-dimensional versions of the lemma in [124].

Example 1.40 is from [325, Remark 5.10] and [333, Section 3]. The fact that $(L_p, \|x - y\|_{L_p}^{p/q})$ admits an isometric embedding into L_q when $1 \leq p \leq q \leq 2$ is due to Bretagnolle, Dacunha-Castelle, and Krivine [80, p. 251]

1.6.3 To Section 1.3

In this section we list some standard definitions of probability theory. Our summary is similar to the one given in [233]. Our recommended source on probability theory is [146],[147]. For more information on Banach space valued martingales we recommend [383]. This book also contains basic material on vector-valued random variables and conditional expectations.

1.6.4 To Section 1.4

The application of low-distortion embeddings described in this section appears in [290] and [33].

Distortion estimates for finite metric spaces have been studied extensively, a survey of this study appears in [208] and [314, Chapter 15]. It is worth mentioning that an updated version of [314, Chapter 15] is available on the internet, see [315]. In my opinion [314, Chapter 15] (both in 2002 and in 2005 versions) is very nice reading with a nice selection of exercises.

Applications of low-distortion bilipschitz embeddings in computer science are so numerous that I do not even try to survey them. Information on some of the developments in the area can be found in [2],[3],[142],[207],[333].

1.6.5 To exercises

Exercise 1.53 is taken from [155, pp. 161–162].

Exercise 1.55 is from [193] and [367, Lemma 3].

Exercise 1.60 is from [441]. The problem of estimating the least dimension $d(n)$ such that all n-element metric spaces are isometric to subsets of $\ell_\infty^{d(n)}$ was further studied in [35] and [371]. The problem obtained if we replace $\ell_\infty^{d(n)}$ with "some $d(n)$-dimensional normed space" was studied in [34].

Exercise 1.61 is a discrete version of Example 1.40. See [8] and [47] for closely related results. See [344] and [354, Theorem 5.1] for a different approach to construction of coarse embeddings of ℓ_2 into other Banach spaces.

1.7 On applications in topology

Usage of coarse embeddings of uniformly discrete metric spaces into sufficiently good Banach spaces in topology was initiated by Gromov. In this section we cite the corresponding paper of Gromov, and state the results which could be considered as very important steps in the realization of Gromov's program. This section requires much more background than the rest of the book. For this reason readers who do not have such background are advised to skip this section and to read only the following summary of it: Some important recent results in topology are proved in two steps. (1) Certain uniformly discrete metric spaces related to the problem in question (for example, homotopy groups of topological spaces with their word metrics) are coarsely embedded into a Hilbert space or a uniformly convex Banach space (see Definition 1.65 below). (2) This embeddability is used to answer the problem.

This approach has been already mentioned in the essay [175], and on September 9, 1993 Gromov suggested the following problems (the enclosed is a citation from [176]):

(4) Does every finitely generated or finitely presented group admit a uniformly metrically proper Lipschitz embedding into a Hilbert space? Even such an embedding into a reflexive uniformly convex Banach space would be interesting. This seems hard.

(5) Can one give a new proof using the above philosophy (of mapping to Euclidean or Hilbert space) of the Strong Novikov conjecture (injectivity of the assembly map for the K-theory of the group C^*-algebra) or the Baum–Connes Conjecture for uniformly discrete subgroups of $SO(n,1)$, $SU(n,1)$?

Remark 1.64. The notion of a 'uniformly metrically proper Lipschitz embedding' coincides with the notion of 'coarse embedding' introduced in Definition 1.36

Somewhat later G. Yu [442] and G. Kasparov and G. Yu [243] obtained strong positive results in the direction of Problem (5).

Because of this the problems of embeddability of uniformly discrete metric spaces into well-structured Banach spaces became important for topology.

So far only embeddings into uniformly convex spaces, or better into ℓ_2, are useful.

Definition 1.65. A Banach space is called *uniformly convex* if for every $\varepsilon > 0$ there is some $\delta > 0$ so that for any two vectors with $\|x\| \le 1$ and $\|y\| \le 1$, the inequality

$$\|x + y\| > 2 - \delta$$

implies

$$\|x - y\| < \varepsilon.$$

Intuitively, the center of a line segment inside the unit ball must lie deep inside the unit ball unless the segment is short. Uniformly convex Banach spaces form a rather wide and very complicated (from the Banach-space-theoretical point of view) class of Banach spaces.

Metric spaces which are the most interesting for topology are finitely generated infinite groups with their word metrics (see the definitions on page 3). In problems related to the Baum–Connes conjecture, the following class of metric spaces is also of importance:

Definition 1.66. A uniformly discrete metric space A is said to have a *bounded geometry* if for each $r > 0$ there exist a positive integer $M(r)$ such that each ball in A of radius r contains at most $M(r)$ elements.

In the problems cited above M. Gromov suggested to use uniform (coarse) embeddings into a Hilbert space or into a uniformly convex space as a tool for approaching some of the well-known problems. G. Yu [442] and G. Kasparov and G. Yu [243] have shown that this is indeed a very powerful tool. They proved (we warn the reader that the term "uniform embedding" in the theorems below is used for what we call a "coarse embedding"):

Theorem 1.67 (G. Yu, [442]). *Let Γ be a discrete metric space with bounded geometry. If Γ admits a uniform embedding into Hilbert space, then the coarse Baum–Connes conjecture holds for Γ.*

Theorem 1.68 (G. Kasparov and G. Yu, [243]). *Let Γ be a discrete metric space with bounded geometry. If Γ is uniformly embeddable into a uniformly convex Banach space, then the coarse geometric Novikov conjecture holds for Γ, i.e., the index map from $\lim_{d\to\infty} K_*(P_d(\Gamma))$ to $K_*(C^*(\Gamma))$ is injective, where $P_d(\Gamma)$ is the Rips complex of Γ and $C^*(\Gamma)$ is the Roe algebra associated to Γ.*

Theorems 1.67 and 1.68 made Gromov's problem (4) (cited above) and its generalization for spaces with bounded geometry very important.

In most of the known applications of embeddings in topology coarse embeddings are used. However, recently Chen, Wang, and Yu [99] showed that a strictly weaker notion of embedding (fibered coarse embedding) can be used as well.

In the negative direction, Gromov [179] constructed finitely generated groups which do not admit coarse embeddings into a Hilbert space. Sapir [403] proved that there is a 4-dimensional closed aspherical manifold M such that the fundamental group $\pi_1(M)$ contains weakly a family of expanders (see Definition 7.2), and so $\pi_1(M)$ is not coarsely embeddable into a Hilbert space.

1.8 Hints to exercises

To Exercise 1.54. Use the argument similar to the one used at the end of Example 1.33 to construct a bilipschitz embedding of ℓ_2^2 into ℓ_∞^m.

To Exercise 1.55. For $1 \le p < 2$ and $x, y \in \mathbb{R}$ prove that $|x+y|^p + |x-y|^p \le 2|x|^p + 2|y|^p$ and the equality holds if and only if at least one of the numbers x and y is equal to 0. For $2 < p < \infty$ prove that $|x+y|^p + |x-y|^p \ge 2|x|^p + 2|y|^p$ and the equality holds if and only if at least one of the numbers x and y equals 0.

Then apply these inequalities to each of the coordinates of the images of two unit vectors, in order to show that they are disjoint.

To Exercises 1.57 and 1.59. Use the parallelogram identity:

$$\forall x, y \in \ell_2 \quad \|x+y\|^2 + \|x-y\|^2 = 2\|x\|^2 + 2\|y\|^2.$$

To Exercise 1.60. Do it in three steps:

(a) Start with $n = 4$, let (X, d) be a 4-element metric space. We say that a real-valued function f on X *takes care* of the distance between u and v ($u, v \in X$) if $\text{Lip}(f) = 1$ and $|f(u) - f(v)| = d(u, v)$. Consider the bijective correspondence between distances in X and edges of K_4. Observe that we need to split the set of edges of K_4 into two subsets, S_1, and S_2, such that there exist real-valued functions f_1 and f_2 with f_1 taking care of distances from S_1 and f_2 taking care of distances from S_2.

We can find sets S_1 and S_2 for which this is possible in the following way: We partition edges of K_4 into three pairs of edges without common end vertices, and compute the sums of distances corresponding to the pairs. Then we split the pair with the largest sum, and distribute its edges among the remaining pairs. Finally we prove that the obtained triples of edges can serve as S_1 and S_2.

(b) Prove: If $Y \subset X$, then any function f on Y with $\text{Lip}(f) = 1$ may be extended to a function $\tilde{f}$ on X with $\text{Lip}(\tilde{f}) = 1$.

(c) We need to embed isometrically a $(k+4)$-element metric space X into ℓ_∞^{k+2}. Let Y be a 4-element subset of X and f_1 and f_2 be functions producing an isometric embedding of Y into ℓ_∞^2 (according to (a)). Show that their extensions to X, existing according to (b), together with the functions $d(\cdot, v)$, $v \in X \backslash Y$ produce an isometric embedding into ℓ_∞^{k+2}.

To Exercise 1.61. One of the approaches is to consider, for $0 < \alpha < 2\beta$, the series $\sum_{n=-\infty}^{\infty} \frac{(1-\cos(2^n x))^\beta}{2^{n\alpha}}$ and to mimic steps of the proof in Example 1.40 to the extent to which it is possible.

Chapter 2

Embeddability of locally finite metric spaces into Banach spaces is finitely determined. Related Banach space theory

2.1 Introduction

Definition 2.1. A uniformly discrete metric space X is called *locally finite* if for every $x \in X$ and every $r > 0$ the set $\{u \in X : d_X(x,u) \leq r\}$ is finite.

Remark 2.2. Some authors do not require uniform discreteness of locally finite metric spaces (see in this connection Remark 2.35 and the corresponding comment in Notes and Remarks section). Most of our results hold in this more general setting. We invite readers to check in each case whether the uniform discreteness is needed.

Examples 2.3. 1. Unweighted graphs with finite degrees, that is, with finitely many edges incident with each vertex, are locally finite metric spaces with their graph distances.

2. Important examples of uniformly discrete but *not locally finite* metric spaces are nets in infinite-dimensional Banach spaces. Let us recall the definition:

Definition 2.4. Let $\delta > 0$. By a *δ-net* in a metric space X we mean a collection U of elements of X satisfying the conditions:

(a) $\forall x \in X\ \exists u \in U\ d_X(u,x) \leq \delta$.

(b) $\forall u, v \in U\ d_X(u,v) \geq \delta$.

The existence of nets can be derived from the following result:

Lemma 2.5 (Zorn's lemma). *A partially ordered set has a maximal element if every totally ordered subset of it has an upper bound.*

In this lemma we use the following definitions: A partially ordered set is called *totally ordered* if any two elements in it are comparable (that is, satisfy $a \leq b$ or $b \leq a$). An *upper bound* of a subset A in a partially ordered set B is an element $u \in B$ satisfying $a \leq u$ for each $a \in A$.

We are going to consider Zorn's lemma as a basic set-theoretical principle. In the Notes and Remarks section we provide references to sources where Zorn's lemma is derived from other set-theoretical axioms and results.

To get the existence of a δ-net in an arbitrary metric space X we consider the partially ordered by inclusion set P of all subsets $U \subset X$ satisfying the second condition of Definition 2.4. This set P satisfies the assumptions of Zorn's lemma because if $\{U_\gamma\}_{\gamma\in\Gamma}$ is totally ordered by inclusion, then $\cup_{\gamma\in\Gamma} U_\gamma$ is also in P and is an upper bound for $\{U_\gamma\}_{\gamma\in\Gamma}$. By Zorn's lemma there is a maximal element M in P. It should satisfy the first condition of Definition 2.4 because otherwise we would be able to find $x \in X$ such that $M \cup \{x\}$ still satisfies the second condition of Definition 2.4 contrary to the maximality of M.

This chapter has two main purposes. The first main purpose is to prove the following theorems (we use Definitions 1.28 and 1.46):

Theorem 2.6. *Let A be a locally finite metric space whose finite subsets admit uniformly bilipschitz embeddings into a Banach space X. Then A admits a bilipschitz embedding into X.*

Theorem 2.7. *Let A be a locally finite metric space whose finite subsets admit uniformly coarse embeddings into a Banach space X. Then A admits a coarse embedding into X.*

Proofs are given in Section 2.3 of this chapter.

The second main purpose of this chapter is to develop two important tools of Banach space theory. The first is the theory of ultraproducts and finite representability. The second tool is the theory of type and cotype.

2.2 Banach space theory: ultrafilters, ultraproducts, finite representability

In order to make some of our arguments more readable (both in this chapter and in further chapters) it is convenient to use the limiting procedure based on the notions of an ultrafilter and an ultraproduct.

2.2.1 Ultrafilters

Definition 2.8. Let I be an infinite set. A *filter* on I is a subset $\mathcal{F}$ of $\mathcal{P}(I)$ (where $\mathcal{P}(I)$ is the set of all subsets of I) satisfying the following conditions

(a) $\emptyset \notin \mathcal{F}$.

(b) If $A \subset B$ and $A \in \mathcal{F}$, then $B \in \mathcal{F}$.

(c) If $A, B \in \mathcal{F}$, then $A \cap B \in \mathcal{F}$.

Let $\mathcal{R}$ be a collection of subsets of I satisfying the conditions:

- $\emptyset \notin \mathcal{R}$.
- If $A, B \in \mathcal{R}$, then $\exists C \in \mathcal{R}$ such that $C \subset A \cap B$.

Then the set $\mathcal{S} := \{A \subset I \ : \ \exists B \in \mathcal{R},\ A \supset B\}$ is a filter on I called the *filter generated by* $\mathcal{R}$.

Example 2.9. Let I be an infinite set. The set of all subsets of I with finite complements is a filter on I.

Example 2.10. Let I be any set, $x \in I$. Then the set of all subsets of I containing x is a filter on I.

Let Z be a topological space, $f : I \to Z$ be a function. We say that f *converges to* $z \in Z$ *through* $\mathcal{F}$ and write $\lim_{\mathcal{F}} f(x) = z$, if $f^{-1}(U) \in \mathcal{F}$ for every open set U containing z.

This notion of convergence applied to Example 2.9 in the case where $I = \mathbb{N}$ leads to the standard notion of convergence for sequences.

Definition 2.11. An *ultrafilter* $\mathcal{U}$ (on I) is a maximal filter (on I) with respect to inclusion, that is, a filter which is not properly contained in any larger filter.

Lemma 2.12. *Every filter is contained in an ultrafilter.*

Proof. By Zorn's lemma (see Lemma 2.5 in this section) it suffices to show that for each totally ordered set of filters there is an upper bound. But this immediately follows from the observation that the union of an increasing family of filters is a filter. □

It is easy to see that each filter of the type described in Example 2.10 is an ultrafilter. We are interested mostly in ultrafilters with empty intersection.

Definition 2.13. An ultrafilter is called *free* if the intersection of all sets of the ultrafilter is empty. (Some authors use *nonprincipal* or *nontrivial* instead of 'free'.)

Observation 2.14. If $\mathcal{U}$ is an ultrafilter on I and $E \subset I$, then either $E \in \mathcal{U}$ or $(I \backslash E) \in \mathcal{U}$. The converse is also true: If a filter $\mathcal{U}$ on I is such that for each $E \subset I$ either $E \in \mathcal{U}$ or $(I \backslash E) \in \mathcal{U}$, then $\mathcal{U}$ is an ultrafilter.

Proof. In fact, if $E \notin \mathcal{U}$, then all intersections $A \cap (I \backslash E)$, $A \in \mathcal{F}$, are nonempty. It is easy to verify that these sets generate a filter containing $\mathcal{U}$, hence this filter coincides with $\mathcal{U}$. Hence $(I \backslash E) \in \mathcal{U}$.

To prove the converse observe that if a filter $\mathcal{U}$ satisfies the condition of Observation 2.14 and is such that $E \notin \mathcal{U}$, then the set $\mathcal{U} \cup \{E\}$ cannot be a subset of any filter because $I \backslash E \in \mathcal{U}$ and $(I \backslash E) \cap E = \emptyset$. Thus $\mathcal{U}$ is a maximal filter, that is, an ultrafilter. □

Lemma 2.15. *Let* $\mathcal{U}$ *be an ultrafilter on* I, K *be a compact set, and* $f : I \to K$ *be a function, then* f *converges to some point* $k \in K$ *through* $\mathcal{U}$.

Proof. Assume the contrary. Then each point $k \in K$ has a neighborhood N_k such that $f^{-1}(N_k)$ is not in $\mathcal{U}$. By Observation 2.14

$$(I \backslash f^{-1}(N_k)) \in \mathcal{U}.$$

Since K is compact there are finitely points $k_1, \ldots, k_n$ such that $\cup_{i=1}^n N_{k_i} \supset K$, but then $\cap_{i=1}^n (I \backslash f^{-1}(N_{k_i})) \in \mathcal{U}$. On the other hand, this set is empty. We get a contradiction. □

Corollary 2.16. *Let I be a set, $\mathcal{U}$ be an ultrafilter on I, and f be a bounded real- or complex-valued function on I. Then f converges to some number through $\mathcal{U}$.*

Note. The limit whose existence is claimed in Lemma 2.15 and Corollary 2.16 is denoted by $\lim_{\mathcal{U}} f(i)$.

2.2.2 Ultraproducts

Definition 2.17. Given a family $(X_i)_{i \in I}$ of Banach spaces, the ℓ_∞ *direct sum* of $(X_i)_{i \in I}$ is defined as the space of all bounded collections $(x_i)_{i \in I}$, $x_i \in X_i$ with the vector operations $(x_i)_{i \in I} + (y_i)_{i \in I} = (x_i + y_i)_{i \in I}$, $\alpha(x_i)_{i \in I} = (\alpha x_i)_{i \in I}$, and the norm given by

$$\|(x_i)_{i \in I}\|_\infty = \sup_{i \in I} \|x_i\|_{X_i}.$$

The ℓ_∞ direct sum is denoted by $(\oplus_{i \in I} X_i)_\infty$. It is easy to check that $(\oplus_{i \in I} X_i)_\infty$ is a Banach space.

Remark 2.18. We introduce ℓ_p direct sums for $1 \le p < \infty$ in Definition 3.6. We introduce ℓ_∞ direct sums separately because it is convenient and customary to use a slightly different notation here.

Let $\mathcal{U}$ be a free ultrafilter on I. By Corollary 2.16 the limit $\lim_{\mathcal{U}} \|x_i\|_{X_i}$ exists for each $(x_i)_{i \in I} \in (\oplus_{i \in I} X_i)_\infty$. It is easy to see that $\lim_{\mathcal{U}} \|x_i\|_{X_i}$ is a seminorm on $(\oplus_{i \in I} X_i)_\infty$. (Recall that a *seminorm* is like norm except that $\|x\| = 0 \Rightarrow x = 0$ is not required.) Let $N_{\mathcal{U}}$ be the subspace of $(\oplus_{i \in I} X_i)_\infty$ on which this seminorm is equal to 0.

Lemma 2.19. *$N_{\mathcal{U}}$ is a closed subspace of $(\oplus_{i \in I} X_i)_\infty$.*

Proof. We need to show that each vector $(z_i)_{i \in I}$ for which $\lim_{\mathcal{U}} \|z_i\|_{X_i} = \alpha > 0$ has a neighborhood satisfying $\lim_{\mathcal{U}} \|x_i\|_{X_i} \neq 0$. It is clear that the neighborhood $\{(x_i)_{i \in I} : \|x_i - z_i\| < \alpha/2 \text{ for all } i \in I\}$ satisfies this condition. □

Since $N_{\mathcal{U}}$ is a closed subspace, the seminorm $\lim_{\mathcal{U}} \|x_i\|_{X_i}$ induces a norm on the quotient space $(\oplus_{i \in I} X_i)_\infty / N_{\mathcal{U}}$.

Definition 2.20. The obtained Banach space is called the *ultraproduct* of $(X_i)_{i\in I}$ with respect to the ultrafilter $\mathcal{U}$. We denote it by $(\prod_{i\in I} X_i)_{\mathcal{U}}$ or $(\prod X_i)_{\mathcal{U}}$. If all X_i are the same, the corresponding ultraproduct is also called an *ultrapower* and is denoted $X^{\mathcal{U}}$.

The element corresponding to $(x_i)_{i\in I}$ in the ultraproduct will be written $(x_i)_{\mathcal{U}}$. As we know its norm is given by $\|(x_i)_{\mathcal{U}}\| = \lim_{\mathcal{U}} \|x_i\|$. We also use the notation $\|(x_i)\|_{\mathcal{U}}$ for this norm. From our definitions we see that each element of $(\prod X_i)_{\mathcal{U}}$ or $X^{\mathcal{U}}$ is of the form $(x_i)_{\mathcal{U}}$, but this representation is not unique.

The notion of an ultraproduct is of importance for us because it provides a way of "pasting" bilipschitz and coarse embeddings out of uniformly bilipschitz and uniformly coarse pieces. We mean the following result which will be used in this chapter.

Proposition 2.21. *Let A be a metric space which is represented as a union of metric subspaces $\{A_i\}_{i=1}^{\infty}$ satisfying $A_1 \subset A_2 \subset A_3 \subset \dots$. Suppose that $\{A_i\}_{i=1}^{\infty}$ admit uniformly bilipschitz (uniformly coarse) embeddings $f_i : A_i \to X_i$ into Banach spaces $\{X_i\}_{i=1}^{\infty}$. Then A admits a bilipschitz (coarse) embedding into $(\prod X_i)_{\mathcal{U}}$ for any free ultrafilter $\mathcal{U}$.*

Proof. We pick a point $O \in A_1$. We may assume without loss of generality that $f_i(O) = 0$ (in fact, the conditions in the definitions of bilipschitz and coarse embeddings are shift-invariant).

Next, in the bilipschitz case we may assume that all embeddings $\{f_i\}_{i=1}^{\infty}$ satisfy the inequalities:

$$\forall u, v \in A_i \quad d_A(u,v) \le \|f_i(u) - f_i(v)\| \le C d_A(u,v), \tag{2.1}$$

where C is any number exceeding distortions of all f_i. This can be achieved by multiplying all of the maps by $\frac{1}{r}$, where r is the number in the definition of a bilipschitz embedding (see (1.5) in Definition 1.26).

We claim that the embedding which maps $a \in A$ onto $(f_i(a))_{\mathcal{U}}$ has the desired property. First we need to fix the following problem with this definition: if $a \notin A_i$, then $f_i(a)$ is not defined. The point is that for each $a \in A$ this can happen only for finitely many indices, and the definition of the norm in the ultraproduct is such that no change in finitely many terms affects the norm (for free ultrafilters). So we may assume, for example, that $f_i(a) = 0 \in X_i$ if $a \notin A_i$.

It remains to estimate

$$\|(f_i(u))_{i\in I} - (f_i(v))_{i\in I}\|_{\mathcal{U}}.$$

In the uniformly bilipschitz case we get immediately from (2.1) and the definition of a convergence through an ultrafilter that

$$\forall u, v \in A \quad d_A(u,v) \le \|(f_i(u))_{i\in I} - (f_i(v))_{i\in I}\|_{\mathcal{U}} \le C d_A(u,v).$$

In the uniformly coarse case the assumption (2.1) is replaced with

$$\forall u, v \in A_i \quad \rho_1(d_A(u,v)) \le \|f_i(u) - f_i(v)\| \le \rho_2(d_A(u,v)),$$

and we get

$$\forall u, v \in A \quad \rho_1(d_A(u,v)) \le \|(f_i(u))_{i\in I} - (f_i(v))_{i\in I}\|_{\mathcal{U}} \le \rho_2(d_A(u,v)). \qquad \square$$

Proposition 2.22. *If the sequence $\{\dim X_i\}_{i=1}^{\infty}$ is bounded, then the space $(\prod X_i)_{\mathcal{U}}$ is also finite-dimensional.*

We start by proving a result which is very interesting and important by itself. Let $\{x_i\}_{i=1}^n$ be a basis in an n-dimensional space X, its *biorthogonal functionals* are defined by $x_i^*(x_j) = \delta_{ij}$ (Kronecker delta). The basis $\{x_i\}_{i=1}^n$ is called an *Auerbach basis* if $\|x_i\| = \|x_i^*\| = 1$ for all $i \in \{1,\dots,n\}$.

Lemma 2.23 (Auerbach's lemma). *Every finite-dimensional Banach space has an Auerbach basis.*

Proof. For a vector $x \in X$ by $[-x, x]$ we denote the line segment joining $-x$ and x. For $x_1,\dots,x_k \in X$ by $M(\{x_i\}_{i=1}^k)$ we denote the Minkowski sum of the corresponding line segments, that is,

$$M(\{x_i\}_{i=1}^k) = \{x : x = y_1 + \dots + y_k \text{ for some } y_i \in [-x_i, x_i],\ i = 1,\dots,k\}.$$

Let $n = \dim X$. Consider the set $N(= N(X))$ consisting of all subsets $\{x_i\}_{i=1}^n \subset X$ satisfying $\|x_i\| = 1, i \in \{1,\dots,n\}$. It is a compact set in its natural topology; and the n-dimensional volume of $M(\{x_i\}_{i=1}^n)$ is a continuous function on N. Hence it attains its maximum on N. Let $U \subset N$ be the set of n-tuples on which the maximum is attained. It is easy to see that each $\{x_i\}_{i=1}^n \in U$ is a basis (for linearly dependent sets the volume is zero). Another important observation is that $M(\{x_i\}_{i=1}^n) \supset B_X$ if $\{x_i\}_{i=1}^n \in U$ (where $B_X = \{x \in X : \|x\| \le 1\}$). In fact, if there is $y \in B_X \setminus M(\{x_i\}_{i=1}^n)$ then (since the volume of a parallelepiped is the product of the length of its height and the $(n-1)$-dimensional volume of its base), there is $i \in \{1,\dots,n\}$ such that replacing x_i by y we get a parallelepiped whose volume is strictly greater the volume of $M(\{x_i\}_{i=1}^n)$. Since we may assume that $\|y\| = 1$, this leads to a contradiction with the definition of U.

The following claim shows that each basis in U is an Auerbach basis.

Claim 2.24. A system $\{x_i\}_{i=1}^n \in N$ is an Auerbach basis if and only if

$$M(\{x_i\}_{i=1}^n) \supset B_X.$$

Proof. It is easy to see that

$$M(\{x_i\}_{i=1}^n) = \{x : |x_i^*(x)| \le 1 \text{ for } i = 1,\dots,n\}$$

for each basis $\{x_i\}_{i=1}^n$. Hence $M(\{x_i\}_{i=1}^n) \supset B_X$ if and only if $\|x_i^*\| \le 1$ for each i. It remains to observe that the equality $\|x_i\| = 1$ implies $\|x_i^*\| \ge 1, i = 1, \dots, n$. □

This completes the proof of Lemma 2.23. □

Proof of Proposition 2.22. Let $\{e_{j,i}\}_{j=1}^{\dim X_i}$, $i \in I$, be Auerbach bases in X_i. We may consider each X_i as a linear subspace of ℓ_2^m, where $m = \max_i \dim X_i$. Also we may assume that $e_{j,i}$ is identified with e_j, where $\{e_j\}_{j=1}^m$ is an orthonormal basis in ℓ_2^m. (Of course, in general X_i is not a subspace of ℓ_2^m as a normed space.)

Under these conditions a bounded sequence $(x_i)_{i\in I}$ corresponds to a bounded sequence in ℓ_2^m. In fact, since $\{e_{j,i}\}$ are Auerbach bases, the decomposition $x_i = \sum_j a_{j,i}e_{j,i}$ implies $|a_{j,i}| \le \|x_i\|$, and therefore the norm of the natural image of x_i in ℓ_2^m is $\le \sqrt{m}\|x_i\|$. Thus, by Lemma 2.15, $\lim_{\mathcal{U}} x_i$ exists as an element of ℓ_2^m. It is easy to see that the map $(x_i)_{\mathcal{U}} \mapsto \lim_{\mathcal{U}} x_i$ is a linear embedding of $(\prod X_i)_{\mathcal{U}}$ into ℓ_2^m. This embedding is injective. In fact, $\lim_{\mathcal{U}} x_i = 0$ implies $\lim_{\mathcal{U}} \|x_i\|_{\ell_2^m} = 0$. Therefore $\lim_{\mathcal{U}} a_{j,i} = 0$ for each $j = 1, \dots, m$ and $\lim_{\mathcal{U}} \|x_i\|_{X_i} = 0$. The injectivity of the embedding implies that $(\prod X_i)_{\mathcal{U}}$ is a finite-dimensional space. □

Remark 2.25. In the case where all X_i have the same dimension m, the ultraproduct $(\prod X_i)_{\mathcal{U}}$ is also m-dimensional. In fact, in this case each X_i is identified (as a linear space) with ℓ_2^m (using Auerbach bases as above), and it is clear that each stationary sequence $(x, x, \dots, x, \dots)$ is in $(\prod X_i)_{\mathcal{U}}$ and is identified with the corresponding $x \in \ell_2^m$. Thus $(\prod X_i)_{\mathcal{U}}$ is m-dimensional.

We also use the following important theorem whose proof is beyond the scope of the present book (see Notes and Remarks for references and a proof sketch).

Theorem 2.26. *If $X = L_p(0,1)$ for some $p \in [1, \infty)$ and $\mathcal{U}$ is any ultrafilter, then $X^{\mathcal{U}}$ is isometric to $L_p(\Omega, \Sigma, \mu)$ for some measure space (Ω, Σ, μ), and each separable subspace of $X^{\mathcal{U}}$ is isometric to a subspace of X.*

2.2.3 Finite representability

Definition 2.27. The standard notion of distance between two Banach spaces is the *Banach–Mazur distance* which is defined by

$$d(X, Y) = \inf\{\|T\|\|T^{-1}\| : T : X \to Y \text{ is an isomorphism}\}.$$

If X and Y are not isomorphic, we put $d(X, Y) = \infty$.

This notion is especially useful for pairs of finite-dimensional spaces of the same dimension. The Banach–Mazur distance is not a metric in the sense of the definition given at the beginning of Chapter 1, but it is easy to verify that it satisfies multiplicative analogues of the corresponding conditions:

(a) *Multiplicative triangle inequality:* $\forall X, Y, Z \quad d(X,Z) \le d(X,Y) \cdot d(Y,Z)$.

(b) *Symmetry:* $\forall X, Y \quad d(X,Y) = d(Y,X)$.

(c) $d(X,X) = 1$.

(d) $d(X,Y) \ge 1$.

Therefore, the restriction of $\log d(X,Y)$ to any set of isomorphic Banach spaces is a semimetric. It is not difficult to prove using compactness that in the finite-dimensional case it is a metric, if we identify linearly isometric spaces.

Definition 2.28. Let X and Y be two Banach spaces. The space X is said to be *finitely representable* in Y if for any $\varepsilon > 0$ and any finite-dimensional subspace $F \subset X$ there exists a finite-dimensional subspace $G \subset Y$ such that $d(F,G) < 1 + \varepsilon$.

The space X is said to be *crudely finitely representable* in Y if there exists $1 \le C < \infty$ such that for any finite-dimensional subspace $F \subset X$ there exists a finite-dimensional subspace $G \subset Y$ such that $d(F,G) < C$.

Important example:

Proposition 2.29. *The space L_p is finitely representable in ℓ_p.*

Proof. Let F be a finite-dimensional subspace of L_p, $\dim F = n$. Let $\{e_i\}_{i=1}^n$ be a basis in F. Using compactness we get that there exists a constant $c > 0$ such that

$$\forall \{t_i\}_{i=1}^n \subset \mathbb{R} \quad \left\| \sum_{i=1}^n t_i e_i \right\| \ge c \sum_{i=1}^n |t_i|.$$

(The constant might depend on the subspace, the existence of such a constant is immediate for Auerbach bases.) Recall that a measurable function is called *simple* if it takes only finitely many different values. It is well known (and is easy to see from the definition of $L_p = L_p(0,1)$) that simple functions are dense in L_p. (Even the set of all those simple functions for which pre-images of all values are half-open intervals is dense.)

Let $\varepsilon > 0$ be arbitrary, pick $\omega > 0$ satisfying $(1+\omega)/(1-\omega) < 1+\varepsilon$. Let $f_i \in L_p$ be simple functions satisfying $\|f_i - e_i\| \le c\omega$. There is a partition $\{P_1, \dots, P_m\}$ of $(0,1)$ such that each of the functions f_i is constant on each of the elements of the partition. Consider the subspace of L_p spanned by the indicator functions $\{\mathbf{1}_{P_i}\}_{i=1}^m$. It is easy to see that this space is isometric to ℓ_p^m (see the proof of Proposition 1.19; as we have mentioned, we may assume that each P_i is an interval).

It remains to estimate the norm of the operator $T : F \to L_p$ given by $Te_i = f_i$ and the norm of its inverse. We have

$$\left\| \sum_i t_i e_i - \sum_i t_i f_i \right\| \le \sum_i |t_i| \|e_i - f_i\| \le \omega c \sum_i |t_i| \le \omega \left\| \sum_i t_i e_i \right\|.$$

Thus

$$(1-\omega)\|x\| \le \|Tx\| \le (1+\omega)\|x\|$$

and

$$\|T\| \cdot \|T^{-1}|_{T(F)}\| < 1+\varepsilon. \qquad \square$$

The following theorem is one of the most famous results on finite representability. We do not prove it because we shall use it in some examples only, and also because proofs of this theorem are presented in numerous books and surveys.

Theorem 2.30 (Dvoretzky theorem). *The space ℓ_2 is finitely representable in each infinite-dimensional Banach space.*

The notion of an ultraproduct is closely related to finite representability. One of the results establishing such relations is the following proposition.

Proposition 2.31. *Let X be any infinite-dimensional Banach space and $\mathcal{U}$ be a free ultrafilter. Then $X^{\mathcal{U}}$ is finitely representable in X.*

Proof. Let F be a finite-dimensional subspace of $X^{\mathcal{U}}$. We pick a basis $\{x_j\}_{j=1}^n$ in F. Let $(x_{j,i})_{i\in I}$ be a representative of x_j in $(\oplus_{i\in I}X_i)_\infty$. It is clear that we can extend the map $x_j \mapsto (x_{j,i})_{i\in I}$ to a linear map on $R : F \to (\oplus_{i\in I}X_i)_\infty$ which maps each vector of F to an element in $(\oplus_{i\in I}X_i)_\infty$ representing it.

Now we pick $\delta > 0$ and $\omega > 0$ satisfying the condition

$$\frac{1+\omega}{1-\omega-2\delta} < 1+\varepsilon. \tag{2.2}$$

Let $\{d_k\}_{k=1}^m$ be a δ-net in the unit sphere S_F of F (the *unit sphere* of a normed space F is defined as $S_F = \{x \in F : \|x\| = 1\}$). Let $(d_{k,i})_{i\in I} = Rd_k$.

Since $\|d_k\| = \lim_{\mathcal{U}} \|d_{k,i}\|$, for each k there is a subset $U_k \in \mathcal{U}$ such that

$$\forall i \in U_k \quad 1-\omega \le \|d_{k,i}\| \le 1+\omega$$

Since $\mathcal{U}$ is an ultrafilter, the intersection $\cap_{k=1}^m U_k$ is nonempty. Let $g \in \cap_{k=1}^m U_k$. Then the map given by $d_k \mapsto d_{k,g}$ is a linear map and it satisfies

$$1-\omega \le \|d_{k,g}\| \le 1+\omega \tag{2.3}$$

for each $k = 1,\dots,m$. We denote this linear map by T. Let us estimate the norm of T and $T^{-1}|_{T(F)}$. We use the following general lemma.

Lemma 2.32. *Let $\{d_k\}_{k=1}^m$ be a δ-net in the unit sphere of a finite-dimensional normed space F and $T : F \to X$ be a linear map satisfying*

$$1-\omega \le \|Td_k\| \le 1+\omega.$$

Then $\|T\| \le \frac{1+\omega}{1-\delta}$ and $\|T^{-1}|_{T(F)}\| \le \frac{1-\delta}{1-\omega-2\delta}$.

Proof. Let $f \in S_F$ be such that $\|Tf\| = \|T\|$. Let $d \in \{d_k\}_{k=1}^m$ be such that $\|f - d\| \leq \delta$. We have $\|T\| = \|Tf\| \leq \|Td\| + \|T(f-d)\| \leq 1 + \omega + \|T\|\delta$. Therefore

$$\|T\| \leq \frac{1+\omega}{1-\delta}.$$

Now let f be any element in S_F and $d \in \{d_k\}_{k=1}^m$ be such that $\|f - d\| \leq \delta$. Then

$$\|Tf\| \geq \|Td\| - \|T(f-d)\| \geq 1 - \omega - \delta\|T\| \geq 1 - \omega - \delta\,\frac{1+\omega}{1-\delta}$$

and therefore

$$\|T^{-1}|_{T(F)}\| \leq \frac{1-\delta}{1-\omega-2\delta}. \qquad \square$$

Combining estimates of Lemma 2.32 with the assumption (2.2) we get that the embedding $T{:}F \to X^{\mathcal{U}}$ satisfies the conditions from the definition of finite representability. $\square$

Now we prove a result which is a kind of a converse to Proposition 2.31.

Proposition 2.33. *Let Z be a Banach space which is finitely representable in a Banach space X. Then there exists a free ultrafilter $\mathcal{U}$ such that the ultrapower $X^{\mathcal{U}}$ contains a subspace isometric to Z.*

Proof. Denote by J the set of all finite-dimensional subspaces of Z. Consider the set $I = J \times (0,1)$ as an ordered set: $(j_1, \varepsilon_1) \succeq (j_2, \varepsilon_2)$ if and only if $j_1 \supseteq j_2$ and $\varepsilon_1 \leq \varepsilon_2$. Consider an ultrafilter $\mathcal{U}$ on I containing the filter generated by sets of the form $\{(j,\varepsilon) : (j,\varepsilon) \succeq (j_0, \varepsilon_0)\}$, where $j_0 \in J$, $\varepsilon_0 \in (0,1)$.

Since Z is finitely representable in X, for each pair $(j,\varepsilon) \in J \times (0,1)$ we can pick a linear operator $T_{(j,\varepsilon)} : j \to X$ satisfying

$$\|z\| \leq \|T_{(j,\varepsilon)}(z)\| \leq (1+\varepsilon)\|z\|.$$

It remains to observe that the maps

$$z \mapsto \begin{cases} T_{(j,\varepsilon)}(z) & \text{if } z \in j \\ 0 & \text{if } z \notin j \end{cases}$$

(parameterized by pairs $(j,\varepsilon) \in I$) induce an isometric embedding of Z into $X^{\mathcal{U}}$. $\square$

2.3 Proofs of the main results on relations between embeddability of a locally finite metric space and its finite subsets

2.3.1 Proof in the bilipschitz case

Proof of Theorem 2.6. We pick a point O in A and let $A_i = \{a \in A : d_A(O,a) \leq 2^i\}$. By the assumption there are uniformly bilipschitz maps $f_i : A_i \to X$. We may and shall assume that $f_i(O) = 0$ and that there is a constant $1 \leq C < \infty$ such that

$$\forall u, v \in A_i \quad d_A(u,v) \leq \|f_i(u) - f_i(v)\| \leq C d_A(u,v).$$

(See the proof of Proposition 2.21.)

Let $\mathcal{U}$ be a free ultrafilter on $\mathbb{N}$. The maps $\{f_i\}_{i=1}^{\infty}$ induce a map $f : A \to X^{\mathcal{U}}$ defined by $f(u) = \{\widetilde{f_i}(u)\}_{i=1}^{\infty}$, where

$$\widetilde{f_i}(u) = \begin{cases} f_i(u) & \text{if } u \in A_i \\ 0 & \text{if } u \notin A_i. \end{cases}$$

The definition of an ultraproduct immediately implies that $f : A \to X^{\mathcal{U}}$ is a bilipschitz embedding (the argument is the same as in Proposition 2.21). Let $N = f(A)$. Since the composition of two bilipschitz embeddings is a bilipschitz embedding, it suffices to find a bilipschitz embedding of N (with the metric induced from $X^{\mathcal{U}}$) into X.

Note. This passage from A to its image in $X^{\mathcal{U}}$ is not essential for the proof of Theorem 2.6, it just simplifies some formulas in our proof. A similar step is more essential for other classes of embeddings.

Remark 2.34. We would like to emphasize that the rest of the proof of Theorem 2.6 consists of establishing the fact that a locally finite metric subspace of $X^{\mathcal{U}}$ admits a bilipschitz embedding into X.

Remark 2.25 implies that the case where X is finite-dimensional is trivial. In this connection in the rest of the proof we assume that X is infinite-dimensional.

Let $N_i = \{u \in N : \|u\|_{X^{\mathcal{U}}} \leq 2^i\}$. It is clear that N_i are finite sets. By Proposition 2.31 there exist maps $s_i : N_i \to X$ such that $s_i(0) = 0$ and

$$\forall u, v \in N_i \quad \|u - v\| \leq \|s_i(u) - s_i(v)\| \leq \left(1 + \frac{1}{i}\right)\|u - v\|. \tag{2.4}$$

(The direct proof of this statement is simpler than the proof of Proposition 2.31, since we do not need the argument with nets.)

Since the sets N_i form an increasing sequence, any subsequence $\{s_{n_i}\}_{i=1}^{\infty}$ of the sequence $\{s_i\}_{i=1}^{\infty}$ maps $\{N_i\}_{i=1}^{\infty}$ into X and satisfies (2.4). We are going to construct a bilipschitz embedding of N into X using such subsequences.

Note. We are going to pass to a subsequence in $\{s_i\}_{i=1}^{\infty}$ several times. Each time we keep the notation $\{s_i\}_{i=1}^{\infty}$ for the subsequence.

Recall that a subspace $M \subset X^*$ is called 1-*norming* if

$$\forall x \in X \ \sup\{|f(x)| : f \in M, \ \|f\| \le 1\} = \|x\|.$$

It is clear that we may assume that X is separable (replacing it by the closure of the linear span of $\bigcup_{i=1}^{\infty} s_i(N_i)$, if necessary).

For a separable Banach space X there exists a separable 1-norming subspace $M \subset X^*$. It can be constructed as follows: Let $\{x_i\}_{i=1}^{\infty}$ be a dense sequence in the unit sphere $S_X = \{x \in X : \|x\| = 1\}$. Let $f_i \in S_{X^*}$ be such that $f_i(x_i) = 1$. It is easy to check that the closed linear span M of the sequence $\{f_i\}_{i=1}^{\infty}$ is 1-norming.

Let $M \subset X^*$ be a separable 1-norming subspace. Then the natural embedding of X into M^* is an isometry. We identify X with its image under this embedding. Since M is separable, there is a subsequence in $\{s_i\}$ such that the sequence $\{s_i(a)\}_{i=k}^{\infty}$ is convergent in the weak* topology of M^* for each $a \in N_k$. We denote the weak* limit of this sequence by $m(a)$.

We need to select further subsequences of $\{s_i\}$. We do this in the following two steps.

Step 1. If $m(a) \ne m(b)$ for $a, b \in N_j$, we find and fix $f = f_{a,b} \in S_M$ satisfying

$$f(m(a) - m(b)) \ge \frac{99}{100}\|m(a) - m(b)\|. \tag{2.5}$$

We select a subsequence of $\{s_i\}$ satisfying

$$|f((s_k(a) - s_k(b)) - (m(a) - m(b)))| \le \frac{1}{100}\|m(a) - m(b)\| \tag{2.6}$$

for $k \ge j$. This can be achieved because N_j is finite and the sequence $\{s_k(a)\}$ converges to $m(a)$ in the weak* topology of M^*.

Step 2. If for some $a, b \in N_\ell$ and some $j \ge \ell$ the vector $(s_j(a) - s_j(b)) - (m(a) - m(b))$ is nonzero, we find and fix $f = f_{j,a,b} \in S_M$ such that

$$\begin{aligned} &f((s_j(a) - s_j(b)) - (m(a) - m(b))) \\ &\qquad \ge \frac{99}{100}\|(s_j(a) - s_j(b)) - (m(a) - m(b))\|. \end{aligned} \tag{2.7}$$

In such a case we assume that for all $k > j$ the condition

$$|f((s_k(a) - s_k(b)) - (m(a) - m(b)))| \le \frac{1}{1000}\|a - b\| \tag{2.8}$$

holds. This goal can be achieved because there are finitely many $a, b \in N_\ell$ and because $s_k(a)$ converges to $m(a)$ in the weak* topology of M^*.

We introduce a map $\varphi : N \to X$ by

$$\varphi(a) = \frac{2^i - \|a\|}{2^{i-1}} s_i(a) + \frac{\|a\| - 2^{i-1}}{2^{i-1}} s_{i+1}(a) \tag{2.9}$$

if $2^{i-1} \le \|a\| \le 2^i$. This formula gives two definitions of $\varphi(a)$ if $\|a\| = 2^i$, but one can easily check that they coincide.

We start by considering the case where X is isomorphic to $X \oplus \mathbb{R}$. To make this condition more clear we need to specify the norm on $X \oplus \mathbb{R}$; we do this by $\|(x,t)\| = \|x\|_X + |t|$. In this case we show that the embedding $\widetilde{\varphi} : N \to X \oplus \mathbb{R}$ given by $\widetilde{\varphi}(a) = (\varphi(a), \|a\|)$ is a bilipschitz embedding.

Recall that a *hyperplane* in a Banach space X is defined as the subspace $\{x \in X : f(x) = 0\}$, where f is a linear functional. A *closed hyperplane* is defined in the same way with the additional condition that f is continuous. It is easy to see that a Banach space X is isomorphic to $X \oplus \mathbb{R}$ if and only if X is isomorphic to its closed hyperplane. Gowers and Maurey (1993) proved that Banach spaces which are not isomorphic to their closed hyperplanes exist. Because of this we have to consider also the case where X is not isomorphic to $X \oplus \mathbb{R}$. This is done in Section 2.3.1.

Now we estimate the Lipschitz constants of $\widetilde{\varphi}$ and $(\widetilde{\varphi})^{-1}$. We consider three cases.

Case 1. $2^{i-1} \le \|b\| \le \|a\| \le 2^i$

In this case we have

$$\begin{aligned}
&\varphi(a) - \varphi(b)\\
&\quad= \frac{2^i - \|a\|}{2^{i-1}} s_i(a) + \frac{\|a\| - 2^{i-1}}{2^{i-1}} s_{i+1}(a)\\
&\qquad - \frac{2^i - \|b\|}{2^{i-1}} s_i(b) - \frac{\|b\| - 2^{i-1}}{2^{i-1}} s_{i+1}(b)\\
&\quad= \frac{2^i - \|a\|}{2^{i-1}} (s_i(a) - s_i(b)) + \frac{\|a\| - 2^{i-1}}{2^{i-1}} (s_{i+1}(a) - s_{i+1}(b))\\
&\qquad + \frac{\|b\| - \|a\|}{2^{i-1}} s_i(b) + \frac{\|a\| - \|b\|}{2^{i-1}} s_{i+1}(b).
\end{aligned} \tag{2.10}$$

Using (2.4) we get

$$\begin{aligned}
\|\varphi(a) - \varphi(b)\| \le{}& \frac{2^i - \|a\|}{2^{i-1}} 2\|a - b\| + \frac{\|a\| - 2^{i-1}}{2^{i-1}} 2\|a - b\|\\
&+ 4|\,\|b\| - \|a\|\,| + 4|\,\|a\| - \|b\|\,|,
\end{aligned}$$

therefore $\|\widetilde{\varphi}(a) - \widetilde{\varphi}(b)\| \le C\|a - b\|$ for some absolute constant C.

Now we estimate $\|\widetilde{\varphi}(a)-\widetilde{\varphi}(b)\|$ from below. So we need to estimate $\|\varphi(a)-\varphi(b)\|$ from below. Observe that

$$\begin{aligned}\varphi(a)-\varphi(b) &= m(a)-m(b)\\ &\quad+\frac{2^i-\|a\|}{2^{i-1}}(s_i(a)-s_i(b)-(m(a)-m(b)))\\ &\quad+\frac{\|a\|-2^{i-1}}{2^{i-1}}(s_{i+1}(a)-s_{i+1}(b)-(m(a)-m(b)))\\ &\quad+(\|b\|-\|a\|)\cdot\frac{s_i(b)}{2^{i-1}}+(\|a\|-\|b\|)\frac{s_{i+1}(b)}{2^{i-1}}.\end{aligned}\tag{2.11}$$

First we consider the case where $\|m(a)-m(b)\|\geq\frac{1}{100}\|a-b\|$. Let $f_{a,b}$ be the corresponding functional (see Step 1 above). We have

$$\begin{aligned}&\|\varphi(a)-\varphi(b)\|\\ &\qquad\geq f_{a,b}(\varphi(a)-\varphi(b))\\ &\qquad= f_{a,b}(m(a)-m(b))\\ &\qquad\quad+f_{a,b}\Big(\frac{2^i-\|a\|}{2^{i-1}}(s_i(a)-s_i(b)-(m(a)-m(b)))\Big)\\ &\qquad\quad+f_{a,b}\Big(\frac{\|a\|-2^{i-1}}{2^{i-1}}(s_{i+1}(a)-s_{i+1}(b)-(m(a)-m(b)))\Big)\\ &\qquad\quad+(\|b\|-\|a\|)\cdot\frac{f_{a,b}(s_i(b))}{2^{i-1}}+(\|a\|-\|b\|)\frac{f_{a,b}(s_{i+1}(b))}{2^{i-1}}\\ &\qquad\overset{(2.5)\,\&\,(2.6)}{\geq}\frac{99}{100}\|m(a)-m(b)\|-\frac{1}{100}\|m(a)-m(b)\|\\ &\qquad\quad-4|\,\|b\|-\|a\|\,|-4|\,\|a\|-\|b\|\,|\\ &\qquad\geq\frac{98}{10000}\|a-b\|-8(\|a\|-\|b\|).\end{aligned}\tag{2.12}$$

In the case where $\|a\|-\|b\|<\frac{1}{1000}\|a-b\|$, we get an estimate for $\|\varphi(a)-\varphi(b)\|$ (and thus for $\|\widetilde{\varphi}(a)-\widetilde{\varphi}(b)\|$) from below of the form $c\|a-b\|$, where $c>0$ is an absolute constant.

The estimate $\|\widetilde{\varphi}(a)-\widetilde{\varphi}(b)\|$ from below in the case where $\|a\|-\|b\|\geq\frac{1}{1000}\|a-b\|$ is immediate, we just recall that

$$\widetilde{\varphi}(a)-\widetilde{\varphi}(b)=(\varphi(a)-\varphi(b))\oplus(\|a\|-\|b\|).\tag{2.13}$$

Hence, to finish the lower estimate for $\|\widetilde{\varphi}(a)-\widetilde{\varphi}(b)\|$ in Case 1 it remains to consider the case where $\|m(a)-m(b)\|<\frac{1}{100}\|a-b\|$. In this case we consider two subcases:

$$\frac{2^i-\|a\|}{2^{i-1}}\;\|s_i(a)-s_i(b)-(m(a)-m(b))\|\geq\frac{1}{10}\|a-b\|\tag{2.14}$$

$$\frac{2^i-\|a\|}{2^{i-1}}\;\|s_i(a)-s_i(b)-(m(a)-m(b))\|<\frac{1}{10}\|a-b\|.\tag{2.15}$$

We start with subcase (2.14). Let $f_{i,a,b}$ be the functional found in Step 2. We get

$$\begin{aligned}\|\varphi(a)-\varphi(b)\| &\geq f_{i,a,b}(\varphi(a)-\varphi(b)) \geq f_{i,a,b}(m(a)-m(b))\\ &\quad+ f_{i,a,b}\Big(\frac{2^i-\|a\|}{2^{i-1}}(s_i(a)-s_i(b)-(m(a)-m(b)))\Big)\\ &\quad+ f_{i,a,b}\Big(\frac{\|a\|-2^{i-1}}{2^{i-1}}(s_{i+1}(a)-s_{i+1}(b)-(m(a)-m(b)))\Big)\\ &\quad+(\|b\|-\|a\|)\cdot\frac{f_{i,a,b}(s_i(b))}{2^{i-1}}+(\|a\|-\|b\|)\frac{f_{i,a,b}(s_{i+1}(b))}{2^{i-1}} \qquad (2.16)\\ &\overset{(2.7),(2.8),\ \&\ (2.14)}{>} \frac{99}{1000}\|a-b\|-\frac{1}{1000}\|a-b\|-\frac{1}{100}\|a-b\|\\ &\quad-4|\,\|b\|-\|a\|\,|-4|\,\|a\|-\|b\|\,|\\ &=\frac{88}{1000}\|a-b\|-8(\|a\|-\|b\|).\end{aligned}$$

Now, as above, we consider the case where $\|a\|-\|b\|<\frac{1}{1000}\|a-b\|$ separately, and complete the argument in the same way as above.

We turn to subcase (2.15). Recall (see (2.4)) that $\|s_i(a)-s_i(b)\|\geq\|a-b\|$. Combining this with (2.15) and with the inequality $\|m(a)-m(b)\|<\frac{1}{100}\|a-b\|$, we get $\|s_i(a)-s_i(b)-(m(a)-m(b))\|\geq\frac{99}{100}\|a-b\|$ and $\frac{2^i-\|a\|}{2^{i-1}}<\frac{10}{99}$. (In the same way we get the inequality $\|s_{i+1}(a)-s_{i+1}(b)-(m(a)-m(b))\|\geq\frac{99}{100}\|a-b\|$ which we use below.) Therefore $\frac{\|a\|-2^{i-1}}{2^{i-1}}>\frac{89}{99}$. Applying the triangle inequality we get

$$\begin{aligned}\|\varphi(a)-\varphi(b)\| &\geq \Big\|\frac{\|a\|-2^{i-1}}{2^{i-1}}(s_{i+1}(a)-s_{i+1}(b)-(m(a)-m(b)))\Big\|\\ &\quad-\Big\|\frac{2^i-\|a\|}{2^{i-1}}(s_i(a)-s_i(b)-(m(a)-m(b)))\Big\|\\ &\quad-\|m(a)-m(b)\|-8(\|a\|-\|b\|) \qquad (2.17)\\ &>\frac{89}{99}\cdot\frac{99}{100}\|a-b\|-\frac{1}{10}\|a-b\|-\frac{1}{100}\|a-b\|-8(\|a\|-\|b\|)\\ &=\frac{78}{100}\|a-b\|-8(\|a\|-\|b\|).\end{aligned}$$

Now, as was done already twice, we consider the case where $\|a\|-\|b\|<\frac{1}{100}\|a-b\|$ separately, and complete the argument in the same way as above. This completes the argument in Case 1.

Case 2. $\mathbf{2^{i-1} \le \|b\| \le 2^i \le \|a\| \le 2^{i+1}}$

We have

$$\begin{aligned}\varphi(a) - \varphi(b) = &-\frac{2^i - \|b\|}{2^{i-1}} s_i(b) \\ &+ \frac{2^{i+1} - \|a\|}{2^i} s_{i+1}(a) - \frac{\|b\| - 2^{i-1}}{2^{i-1}} s_{i+1}(b) \\ &+ \frac{\|a\| - 2^i}{2^i} s_{i+2}(a).\end{aligned}$$

Estimate from above: The first and the last terms have norms $\le 4(\|a\| - \|b\|)$. The norm of the two remaining terms can be estimated as follows:

$$\begin{aligned}&\left\| \frac{2^{i+1} - \|a\|}{2^i} s_{i+1}(a) - \frac{\|b\| - 2^{i-1}}{2^{i-1}} s_{i+1}(b) \right\| \\ &= \left\| \frac{2^i - (\|a\| - 2^i)}{2^i} s_{i+1}(a) + \frac{(2^i - \|b\|) - 2^{i-1}}{2^{i-1}} s_{i+1}(b) \right\| \\ &= \left\| (s_{i+1}(a) - s_{i+1}(b)) - \frac{(\|a\| - 2^i)}{2^i} s_{i+1}(a) + \frac{(2^i - \|b\|)}{2^{i-1}} s_{i+1}(b) \right\| \\ &\le 2\|a - b\| + 4(\|a\| - 2^i) + 4(2^i - \|b\|) \le 6\|a - b\|.\end{aligned} \tag{2.18}$$

Now we turn to estimates from below. Rewriting and estimating some of the terms as in (2.18) we get

$$\begin{aligned}&\|\varphi(a) - \varphi(b)\| \\ &\ge \left\| (s_{i+1}(a) - s_{i+1}(b)) - \frac{(\|a\| - 2^i)}{2^i} s_{i+1}(a) + \frac{(2^i - \|b\|)}{2^{i-1}} s_{i+1}(b) \right\| \\ &\quad - \frac{2^i - \|b\|}{2^{i-1}} \|s_i(b)\| - \frac{\|a\| - 2^i}{2^i} \|s_{i+2}(a)\| \\ &\ge \|s_{i+1}(a) - s_{i+1}(b)\| - 12(\|a\| - \|b\|) \\ &\ge \|a - b\| - 12(\|a\| - \|b\|),\end{aligned} \tag{2.19}$$

where in the last line we used (2.4). We complete the proof in this case as three times before. If $\|a\| - \|b\| < \frac{1}{20}\|a - b\|$, we get an estimate using (2.19). Otherwise we use (2.13).

Case 3. $\mathbf{2^{i-1} \le \|b\| \le 2^i < 2^{k-1} \le \|a\| \le 2^k}$

In this case we have

$$3(2^k + 2^i) \ge 3(\|a\| + \|b\|) \ge \|\tilde{\varphi}(a) - \tilde{\varphi}(b)\| \ge \|a\| - \|b\| \ge 2^{k-1} - 2^i.$$

Since

$$\frac{3(2^k+2^i)}{2^{k-1}-2^i} \le \frac{3\cdot 2^{k+1}}{2^{k-2}} \le 24$$

and

$$\|a\|+\|b\| \ge \|a-b\| \ge \|a\|-\|b\|,$$

it follows that $\widetilde{\varphi}$ is bilipschitz.

Completion of the proof for spaces non-isomorphic to their closed hyperplanes

We have proved Theorem 2.6 in the case where X is isomorphic to its closed hyperplanes. To prove Theorem 2.6 in the general case we find a Lipschitz map $\tau : \mathbb{R}_+ \to X$ such that the map $\widehat{\varphi} : N \to X$ given by $\widehat{\varphi}(a) = \tau(\|a\|) + \varphi(a)$ works just in the same way as $\widetilde{\varphi}$. It is easy to see that for this to be true we need the inequality

$$\|\widehat{\varphi}(a) - \widehat{\varphi}(b)\| \ge \alpha(\|a\| - \|b\|)$$

to hold for some $\alpha > 0$. We rewrite this inequality as

$$\|\tau(\|a\|) - \tau(\|b\|) + (\varphi(a) - \varphi(b))\| \ge \alpha(\|a\| - \|b\|). \tag{2.20}$$

Let $T_i = \{\varphi(u) : u \in N,\ \|u\| \le 3^{i+1}\}$. It is clear that all these sets are finite. We construct inductively a sequence $\{F_i\}_{i=1}^\infty$ of finite-dimensional subspaces of X and a sequence $\{p_i\}_{i=1}^\infty$ of vectors. We let $F_1 = \operatorname{lin}(T_1)$. (Here and below $\operatorname{lin}(T)$ denotes the linear span of a set T in a vector space.) Recall that we have assumed that X is infinite-dimensional. Therefore there is $p_1 \in S_X$ such that $\operatorname{dist}(p_1, F_1) = 1$. Let $F_2 = \operatorname{lin}(T_2 \cup \{p_1\})$ and $p_2 \in S_X$ be such that $\operatorname{dist}(p_2, F_2) = 1$. Let $F_3 = \operatorname{lin}(T_3 \cup \{p_i\}_{i=1}^2)$, we continue in an obvious way.

We introduce the map $\tau : \mathbb{R}_+ \to X$ in the following way:

$$\tau(t) = \begin{cases} tp_1 & \text{if } 0 \le t \le 3^1 \\ 3^1 p_1 + (t-3^1)p_2 & \text{if } 3^1 \le t \le 3^2 \\ 3^1 p_1 + (3^2-3^1)p_2 + (t-3^2)p_3 & \text{if } 3^2 \le t \le 3^3 \\ \dots & \dots \\ 3^1 p_1 + (3^2-3^1)p_2 + \cdots & \\ +(3^k-3^{k-1})p_k + (t-3^k)p_{k+1} & \text{if } 3^k \le t \le 3^{k+1} \\ \dots & \dots \end{cases}$$

Since $\|p_i\| = 1$, the map τ is 1-Lipschitz. It remains to show that the inequality (2.20) holds. We consider three cases:

(a) $3^i \le \|b\| \le \|a\| \le 3^{i+1}$. The argument used in this case can be used also in the case $0 \le \|b\| \le \|a\| \le 3^1$. Minor adjustments of the other cases are needed if $0 \le \|b\| \le 3^1 \le \|a\|$.

(b) $3^{i-1} \leq \|b\| \leq 3^i \leq \|a\| \leq 3^{i+1}$

(c) $3^{k-1} \leq \|b\| \leq 3^k \leq 3^i \leq \|a\| \leq 3^{i+1}$, where $k < i$.

In case (a) we have

$$\begin{aligned}\|\tau(\|a\|) - \tau(\|b\|) + (\varphi(a) - \varphi(b))\| &= \|(\|a\| - \|b\|)p_{i+1} + (\varphi(a) - \varphi(b))\| \\ &\geq \|a\| - \|b\|.\end{aligned}$$

The last inequality follows from $\operatorname{dist}(p_{i+1}, F_{i+1}) = \|p_{i+1}\|$ and $\varphi(a), \varphi(b) \in T_i$, therefore $\varphi(a) - \varphi(b) \in F_i \subset F_{i+1}$.

In case (b) we consider two subcases:

$$\|a\| - 3^i \geq \frac{1}{3}(\|a\| - \|b\|). \tag{2.21}$$

$$\|a\| - 3^i < \frac{1}{3}(\|a\| - \|b\|). \tag{2.22}$$

In subcase (2.21) we get

$$\begin{aligned}&\|\tau(\|a\|) - \tau(\|b\|) + (\varphi(a) - \varphi(b))\| \\ &\qquad = \left\|(\|a\| - 3^i)p_{i+1} + (3^i - \|b\|)p_i + (\varphi(a) - \varphi(b))\right\| \\ &\qquad \geq \|a\| - 3^i \geq \frac{1}{3}(\|a\| - \|b\|),\end{aligned}$$

where we use (2.21), $\operatorname{dist}(p_{i+1}, F_{i+1}) = \|p_{i+1}\|$, and $p_i, \varphi(a), \varphi(b) \in F_{i+1}$.

In subcase (2.22) we have

$$\begin{aligned}&\|\tau(\|a\|) - \tau(\|b\|) + (\varphi(a) - \varphi(b))\| \\ &\qquad = \left\|(\|a\| - 3^i)p_{i+1} + (3^i - \|b\|)p_i + (\varphi(a) - \varphi(b))\right\| \\ &\qquad \geq \left\|(3^i - \|b\|)p_i + (\varphi(a) - \varphi(b))\right\| - (\|a\| - 3^i) \\ &\qquad \geq (3^i - \|b\|) - (\|a\| - 3^i) \geq \frac{1}{3}(\|a\| - \|b\|).\end{aligned}$$

In this chain of inequalities we use the fact that $\varphi(a), \varphi(b) \in T_i$, $\operatorname{dist}(p_i, F_i) = \|p_i\|$, in the last line we use the inequality $\|b\| \leq 3^i \leq \|a\|$ and (2.22).

Now we consider case (c). In this case we consider two subcases:

$$\|a\| - 3^i \geq \frac{1}{3}(3^i - 3^{i-1}). \tag{2.23}$$

$$\|a\| - 3^i < \frac{1}{3}(3^i - 3^{i-1}). \tag{2.24}$$

In subcase (2.23) our argument is very close to the argument above:

$$\begin{aligned}
&\|\tau(\|a\|) - \tau(\|b\|) + (\varphi(a) - \varphi(b))\| \\
&\qquad = \big\|(\|a\| - 3^i)p_{i+1} + (3^i - 3^{i-1})p_i \\
&\qquad\quad + \cdots + (3^k - \|b\|)p_k + (\varphi(a) - \varphi(b))\big\| \\
&\qquad \geq \|a\| - 3^i \\
&\qquad \geq \frac{1}{3}(3^i - 3^{i-1}) \\
&\qquad = \frac{2}{27}3^{i+1} \\
&\qquad \geq \frac{2}{27}(\|a\| - \|b\|).
\end{aligned}$$

Now we consider subcase (2.24). In this subcase we get (from the previous formula)

$$\tau(\|a\|) - \tau(\|b\|) = (\|a\| - 3^i)p_{i+1} + (3^i - 3^{i-1})p_i + r,$$

where r is a vector contained in F_i. Thus

$$\begin{aligned}
&\|\tau(\|a\|) - \tau(\|b\|) + (\varphi(a) - \varphi(b))\| \\
&\qquad \geq \big\|(3^i - 3^{i-1})p_i + r + (\varphi(a) - \varphi(b))\big\| - \|(\|a\| - 3^i)p_{i+1}\| \\
&\qquad \geq (3^i - 3^{i-1}) - \frac{1}{3}(3^i - 3^{i-1}) = \frac{4}{3}\cdot 3^{i-1} \geq \frac{4}{27}(\|a\| - \|b\|),
\end{aligned}$$

where we use the fact that r and $\varphi(a) - \varphi(b)$ are in F_i and $\operatorname{dist}(p_i, F_i) = \|p_i\|$. This completes the proof of (2.20) and thus Theorem 2.6. □

2.3.2 Proof in the coarse case

Proof of Theorem 2.6 contains almost everything we need for the proof of Theorem 2.7, we need just to modify the beginning of the proof.

Proof of Theorem 2.7. We pick a point O in A and let $A_i = \{a \in A : d_A(O,a) \leq 2^i\}$. By the assumption there are uniformly coarse maps $f_i : A_i \to X$. We may and shall assume that $f_i(O) = 0$. Let $\mathcal{U}$ be a free ultrafilter on $\mathbb{N}$. The maps $\{f_i\}_{i=1}^{\infty}$ induce a map $f : A \to X^{\mathcal{U}}$ defined by $f(u) = \{\widetilde{f_i}(u)\}_{i=1}^{\infty}$, where

$$\widetilde{f_i}(u) = \begin{cases} f_i(u) & \text{if } u \in A_i \\ 0 & \text{if } u \notin A_i. \end{cases}$$

The definition of an ultraproduct immediately implies that $f : A \to X^{\mathcal{U}}$ is a coarse embedding (see the proof of Proposition 2.21). Let $N = f(A)$, it is easy to check that in N with the metric induced from $X^{\mathcal{U}}$ all balls of finite radius have finite cardinality.

The argument of the proof of Theorem 2.6 shows that there is a bilipschitz embedding of N into X (see Remarks 2.34 and 2.35). Since the composition of a coarse and a bilipschitz embedding is a coarse embedding, the proof is completed. □

2.3.3 Remarks on extensions of finite determination results

Theorems 2.6 and 2.7 can be extended in two directions:

Remark 2.35. In the proofs of Theorems 2.6 and 2.7 we never used the condition that the metric space A is uniformly discrete. For this reason these results remain true if the only assumption on A is that for every $a \in A$ and every $r > 0$ the set $\{u \in A : d_X(a,u) \le r\}$ is finite.

Remark 2.36. It is worth mentioning that our argument implies that a result similar to Theorems 2.6 and 2.7 holds for any class $\mathcal{E}$ of embeddings provided that

(a) There is a notion of being *uniformly in* $\mathcal{E}$ for a collection of maps of finite metric spaces into a Banach space.

(b) The notion in (a) is such that if all finite subspaces of a metric space A admit uniformly-in-$\mathcal{E}$ embeddings into a Banach space X, then there is an embedding of the class $\mathcal{E}$ of A into an ultraproduct $X^{\mathcal{U}}$ where $\mathcal{U}$ is a free ultrafilter (see the construction below).

(c) The image of a locally finite metric space under an embedding of the class $\mathcal{E}$ is locally finite.

(d) A composition of an embedding of the class $\mathcal{E}$ and a bilipschitz embedding is in $\mathcal{E}$.

2.4 Banach space theory: type and cotype of Banach spaces, Khinchin and Kahane inequalities

2.4.1 Definitions: Rademacher type and cotype, parallelogram identity

One of the most important directions of research in Banach space theory is to find embeddings of Banach spaces, one into another; and to find obstructions for such embeddings. Of course the results depend heavily on the class of "admissible" embeddings. In classical linear theory the most important class of embeddings is the class of *linear isomorphic embeddings*, that is, linear maps $T : X \to Y$ satisfying $c\|x\|_X \le \|Tx\|_Y \le C\|x\|_X$ for some $0 < c \le C < \infty$. Such a T is called K-*isomorphic*, where $K = \frac{C}{c}$. Sometimes linear isomorphic embeddings are called just *isomorphic embeddings*. One of the important tools and outstanding achievements in the linear isomorphic theory is the type-cotype theory for Banach spaces.

Definition 2.37. *Rademacher functions* $\{r_k(t)\}_{k=0}^{\infty}$ on $[0,1]$ are defined by

$$r_k(t) = \begin{cases} 1 & \text{for all } t \in [0,1], \text{ if } k = 0 \\ \operatorname{sign}(\sin 2^k \pi t) & \text{if } k \geq 1. \end{cases}$$

A Banach space X is said to have *Rademacher type* p, $1 \leq p \leq 2$ if there exists a constant $T_p(X) < \infty$ such that for every n and for every $x_1, \dots, x_n \in X$,

$$\left(\int_0^1 \left\| \sum_{i=1}^n r_i(t) x_i \right\|_X^p dt \right)^{1/p} \leq T_p(X) \left(\sum_{i=1}^n \|x_i\|_X^p \right)^{1/p}, \tag{2.25}$$

where $\{r_i\}$ are Rademacher functions. The Banach space X is said to have *Rademacher cotype* q, $2 \leq q \leq \infty$ if there exists a constant $0 < C_q(X) < \infty$ such that for every n and for every $x_1, \dots, x_n \in X$,

$$\left(\int_0^1 \left\| \sum_{i=1}^n r_i(t) x_i \right\|_X^q dt \right)^{1/q} \geq \frac{1}{C_q(X)} \left(\sum_{i=1}^n \|x_i\|_X^q \right)^{1/q}. \tag{2.26}$$

In the case where $q = \infty$ we understand inequality (2.26) as

$$\max_{0 \leq t \leq 1} \left\| \sum_{i=1}^n r_i(t) x_i \right\|_X \geq \frac{1}{C_\infty(X)} \max_i \|x_i\|_X.$$

Let us explain the restrictions on p and q in this definition and give some simple examples.

Observation 2.38. The triangle inequality implies that each Banach space has type 1 with constant 1, and cotype ∞ with constant 1.

In fact, to show the first statement we observe that the triangle inequality implies

$$\left\| \sum_{i=1}^n r_i(t) x_i \right\|_X \leq \sum_{i=1}^n \|x_i\|_X$$

for each $t \in [0,1]$.

The second statement can be proved using induction on n and the following two observations. We may assume that $\max_i \|x_i\| = \|x_1\|$. For every $z \in X$ and $i = 1, \dots, n$ we have $z = \frac{1}{2}((z + x_i) + (z - x_i))$. Therefore $\max\{\|z + x_i\|, \|z - x_i\|\} \geq \|z\|$.

Definition 2.39. If a Banach space X has some type $p > 1$, we say that X has *nontrivial type*. If a Banach space X has some cotype $q < \infty$, we say that X has *nontrivial cotype*.

Proposition 2.40. *Any Hilbert space (finite- or infinite-dimensional) has type* 2 *with constant* 1 *and cotype* 2 *with constant* 1.

Proof. Let $x_1, \dots, x_n$ be elements of a Hilbert space H. Then

$$\Big\| \sum_{i=1}^{n} r_i(t)x_i \Big\|_H^2 = \Big\langle \sum_{i=1}^{n} r_i(t)x_i, \sum_{j=1}^{n} r_j(t)x_j \Big\rangle = \sum_{i=1}^{\infty}\sum_{j=1}^{\infty} r_i(t)r_j(t)\langle x_i, x_j\rangle.$$

Now, when we integrate with respect to t, we use the fact that

$$\int_0^1 r_i(t)r_j(t)dt = \delta_{i,j} \quad \text{(Kronecker delta)},$$

and get

$$\Big(\int_0^1 \Big\| \sum_{i=1}^{n} r_i(t)x_i \Big\|_H^2 dt \Big)^{1/2} = \Big(\sum_{i=1}^{n} \|x_i\|_H^2 \Big)^{1/2}. \tag{2.27}$$

This proves the proposition. □

To develop some theory of type-cotype we need an important inequality showing that we can replace inequalities (2.25) and (2.26) in the definitions of Rademacher type and cotype with

$$\Big(\int_0^1 \Big\| \sum_{i=1}^{n} r_i(t)x_i \Big\|_X^2 dt \Big)^{1/2} \le \widetilde{T}_p(X) \Big(\sum_{i=1}^{n} \|x_i\|_X^p \Big)^{1/p} \tag{2.28}$$

and

$$\Big(\int_0^1 \Big\| \sum_{i=1}^{n} r_i(t)x_i \Big\|_X^2 dt \Big)^{1/2} \ge \frac{1}{\widetilde{C}_q(X)} \Big(\sum_{i=1}^{n} \|x_i\|_X^q \Big)^{1/q}, \tag{2.29}$$

respectively, and this will affect the constants only. Even more, we can replace 2 in the left-hand sides of these inequalities with any number in $[1, \infty)$, and again, this change will affect the constants only: the obtained inequalities will characterize the same class of Banach spaces. This goal will be achieved in the next section.

Remark 2.41. Now we can explain why in the definition of type we restrict to $p \le 2$ and in the definition of cotype to $q \ge 2$. The reason is that any space which has dimension at least 1 does not have any type $p > 2$ and does not have any cotype $q < 2$. Since type and cotype are obviously inherited by subspaces and each Banach space, except for the space $\{0\}$, contains a one-dimensional subspace, it suffices to show that a one-dimensional space (which can be assumed to be a Hilbert space) does not have any type $p > 2$ and does not have any cotype $q < 2$. To see this we combine equality (2.27) with the equivalent definitions of type and cotype in (2.28) and (2.29). In fact for a sequence $\{x_i\}_{i=1}^n$ consisting of vectors of norm 1 (it can be the same vector repeated n times) in the space $X = \mathbb{R}^1$, we get $\sqrt{n} \le \widetilde{T}_p(X) \cdot n^{1/p}$ and $\sqrt{n} \ge \frac{1}{\widetilde{C}_q(X)} n^{1/q}$. For $p > 2$ or $q < 2$ we get a contradiction for sufficiently large n.

The converse to Proposition 2.40 is also true, spaces having type 2 with constant 1 and cotype 2 with constant 1 are isometric to a Hilbert space. It is sufficient even to require the corresponding inequalities for $n = 2$ only.

Definition 2.42. We say that vectors x and y in a Banach space satisfy the *parallelogram identity* if

$$\|x + y\|^2 + \|x - y\|^2 = 2\|x\|^2 + 2\|y\|^2 \tag{2.30}$$

Observe that the inequality $\|x_1 + x_2\|^2 + \|x_1 - x_2\|^2 \le 2\|x_1\|^2 + 2\|x_2\|^2$ can be written as $\frac{1}{2}(\|x_1 + x_2\|^2 + \|x_1 - x_2\|^2) \le \|x_1\|^2 + \|x_2\|^2$ and so it is just the type-2 inequality (2.25) with constant 1 and $n = 2$.

Theorem 2.43. *A Banach space is a Hilbert space if and only if any two vectors in it satisfy the parallelogram identity. The corresponding inner product is uniquely determined by the norm.*

Proof. Properties of inner product give immediately

$$\begin{aligned}\|x + y\|^2 + \|x - y\|^2 &= \langle x + y, x + y\rangle + \langle x - y, x - y\rangle \\ &= \langle x, x\rangle + \langle y, x\rangle + \langle x, y\rangle + \langle y, y\rangle + \langle x, x\rangle - \langle y, x\rangle - \langle x, y\rangle + \langle y, y\rangle \\ &= 2\langle x, x\rangle + 2\langle y, y\rangle = 2\|x\|^2 + 2\|y\|^2.\end{aligned}$$

Therefore any pair of vectors in a Hilbert space satisfies the parallelogram identity.

Now we assume that any two vectors in a Banach space X satisfy the parallelogram identity. Observe that any two vectors in a Hilbert space satisfy also the following identity:

$$\begin{aligned}\|x + y\|^2 - \|x - y\|^2 &= \langle x + y, x + y\rangle - \langle x - y, x - y\rangle \\ &= \langle x, x\rangle + \langle y, x\rangle + \langle x, y\rangle + \langle y, y\rangle - \langle x, x\rangle + \langle y, x\rangle + \langle x, y\rangle - \langle y, y\rangle \\ &= 2\langle x, y\rangle + 2\langle y, x\rangle.\end{aligned}$$

We continue our proof in the real case only (see Notes and Remarks for the complex case). In the real case $\langle x, y\rangle = \langle y, x\rangle$, and we get that the only possibility to define the inner product corresponding to the norm is

$$\langle x, y\rangle = \frac{1}{4}(\|x + y\|^2 - \|x - y\|^2). \tag{2.31}$$

It remains to check that under the assumption that the parallelogram identity holds, this formula defines an inner product on X, and that the norm of X corresponds to this inner product. Observe that $\langle x, y\rangle = \langle y, x\rangle$, $\|x\|^2 = \langle x, x\rangle$, and $\langle x, 0\rangle = 0$ are obvious. So we need to verify that $\langle x, y\rangle$ given by (2.31) satisfies

(a) $\langle x_1 + x_2, y\rangle = \langle x_1, y\rangle + \langle x_2, y\rangle$

(b) $\forall \alpha \in \mathbb{R} \quad \langle \alpha x, y\rangle = \alpha\langle x, y\rangle$

Observe that the parallelogram identity implies that for all $u, v, y \in X$ the following equality holds (the first and the last equalities come from our definition of the inner product, the second from the parallelogram identity):

$$\begin{aligned}
&\langle u+v, y\rangle + \langle u-v, y\rangle \\
&= \frac{1}{4}(\|u+v+y\|^2 - \|u+v-y\|^2 + \|u-v+y\|^2 - \|u-v-y\|^2) \\
&= \frac{1}{2}(\|u+y\|^2 + \|v\|^2 - \|u-y\|^2 - \|v\|^2) \\
&= 2\langle u, y\rangle.
\end{aligned}$$

In particular, $\langle 2u, y\rangle = 2\langle u, y\rangle$. Replacing $u+v$ by x_1 and $u-v$ by x_2, we get

$$\langle x_1, y\rangle + \langle x_2, y\rangle = 2\Big\langle \frac{x_1+x_2}{2}, y\Big\rangle = \langle x_1 + x_2, y\rangle.$$

Using this equality and induction we get that $\langle mx, y\rangle = m\langle x, y\rangle$ for all $m \in \mathbb{N}$. Letting $u = mx$, we get $\frac{1}{m}\langle u, y\rangle = \langle \frac{1}{m}u, y\rangle$ for all $m \in \mathbb{N}$. Since it is also clear that (2.31) implies $\langle -x, y\rangle = -\langle x, y\rangle$, we get $\forall \alpha \in \mathbb{Q}\quad \langle \alpha x, y\rangle = \alpha\langle x, y\rangle$. Now item (b) above follows because it can be written as

$$\frac{1}{4}(\|\alpha x + y\|^2 - \|\alpha x - y\|^2) = \frac{\alpha}{4}(\|x+y\|^2 - \|x-y\|^2),$$

the identity is known to hold for rational α, and both sides are continuous with respect to α. □

We add some more observations related to Theorem 2.43 in Exercise 2.68. In this connection we also would like to mention the following very interesting result (we do not prove it because we are not going to use it in this book).

Theorem 2.44 (Kwapień theorem). *A Banach space is isomorphic to a Hilbert space if and only if it has both type* 2 *and cotype* 2.

2.4.2 Kahane–Khinchin inequality

Theorem 2.45 (Kahane–Khinchin inequality). *For each* $1 \le p < \infty$ *there exists a constant* C_p *such that, for any Banach space* X *and any finite sequence* $\{x_i\}_{i=1}^n$ *in* X *the inequality*

$$\begin{aligned}
\int_0^1 \Big\|\sum_{i=1}^n r_i(t)x_i\Big\|_X dt &\le \Big(\int_0^1 \Big\|\sum_{i=1}^n r_i(t)x_i\Big\|_X^p dt\Big)^{1/p} \\
&\le C_p \int_0^1 \Big\|\sum_{i=1}^n r_i(t)x_i\Big\|_X dt
\end{aligned} \tag{2.32}$$

holds.

Note. This inequality in the case where $X = \mathbb{R}$ is called the *Khinchin inequality* and in the vector-valued case the inequality is called the *Kahane inequality*.

It will be convenient to rewrite and to prove inequality (2.32) using probabilistic language, that is, considering instead of the sequence of Rademacher functions a sequence $\{\varepsilon_i\}_{i=1}^{\infty}$ of independent random variables on a probability space $(\Omega, \Sigma, \mathbb{P})$, such that each ε_i takes values 1 and -1, each with probability $\frac{1}{2}$. The name *Bernoulli random variables* is historically more appropriate for such random variables. However, we shall follow the Banach-space-theoretical tradition, and call the sequence $\{\varepsilon_i\}_{i=1}^{\infty}$ a *Rademacher sequence*. With this notation inequality (2.32) becomes:

$$\begin{aligned}\mathbb{E}\left\|\sum_{i=1}^{n}\varepsilon_i(\omega)x_i\right\|_X &\le \left(\mathbb{E}\left\|\sum_{i=1}^{n}\varepsilon_i(\omega)x_i\right\|_X^p\right)^{1/p} \\ &\le C_p\,\mathbb{E}\left\|\sum_{i=1}^{n}\varepsilon_i(\omega)x_i\right\|_X\end{aligned}. \tag{2.33}$$

We make some remarks before proving (2.33). To be more precise, only the right-hand side inequalities in (2.32) and (2.33) are attributed to Khinchin and Kahane. The left-hand side inequalities are easy consequences of the *Hölder inequality* which we recall now.

Theorem 2.46 (Hölder inequality). *Suppose that $1 < p < \infty$ and $\frac{1}{p} + \frac{1}{q} = 1$. Let $f \in L_p(\Omega, \Sigma, \mathbb{P})$ and $g \in L_q(\Omega, \Sigma, \mathbb{P})$. Then $fg \in L_1(\Omega, \Sigma, \mathbb{P})$, and*

$$|\mathbb{E}(fg)| \le \mathbb{E}|fg| \le \left(\mathbb{E}|f|^p\right)^{1/p}\left(\mathbb{E}|g|^q\right)^{1/q}. \tag{2.34}$$

Remark 2.47. A similar inequality holds for $f \in L_p(0,1)$ and $g \in L_q(0,1)$. In this case the inequality is written as

$$\left|\int_0^1 (f(t)g(t))dt\right| \le \left(\int_0^1 |f(t)|^p dt\right)^{1/p}\left(\int_0^1 |g(t)|^q dt\right)^{1/q}. \tag{2.35}$$

A similar inequality holds for $f \in L_p(\Omega, \Sigma, \mu)$ and $g \in L_q(\Omega, \Sigma, \mu)$, where (Ω, Σ, μ) is any measure space:

$$\left|\int_\Omega (f(t)g(t))\,d\mu\right| \le \left(\int_\Omega |f(t)|^p d\mu\right)^{1/p}\left(\int_\Omega |g(t)|^q d\mu\right)^{1/q}. \tag{2.36}$$

To get the left-hand side of (2.33) from (2.34) we apply the latter to the functions $f(\omega) = \left\|\sum_{i=1}^n \varepsilon_i(\omega)x_i\right\|_X$ and $g(\omega) = 1$ for each ω (observe that the function f is bounded and hence satisfies the condition $f \in L_p(\Omega, \Sigma, \mathbb{P})$). We get

$$\begin{aligned}\mathbb{E}\Big\| \sum_{i=1}^{n} \varepsilon_i(\omega)x_i \Big\|_X &\le \Big(\mathbb{E}\Big\| \sum_{i=1}^{n} \varepsilon_i(\omega)x_i \Big\|_X^p\Big)^{1/p} \big(\mathbb{E}|1|^q\big)^{1/q} \\ &= \Big(\mathbb{E}\Big\| \sum_{i=1}^{n} \varepsilon_i(\omega)x_i \Big\|_X^p\Big)^{1/p}.\end{aligned}$$

So we turn to the right-hand side inequality of (2.33) (the main point in the proof). Recall that a random variable $f : \Omega \to X$ is called *symmetric* if $\mathbb{P}(f \in B) = \mathbb{P}(-f \in B)$ for each Borel set $B \subset X$.

Lemma 2.48. *Let $f : \Omega \to X$ be a symmetric random variable, and $x \in X$. Then*

$$\mathbb{P}\{\omega : \|f(\omega) + x\| \ge \|x\|\} \ge \frac{1}{2}.$$

Proof. Observe that for each ω

$$\|x\| = \frac{\|x + f(\omega) + x - f(\omega)\|}{2} \le \max\{\|x + f(\omega)\|, \|x - f(\omega)\|\}.$$

Therefore $\Omega \subset (\{\omega : \|x + f(\omega)\| \ge \|x\|\} \cup \{\omega : \|x - f(\omega)\| \ge \|x\|\})$. If we introduce the notation $B = \{y \in X : \|x + y\| \ge \|x\|\}$, we can write

$$\Omega \subset (\{\omega : f(\omega) \in B\} \cup \{\omega : -f(\omega) \in B\}).$$

Therefore $\mathbb{P}(\{\omega : f(\omega) \in B\}) \ge \frac{1}{2}$ or $\mathbb{P}(\{\omega : -f(\omega) \in B\}) \ge \frac{1}{2}$. It remains to observe that since the random variable f is symmetric, these probabilities are the same. □

For given $n \in \mathbb{N}$ and $x_1, \dots, x_n \in X$ we introduce random variables $Y_m : \Omega \to X$ $(1 \le m \le n)$ given by

$$Y_m(\omega) = \sum_{i=1}^{m} \varepsilon_i(\omega)x_i.$$

Lemma 2.49. $\mathbb{P}\{\omega : \max_{m \le n} \|Y_m(\omega)\| > r\} \le 2\mathbb{P}\{\omega : \|Y_n(\omega)\| > r\}$ *for all* $r > 0$.

Proof. Let

$$\begin{aligned}B &= \{\omega : \|Y_n(\omega)\| > r\} \\ A &= \{\omega : \max_{m \le n} \|Y_m(\omega)\| > r\} \\ A_1 &= \{\omega : \|Y_1(\omega)\| > r\} \\ A_2 &= \{\omega : \|Y_1(\omega)\| \le r, \|Y_2(\omega)\| > r\} \\ A_3 &= \{\omega : \|Y_1(\omega)\| \le r, \|Y_2(\omega)\| \le r, \|Y_3(\omega)\| > r\} \\ &\dots \\ A_n &= \{\omega : \|Y_1(\omega)\| \le r, \dots, \|Y_{n-1}(\omega)\| \le r, \|Y_n(\omega)\| > r\}.\end{aligned}$$

It is clear that $A = \cup_{m=1}^n A_m$, that the sets A_m, $m = 1\dots,n$, are disjoint, and that $B \subset \cup_{m=1}^n A_m$. Therefore $\mathbb{P}(A) = \sum_{m=1}^n \mathbb{P}(A_m)$ and

$$\mathbb{P}(B) = \sum_{m=1}^{n} \mathbb{P}(A_m \cap B). \tag{2.37}$$

Notice that

$$A_m = \bigcup_{\theta_1,\dots,\theta_m\in\{1,-1\}} \{\omega \in A_m : \varepsilon_j(\omega) = \theta_j, j \in \{1,\dots,m\}\}.$$

(We partition A_m according to the values of $\theta_1,\dots,\theta_m$.) On the other hand, by Lemma 2.48

$$\mathbb{P}\left(\left\|\sum_{j=1}^{m}\theta_j x_j + \sum_{j=m+1}^{n}\varepsilon_j(\omega)x_j\right\| \geq \left\|\sum_{j=1}^{m}\theta_j x_j\right\|\right) \geq \frac{1}{2}$$

for each collection $\theta_1,\dots,\theta_m$ of ± 1. By independence of $\{\varepsilon_i\}_{i=1}^n$ we get

$$\begin{aligned}
&\mathbb{P}\left(\{\omega \in A_m : \varepsilon_j(\omega) = \theta_j, j \in \{1,\dots,m\}\} \cap \{\omega : \|Y_n(\omega)\| \geq \|Y_m(\omega)\|\}\right)\\
&= \mathbb{P}\Bigg(\{\omega \in A_m : \varepsilon_j(\omega) = \theta_j, j \in \{1,\dots,m\}\}\\
&\qquad \cap \left\{\omega : \left\|\sum_{j=1}^{m}\theta_j x_j + \sum_{j=m+1}^{n}\varepsilon_j(\omega)x_j\right\| \geq \left\|\sum_{j=1}^{m}\theta_j x_j\right\|\right\}\Bigg)\\
&\geq \frac{1}{2}\mathbb{P}\left(\{\omega \in A_m : \varepsilon_j(\omega) = \theta_j, j \in \{1,\dots,m\}\}\right).
\end{aligned}$$

Summing these inequalities over all collections $\theta_1,\dots,\theta_m$ of ± 1, we get

$$\mathbb{P}\left(A_m \cap \{\omega : \|Y_n(\omega)\| \geq \|Y_m(\omega)\|\}\right) \geq \frac{1}{2}\mathbb{P}(A_m).$$

Therefore

$$\mathbb{P}\left(A_m \cap B\right) \geq \frac{1}{2}\mathbb{P}(A_m).$$

Summing these inequalities over m and using (2.37) we get $P(B) \geq \frac{1}{2}P(A)$, which is the desired inequality. □

We use the notation introduced above.

Lemma 2.50. *$\mathbb{P}\{\omega : \|Y_n(\omega)\| > 2r\} \leq 4(\mathbb{P}\{\omega : \|Y_n(\omega)\| > r\})^2$ for all $r > 0$.*

Proof. For each $m \in \{1,\dots,n\}$, the random variables $\|\sum_{i=m}^n \varepsilon_i x_i\|, \varepsilon_1,\dots,\varepsilon_m$ are independent. We would like to emphasize that this is true despite the fact that ε_m is included into the first random variable, too. One of the ways of showing the independence is to observe that the first random variable is equal to $\|x_m + \sum_{i=m+1}^n \varepsilon_i \varepsilon_m x_i\|$, and that the variables $\varepsilon_1, \dots, \varepsilon_m, \varepsilon_{m+1}\varepsilon_m, \varepsilon_{m+2}\varepsilon_m, \dots, \varepsilon_n\varepsilon_m$ are independent. Hence the events $\{\omega : \|\sum_{i=m}^n \varepsilon_i x_i\| > r\}$ and A_m are independent. Observe that if $\omega \in A_m$ satisfies $\|Y_n(\omega)\| > 2r$, then $\|Y_n(\omega) - Y_{m-1}(\omega)\| > r$, where we let $Y_0(\omega) = 0$ for all $\omega \in \Omega$.

Observe that Lemma 2.49 implies

$$\mathbb{P}\left\{\omega : \left\|\sum_{i=m}^n \varepsilon_i x_i\right\| > r\right\} \le 2\mathbb{P}\{\omega : \|Y_n(\omega)\| > r\}.$$

In fact, one can see that the result of Lemma 2.49 does not depend on the ordering of indices i. Hence we have

$$\begin{aligned}\mathbb{P}(A_m \cap \{\omega : \|Y_n(\omega)\| > 2r\}) &\le \mathbb{P}(A_m \cap \{\omega : \left\|\sum_{i=m}^n \varepsilon_i x_i\right\| > r\}) \\ &= \mathbb{P}(A_m)\mathbb{P}\{\omega : \left\|\sum_{i=m}^n \varepsilon_i x_i\right\| > r\} \\ &\le 2\mathbb{P}(A_m)\mathbb{P}\{\omega : \|Y_n(\omega)\| > r\}.\end{aligned}$$

Summing in m and using the inclusion $\{\omega : \|Y_n(\omega)\| > 2r\} \subset A$ and Lemma 2.49 again, we get

$$\begin{aligned}\mathbb{P}\{\omega : \|Y_n(\omega)\| > 2r\} &\le 2\mathbb{P}(A)\mathbb{P}\{\omega : \|Y_n(\omega)\| > r\} \\ &\le 4(\mathbb{P}\{\omega : \|Y_n(\omega)\| > r\})^2.\end{aligned}$$ □

Proof of (2.33) *and of Theorem* 2.45. Let $1 \le p < \infty$ and let $\{x_i\}_{i=1}^n$ be a finite set of vectors in X. We may assume without loss of generality that

$$\mathbb{E}\left\|\sum_{m=1}^n \varepsilon_i x_i\right\| = \mathbb{E}\|Y_n\| = 1.$$

Therefore

$$\mathbb{P}\{\omega : \|Y_n(\omega)\| > 8\} \le \frac{1}{8} \tag{2.38}$$

Using Lemma 2.50 repeatedly we get

$$\mathbb{P}\{\omega : \|Y_n(\omega)\| > 2 \cdot 8\} \le 4\left(\frac{1}{8}\right)^2,$$

$$\mathbb{P}\{\omega : \|Y_n(\omega)\| > 2^2 \cdot 8\} \le 4^3\left(\frac{1}{8}\right)^4,$$

$$\mathbb{P}\{\omega : \|Y_n(\omega)\| > 2^3 \cdot 8\} \le 4^7\left(\frac{1}{8}\right)^8,$$

and so on. Now we use the inequality which is easy to get from the definition of expectation:

$$\begin{aligned}\mathbb{E}\left(\|Y_n\|^p\right) &\le 8^p \cdot \mathbb{P}\{\omega : \|Y_n(\omega)\| \le 8\} \\ &\quad + (2\cdot 8)^p \cdot \mathbb{P}\{\omega : \|Y_n(\omega)\| > 8\} \\ &\quad + (2^2\cdot 8)^p \cdot \mathbb{P}\{\omega : \|Y_n(\omega)\| > 2\cdot 8\} \\ &\quad + \cdots \\ &\le 8^p + \sum_{j=0}^{\infty}(2^{j+1}\cdot 8)^p \cdot 4^{2^j-1}\left(\frac{1}{8}\right)^{2^j}.\end{aligned}$$

It remains to observe that this series is convergent. Denoting its sum by $(C_p)^p$, we get (2.33). This completes the proof of Theorem 2.45. □

Usually the Khinchin inequality is written in an equivalent but slightly different form than (2.32):

Theorem 2.51 (Khinchin inequality). *For each $1 \le p < \infty$ there exist constants A_p and B_p such that for any finite sequence $\{x_i\}_{i=1}^n$ of scalars the inequality*

$$A_p\left(\sum_{i=1}^n |x_i|^2\right)^{1/2} \le \left(\int_0^1 \left|\sum_{i=1}^n r_i(t)x_i\right|^p dt\right)^{1/p} \le B_p\left(\sum_{i=1}^n |x_i|^2\right)^{1/2} \tag{2.39}$$

holds.

Theorem 2.51 is an immediate consequence of Theorem 2.45 because the Rademacher sequence is orthonormal and therefore

$$\int_0^1 \left|\sum_{i=1}^n r_i(t)x_i\right|^2 dt = \sum_{i=1}^n |x_i|^2.$$

We shall also use the Khinchin inequality in probabilistic notation:

$$A_p\left(\sum_{i=1}^n |x_i|^2\right)^{1/2} \le \left(\mathbb{E}\left|\sum_{i=1}^n \varepsilon_i(\omega)x_i\right|^p\right)^{1/p} \le B_p\left(\sum_{i=1}^n |x_i|^2\right)^{1/2}. \tag{2.40}$$

Remark 2.52. Theorem 2.51 (that is, the Khinchin inequality) implies that the space ℓ_2 admits an isomorphic embedding into any of the spaces L_p ($1 \le p < \infty$) (see Observation 4.12 for more details). For L_∞ this result also holds, but the reason is different: first we partition $[0, 1]$ into infinitely many measurable subsets of nonzero measure and show that functions which are constant on each of the subsets form a subspace isometric to ℓ_∞. After that we prove a linear analogue of Proposition 1.17, namely we prove that each separable Banach space X admits a linear isometric embedding into ℓ_∞ and that this embedding can be constructed as follows. Let $\{x_i\}$ be a dense sequence in X. Let $\{f_i\}$ be a sequence of functionals satisfying the conditions $\|f_i\| = 1$ and $f_i(x_i) = \|x_i\|$. Then the mapping $X \mapsto \ell_\infty$ given by $x \mapsto \{f_i(x)\}_{i=1}^\infty$ is a linear isometric embedding.

Theorem 2.53. *The space* $L_p = L_p(0,1)$, $1 \le p < \infty$ *has type* $\min\{p,2\}$ *and cotype* $\max\{p,2\}$.

Remark 2.54. The same proof works for $L_p(\Omega, \Sigma, \mu)$, where (Ω, Σ, μ) is an arbitrary measure space.

Proof. First we consider the case $1 \le p \le 2$. In this case we need to show that L_p has cotype 2. Let $x_1, \dots, x_n \in L_p$. By the Kahane inequality (2.33), it suffices to prove the inequality

$$\left(\mathbb{E}\left\|\sum_{i=1}^n \varepsilon_i(\omega)x_i\right\|^p\right)^{1/p} \ge \frac{1}{\widetilde{C}_p}\left(\sum_{i=1}^n \|x_i\|^2\right)^{1/2},$$

where $\widetilde{C}_p$ is a constant which does not depend on the choice of n and $\{x_i\}_{i=1}^n$, but, possibly, depends on p. We have

$$\begin{aligned}\left(\mathbb{E}\left\|\sum_{i=1}^n \varepsilon_i(\omega)x_i\right\|^p\right)^{1/p} &= \left(\mathbb{E}\left(\int_0^1\left|\sum_{i=1}^n \varepsilon_i(\omega)x_i(t)\right|^p dt\right)\right)^{1/p} \\ &= \left(\int_0^1 \mathbb{E}\left|\sum_{i=1}^n \varepsilon_i(\omega)x_i(t)\right|^p dt\right)^{1/p} \qquad (2.41) \\ &\ge A_p\left(\int_0^1\left(\sum_{i=1}^n |x_i(t)|^2\right)^{p/2} dt\right)^{1/p},\end{aligned}$$

where we used the Khinchin inequality (2.40). Comparing the right-hand side of the inequality (2.41) with the desired result, we see that it suffices to prove the inequality

$$\left(\int_0^1\left(\sum_{i=1}^n |x_i(t)|^2\right)^{p/2} dt\right)^{1/p} \ge \left(\sum_{i=1}^n\left(\int_0^1 |x_i(t)|^p\,dt\right)^{2/p}\right)^{1/2}.$$

To prove this inequality we use the fact that $r := 2/p \ge 1$. Therefore there is a supporting functional $g = \{g_i\}_{i=1}^n \in \ell_{r'}^n$, where r' is determined by $\frac{1}{r} + \frac{1}{r'} = 1$, such that $\|g\| = 1$ (the norm in $\ell_{r'}^n$) and

$$\left(\sum_{i=1}^n\left(\int_0^1 |x_i(t)|^p\,dt\right)^{2/p}\right)^{p/2} = \sum_{i=1}^n\left(\int_0^1 |x_i(t)|^p\,dt\right)g_i.$$

We have

$$\left(\sum_{i=1}^{n}\left(\int_0^1 |x_i(t)|^p dt\right)^{2/p}\right)^{1/2} = \left(\sum_{i=1}^{n}\left(\int_0^1 |x_i(t)|^p dt\right) g_i\right)^{1/p}$$
$$= \left(\int_0^1 \left(\sum_{i=1}^{n} |x_i(t)|^p g_i\right) dt\right)^{1/p}.$$

Now we apply the inequality $\sum_{i=1}^{n} |x_i(t)|^p g_i \le \|\{|x_i(t)|^p\}_{i=1}^{n}\|_r \cdot \|g\|_{r'}$, and get

$$\sum_{i=1}^{n} |x_i(t)|^p g_i \le \left(\sum_{i=1}^{n} |x_i(t)|^2\right)^{p/2} \|g\| = \left(\sum_{i=1}^{n} |x_i(t)|^2\right)^{p/2},$$

the desired inequality follows.

Our next purpose is to show that the space L_p, $1 \le p \le 2$, has type p. Let $x_1, \dots, x_n \in L_p$, we have

$$\left(\mathbb{E}\left\|\sum_{i=1}^{n} \varepsilon_i(\omega) x_i\right\|^p\right)^{1/p} = \left(\mathbb{E}\left(\int_0^1 \left|\sum_{i=1}^{n} \varepsilon_i(\omega) x_i(t)\right|^p dt\right)\right)^{1/p}$$
$$= \left(\int_0^1 \mathbb{E}\left|\sum_{i=1}^{n} \varepsilon_i(\omega) x_i(t)\right|^p dt\right)^{1/p} \tag{2.42}$$
$$\le B_p\left(\int_0^1 \left(\sum_{i=1}^{n} |x_i(t)|^2\right)^{p/2} dt\right)^{1/p}.$$

At this point we use the fact that for $p \le 2$ the following inequality holds:

$$\left(\sum_{i=1}^{n} |x_i(t)|^2\right)^{1/2} \le \left(\sum_{i=1}^{n} |x_i(t)|^p\right)^{1/p}. \tag{2.43}$$

This inequality can be proved in the following way. It is clear that it suffices to consider the case where $\sum_{i=1}^{n} |x_i(t)|^2 = 1$. In this case $|x_i(t)| \le 1$ for each i and t. Therefore $|x_i(t)|^2 \le |x_i(t)|^p$, and the proof of (2.43) is completed. Combining (2.42) and (2.43) we get

$$\left(\mathbb{E}\left\|\sum_{i=1}^{n} \varepsilon_i(\omega) x_i\right\|^p\right)^{1/p} \le B_p\left(\int_0^1 \sum_{i=1}^{n} |x_i(t)|^p dt\right)^{1/p}$$
$$= B_p\left(\sum_{i=1}^{n} \|x_i\|^p\right)^{1/p}.$$

Now we turn to L_p with $p \geq 2$. First we prove the cotype inequality. We have

$$\begin{aligned}\left(\mathbb{E}\left\|\sum_{i=1}^{n}\varepsilon_i(\omega)x_i\right\|^p\right)^{1/p} &= \left(\mathbb{E}\left(\int_0^1\left|\sum_{i=1}^{n}\varepsilon_i(\omega)x_i(t)\right|^p dt\right)\right)^{1/p}\\ &= \left(\int_0^1 \mathbb{E}\left|\sum_{i=1}^{n}\varepsilon_i(\omega)x_i(t)\right|^p dt\right)^{1/p}\\ &\geq A_p\left(\int_0^1\left(\sum_{i=1}^{n}|x_i(t)|^2\right)^{p/2} dt\right)^{1/p}.\end{aligned}$$

At this point we use the fact that for $p \geq 2$ the inequality

$$\left(\sum_{i=1}^{n}|x_i(t)|^2\right)^{1/2} \geq \left(\sum_{i=1}^{n}|x_i(t)|^p\right)^{1/p}$$

holds (it can be proved in the same way as (2.43)). We get

$$\begin{aligned}\left(\mathbb{E}\left\|\sum_{i=1}^{n}\varepsilon_i(\omega)x_i\right\|^p\right)^{1/p} &\geq A_p\left(\int_0^1\sum_{i=1}^{n}|x_i(t)|^p dt\right)^{1/p}\\ &= A_p\left(\sum_{i=1}^{n}\|x_i\|^p\right)^{1/p}.\end{aligned}$$

Now we turn to proving that L_p $(2 \leq p < \infty)$ has type 2. Let $x_1, \dots, x_n \in L_p$. By the Kahane inequality (2.33), it suffices to prove the inequality

$$\left(\mathbb{E}\left\|\sum_{i=1}^{n}\varepsilon_i(\omega)x_i\right\|^p\right)^{1/p} \leq \widetilde{T}_p\left(\sum_{i=1}^{n}\|x_i\|^2\right)^{1/2},$$

where $\widetilde{T}_p$ is a constant which does not depend on the choice of n and $\{x_i\}_{i=1}^n$, but, possibly, depends on p.

Using the Khinchin inequality as in (2.42) we get

$$\left(\mathbb{E}\left\|\sum_{i=1}^{n}\varepsilon_i(\omega)x_i\right\|^p\right)^{1/p} \leq B_p\left(\int_0^1\left(\sum_{i=1}^{n}|x_i(t)|^2\right)^{p/2} dt\right)^{1/p}$$

We use the same trick as in the first part of the proof, now with $r = p/2 \geq 1$, and with respect to the interval $[0, 1]$ and not the set $\{1, \dots, n\}$. There is a supporting functional $g = g(t) \in L_{r'}$, where r' is determined by $\frac{1}{r} + \frac{1}{r'} = 1$, such that $\|g\| = 1$ (the norm is in the space $L_{r'}$) and

$$\left(\int_0^1\left(\sum_{i=1}^{n}|x_i(t)|^2\right)^{p/2} dt\right)^{2/p} = \int_0^1\left(\sum_{i=1}^{n}|x_i(t)|^2\right)g(t)\,dt.$$

We have

$$\left(\int_0^1\left(\sum_{i=1}^n|x_i(t)|^2\right)^{p/2}dt\right)^{1/p}=\left(\int_0^1\left(\sum_{i=1}^n|x_i(t)|^2\right)g(t)\,dt\right)^{1/2}$$
$$=\left(\sum_{i=1}^n\int_0^1|x_i(t)|^2g(t)\,dt\right)^{1/2}.$$

Now we apply the inequality $\int_0^1|x_i(t)|^2g(t)\,dt\le\|(|x_i(t)|^2)\|_r\cdot\|g\|_{r'}$, and get

$$\int_0^1|x_i(t)|^2g(t)\,dt\le\|x_i\|^2\|g\|=\|x_i\|^2,$$

the desired inequality follows. □

2.4.3 Characterization of spaces with trivial type or cotype

The following result shows an important relation between type, cotype, and finite representability.

Theorem 2.55 (Maurey–Pisier theorem). *Let X be an infinite-dimensional Banach space, $p(X)=\sup\{p:\ X$ has type $p\}$ and $q(X)=\inf\{q:\ X$ has cotype $q\}$. Then the spaces $\ell_{p(X)}$ and $\ell_{q(X)}$ are finitely representable in X.*

Theorem 2.55 is beyond the scope of this book. We are going to prove only the following special cases of it (only these cases will be used in this book).

Theorem 2.56. *A Banach space X has no nontrivial type if and only if ℓ_1 is finitely representable in X. A Banach space X has no nontrivial cotype if and only if ℓ_∞ is finitely representable in X.*

Proof. Let X be an infinite-dimensional Banach space, and $\{\varepsilon_i\}_{i=1}^n$ be a Rademacher sequence. For each $n\in\mathbb{N}$, define $\alpha_n(X)$ to be the least constant α so that

$$\left(\mathbb{E}\left\|\sum_{i=1}^n\varepsilon_i(\omega)x_i\right\|^2\right)^{1/2}\le\alpha\left(\sum_{i=1}^n\|x_i\|^2\right)^{1/2}$$

for all sequences $\{x_i\}_{i=1}^n$ in X. We define $\beta_n(X)$ to be the least constant β so that

$$\left(\sum_{i=1}^n\|x_i\|^2\right)^{1/2}\le\beta\left(\mathbb{E}\left\|\sum_{i=1}^n\varepsilon_i(\omega)x_i\right\|^2\right)^{1/2}$$

for all sequences $\{x_i\}_{i=1}^n$ in X. We have $1\le\alpha_n(X)\le\sqrt{n}$, $1\le\beta_n(X)\le\sqrt{n}$. The lower bounds should hold because the inequalities should be satisfied for sequences with exactly one nonzero vector. The upper bounds follow from the following inequality:

$$\max_i \|x_i\| \leq \left(\mathbb{E}\left\|\sum_{i=1}^n \varepsilon_i(\omega)x_i\right\|^2\right)^{1/2} \leq \sum_{i=1}^n \|x_i\|.$$

The right-hand-side inequality is an immediate consequence of the triangle inequality. The left-hand-side inequality follows from convexity of the function $\|\cdot\|^2$, which implies $\|x-y\|^2 + \|x+y\|^2 \geq 2\|x\|^2$. It is also easy to see that $\alpha_n(X)$ and $\beta_n(X)$ are non-decreasing.

Lemma 2.57. *The parameters $\alpha_n(X)$ and $\beta_n(X)$ are submultiplicative in the sense that they satisfy*

$$\forall m,n \in \mathbb{N} \quad \alpha_{mn}(X) \leq \alpha_m(X)\alpha_n(X) \tag{2.44}$$

and

$$\forall m,n \in \mathbb{N} \quad \beta_{mn}(X) \leq \beta_m(X)\beta_n(X). \tag{2.45}$$

Proof. Let $\{\varepsilon_{i,j}(\omega) : i = 1,\dots,m;\ j = 1,\dots,n\}$ be a Rademacher sequence on a probability space $(\Omega, \Sigma_\Omega, \mathbb{P}_\Omega)$, numbered as a double-parametric sequence (but still being an independent set of random variables with values 1 and -1, taking each of the values with probability $\frac{1}{2}$). Let $\{\varepsilon_i(\theta) : i = 1,\dots,m\}$ be a Rademacher sequence on a probability space $(\Theta, \Sigma_\Theta, \mathbb{P}_\Theta)$.

The products

$$\{\varepsilon_i(\theta)\varepsilon_{i,j}(\omega) : i = 1,\dots,m;\ j = 1,\dots,n\} \tag{2.46}$$

form a Rademacher sequence on the probability space $(\Omega \times \Theta, \Sigma_{\mathrm{prod}}, \mathbb{P}_{\mathrm{prod}})$, where $\Omega \times \Theta$ is the Cartesian product of the sets Ω and Θ, Σ_{prod} is the smallest with respect to the inclusion σ-algebra of subsets of $\Omega \times \Theta$ containing all subsets of the form $A \times B$ with $A \in \Sigma_\Omega$ and $B \in \Sigma_\Theta$; the probability $\mathbb{P}_{\mathrm{prod}}$ is determined by $\mathbb{P}_{\mathrm{prod}}(A \times B) = \mathbb{P}_\Omega(A) \cdot \mathbb{P}_\Theta(B)$. Let us denote $\mathbb{E}_\omega$, $\mathbb{E}_\theta$, and $\mathbb{E}_{\omega\times\theta}$ the expectations corresponding to the spaces $(\Omega, \Sigma_\Omega, \mathbb{P}_\Omega)$, $(\Theta, \Sigma_\Theta, \mathbb{P}_\Theta)$, and $(\Omega \times \Theta, \Sigma_{\mathrm{prod}}, \mathbb{P}_{\mathrm{prod}})$, respectively.

Since the sequence (2.46) is a Rademacher sequence, we have

$$\mathbb{E}_\omega \left\|\sum_{i,j} \varepsilon_{i,j}(\omega)x_{i,j}\right\|^2 = \mathbb{E}_{\omega\times\theta}\left\|\sum_{i=1}^m \varepsilon_i(\theta)\sum_{j=1}^n \varepsilon_{i,j}(\omega)x_{i,j}\right\|^2$$

for any collection $\{x_{i,j}\}_{i,j}$ in X.

We use the well-known fact:

$$\mathbb{E}_{\omega\times\theta}(f) = \mathbb{E}_\omega(\mathbb{E}_\theta(f))$$

for every function on $\Omega \times \Theta$. It is worth mentioning that in our case if suffices to consider finite probability spaces. In this case we just use the fact that the finite sum does not depend on the order of summation.

By the definition of $\alpha_m(X)$ we have

$$\mathbb{E}_\theta\left\|\sum_{i=1}^m \varepsilon_i(\theta)\sum_{j=1}^n \varepsilon_{i,j}(\omega)x_{i,j}\right\|^2 \le (\alpha_m(X))^2\sum_i\left\|\sum_j \varepsilon_{i,j}(\omega)x_{i,j}\right\|^2.$$

Thus

$$\mathbb{E}_\omega\mathbb{E}_\theta\left\|\sum_{i=1}^m \varepsilon_i(\theta)\sum_{j=1}^n \varepsilon_{i,j}(\omega)x_{i,j}\right\|^2 \le (\alpha_m(X))^2\sum_i\mathbb{E}_\omega\left\|\sum_j \varepsilon_{i,j}(\omega)x_{i,j}\right\|^2.$$

Now we observe that for each fixed i the sequence $\{\varepsilon_{i,j}(\omega)\}_{j=1}^n$ is a Rademacher sequence. Therefore

$$\mathbb{E}_\omega\left\|\sum_j \varepsilon_{i,j}(\omega)x_{i,j}\right\|^2 \le (\alpha_n(X))^2\sum_{j=1}^n \|x_{i,j}\|^2,$$

and we get

$$\mathbb{E}_\omega\left\|\sum_{i,j} \varepsilon_{i,j}(\omega)x_{i,j}\right\|^2 \le (\alpha_m(X)\alpha_n(X))^2\sum_{i,j} \|x_{i,j}\|^2.$$

Therefore $\alpha_{mn}(X) \le \alpha_m(X)\alpha_n(X)$.

Similarly we get the results for $\beta_{mn}(X)$ (we do not repeat all steps of the argument):

$$\begin{aligned}\sum_{i,j}\|x_{i,j}\|^2 &\le (\beta_n(X))^2\sum_{i=1}^m\mathbb{E}_\omega\left\|\sum_{j=1}^n \varepsilon_{i,j}(\omega)x_{i,j}\right\|^2\\ &\le (\beta_m(X)\beta_n(X))^2\mathbb{E}_\omega\left\|\sum_{i,j}\varepsilon_{i,j}(\omega)x_{i,j}\right\|^2.\end{aligned}$$

Therefore $\beta_{mn}(X) \le \beta_m(X)\beta_n(X)$. □

To prove Theorem 2.56 for spaces with no nontrivial type we need the following lemma.

Lemma 2.58. *A Banach space X has nontrivial type if and only if $\alpha_n(X) < \sqrt{n}$ for some $n \in \mathbb{N}$.*

Proof. Suppose that X has type $p > 1$ (as we know $p \le 2$). Combining the definition of type with the Kahane inequality we get

$$\left(\mathbb{E}\left\|\sum_{i=1}^n \varepsilon_i(\omega)x_i\right\|^2\right)^{1/2} \le C\left(\sum_{i=1}^n \|x_i\|^p\right)^{1/p}. \qquad (2.47)$$

Now we use the inequality

$$\left(\sum_{i=1}^{n}\|x_i\|^p\right)^{1/p} \le n^{\frac{1}{p}-\frac{1}{2}}\left(\sum_{i=1}^{n}\|x_i\|^2\right)^{1/2}.$$

This inequality can be obtained by using

$$\sum_{i=1}^{n} a_i b_i \le \|\{a_i\}\|_r \cdot \|\{b_i\}\|_{r'}$$

with $a_i = \|x_i\|^p$, $r = \frac{2}{p}$, $b_i = 1$, and r' given by $\frac{1}{r} + \frac{1}{r'} = 1$.

We get $\alpha_n(X) \le Cn^{\frac{1}{p}-\frac{1}{2}}$. Since $p > 1$, this inequality implies that $\alpha_n(X) < \sqrt{n}$ for some n.

Now we prove the other direction. Suppose that $\alpha_{n_0}(X) < \sqrt{n_0}$ for some $n_0 \in \mathbb{N}$. Then $\alpha_{n_0}(X) = n_0^{\frac{1}{s}-\frac{1}{2}}$ for some $s > 1$. Using the submultiplicativity of $\alpha_n(X)$ we get $\alpha_n(X) \le n^{\frac{1}{s}-\frac{1}{2}}$, for each n which is a positive integer power of n_0. Now we recall the fact that $\alpha_n(X)$ is a non-decreasing sequence and use the observation that the interval $[n, n_0 n]$ always contains a positive integer power of n_0. Therefore

$$\alpha_n(X) \le (nn_0)^{\frac{1}{s}-\frac{1}{2}} = Dn^{\frac{1}{s}-\frac{1}{2}} \tag{2.48}$$

for all $n \in \mathbb{N}$, where D is some constant.

Now we show that the estimate (2.48) implies that X has type p for each $p < s$. By the Kahane inequality, it suffices to prove (2.47) for some constant C.

To do so, we assume that $\|x_1\| \ge \|x_2\| \ge \cdots \ge \|x_n\|$. With this assumption we have (writing $\sum_{i=1}^{\infty} x_i$ we mean $x_i = 0$ for $i > n$)

$$\begin{aligned}
\left(\mathbb{E}\left\|\sum_{i=1}^{n}\varepsilon_i(\omega)x_i\right\|^2\right)^{1/2} &\le \sum_{k=1}^{\infty}\left(\mathbb{E}\left\|\sum_{i=2^{k-1}}^{2^k-1}\varepsilon_i(\omega)x_i\right\|^2\right)^{1/2} \\
&\le \sum_{k=1}^{\infty} D(2^{k-1})^{\frac{1}{s}-\frac{1}{2}}\left(\sum_{i=2^{k-1}}^{2^k-1}\|x_i\|^2\right)^{1/2} \\
&\le \sum_{k=1}^{\infty} D(2^{k-1})^{\frac{1}{s}}\|x_{2^{k-1}}\| \\
&\le \sum_{k=1}^{\infty} D(2^{k-1})^{\frac{1}{s}}(2^{k-1})^{-\frac{1}{p}}\left(\sum_{i=1}^{n}\|x_i\|^p\right)^{1/p} \\
&\le C\left(\sum_{i=1}^{n}\|x_i\|^p\right)^{1/p}.
\end{aligned} \tag{2.49}$$

(We used the inequality $\frac{1}{s} - \frac{1}{p} < 0$, which implies that the series

$$\sum_{k=1}^{\infty} D(2^{k-1})^{\frac{1}{s}}(2^{k-1})^{-\frac{1}{p}}$$

is convergent.) We have proved (2.47). □

The last step in the proof of the statement of Theorem 2.56 regarding type is the following.

Lemma 2.59. *Let X be a Banach space. The space ℓ_1 is finitely representable in X if and only if $\alpha_n(X) = \sqrt{n}$ for all $n \in \mathbb{N}$.*

Proof. First we observe that $\alpha_n(\ell_1) = \sqrt{n}$ for all n. In fact, if we let $x_1, \dots, x_n$ be vectors of the unit vector basis of ℓ_1, then

$$\left(\mathbb{E}\left\|\sum_{i=1}^{n} \varepsilon_i(\omega)x_i\right\|^2\right)^{1/2} = n$$

and $\left(\sum_{i=1}^{n} \|x_i\|^2\right)^{1/2} = \sqrt{n}$. Therefore if ℓ_1 is finitely representable in X, we easily get $\alpha_n(X) \ge \sqrt{n}/(1+\varepsilon)$ for every $\varepsilon > 0$, therefore $\alpha_n(X) = \sqrt{n}$.

The assumption $\alpha_n(X) = \sqrt{n}$ implies that for each $k \in \mathbb{N}$ we can find a sequence $\{x_{k,j}\}_{j=1}^{n}$ in X such that

$$\left(\sum_{j=1}^{n} \|x_{k,j}\|^2\right)^{1/2} = \sqrt{n}$$

and

$$n - \frac{1}{k} \le \left(\mathbb{E}\left\|\sum_{j=1}^{n} \varepsilon_j(\omega)x_{j,k}\right\|^2\right)^{1/2} \le \sum_{j=1}^{n} \|x_{k,j}\| \le n.$$

Let $\mathcal{U}$ be a free ultrafilter on $\mathbb{N}$. Consider elements $\{x_j\}_{j=1}^{n}$ represented in the ultraproduct $X^{\mathcal{U}}$ by the sequences $\{x_{k,j}\}_{k=1}^{\infty}$. We have

$$\left(\sum_{j=1}^{n} \|x_j\|^2\right)^{1/2} = \sqrt{n},$$

$$\left(\mathbb{E}\left\|\sum_{j=1}^{n} \varepsilon_j(\omega)x_j\right\|^2\right)^{1/2} = n,$$

and

$$\sum_{j=1}^{n} \|x_j\| = n.$$

It is easy to see that these equalities imply that $\|x_j\| = 1$ for each j and that

$$\left\| \sum_{j=1}^{n} \varepsilon_j(\omega) x_j \right\| = n$$

for almost each $\omega \in \Omega$ (that is, for each collection of $\varepsilon_j(\omega) = \pm 1$).

Now let $\{a_j\}_{j=1}^n$ be a sequence of real numbers. Assume that $0 < |a_j| \le 1$ for each j. We have

$$\begin{aligned} \sum_{j=1}^{n} |a_j| \ge \left\| \sum_{j=1}^{n} a_j x_j \right\| &\ge \left\| \sum_{j=1}^{n} \operatorname{sign}(a_j) x_j \right\| - \left\| \sum_{j=1}^{n} \operatorname{sign}(a_j)(1 - |a_j|) x_j \right\| \\ &\ge n - \Big(n - \sum_{j=1}^{n} |a_j|\Big) = \sum_{j=1}^{n} |a_j|. \end{aligned}$$

It follows immediately that the subspace spanned by $\{x_j\}_{j=1}^n$ in $X^{\mathcal{U}}$ is isometric to ℓ_1^n. We combine this statement with the following results.

- $X^{\mathcal{U}}$ is finitely representable in X (Proposition 2.31).
- For each finite-dimensional subspace F of ℓ_1 and each $\varepsilon > 0$ there exists $n \in \mathbb{N}$ and an operator $T : F \to \ell_1^n$ such that $\|T\| \le 1$ and $\|T^{-1}\| \le 1 + \varepsilon$, where T^{-1} is the operator defined on $T(F)$. (The necessary argument is contained in the proof of Proposition 2.29.)
- Composition S of two operators satisfying the inequalities of the previous item satisfies $\|S\| \le 1$ and $\|S^{-1}\| \le (1 + \varepsilon)^2$.

We conclude that ℓ_1 is finitely representable in X. □

To prove Theorem 2.56 for spaces with no nontrivial cotype we need the following lemma.

Lemma 2.60. *A Banach space X has nontrivial cotype if and only if $\beta_n(X) < \sqrt{n}$ for some $n \in \mathbb{N}$.*

Proof. Suppose that X has cotype $q < \infty$ (as we know $q \ge 2$). Combining the definition of cotype with the Kahane inequality we get

$$\left(\mathbb{E} \left\| \sum_{i=1}^{n} \varepsilon_i(\omega) x_i \right\|^2 \right)^{1/2} \ge C \left(\sum_{i=1}^{n} \|x_i\|^q \right)^{1/q}. \tag{2.50}$$

Now we use the inequality

$$\left(\sum_{i=1}^{n} \|x_i\|^2 \right)^{1/2} \le n^{\frac{1}{2} - \frac{1}{q}} \left(\sum_{i=1}^{n} \|x_i\|^q \right)^{1/q}.$$

This inequality can be obtained by using the inequality

$$\sum_{i=1}^{n} a_i b_i \le \|\{a_i\}\|_r \cdot \|\{b_i\}\|_{r'}$$

with $a_i = \|x_i\|^2$, $r = \frac{q}{2}$, $b_i = 1$, and r' given by $\frac{1}{r} + \frac{1}{r'} = 1$.

We get $\beta_n(X) \le \frac{1}{C} n^{\frac{1}{2}-\frac{1}{q}}$. Since $q < \infty$, this inequality implies that $\beta_n(X) < \sqrt{n}$ for some n.

Now we prove the other direction. Suppose that $\beta_{n_0}(X) < \sqrt{n_0}$ for some $n_0 \in \mathbb{N}$. Then $\beta_{n_0}(X) = n_0^{\frac{1}{2}-\frac{1}{t}}$ for some $2 \le t < \infty$. Using the submultiplicativity of $\beta_n(X)$ we get $\beta_n(X) \le n^{\frac{1}{2}-\frac{1}{t}}$ for each n which is a positive integer power of n_0. Now we recall the fact that $\beta_n(X)$ is a non-decreasing sequence. Since each interval of the form $[n, nn_0]$ contains a positive integer power of n_0, we get

$$\beta_n(X) \le (nn_0)^{\frac{1}{2}-\frac{1}{t}} = Dn^{\frac{1}{2}-\frac{1}{t}} \tag{2.51}$$

for all $n \in \mathbb{N}$, where D is some constant.

Now we show that the estimate (2.51) implies that X has cotype q for each $q > t$. By the Kahane inequality, it suffices to prove (2.50) for some constant C.

To do so, we assume that $\|x_1\| \ge \|x_2\| \ge \cdots \ge \|x_n\|$. With this assumption we have

$$\begin{aligned}\Big(\sum_{i=2^{k-1}}^{2^k-1} \|x_i\|^2\Big)^{1/2} &\le D(2^{k-1})^{\frac{1}{2}-\frac{1}{t}}\Big(\mathbb{E}\Big\|\sum_{i=2^{k-1}}^{2^k-1} \varepsilon_i(\omega)x_i\Big\|^2\Big)^{1/2}\\ &\le D(2^{k-1})^{\frac{1}{2}-\frac{1}{t}}\Big(\mathbb{E}\Big\|\sum_{i=1}^{n} \varepsilon_i(\omega)x_i\Big\|^2\Big)^{1/2}.\end{aligned}$$

Since

$$\sum_{i=2^{k-1}}^{2^k-1} \|x_i\|^q \le \|x_{2^{k-1}}\|^{q-2} \sum_{i=2^{k-1}}^{2^k-1} \|x_i\|^2,$$

we have

$$\begin{aligned}\sum_{i=1}^{n} \|x_i\|^q &\le D^2\Big(\sum_{k=1}^{\infty}(2^{k-1})^{1-\frac{2}{t}}\|x_{2^{k-1}}\|^{q-2}\Big)\mathbb{E}\Big\|\sum_{i=1}^{n}\varepsilon_i(\omega)x_i\Big\|^2\\ &= D^2\Big(\sum_{k=1}^{\infty}(2^{k-1})^{1-\frac{2}{t}-(1-\frac{2}{q})}(2^{k-1})^{1-\frac{2}{q}}\|x_{2^{k-1}}\|^{q-2}\Big)\mathbb{E}\Big\|\sum_{i=1}^{n}\varepsilon_i(\omega)x_i\Big\|^2.\end{aligned}$$

Now we apply $\sum_{k=1}^{\infty} a_k b_k \le \|\{a_k\}\|_r \cdot \|\{b_k\}\|_{r'}$, with $a_k = (2^{k-1})^{\frac{2}{q}-\frac{2}{t}}$, $r = \frac{q}{2}$, $b_i = (2^{k-1})^{1-\frac{2}{q}}\|x_{2^{k-1}}\|^{q-2}$, and r' given by $\frac{1}{r}+\frac{1}{r'}=1$, so $r' = \frac{q}{q-2}$. We get

$$\begin{aligned}\sum_{i=1}^{n}\|x_i\|^q &\le D^2\Big(\sum_{k=1}^{\infty}(2^{k-1})^{\left(\frac{2}{q}-\frac{2}{t}\right)\frac{q}{2}}\Big)^{\frac{2}{q}} \\ &\quad\times\Big(\sum_{k=1}^{\infty}2^{k-1}\|x_{2^{k-1}}\|^q\Big)^{\frac{q-2}{q}}\mathbb{E}\Big\|\sum_{i=1}^{n}\varepsilon_i(\omega)x_i\Big\|^2 \\ &\le D_1\Big(\sum_{i=1}^{n}\|x_i\|^q\Big)^{\frac{q-2}{q}}\mathbb{E}\Big\|\sum_{i=1}^{n}\varepsilon_i(\omega)x_i\Big\|^2.\end{aligned} \tag{2.52}$$

We use the inequalities

$$\sum_{k=1}^{\infty}2^{k-1}\|x_{2^{k-1}}\|^q \le 2\sum_{i=1}^{n}\|x_i\|^q$$

and $\frac{2}{q}-\frac{2}{t}<0$. The latter inequality implies that the series

$$\sum_{k=1}^{\infty}(2^{k-1})^{\left(\frac{2}{q}-\frac{2}{t}\right)\frac{q}{2}}$$

converges. Inequality (2.52) can be easily transformed into the desired inequality. □

The last step in the proof of the statement of Theorem 2.56 regarding cotype is the following.

Lemma 2.61. *Let X be a Banach space. The space ℓ_∞ is finitely representable in X if and only if $\beta_n(X) = \sqrt{n}$ for all $n \in \mathbb{N}$.*

Proof. First we observe that $\beta_n(\ell_\infty) = \sqrt{n}$ for all n. In fact, let $x_1, \dots, x_n$ be vectors of the unit vector basis of ℓ_∞, then

$$\Big(\mathbb{E}\Big\|\sum_{i=1}^{n}\varepsilon_i(\omega)x_i\Big\|^2\Big)^{1/2} = 1$$

and $\left(\sum_{i=1}^{n}\|x_i\|^2\right)^{1/2} = \sqrt{n}$. As for X, since ℓ_∞ is finitely representable in X, we easily get $\beta_n(X) \ge \sqrt{n}/(1+\varepsilon)$ for every $\varepsilon > 0$, therefore $\beta_n(X) = \sqrt{n}$.

The assumption $\beta_n(X) = \sqrt{n}$ implies that for each $k \in \mathbb{N}$ we can find a sequence $\{x_{k,j}\}_{j=1}^{n}$ in X such that

$$\Big(\sum_{j=1}^{n}\|x_{k,j}\|^2\Big)^{1/2} = \sqrt{n}$$

and

$$1 \le \left(\mathbb{E}\left\|\sum_{j=1}^{n} \varepsilon_j(\omega)x_{j,k}\right\|^2\right)^{1/2} \le 1 + \frac{1}{k}.$$

Let $\mathcal{U}$ be a free ultrafilter of $\mathbb{N}$. Consider elements $\{x_j\}_{j=1}^n$ represented in the ultraproduct $X^{\mathcal{U}}$ by the sequences $\{x_{k,j}\}_{k=1}^{\infty}$. We have

$$\left(\sum_{j=1}^{n} \|x_j\|^2\right)^{1/2} = \sqrt{n} \tag{2.53}$$

and

$$\left(\mathbb{E}\left\|\sum_{j=1}^{n} \varepsilon_j(\omega)x_j\right\|^2\right)^{1/2} = 1.$$

Now we prove that these inequalities imply that $\|x_j\| = 1$ for each j and that $\|\sum_{j=1}^n \varepsilon_j(\omega)x_j\| = 1$ for almost each $\omega \in \Omega$ (that is, for each collection of $\varepsilon_j(\omega) = \pm 1$). In fact, it is clear that (2.53) implies that $\|x_{j_0}\| \ge 1$ for some j_0. If

$$\left\|\sum_{j=1}^{n} \varepsilon_j(\omega)x_j\right\| = 1 - \delta(\omega) \tag{2.54}$$

for some ω and $\delta(\omega) \ge 0$, applying the triangle inequality we get

$$\left\|\sum_{j=1}^{n} \varepsilon_j(\omega)x_j - 2\varepsilon_{j_0}(\omega)x_{j_0}\right\| \ge 1 + \delta(\omega),$$

so each $\omega \in \Omega$ satisfying (2.54) can be coupled with $\widetilde{\omega} \in \Omega$ such that

$$\left\|\sum_{j=1}^{n} \varepsilon_j(\widetilde{\omega})x_j\right\| \ge 1 + \delta(\omega).$$

This shows that $\delta(\omega)$ is always 0, because otherwise

$$\mathbb{E}\left\|\sum_{j=1}^{n} \varepsilon_j(\omega)x_j\right\|^2 > 1.$$

The equalities $\|x_i\| = 1$ also follow.

Now let $\{a_j\}_{j=1}^n$ be a sequence of real numbers satisfying $|a_j| \le 1$ for all $j = 1, \dots, n$. We have

(1) $\sum_{j=1}^n a_j x_j$ is a convex combination of $\sum_{j=1}^n \theta_j x_j$, where $\theta_j = \pm 1$. Therefore

$$\left\|\sum_{j=1}^{n} a_j x_j\right\| \le 1.$$

(2) Assume $a_{j_0} = 1$ and other coefficients are ≤ 1. Our purpose is to show that in such a case $\left\| \sum_{j=1}^{n} a_j x_j \right\| = 1$. As we have shown, $\left\| \sum_{j=1}^{n} a_j x_j \right\| \leq 1$ and $\left\| 2a_{j_0}x_{j_0} - \sum_{j=1}^{n} a_j x_j \right\| \leq 1$ (because the absolute values of the coefficients of the latter vector are also ≤ 1). Hence

$$1 = \|x_{j_0}\| \leq \frac{\left\| \sum_{j=1}^{n} a_j x_j \right\| + \left\| 2a_{j_0}x_{j_0} - \sum_{j=1}^{n} a_j x_j \right\|}{2}.$$

Therefore $\left\| \sum_{j=1}^{n} a_j x_j \right\| = 1$.

It follows that the subspace spanned by $\{x_j\}_{j=1}^n$ in $X^{\mathcal{U}}$ is isometric to ℓ_∞^n. We combine this statement with the following results.

- $X^{\mathcal{U}}$ is finitely representable in X (Proposition 2.31).
- For each finite-dimensional normed space F and each $\varepsilon > 0$ there exist $n \in \mathbb{N}$ and an operator $T : F \to \ell_\infty^n$ such that $\|T\| \leq 1$ and $\|T^{-1}\| \leq 1 + \varepsilon$, where T^{-1} is the operator defined on $T(F)$. We sketched proofs of similar results in Example 1.33 and Remark 2.52.
- Composition S of two operators satisfying the inequalities of the previous item satisfies $\|S\| \leq 1$ and $\|S^{-1}\| \leq (1 + \varepsilon)^2$.

We conclude that ℓ_∞ is finitely representable in X. □

This completes the proof of Theorem 2.56. □

2.5 Some corollaries of the theorems on finite determination of embeddability of locally finite metric spaces

Theorem 2.62. *Let M be a locally finite subset of a Hilbert space. Then M admits a bilipschitz embedding into an arbitrary infinite-dimensional Banach space.*

Follows from the Dvoretzky theorem (Theorem 2.30) and Theorem 2.6.

Theorem 2.63. *Each locally finite metric space admits a bilipschitz embedding into any Banach space with no nontrivial cotype.*

Follows from the result of Fréchet (Proposition 1.17), characterization of spaces with no nontrivial cotype (Theorem 2.56), and Theorem 2.6.

2.6 Exercises

Exercise 2.64. (Applying Zorn's lemma) Let X be a linear space. A set $A \subset X$ is called a *Hamel basis* for X if each vector $x \in X$ has a unique (up to an order of terms) representation $x = \alpha_1 a_1 + \cdots + \alpha_n a_n$ where $n \in \mathbb{N}$, $\alpha_1, \dots, \alpha_n$ are nonzero scalars, and $a_1, \dots, a_n \in A$. Use Zorn's lemma to show that each linear space has a Hamel basis.

Recall that a metric space is called *nonseparable* if it does not have a countable dense subset.

Exercise 2.65. Show that for any infinite-dimensional Banach space X and any free ultrafilter $\mathcal{U}$ on $\mathbb{N}$ the ultrapower $X^{\mathcal{U}}$ is a nonseparable space.

If needed, see a hint in the Hints to exercises section below.

Exercise 2.66. Give an example of two isomorphic but not isometric Banach spaces X and Y with $d(X, Y) = 1$, where d is the Banach–Mazur distance.

There is a hint to this exercise.

A Banach space X is called *strictly convex* if its unit sphere does not contain line segments. (We also say that the norm of the space X is *strictly convex*.)

Exercise 2.67. Prove that a Banach space Y is uniformly convex if and only if each Banach space X which is finitely representable in Y is strictly convex.

Exercise 2.68. (a) Show that each Banach space having type 2 with constant 1 is a Hilbert space. (b) Show that each Banach space having cotype 2 with constant 1 is a Hilbert space.

There is a hint to this exercise.

Let $A \subset S_{\ell_1^n}$ ($S_{\ell_1^n}$ is the unit sphere of ℓ_1^n). For a Banach space X we define the *index* $h(X, A)$ *of* A *in* X by

$$h(X, A) = \sup_{\{x_i\}_{i=1}^n \subset B_X} \inf_{\{a_i\} \in A} \left\| \sum_{i=1}^n a_i x_i \right\|.$$

Exercise 2.69. Let $\alpha \in (0, 1)$. Prove

$$h(X, S_{\ell_1^{n^2}}) \ge \alpha \;\Rightarrow h(X, S_{\ell_1^n}) \ge \sqrt{\alpha}.$$

2.7 Notes and remarks

Zorn's lemma is from [445]. See [130, Chapter 1, Sections 2 and 15] for relations of Zorn's lemma with the axiom of choice and other statements of set theory, and for related references.

Theorems 2.6 and 2.7 were proven in [360] and our presentation follows this paper. The results of [360] provide a common generalization of different theorems showing that one can paste an embedding of a locally finite metric space into a Banach space from embeddings of finite pieces of the metric space. Such results were obtained in [45], [46], [48], [352], [353], [354]. Some of these results, in turn, were proven as generalizations and analogues of the result of [85] stating that each space of bounded geometry admits a coarse embedding into a reflexive Banach space. A generalization of the mentioned result of [85] in a different direction was obtained in [236, Theorem 2.1].

Formula (2.9) for pasting pieces was suggested in [48] in the proof of Theorem 2.63. The idea of using norming subspaces for selection of good-behaving sequences in Banach spaces goes back to [230].

The result of Gowers and Maurey on the existence of Banach spaces which are not isomorphic to their closed hyperplanes can be found in [173].

The extended notion of a locally finite space mentioned in Remark 2.35 is used in some papers, e.g. [247]. Our definition of a locally finite metric space follows [345].

Filters and ultrafilters were introduced by H. Cartan [89],[90]. A detailed presentation of Cartan's results on filters and ultrafilters was given in [73]. Our summary of basic results on ultrafilters is similar to the summary in [130, pp. 30–31]. Ultraproducts were introduced to Banach space theory by Dacuhna-Castelle and Krivine [107]. An important contribution to the development and dissemination of these ideas is the paper by Heinrich [199]. Our presentation of the basic results on ultraproducts is similar to those of [120] and [383]. Theorem 2.26 has its roots in [107]. It was explicitly formulated in [199], where the following proof was suggested: to show that an ultraproduct of Banach lattices (see the definition of a Banach lattice in [288]) is a Banach lattice. As Banach lattices, L_p-spaces admit a characterization [288, Theorem 1.b.2] which goes back to Kakutani [234]. It is not difficult to check that this characterization passes from $(X_i)_{i\in I}$ to their ultraproducts, see [199] or [120, p. 172]. The first part of Theorem 2.26 follows. The second part follows if we combine the first part with Fact 1.20 mentioned in Chapter 1.

According to [40, Remarks to Chapter VII] Auerbach proved the existence of Auerbach bases for each finite-dimensional Banach space X. The book [40] does not contain any proofs of the existence of Auerbach bases. The two-dimensional case of Auerbach's result was proved in [32]. During the Second World War Herman Auerbach (1901–1942) fell victim to the Nazi regime (see [373, p. 590]), and his original proof of the general case seems to have been lost. Apparently the first published proofs of Auerbach's lemma appeared in [116] and [424]. The paper [385] contains interesting results on the relations between Auerbach bases obtained using different constructions, and some related references, see also [358].

The notion of finite representability was introduced in [213]. The Dvoretzky theorem is from [132]. It preceded the introduction of the notion of finite representability and was proved in terms of finite-dimensional normed spaces: for each $n \in \mathbb{N}$ and $\varepsilon > 0$ a normed space of sufficiently high dimension contains an n-dimensional subspace whose Banach–Mazur distance to ℓ_2^n is $\leq 1 + \varepsilon$. The Dvoretzky theorem and its ramifications form a very important part of Banach space theory, see [332] and [381].

The precursors to the type and cotype theory first appear in papers [348],[349]. Various versions of the notions of type and cotype appear in the papers [129], [204], [376], [395], all of which were published in the early 1970's. Apparently the term 'type' appeared for the first time in [376]. One of the most important contributions to the type-cotype theory and the first paper where this theory was systematically developed is [319]. This paper contains the Maurey–Pisier Theorem 2.55. Theorem 2.56 is a special case of Theorem 2.55, its first part was proved earlier in [375]. Our presentation of the proof of Theorem 2.56 is based on [9, Chapters 7 and 11] and [375].

Proofs of the full version of Maurey–Pisier theorem (Theorem 2.55) and of other important results about type and cotype can be found in [332]. See [318] for a historical account of the type-cotype theory written by one of its main contributors. Among predecessors which were not mentioned in [318] we would like to mention [133] (nonlinear versions of type) and [365] (Fourier-analytic versions of type-cotype).

Theorem 2.43 is from [227]. Jordan and von Neumann proved the result in the complex case too. Many other isometric geometric characterizations of Hilbert spaces are known, interested readers should consult the book [18].

The Kwapień theorem 2.44 is from [260] (another recommended source is the book [380, Chapter 3]). The part of Theorem 2.53 concerning cotype appears in [348],[349] and the part concerning type appears in [343]. However these papers were written without the type-cotype terminology.

The Khinchin inequality goes back to [248], in the present form it was stated and proved in Littlewood [295], apparently Littlewood was unaware of [248]. The Kahane inequality goes back to [232] and [233]. None of these sources contains an explicit statement of the Kahane inequality, the closest statement is [233, Theorem 4 in Chapter II]. Because of its importance several different proofs of the Kahane inequality were found and presented in many texts and monographs. Here is the list of some of the proofs of the Kahane inequality. *Probabilistic proofs:* Most of them are close to the original original argument sketched in [232] and presented in detail in [233]. See [9, Section 6.2] and [160, Section 12.8] for similar proofs. Our presentation is very close to the one given in [9]. Somewhat different are the tail estimates proofs, see [272, pp. 91, 100–101]. *Geometric proof:* A proof based on the Brunn–Minkowski inequality, see [332, pp. 134–136]. *Analytic proof:* See [72], [288, pp. 74–77], [378], and [383]. *Latała–Oleszkiewicz* [268] suggested a very elegant proof in the case $p = 2$ with the best constant, see also [160, Section 13.3].

The Hölder inequality is presented in numerous analysis books, see [194, Sections 2.7-2.8], [160, Section 5.4]. Historical studies show that it would be more natural to attribute this inequality to L. J. Rogers, see the footnote to Theorem 13 in [194] and [414, Chapter 9] and references therein for more on this matter. Lemma 2.49 is attributed to P. Levy in [233, Chapter II, Section 3].

Theorem 2.62 is from [354], Theorem 2.63 is from [48]. Some more results in the same spirit can be found in [46].

Exercise 2.69 is the well-known Giesy lemma (see [161, Lemmas I.4 and I.6]), it is a more precise finite-dimensional version of the result of James [211, Lemma 2.1]. It is worth mentioning that there exists a version of the Giesy lemma related to superreflexivity which we are going to study in Section 9.3.1. To state it we denote the subset of $S_{\ell_1^n}$ consisting of all sequences with at most one change of signs by J_n. With this notation one of the results of James [213] can be stated as

Theorem 2.70. *For a Banach space X the following three conditions are equivalent.*

- *The space X is non-superreflexive.*
- $\inf_n h(X, J_n) > 0$.
- $h(X, J_n) = 1 \ \forall n \in \mathbb{N}$.

Using the standard ultraproduct argument it can be shown that Theorem 2.70 implies the following result.

Theorem 2.71. *For each $n \in \mathbb{N}$ and every real number α, β satisfying $1 > \alpha > \beta > 0$ there is $N = N(n, \alpha, \beta) \in \mathbb{N}$ such that $h(X, J_N) \geq \beta$ implies $h(X, J_n) \geq \alpha$.*

It turns out, however that in this case the dependence of $N = N(n, \alpha, \beta)$ on n, α, β is much worse than in the Giesy lemma, see [350], [437].

2.8 Hints to exercises

To Exercise 2.65. Show that X contains a bounded infinite sequence $\{x_k\}_{k=1}^\infty$ satisfying $\|x_k - x_m\| \geq 1$ for $k \neq m$. Show that this sequence contains uncountably many subsequences with any two of them having only finitely many common elements. Consider the elements of $X^{\mathcal{U}}$ corresponding to these subsequences.

To Exercise 2.66. Consider the ℓ_2 direct sums of two-dimensional $\{\ell_{p_n}^2\}_{n=1}^\infty$ and $\{\ell_{q_n}^2\}_{n=1}^\infty$, where $\{p_n\}_{n=1}^\infty$ and $\{q_n\}_{n=1}^\infty$ are two dense subsets of $[1, 2]$, but only one of them contains 1. (The definition of an ℓ_2 direct sum is similar to the definition of the ℓ_∞ direct sum given in this chapter, see Definition 3.6 in the next chapter.)

To Exercise 2.68. Show that each of the assumptions, used for different pairs of vectors, implies the parallelogram identity.

Chapter 3

Constructions of embeddings

Let (X, d_X) and (Y, d_Y) be metric spaces. The infimum of distortions of bilipschitz embeddings of X into Y is denoted $c_Y(X)$. We let $c_Y(X) = \infty$ if there are no bilipschitz embeddings of X into Y. When $Y = L_p$ we use the notation $c_Y(\cdot) = c_p(\cdot)$ and call this number the *L_p-distortion* of X. The parameter $c_2(X)$ is called the *Euclidean distortion* of X.

We are going to study some constructions of low-distortion embeddings of finite metric spaces (X, d) into ℓ_p $(1 \le p < \infty)$. The embeddings which we are going to construct actually will be into ℓ_p^n with large n. To construct such an embedding means to construct a finite or infinite sequence of Lipschitz real-valued functions on X, so that the embedding will be $u \mapsto \{f_i(u)\}_{i=1}^n \in \ell_p^n$ or $u \mapsto \{f_i(u)\}_{i=1}^\infty \in \ell_p$. We start with embeddings called *Fréchet embeddings*. They are defined as embeddings into ℓ_p^n or ℓ_p for which the functions f_i are defined as scalar multiples of functions of the form $d(u, A)$ where A is a subset of X and

$$d(u, A) = \inf\{d(u, v) : v \in A\}.$$

Some of the constructions of suitable functions of this form are based on a very useful notion of a *padded decomposition* of a metric space X.

3.1 Padded decompositions and their applications to constructions of embeddings

Definition 3.1. A *decomposition* of a metric space (X, d) is a partition of X into disjoint subsets. Given a decomposition $P = \{C_1, \dots, C_m\}$ of X, we refer to the sets C_i as *clusters*. We write $\mathcal{P}_X$ for the set of all partitions of X. For $x \in X$ and a partition $P \in \mathcal{P}_X$ we denote by $P(x)$ the unique cluster of P containing x.

Recall that the *diameter* of a set A in a metric space X is $\sup_{u,v \in A} d(u, v)$. We are interested in collections of decompositions for which the diameters of clusters are bounded by some number Δ, but each $x \in X$ is, on average over the collection, reasonably far from the complement of the cluster $P(x)$. In this connection we consider the set $\mathcal{D}_X$ of all probability measures on $\mathcal{P}_X$. We are interested mostly in the case where X is a finite set. In such a case $\mathcal{P}_X$ is also finite, each element of $\mathcal{D}_X$ is a measure on a finite set. *Support* of such a measure is defined as the set of decompositions having nonzero probability.

Definition 3.2. A *stochastic decomposition* of a finite metric space (X, d) is a probability measure $\Psi \in \mathcal{D}_X$. Let $\varepsilon : X \to (0, 1]$ and $\Delta > 0$. The measure Ψ is called an *ε-padded Δ-bounded decomposition* of X (or an *(ε, Δ)-padded decomposition*) if it satisfies the following two conditions:

(a) *Diameter condition:* $\operatorname{diam}(C) < \Delta$ for all $P \in \operatorname{supp}(\Psi)$ and all $C \in P$.

(b) *Pad condition:* For all $x \in X$, $\Psi\{P : d(x, X \backslash P(x)) \geq \varepsilon(x)\Delta\} \geq \frac{1}{2}$.

Remark 3.3. The number $\frac{1}{2}$ in the pad condition can be replaced by any other number in the interval $(0, 1)$. This will change the class of (ε, Δ)-padded decompositions, but the theory could be developed in the same way.

Now we describe a way in which padded decompositions can be used to construct bilipschitz embeddings with relatively small distortions. (Later we shall prove stronger results, but some of the main ideas are more transparent in the proof of Theorem 3.4.)

Theorem 3.4. *Let X be a finite metric space with* $\operatorname{diam}(X) < 2^t$ $(t \in \mathbb{N})$ *and* $d(u, v) \geq 1$ *for all $u, v \in X$, $u \neq v$. Suppose that there is $\alpha > 0$ such that the metric space X has an (ε, Δ)-padded decomposition for each $\Delta \in L = \{1, 2, 2^2 \dots, 2^t\}$ with $\varepsilon(x) = \alpha$ for each $x \in X$. Then $c_2(X) \leq \frac{4\sqrt{2}(t+1)^{1/2}}{\alpha}$.*

Proof. For each $\Delta \in L$ let $\Psi(\Delta)$ be an (ε, Δ)-padded decomposition of X with $\varepsilon(x) = \alpha$ for each $x \in X$. Let $P = \{C_1, \dots, C_{m(P)}\} \in \operatorname{supp}(\Psi(\Delta))$. Denote by $U_1, \dots, U_{2^{m(P)}}$ unions of all possible collections of clusters of P. For each $P \in \operatorname{supp}(\Psi(\Delta))$ and each $j \in \{1, 2, 3, \dots, 2^{m(P)}\}$ we introduce the function $f_{P,j} : X \to \mathbb{R}$ by

$$f_{P,j}(u) = \begin{cases} \min\{d(u, X \backslash P(u)),\ \Delta\}, & \text{if } u \in U_j \\ 0, & \text{if } u \notin U_j. \end{cases} \tag{3.1}$$

Remark 3.5. Here we assume that $d(u, X \backslash P(u)) = \infty$ if $X \backslash P(u) = \emptyset$. We use minimum with Δ in (3.1) for two reasons: (1) To avoid infinite values if the decomposition consists of one cluster; (2) To achieve suitable bounds on the functions which we construct. Although we do not need these bounds for Theorem 3.4, we shall need them in Section 3.4.

Let f_P be the embedding of X into $\ell_2^{2^{m(P)}}$ defined by the formula

$$f_P(u) = \{2^{-m(P)/2} f_{P,j}(u)\}_{j=1}^{2^{m(P)}}.$$

Let us show that $\operatorname{Lip}(f_P) \leq 1$. We start by showing that

$$|f_{P,j}(u) - f_{P,j}(v)| \leq d(u, v).$$

In the case where $P(u) = P(v)$ this inequality follows from the well-known simple fact: the function $d(u, A)$, as a function of u, has Lipschitz constant 1. In the case where $P(u) \neq P(v)$ we have

$$0 \leq d(u, X \backslash P(u)) \leq d(u, v) \text{ and } 0 \leq d(v, X \backslash P(v)) \leq d(u, v),$$

and the conclusion follows in this case also. Therefore we have

$$\|f_P(u) - f_P(v)\| \leq \left(\sum_{j=1}^{2^{m(P)}} 2^{-m(P)} (d(u, v))^2 \right)^{1/2} = d(u, v).$$

On the other hand, the construction does not imply any estimates for $\|f_P(u) - f_P(v)\|$ from below. To get an embedding admitting estimates from below we concatenate the embeddings f_P. To define concatenation we need the following definition.

Definition 3.6. Let $\{X_\gamma\}_{\gamma \in \Gamma}$ be a family of Banach spaces. Their ℓ_p *direct sum*, where $1 \leq p < \infty$, is defined as the Banach space of all functions $F : \Gamma \to \cup_{\gamma \in \Gamma} X_\gamma$ satisfying the conditions $F(\gamma) \in X_\gamma$ and $\sum_{\gamma \in \Gamma} \|F(\gamma)\|_{X_\gamma}^p < \infty$. The sum and multiplication by scalars of such functions are defined pointwise, the norm is defined by $\|F\| = \left(\sum_{\gamma \in \Gamma} \|F(\gamma)\|_{X_\gamma}^p\right)^{1/p}$. The ℓ_p direct sum is denoted $\left(\oplus_{\gamma \in \Gamma} X_\gamma\right)_p$. The function F is also denoted $\oplus_{\gamma \in \Gamma} F(\gamma)$ or $(F(\gamma))_{\gamma \in \Gamma}$.

Remark 3.7. Observe that each $F \in \left(\oplus_{\gamma \in \Gamma} X_\gamma\right)_p$, $1 \leq p < \infty$, can satisfy $F(\gamma) \neq 0$ only for countably many elements $\gamma \in \Gamma$.

Remark 3.8. ℓ_∞ direct sums were introduced in Definition 2.17. In that definition we used slightly different notation.

Observation 3.9. Any ℓ_2 direct sum of Hilbert spaces (finite or infinite-dimensional) is a Hilbert space of the corresponding dimension. In fact, in such a case the norm $\left(\sum_{\gamma \in \Gamma} \|F(\gamma)\|_{\mathcal{H}_\gamma}^2\right)^{1/2}$ is induced by the inner product $\sum_{\gamma \in \Gamma} \langle F(\gamma), G(\gamma) \rangle_{\mathcal{H}_\gamma}$, where $\langle \cdot, \cdot \rangle_{\mathcal{H}_\gamma}$ is the inner product of $\mathcal{H}_\gamma$.

Now we return to the proof of Theorem 3.4. First we concatenate the embeddings f_P for $P \in \text{supp}(\Psi(\Delta))$ in the following way: We consider an embedding f_Δ of X into the direct sum

$$\mathcal{H}_\Delta = \left(\oplus_{P \in \text{supp}(\Psi(\Delta))} \ell_2^{2^{m(P)}} \right)_2$$

defined by

$$f_\Delta(u) = \oplus_{P \in \text{supp}(\Psi(\Delta))} \sqrt{\Psi(P)} f_P(u).$$

The choice of coefficients $\sqrt{\Psi(P)}$ implies $\mathrm{Lip}(f_\Delta) \le 1$. In fact,

$$\begin{aligned}\|f_\Delta(u) - f_\Delta(v)\| &= \Big(\sum_{P \in \mathrm{supp}(\Psi(\Delta))} \Psi(P)\|f_P(u) - f_P(v)\|^2\Big)^{1/2} \\ &\le \Big(\sum_{P \in \mathrm{supp}(\Psi(\Delta))} \Psi(P)(d(u,v))^2\Big)^{1/2} = d(u,v).\end{aligned} \tag{3.2}$$

The pad condition can be used to estimate $\|f_\Delta(u) - f_\Delta(v)\|$ from below if $d(u,v) \ge \Delta$. In fact, if $d(u,v) \ge \Delta$, the diameter condition implies that $P(u) \ne P(v)$ for each $P \in \mathrm{supp}(\Psi(\Delta))$. Therefore a quarter of all sets U_j (corresponding to such P) satisfies the conditions $u \in U_j$ and $v \notin U_j$. Thus

$$\begin{aligned}\|f_P(u) - f_P(v)\| &= \Big(\sum_{j=1}^{2^{m(P)}} 2^{-m(P)}|f_{P,j}(u) - f_{P,j}(v)|^2\Big)^{1/2} \\ &\ge \Big(\Big(\frac{1}{4}\cdot 2^{m(P)}\Big)2^{-m(P)} \min\{d(u, X\backslash P(u)), \Delta\}^2\Big)^{1/2} \\ &= \frac{1}{2}\min\{d(u, X\backslash P(u)), \Delta\}.\end{aligned}$$

Now we use the pad condition (see also (3.2)) and get (since $\alpha \le 1$)

$$\|f_\Delta(u) - f_\Delta(v)\| \ge \sqrt{\Psi\{P : d(u, X\backslash P(u)) \ge \alpha\Delta\}} \cdot \frac{1}{2}\alpha\Delta \ge \frac{\alpha\Delta}{2\sqrt{2}}. \tag{3.3}$$

Remark 3.10. The embedding $f_\Delta : X \to \mathcal{H}_\Delta$ has the following properties

(a) $\mathrm{Lip}(f_\Delta) \le 1$.

(b) $d(u,v) \ge \Delta \quad\Rightarrow\quad \|f_\Delta(u) - f_\Delta(v)\| \ge \frac{\alpha}{2\sqrt{2}}\Delta$

(c) $\|f_\Delta(u)\| \le \Delta$ for each $u \in X$. (This property can be verified one-by-one for maps $f_{P,j}$, f_P, f_Δ.)

Later (in Section 3.4 of this chapter) we introduce a class of embeddings sharing similar properties, and present a way of combining them which is in a certain sense optimal. Now we finish the proof of Theorem 3.4 using a *naïve concatenation.*

Now we concatenate f_Δ over $\Delta \in L$, that is, we consider $f : X \to (\oplus_{\Delta \in L}\mathcal{H}_\Delta)_2$ given by $f(u) = \oplus_{\Delta \in L} f_\Delta(u)$. This concatenation is a bilipschitz embedding and it is easy to estimate its distortion. In fact, on the one hand

$$\begin{aligned}\|f(u) - f(v)\| &= \Big(\sum_{\Delta\in L} \|f_\Delta(u) - f_\Delta(v)\|^2\Big)^{1/2}\\ &\le |L|^{1/2} d(u,v)\\ &= (t+1)^{1/2}\cdot d(u,v).\end{aligned}$$

On the other hand, since X is such that $1 \le d(u,v) < 2^t$ ($\forall u, v \in X,\ u \ne v$), for each pair u, v we can find $\Delta \in L$ such that $\Delta \le d(u,v) < 2\Delta$. By (3.3), for that value of Δ we have $\|f_\Delta(u) - f_\Delta(v)\| \ge \frac{\alpha\Delta}{2\sqrt{2}}$. Thus $\|f_\Delta(u) - f_\Delta(v)\| > \frac{\alpha}{4\sqrt{2}}\, d(u,v)$ and $\|f(u) - f(v)\| > \frac{\alpha}{4\sqrt{2}}\, d(u,v)$. We conclude that $c_2(X) \le \frac{4\sqrt{2}(t+1)^{1/2}}{\alpha}$. □

Remark 3.11. The proof of Theorem 3.4 can be easily modified to prove, under the same assumptions on X, the inequality $c_p(X) \le 2(8(t+1))^{1/p}/\alpha$. It is worth mentioning that for $1 \le p < 2$ we get a better inequality if use the embedding into ℓ_2 and the Dvoretzky theorem 2.30.

3.2 Padded decompositions of minor-excluded graphs

Now we are going to construct a padded decomposition of the type needed for Theorem 3.4. We need the notion of a minor of a graph.

Definition 3.12. Let H be a graph with the vertex set $V(H) = \{v_1, \dots, v_k\}$. We say that H is a *minor* of a graph G if there are disjoint connected sets of vertices $V_i \subset V(G)$, $i = 1, \dots, k$, such that for each edge $e \in E(H)$ with end vertices v_i and v_j there is a corresponding edge $\widetilde{e} \in E(G)$ with end vertices in V_i and V_j. The sets V_i are called *branch sets*.

Remark 3.13. Minors play a significant role in the structure theory of graphs. One of the most well-known results of this type is the Kuratowski theorem: a graph G is planar (that is, can be sketched in the plane without edge crossings) if and only if the complete graph K_5 and the complete bipartite graph $K_{3,3}$ are not minors of G.

Theorem 3.14. *Suppose that an unweighted connected graph G is such that the complete graph K_r is not a minor of G. Then G, considered as a metric space with its graph distance, admits an (ε, Δ)-padded decomposition for each Δ with $\varepsilon(x) = c(r)$, where $c(r) > 0$ depends on r only (and does not depend on G, x, or Δ).*

Proof. We prove the theorem with $c(r) = \frac{1}{16r(8r+2)}$. We do not claim that this bound is close to being optimal.

Observe that there is nothing to prove in the case where $c(r)\Delta \le 1$, because in this case all the conditions of Definition 3.2 on (ε, Δ)-padded decomposition are satisfied for the decomposition of the vertex set of G onto 1-element sets. (This decomposition can be considered as a stochastic decomposition for which the probability measure is supported on one decomposition.)

In the case where $c(r)\Delta > 1$, that is, where $\frac{\Delta}{8r+2} > 16r$, the desired decomposition is constructed in the following way. We let $\delta = \lfloor \frac{\Delta}{8r+2} \rfloor$ and do the following:

We pick a vertex $v(G)$ in G and call it the *center* of G. We pick $r_1 \in \{0, 1, \dots, \delta-1\}$ and let $S_1 \subset E(G)$ be the set of all edges uv in G for which $d_G(u, v(G)) = r_1 \pmod{\delta}$ and $d_G(v, v(G)) = r_1 + 1 \pmod{\delta}$, where d_G is the shortest path metric of G.

We call the components of the graph $G_2 = G \backslash S_1$ (the graph obtained from G after the deletion of edges of S_1) *components of level* 2. For uniformity we call G a *component of level* 1. We split the set of all components of level 2 into two classes which will be handled differently.

We say that a component C of level k satisfies the FC-*condition* (*far-from-centers condition*) if there are vertices in C which are at d_G-distance at least $4r\delta$ from the centers of all components of lower levels containing C. Components satisfying the FC-condition are called FC-*components*.

In application to components of level 2 the FC-condition means that the component is not contained in a ball of radius $< 4r\delta$ centered at $v(G)$. It is clear that in this case non-FC-components already satisfy the diameter condition, the diameter of each of them is $< 8r\delta \leq \Delta$. In general, one can check that as soon as we get a component which satisfies the diameter condition of Definition 3.2, we do not need to decompose it further; but we do this for the uniformity of presentation.

We continue decomposing FC-components. In each FC-component C of level 2 we pick a vertex $v(C)$ satisfying $d_G(v(C), v(G)) \geq 4r\delta$ and call it the *center* of C. We pick $r_2 \in \{0, 1, \dots, \delta - 1\}$ and let S_2 be the set of all edges uv (in all FC-components of level 2) for which $d_C(u, v(C)) = r_2 \pmod{\delta}$ and $d_C(v, v(C)) = r_2 + 1 \pmod{\delta}$. We call the components of the graph $G_3 = G_2 \backslash S_2$ *components of level* 3. It is important that in this step we use the distance d_C, that is, the shortest path metric of C (the FC-component of level 2 which we partition into components of level 3).

Observe that non-FC-components of level 3 do not have to have diameters $< 8r\delta$, and we have to decompose them further to be sure that the diameter condition is satisfied. We are going to do this somewhat later. Now we work on decomposition of FC-components.

FC-components of level 3 and all further levels are decomposed in the same way as in the case of level 2. Here is the general description for level k. In each FC-component C of level k we pick a vertex $v(C)$ satisfying the condition

$$d_G(v(C), v(\widetilde{C})) \geq 4r\delta$$

for each component $\widetilde{C}$ of one of the preceding levels containing C.

We pick a number $r_k \in \{0, 1, \dots, \delta-1\}$ and let S_k be the set of all edges uv in all FC-components C of level k for which $d_C(u, v(C)) = r_k \pmod{\delta}$ and $d_C(v, v(C)) =$

$r_k + 1 \pmod{\delta}$. (Here in each component C of level k the distance d_C is its shortest path metric.) We let $G_{k+1} = G_k \backslash S_k$. We call components of the graph G_{k+1} *components of level* $k+1$. The set of edges uv in an FC-component C of level k for which $d_C(u, v(C)) = r_2 + s\delta$ and $d_C(v, v(C)) = r_2 + 1 + s\delta$ for some $s \in \mathbb{N} \cup \{0\}$ will be called *a cut of level* k and denoted T_k^s.

The following observation is simple but crucial:

Observation 3.15. A shortest path joining a vertex u in a component C of level k with the center $v(C)$ cannot contain more than δ vertices in any of the components of level $k+1$ contained in C.

Proof. Let P be a shortest path joining a vertex of C with $v(C)$. When we form components of level $k+1$, one out of each δ consecutive edges of the path P is deleted. The conclusion follows. □

Claim 3.16. If G has no K_r-minors, then there are no FC-components of level r.

Proof. Assume the contrary. Let C_r be an FC-component of level r in G and let C_j $(j = 1, \dots, r-1)$ be the component of level j containing C_r. Denote by a_j $(j = 1, \dots, r)$ the center of C_j.

We are going to use $\{a_1, \dots, a_r\}$ to find a K_r-minor in G. The branch sets of the K_r-minor which we construct contain $a_1, \dots, a_r$, respectively.

Observe that our construction implies $d_G(a_i, a_j) \geq 4r\delta$. The most natural idea is to join each pair a_i, a_j by a shortest path in G, and then to cut these paths somewhere for each pair, attaching each part to the corresponding branch set. However, this straightforward construction does not always work: no matter how we cut the paths, we cannot claim the disjointness of the obtained sets. Our construction can be regarded as a more complicated modified version of this straightforward construction, the purpose of the modification is to achieve disjointness.

We construct inductively K_j-minors $(j = 2, \dots, r)$ with branch sets

$$B_{r-j+1}(j), \dots, B_r(j)$$

containing vertices $a_{r-j+1}, \dots, a_r$, respectively. These branch sets get some additional vertices in each step of the induction. One of the important ideas is to use C_{r-j}-shortest paths (that is, paths which are shortest in the graph distance of C_{r-j}) instead of shortest paths in G. This helps because in this case the branch sets $B_{r-j+1}(j), \dots, B_r(j)$ are subsets of C_{r-j+1}. On the other hand, by Observation 3.15, each C_{r-j}-shortest path can contain at most $\delta - 1$ edges in C_{r-j+1}. We need to start this path at such a vertex of $B_m(j)$ $(m = r-j+1, \dots, r)$ that within this $(\delta - 1)$-piece the C_{r-j}-shortest path does not intersect components of the collection

$$\{B_{r-j+1}(j), \dots, B_r(j)\} \backslash \{B_m(j)\}.$$

This is the main idea behind the formal description of the construction which we present now.

Remark 3.17. K_2-minors are easy to find in each graph having some edges. Our construction is more complicated than is necessary because we are going to use it in the inductive construction of K_r-minors for larger r.

Constructing a K_2-minor. By construction, the component C_r is contained between two cuts of level $r-1$, say T_{r-1}^{s} and T_{r-1}^{s+1} (the cut T_{r-1}^{s+1} can be empty). We join a_r and a_{r-1} with a C_{r-1}-shortest path $P_r(1)$. We cut this path into two pieces: one of the pieces consists of vertices which are closer to a_r than the 4th cut of level $r-1$ crossed by the path on the way to a_{r-1}, that is, than the cut T_{r-1}^{s-3}. The vertices of this piece form the branch set $B_r(2)$, and all of the remaining vertices of the path form the branch set $B_{r-1}(2)$ (we use this terminology and notation for these sets because they are going to be the branch sets for the K_2-minor). We pick a vertex on $P_r(1)$ between the cuts T_{r-1}^{s-2} and T_{r-1}^{s-1} and denote it $c_r(2)$, we may assume that $d_G(c_r(2), a_r) \leq 2\delta$. We let $c_{r-1}(2) = a_{r-1}$. It is clear that $B_{r-1}(2)$ and $B_r(2)$ are branch sets for a K_2-minor. It is also clear (and will be used below) that $B_{r-1}(2)$ and $B_r(2)$ are subsets of C_{r-1}. (We have much more freedom in the first step. We have chosen the described construction of $B_{r-1}(2)$ and $B_r(2)$ to have uniformity with further steps in the construction.)

Constructing a K_3-minor (assuming $r \geq 3$). By construction, the component C_{r-1} is contained between two cuts of level $r-2$, say T_{r-2}^{s} and T_{r-2}^{s+1} (the letter s here denotes, in general, a different number than in the previous step, we just do not want to use a more complicated notation). We join $c_{r-1}(2)$ and $c_r(2)$ with a_{r-2} using C_{r-2}-shortest paths $P_{r-1}(2)$ and $P_r(2)$, respectively. The conditions $d_G(c_r(2), a_r) \leq 2\delta$ and $d_G(a_i, a_j) \geq 4\delta r$ imply that each of these paths is of length $\geq 4\delta r - 2\delta$.

We decompose each of the paths $P_{r-1}(2)$ and $P_r(2)$ into two pieces. One of the pieces is determined by the condition: it consists of vertices which are closer to $c_{r-1}(2)$ and $c_r(2)$, respectively, than the 4th cut crossed by the path on the way to a_{r-2}, that is, than the cut T_{r-2}^{s-3}. We add those vertices of $P_{r-1}(2)$ and $P_r(2)$ which are closer than the cut T_{r-2}^{s-3} to $c_{r-1}(2)$ and $c_r(2)$, respectively, to the sets $B_{r-1}(2)$ and $B_r(2)$, and denote the obtained sets $B_{r-1}(3)$ and $B_r(3)$. All of the remaining vertices of the paths, including a_{r-2}, form the set $B_{r-2}(3)$, which we create in this step.

We pick vertices between the cuts T_{r-2}^{s-1} and T_{r-2}^{s-2} on each of the paths $P_{r-1}(2)$ and $P_r(2)$ and denote them $c_{r-1}(3)$ and $c_r(3)$, respectively. We may assume that $d_G(c_{r-1}(3), a_{r-1}) \leq 2\delta$ and $d_G(c_r(3), a_r) \leq 4\delta$. We let $c_{r-2}(3) = a_{r-2}$.

Let us show that $B_{r-2}(3)$, $B_{r-1}(3)$, and $B_r(3)$ are branch sets of a K_3-minor. It is clear that only disjointness requires verification. The first step in the verification is easy: The set $B_{r-2}(3)$ cannot intersect any of the other sets because it is separated from them by the cut T_{r-2}^{s-3}.

The added to $B_{r-1}(2)$ and $B_r(2)$ pieces of paths $P_{r-1}(2)$ and $P_r(2)$ cannot intersect each other by the triangle inequality: the d_G-distance between them is $\geq d_G(c_{r-1}(2), c_r(2)) - 2 \cdot 4\delta \geq d_G(a_{r-1}, a_r) - 2 \cdot 4\delta - 2\delta > 0$.

In addition, by Observation 3.15, the path $P_r(2)$ leaves C_{r-1} before it can intersect $B_{r-1}(2)$ because the C_{r-1}-distance between $c_r(2)$ and $B_{r-1}(2)$ is $> \delta$. A similar statement holds for the path $P_{r-1}(2)$ and $B_r(2)$. We emphasize that this statement holds because $P_{r-1}(2)$ and $P_r(2)$ are C_{r-2}-shortest paths; this is one of the main ideas in the proof.

We complete the inductive construction in the following way. When we construct a K_j-minor we introduce the following objects:

- Branch sets $B_{r-j+1}(j), \dots, B_r(j)$.
- Connection vertices $c_{r-j+1}(j), \dots, c_r(j)$ satisfying $c_{r-j+1}(j) \in B_{r-j+1}(j)$, $\dots$, $c_r(j) \in B_r(j)$. (Let us mention the role of these vertices: they are sufficiently distant from other branch sets, and therefore not-too-large portions of C_{r-j}-shortest paths joining connection vertices with the center of C_{r-j} do not intersect other branch sets.)

These objects satisfy the following conditions:

(a) All of them are contained in C_{r-j+1}.

(b) The C_{r-j+1}-distance between $c_m(j)$ and $B_n(j)$ $(m \neq n)$ is $> \delta$.

(c) $d_G(a_m, c_m(j)) \leq 2(j-1)\delta$, and the inequality is strict for all m, except, possibly $m = r$.

Now we describe the inductive step. By construction, the component C_{r-j+1} is contained between two cuts of level $r - j$, say T^s_{r-j} and T^{s+1}_{r-j} (the number s here can be different from the numbers denoted s in the previous steps). We consider C_{r-j}-shortest paths $P_{r-j+1}(j)$, $\dots$, $P_r(j)$ connecting a_{r-j} and the vertices $c_{r-j+1}(j), \dots, c_r(j)$, respectively. We add the vertices of pieces of these paths which are closer to $c_{r-j+1}(j), \dots, c_r(j)$ than the cut T^{s-3}_{r-j} to $B_{r-j+1}(j), \dots, B_r(j)$, respectively. In this way we form the branch sets $B_{r-j+1}(j+1), \dots, B_r(j+1)$. The remaining vertices of the paths $P_{r-j+1}(j), \dots, P_r(j)$ form a branch set $B_{r-j}(j+1)$, which is created in this step. Let $c_{r-j+1}(j+1), \dots, c_r(j+1)$ be vertices in the branch sets $B_{r-j+1}(j+1), \dots, B_r(j+1)$ satisfying the condition: they belong to the pieces of the paths $P_{r-j+1}(j)$, $\dots$, $P_r(j)$ which are between the cuts T^{s-1}_{r-j} and T^{s-2}_{r-j}, and satisfy the conditions

$$d_G(c_m(j+1), c_m(j)) \leq 2\delta. \tag{3.4}$$

The existence of such vertices is obvious. We let $c_{r-j}(j+1) = a_{r-j}$.

We need to show that $B_{r-j}(j+1), \dots, B_r(j+1)$ form a collection of branch sets for a K_{j+1}-minor. The only property which is not obvious is their disjointness. We also have to check that analogues of the conditions in items (b) and (c) for j replaced by $j+1$ are satisfied. The condition in (a) follows from the construction.

The branch set $B_{r-j}(j+1)$ does not intersect the other branch sets because they are separated by the cut T^{s-3}_{r-j}.

By Observation 3.15, the paths $P_{r-j+1}(j), \ldots, P_r(j)$ have at most δ vertices in C_{r-j+1}. Therefore, by condition (b), $P_m(j)$ cannot intersect $B_n(j)$ for $m \neq n$. The pieces of $P_m(j)$ and $P_n(j)$, $m \neq n$ cannot intersect each other because their lengths are $\leq 4\delta$, and therefore, by condition (c) and the triangle inequality, the closest vertices on them are at distance

$$> d_G(a_m, a_n) - 2 \cdot 2(j-1)\delta - 2 \cdot 4\delta = d_G(a_m, a_n) - 4(j+1)\delta,$$

and the last number is assumed to be nonnegative.

Finally, we check the validity of conditions (b) and (c) with j replaced by $j+1$. Condition (b) follows because $c_n(j+1)$ is separated from $B_n(j+1)$ by two cuts, T^s_{r-j} and T^{s-1}_{r-j}. Condition (c) follows immediately from (3.4) and the definition of $c_{r-j}(j+1)$. □

Claim 3.16 implies that the graph G (which has no K_r-minors) does not have FC-components of level r. Therefore to complete the proof of Theorem 3.14 we need to show how to decompose non-FC-components in such a way that the overall decomposition satisfies the diameter and pad conditions of Definition 3.2.

Handling the non-FC-components. For each non-FC-component we define the set of at most $r-1$ *relevant centers*, as the set of centers of FC-components containing it. We pick independently $r-1$ integers $s_1, \ldots, s_{r-1}$ from the set $\{4r\delta, 4r\delta+1, \ldots, 4r\delta+\delta-1\}$.

Let N be a non-FC-component, $c_1, \ldots, c_k$ be relevant centers; where c_i is a center of an FC-component of level i containing N. We consider the following sets of edges in N:

$$R_1 = \{uv : d_G(u, c_1) = s_1, d_G(v, c_1) = s_1 + 1\}$$
$$R_2 = \{uv : d_G(u, c_2) = s_2, d_G(v, c_2) = s_2 + 1\}$$
$$\ldots$$
$$R_k = \{uv : d_G(u, c_k) = s_k, d_G(v, c_k) = s_k + 1\}.$$

We remove these sets of edges from the edge set of N. The components of the obtained graph form the final decomposition. It is clear that the d_G-diameters of the obtained components are $\leq 2(4r\delta + \delta - 1) < (8r+2)\delta \leq \Delta$.

Now we assume that each of the choices $r_1, \ldots, r_{r-1}$ and $s_1, \ldots, s_{r-1}$ was a random choice and that the probabilities of choosing each of the $2(r-1)$-tuples of numbers are the same. We get a stochastic decomposition in which each of the decompositions obtained in the described way has probability $\left(\frac{1}{\delta}\right)^{2(r-1)}$ (if it is not repeated for different choices of the integers $r_1, \ldots, r_{r-1}$ and $s_1, \ldots, s_{r-1}$, and the probability $\left(\frac{1}{\delta}\right)^{2(r-1)} \times$(the number of repetitions) otherwise). It seems worthwhile to add some justification to this statement. In fact, the resulting decomposition depends not only on the choice of numbers $r_1, \dots, r_{r-1}$ and $s_1, \dots, s_{r-1}$, but also on the choice of centers. However we can make the choice of centers uniquely determined by the choice of $r_1, \dots, r_{r-1}$. This can be done as follows. We label vertices of G using positive integers $1, 2, \dots$. Now we let each of the centers to be the vertex with the smallest label satisfying the corresponding conditions.

Verifying the pad property. As we mentioned at the beginning of the proof, it suffices to consider the case where $\frac{\Delta}{8r+2} > 16r$. It is clear that in this case $c(r)\Delta \le \frac{1}{16r}(\delta+1)$.

Consider a vertex v in G. Let A be the event that v is d_G-closer than $\frac{1}{16r}(\delta + 1)$ to any vertex from which v was separated by removal of one of the sets S_i or one of the sets R_i. We can write $A = \cup_{i=1}^{2(r-1)} A_i$, where A_i is the event that the first separation happened when the i-th removal was made (we order the edge sets as $S_1, \dots, S_{r-1}, R_1, \dots, R_{r-1}$).

Let us show that the probability of A_1 is $\le \frac{2}{\delta} \cdot \left\lfloor \frac{\delta+1}{16r} \right\rfloor \le \frac{1}{4r}$. Let d_v be the residue of the number $d(v(G), v)$ modulo δ. It is easy to see that the separation described in the definition of A_1 can happen only if r_1 is the residue of one of the numbers

$$d_v - \left\lfloor \frac{\delta+1}{16r} \right\rfloor, \dots, d_v - 1, d_v, d_v + 1, \dots, d_v + \left\lfloor \frac{\delta+1}{16r} \right\rfloor - 1$$

modulo δ (otherwise each of the deleted edges is too far from v to separate a vertex which is close to v). The claim about probability of A_1 follows immediately. Similar argument works for each of the events A_k, $k = 2, \dots, 2(r-1)$. Therefore the probability of A is $\le \frac{1}{4r}(2r-1) \le \frac{1}{2}$. □

3.3 Padded decompositions in terms of ball growth

As we have already seen in the proof of Theorem 3.4, collections of padded decompositions with the diameter bound Δ ranging over all integral powers of 2 (of course, 2 can be replaced by any other positive number different from 1, but we are going to use 2) can be used to construct embeddings into ℓ_p. (Later we shall see that some of the embeddings constructed in this way are close to being optimal, that is, having the distortion close to the minimal possible.) In this connection it is natural to introduce the following definition:

Definition 3.18 (Decomposition bundle). Let X be a metric space. Given a function $\varepsilon : X \times \mathbb{Z} \to (0, 1]$, an *$\varepsilon$-padded decomposition bundle* on X is a function $\beta : \mathbb{Z} \to \mathcal{D}_X$, where for every $\ell \in \mathbb{Z}$, $\beta(\ell)$ is a 2^ℓ-bounded $\varepsilon(\cdot, \ell)$-padded decomposition of X.

Our next purpose is to construct a reasonably good (as we shall see later) decomposition bundle in an arbitrary finite metric space. Let (X, d) be a finite metric space, we use the notation $B(x, r)$ for the set $\{y \in X : d(x, y) \le r\}$ which is called a *closed ball* in X. In the case where no confusion can occur, we call $B(x,r)$ a *ball* in X. We use the notation $|B(x,r)|$ to denote the cardinality of the ball.

Theorem 3.19. *On an arbitrary finite metric space X there exists an ε-padded decomposition bundle, where*

$$\varepsilon(x,\ell) = \left(16 + 16\ln\left(\frac{|B(x,2^{\ell})|}{|B(x,2^{\ell-3})|}\right)\right)^{-1}. \tag{3.5}$$

Proof. Let $\Delta = 2^{\ell}$ for some $\ell \in \mathbb{Z}$. The corresponding stochastic decomposition $\beta(\ell)$ is constructed in the following way. Choose, uniformly at random, a permutation π of X and a value $\alpha \in \left(\frac{1}{4},\frac{1}{2}\right)$. Denote by $\mathbb{P}$ the corresponding probability measure (the product of two measures: the uniform measure on all permutations and the Lebesgue measure on $\left(\frac{1}{4},\frac{1}{2}\right)$). We may assume without loss of generality that $X = \{1,\dots,n\}$, where n is the cardinality of X, and so the inequality $\pi(y) < \pi(z)$ for $y,z \in X$ makes sense. For each pair (π,α) in the constructed probability space we define the following decomposition of X. The number of clusters in each decomposition is n, but some (maybe many) of the clusters are empty. Clusters are labeled using points of X. For every point $y \in X$, define a cluster

$$C_y = B(y,\alpha\Delta) \setminus \Bigg(\bigcup_{z\colon \pi(z)<\pi(y)} B(z,\alpha\Delta)\Bigg).$$

In words: a point $x \in X$ is assigned to C_y where y is the minimal point according to π that is within distance $\alpha\Delta$ from x.

Clearly $P = \{C_y\}_{y\in X}$ constitutes a decomposition of X. Furthermore, $C_y \subseteq B(y,\alpha\Delta)$, thus $\operatorname{diam}(C_y) < \Delta$, so the diameter condition (a) in Definition 3.2 is satisfied for every decomposition P arising from this process. It remains to prove that the pad condition (b) in Definition 3.2 is also satisfied.

Fix a point $x \in X$ and some value $t \le \Delta/8$. We need to estimate from above the probability $\mathbb{P}$ of the event $\mathcal{B}_{x,t}$ consisting of (bad) decompositions P obtained in the described way and satisfying $B(x,t) \not\subseteq P(x)$. It is clear that the event $\mathcal{B}_{x,t}$ occurs if and only if the decomposition $\{C_y\}_{y\in X}$ contains a cluster which intersects $B(x,t)$, but does not contain $B(x,t)$. Observe that clusters C_y cannot intersect $B(x,t)$ if $y \notin B(x,\Delta)$.

Let $a = |B(x,\Delta/8)|$, $b = |B(x,\Delta)|$. We arrange the points $w_1,\dots,w_b \in B(x,\Delta)$ in order of increasing distance to x, resolving ties arbitrarily. Let $\mathcal{E}_k$ be the event consisting of decompositions for which the following two conditions are satisfied:

(a) w_k is the element with the minimal $\pi(w_k)$ among those satisfying $\alpha\Delta \ge d(x,w_k) - t$.

(b) $\alpha\Delta < d(x,w_k) + t$.

Condition (a) means that C_{w_k} is the first cluster which *could* intersect $B(x,t)$, in the sense that there is no contradiction for the existence of such a nonempty intersection with the distance between the points, the values of t and α, and the triangle inequality. (Note that the intersection can still be empty.)

Condition (b) is necessarily satisfied if C_{w_k} does not contain $B(x,t)$. So informally (b) means that C_{w_k} *does not necessarily contain* $B(x,t)$.

To show that $\mathcal{B}_{x,t}$ is contained in $\cup_{i=k}^{b}\mathcal{E}_k$ it suffices to establish that all decompositions in the complement of $\cup_{i=k}^{b}\mathcal{E}_k$ are "good". This immediately follows from the following description of this complement: it consists of partitions for which the set C_{w_k} with the minimal $\pi(w_k)$ among those which *could* intersect $B(x,t)$ contains $B(x,t)$.

Note that if $w_k \in B(x,\Delta/8)$, then $\mathbb{P}[\mathcal{E}_k] = 0$ since in this case $d(x,w_k)+t \le \Delta/8 + t \le \Delta/4 < \alpha\Delta$. Therefore letting $I_k = [d(x,w_k)-t, d(x,w_k)+t)$ and denoting by $\mathbb{P}[\mathcal{E}_k \mid \alpha\Delta \in I_k]$ the corresponding conditional probabilities, we get

$$\begin{aligned}\mathbb{P}[\mathcal{B}_{x,t}] &\le \sum_{k=a+1}^{b} \mathbb{P}[\mathcal{E}_k] = \sum_{k=a+1}^{b} \mathbb{P}[\alpha\Delta \in I_k]\cdot \mathbb{P}[\mathcal{E}_k \mid \alpha\Delta \in I_k] \\ &\le \sum_{k=a+1}^{b} \frac{2t}{\Delta/4}\cdot\frac{1}{k} \le \frac{8t}{\Delta}\left(1+\ln\frac{b}{a}\right),\end{aligned} \tag{3.6}$$

where we used the fact that the event $\mathcal{E}_k$ does not contains pairs (π,α) for which π is such that $\pi(w_j) < \pi(w_k)$ for some $j < k$. Therefore $\mathbb{P}[\mathcal{E}_k \mid \alpha\Delta \in I_k]$ does not exceed the ratio of the number of permutations of $\{w_1,\dots,w_b\}$ which place numbers of the set $\{w_1,\dots,w_{k-1}\}$ above $\pi(w_k)$ and the number of all permutations. Since each of the numbers $\{\pi(w_1),\dots,\pi(w_k)\}$ has equal chances of being the least, the ratio is equal to $1/k$.

Recall that $\Delta = 2^{\ell}$. Setting $t = \varepsilon(x,\ell)\Delta \le \Delta/8$, where $\varepsilon(x,\ell)$ is as in (3.5), the right-hand side of (3.6) is at most $\frac{1}{2}$, proving the pad condition (b) of Definition 3.2. □

Corollary 3.20. *There exists an absolute constant* $0 < C < \infty$ *such that for every* $\ell \in \mathbb{Z}$ *there is a* 1*-Lipschitz map* $f_\ell : X \to \ell_2$ *satisfying* $\|f_\ell(x)\| \le 2^{\ell}$ *for all* $x \in X$ *and*

$$\|f_\ell(x) - f_\ell(y)\| \ge \frac{d(x,y)}{C\left(1+\ln\frac{|B(x,2^{\ell})|}{|B(x,2^{\ell-3})|}\right)} \tag{3.7}$$

for all $x, y \in X$ *with* $d(x,y) \in [2^{\ell}, 2^{\ell+1}]$.

In this proof one can recognize a modification of the proof of Theorem 3.4.

Proof. Let $\Psi(\ell)$ be an $(\varepsilon(x,\ell), 2^{\ell})$-padded decomposition of X, where $\varepsilon(x,\ell)$ is as in (3.5). Let $P = \{C_1,\dots,C_{m(P)}\} \in \operatorname{supp}(\Psi(\ell))$. Denote by $U_1,\dots,U_{2m(P)}$ unions of all possible collections of clusters of P. For each $P \in \operatorname{supp}(\Psi(\ell))$ and each $j \in \{1, 2, 3, \ldots, 2^{m(P)}\}$ we introduce the function $f_{P,j} : X \to \mathbb{R}$ by

$$f_{P,j}(u) = \begin{cases} \min\{d(u, X \setminus P(u)),\ 2^\ell\}, & \text{if } u \in U_j \\ 0, & \text{if } u \notin U_j. \end{cases}$$

Let f_P be the embedding of X into $\ell_2^{2^{m(P)}}$ defined by the formula

$$f_P(u) = \{2^{-m(P)/2} f_{P,j}(u)\}_{j=1}^{2^{m(P)}}.$$

In the same way as in Theorem 3.4 we check that $\operatorname{Lip}(f_P) \le 1$. We concatenate the embeddings f_P for $P \in \operatorname{supp}(\Psi(\Delta))$ in the following way: We consider an embedding f_ℓ of X into the direct sum

$$\mathcal{H}_\ell = \big(\oplus_{P \in \operatorname{supp}(\Psi(\ell))} \ell_2^{2^{m(P)}} \big)_2$$

defined by

$$f_\ell(u) = \oplus_{P \in \operatorname{supp}(\Psi(\ell))} \sqrt{\Psi(P)} f_P(u).$$

As in Theorem 3.4 we get $\operatorname{Lip}(f_\ell) \le 1$ and $\| f_\ell(x)\| \le 2^\ell$.

Let $x, y \in X$ be such that $d(x, y) \in [2^\ell, 2^{\ell+1}]$. For the same reason as in Theorem 3.4, the pad condition implies

$$\begin{aligned} \| f_\ell(x) - f_\ell(y)\| &\ge \sqrt{\Psi\{P : d(x, X \setminus P(x)) \ge \varepsilon(x,\ell)2^\ell\}} \cdot \frac{1}{2} \varepsilon(x,\ell)2^\ell \\ &\ge \frac{\varepsilon(x,\ell)2^\ell}{2\sqrt{2}} \\ &\ge \frac{2^\ell}{2\sqrt{2}\left(16 + 16 \ln \frac{|B(x,2^\ell)|}{|B(x,2^{\ell-3})|}\right)} \\ &\ge \frac{d(x,y)}{64\sqrt{2}\left(1 + \ln \frac{|B(x,2^\ell)|}{|B(x,2^{\ell-3})|}\right)}. \end{aligned}$$

□

Remark 3.21. The same proof works to show the existence of a 1-Lipschitz map $f_\ell : X \to \ell_p$ ($1 \le p \le \infty$) satisfying the conditions of Corollary 3.20.

3.4 Gluing single-scale embeddings

In this section we are going to show that the problem of constructing a nontrivial embedding of a finite metric space to some extent can be reduced to that of *handling a single scale* in the following sense. Let (X, d_X) and (Y, d_Y) be metric spaces, and let τ be a positive real number.

Definition 3.22. If $f : X \to Y$ is a 1-Lipschitz map such that for every $x, y \in X$ with $d(x, y) \in [\tau, 2\tau)$ the inequality

$$d_Y(f(x), f(y)) \geq \frac{\tau}{K}$$

holds, then we will call f a *scale-τ embedding* with *deficiency* K.

Mappings constructed using padded decompositions, see (3.3) and (3.7), are of this type.

Theorem 3.23 (Gluing lemma). *Let X be a metric space with n elements. Suppose that for each $m \in \mathbb{Z}$, there exists a map $\varphi_m : X \to \ell_p$, $1 \leq p < \infty$, satisfying $\|\varphi_m\| \leq 2^m$, which is a scale-2^m embedding with deficiency K. Then $c_p(X) \leq C\, K^{1-\frac{1}{p}} \log_2^{1/p} n$ for some absolute constant C.*

Remark 3.24. At the end of this section we shall show that the boundedness condition $\|\varphi_m\| \leq 2^m$ is not needed in the Hilbert space case $p = 2$.

Remark 3.25. It is worth mentioning that mappings constructed in the proofs of Theorem 3.4 and Corollary 3.20 satisfy the boundedness condition.

Remark 3.26. In this section we use $c, C, C_1, C_2, \ldots$ to denote absolute constants contained in $(0, \infty)$. The constants denoted in the same way could be different in different places, sometimes we emphasize the fact that the constants are different by adding subscripts.

Remark 3.27. Observe that for a finite metric space X the condition of being a scale-2^m embedding with deficiency K is restrictive only for finitely many values of m. We mean values of m for which there exist pairs x, y such that $d(x, y) \in [2^m, 2^{m+1})$. We call such values of m (or of 2^m) *relevant scales*. For other values of m any map, even having the same value at all points of X, is a scale-2^m embedding with deficiency K. Let R be the number of relevant scales for X. Then the naïve concatenation used at the end of the proof of Theorem 3.4 (see page 83), applied to φ_m corresponding to relevant scales, produces a map showing that $c_p(X) \leq CK \cdot R^{1/p}$. This estimate is not satisfactory for our purposes in two respects (here we restrict our attention to the Euclidean distortion):

(a) One can easily observe that the amount of relevant scales can be much larger than $\log_2 n$ (see Example 3.28 below and Exercise 3.37 for more on this). So replacement of R by $\log_2 n$ is essential.

(b) We would like to have $\sqrt{K}$ rather than K in our estimate for $c_2(X)$.

Example 3.28. Consider the following n-element subset of $\mathbb{R}$:

$$\{0, 1, 1+2, 1+2+4, \dots, 1+2+\dots+2^{n-2}\},$$

endowed with the usual metric inherited from $\mathbb{R}$. It is easy to see that there are $n-1$ relevant scales.

To deal with the issues mentioned above (in items (a) and (b)) we use the following two tools:

(a) Instead of naïve (straightforward) concatenation used in the proof of Theorem 3.4, we form $\lceil \log_2 n \rceil - 1$ different weighted concatenations $\{\psi_t\}$ of $\{\varphi_m\}$. By a *weighted concatenation* we mean a concatenation of maps in which maps are multiplied by some coefficients (usually between 0 and 1) which are functions of x (x is an element of X). The weights are distributed in such a way that

- Each of the weighted concatenations ψ_t is C-Lipschitz where C is an absolute constant.
- The weighted concatenations ψ_t are such that

$$\|\psi_t(x) - \psi_t(y)\| \geq \|\varphi_m(x) - \varphi_m(y)\|$$

if $d(x, y) \in [2^m, 2^{m+1})$ and t is in the intersection of the intervals

$$\left[\log_2 |B(x, 2^{m-3})|, \log_2 |B(x, 2^{m+3})|\right) \bigcap \left[\log_2 |B(y, 2^{m-3})|, \log_2 |B(y, 2^{m+3})|\right).$$

Remark 3.29. One can easily check that the intersection is nonempty, but simple examples show that it can contain no integers. Such situations are handled in Lemma 3.31.

(b) We form similar concatenation for the mappings constructed in Corollary 3.20. This allows us to replace K by $K^{1-\frac{1}{p}}$ in the estimate for the distortion.

As we see from the last item, we plan to construct suitable weighted concatenations for maps which are not necessarily scale-2^m embeddings with deficiency K. This is the purpose of our next lemma.

For $x, y \in X$, define

$$\gamma_m(x, y) = \begin{cases} x & \text{if } |B(x, 2^m)| \geq |B(y, 2^m)| \\ y & \text{otherwise.} \end{cases}$$

Lemma 3.30. *Let $1 \leq p < \infty$. Given, for every $m \in \mathbb{Z}$ a 1-Lipschitz map $h_m : X \to \ell_p$ satisfying $\|h_m(x)\| \leq 2^m$ for each $x \in X$, there exists a map $H : X \to \ell_p$ which satisfies*

(a) $\operatorname{Lip}(H) \leq C(\log n)^{\frac{1}{p}}$.

(b) *For every $m \in \mathbb{Z}$ and every $x, y \in X$ with $d(x,y) \in [2^m, 2^{m+1})$, we have*

$$\|H(x) - H(y)\| \geq \left(\left\lfloor \log_2 \frac{|B(\gamma_{m-3}(x,y), 2^{m+1})|}{|B(\gamma_{m-3}(x,y), 2^{m-3})|} \right\rfloor \right)^{\frac{1}{p}} \cdot \|h_m(x) - h_m(y)\|. \tag{3.8}$$

Proof. For each $t \in \{1, 2, \ldots, \lceil \log_2 n \rceil - 1\}$ (recall that n is the cardinality of the metric space X) define $R(x,t) = \sup\{R : |B(x,R)| \leq 2^t\}$. Let $\rho : \mathbb{R} \to \mathbb{R}_+$ be a Lipschitz map satisfying (i) $\mathrm{supp}(\rho) \subset [2^{-4}, 2^4]$, (ii) $\rho(\tau) = 1$ for $\tau \in [2^{-3}, 2^3]$, and (iii) $0 \leq \rho(\tau) \leq 1$ for each τ. We may assume that ρ is a piecewise linear map with Lipschitz constant 2^4. Define

$$\rho_{m,t} = \rho\left(\frac{R(x,t)}{2^m}\right).$$

Consider the ℓ_p direct sum $\oplus_{m\in\mathbb{Z}} \ell_p$ (we mean that $X_m = \ell_p$ for each $m \in \mathbb{Z}$), and define the weighted concatenation $\psi_t : X \to \oplus_{m\in\mathbb{Z}} \ell_p$ by

$$\psi_t(x) = \oplus_{m\in\mathbb{Z}} \rho_{m,t}(x) h_m(x),$$

and the map H as the straightforward concatenation of $\{\psi_t : t \in \{1, 2, \ldots, \lceil \log_2 n \rceil - 1\}\}$.

Bounding the Lipschitz constant. Observe that for every t the map $x \mapsto R(x,t)$ is 1-Lipschitz. Since the Lipschitz constant of a composition does not exceed the product of Lipschitz constants, we have $\mathrm{Lip}(\rho_{m,t}) \leq 2^{4-m}$. We also have

$$\|\psi_t(x) - \psi_t(y)\|^p = \sum_{m\in\mathbb{Z}} \|\rho_{m,t}(x) h_m(x) - \rho_{m,t}(y) h_m(y)\|^p. \tag{3.9}$$

For a fixed t the value of $R(x,t)$ is just a number and therefore the numbers $\left\{\frac{R(x,t)}{2^m}\right\}_{m\in\mathbb{Z}}$ are in the support of ρ for at most 9 different values of m. Therefore for each t there are at most 18 nonzero terms in the sum (3.9). To bound a nonzero summand in (3.9) we use

$$\begin{aligned} &\|\rho_{m,t}(x) h_m(x) - \rho_{m,t}(y) h_m(y)\| \\ &\quad \leq \|h_m(x)\| \cdot |\rho_{m,t}(x) - \rho_{m,t}(y)| + |\rho_{m,t}(y)| \cdot \|h_m(x) - h_m(y)\| \\ &\quad \leq 2^m \cdot 2^{4-m} d(x,y) + d(x,y) \leq 17 d(x,y). \end{aligned}$$

Since there are at most 18 nonzero terms in the sum (3.9) we get $\|\psi_t(x) - \psi_t(y)\| \leq C d(x,y)$. The estimate $\mathrm{Lip}(H) \leq C(\log_2 n)^{\frac{1}{p}}$ follows.

Proving (3.8). Let $x, y \in X$ be such that $d(x,y) \in [2^m, 2^{m+1})$. Notice that if $\rho_{m,t}(x) = \rho_{m,t}(y) = 1$, then

$$\|\psi_t(x) - \psi_t(y)\| \geq \|h_m(x) - h_m(y)\|.$$

So to get a lower estimate for $\|H(x) - H(y)\|$ we need to estimate from below the number of values $t \in \{1, 2, \ldots, \lceil \log_2 n \rceil - 1\}$ for which $\rho_{m,t}(x) = \rho_{m,t}(y) = 1$.

Observe that $\rho_{m,t}(x) = 1$ if $R(x,t) \in [2^{m-3}, 2^{m+3}]$, and this happens if

$$t \in \big[\log_2 |B(x, 2^{m-3})|, \log_2 |B(x, 2^{m+3})|\big).$$

Similarly, $\rho_{m,t}(y) = 1$ if

$$t \in \big[\log_2 |B(y, 2^{m-3})|, \log_2 |B(y, 2^{m+3})|\big)$$

(the right ends of the intervals are not included).

Assume without loss of generality that $x = \gamma_{m-3}(x, y)$ (γ is defined in the paragraph preceding the statement of Lemma 3.30), so that $|B(x, 2^{m-3})| \geq |B(y, 2^{m-3})|$. Then

$$t \in \big[\log_2 |B(x, 2^{m-3})|, \log_2 |B(x, 2^{m+1})|\big)$$

implies $\rho_{m,t}(x) = \rho_{m,t}(y) = 1$. To see this recall that $R(\cdot, t)$ is 1-Lipschitz and $d(x, y) < 2^{m+1}$. Therefore $R(x,t) \leq 2^{m+1}$ implies $R(y,t) < 2^{m+3}$ (and even $R(y,t) < 2^{m+2}$). Hence the number of values of $t \in \{1, \ldots, \lceil \log_2 n \rceil - 1\}$ for which $\rho_{m,t}(x) = \rho_{m,t}(y) = 1$ is at least

$$\left\lfloor \log_2 \frac{|B(x, 2^{m+1})|}{|B(x, 2^{m-3})|} \right\rfloor.$$

We conclude that

$$\|H(x) - H(y)\|^p \geq \left\lfloor \log_2 \frac{|B(x, 2^{m+1})|}{|B(x, 2^{m-3})|} \right\rfloor \|h_m(x) - h_m(y)\|^p. \qquad \square$$

It is clear that inequality (3.8) is of no significance in the case where

$$\log_2 \left(\frac{|B(x, 2^{m+1})|}{|B(x, 2^{m-3})|} \right) < 1,$$

and in this case we have to do something different. We do this under a weaker assumption $\log_2 \big(\frac{|B(x, 2^{m-1})|}{|B(x, 2^{m-2})|}\big) < 1$ in our next lemma.

Lemma 3.31. *For each $p \in [1, \infty)$ there exists a map $G : X \to \ell_p$ such that*

(a) $\mathrm{Lip}(G) \leq \lceil \log_2 n \rceil^{\frac{1}{p}}$.

(b) *The condition*

$$\|G(x) - G(y)\| \geq c d(x, y)$$

holds for all $m \in \mathbb{Z}$ and all $x, y \in X$ satisfying $d(x, y) \in [2^m, 2^{m+1})$ and $\log_2 \big(\frac{|B(x, 2^{m-1})|}{|B(x, 2^{m-2})|}\big) < 1$,

Proof of Lemma 3.31. For each $t \in \{1, 2, \ldots, \lceil \log_2 n \rceil\}$, consider a random subset of X which contains each point of X independently with probability 2^{-t}. So the probability of a subset $W \subset X$ with k elements is $2^{-tk}\left(1-2^{-t}\right)^{|X|-k}$. Denote by $(\mathcal{P}(X), \mu_t)$ the probability space, where $\mathcal{P}(X)$ is the set of all subsets of X and μ_t is the measure described above. For each $t \in \{1, 2, \ldots, \lceil \log_2 n \rceil\}$ we introduce a map $g_t : X \to L_p(\mathcal{P}(X), \mu_t)$ given by $(g_t(x))(W) = d(x, W)$ for a subset $W \in \mathcal{P}(X)$.

We concatenate all of these maps, that is, we consider the map

$$G : X \to (\oplus_t L_p(\mathcal{P}(X), \mu_t))_p$$

given by $G(x) = \oplus_t g_t(x)$. Straightforward verification shows that $\mathrm{Lip}(g_t) \le 1$ and $\mathrm{Lip}(G) \le \lceil \log_2 n \rceil^{\frac{1}{p}}$.

Fix $x, y \in X$ such that $d(x, y) \in [2^m, 2^{m+1})$ and $\log_2\left(\frac{|B(x,2^{m-1})|}{|B(x,2^{m-2})|}\right) < 1$. Let $s \in \{0, 1, 2, \ldots, \lceil \log_2 n \rceil - 1\}$ be such that $2^s \le |B(x, 2^{m-1})| \le 2^{s+1}$.

We introduce the following four subsets of $\mathcal{P}(X)$:

- $\mathcal{E}^x_{\mathrm{far}} = \{W \in \mathcal{P}(X) : d(x, W) \ge 2^{m-1}\}$
- $\mathcal{E}^x_{\mathrm{close}} = \{W \in \mathcal{P}(X) : d(x, W) \le 2^{m-2}\}$.
- $\mathcal{E}^y_{\mathrm{close}} = \{W \in \mathcal{P}(X) : d(y, W) \le \frac{3}{2} \cdot 2^{m-2}\}$.
- $\mathcal{E}^y_{\mathrm{far}} = \mathcal{P}(X) \backslash \mathcal{E}^y_{\mathrm{close}}$.

Observe that for each of the probability measures μ_t, each of the events $\mathcal{E}^x_{\mathrm{far}}$, $\mathcal{E}^x_{\mathrm{close}}$ is independent of the event $\mathcal{E}^y_{\mathrm{close}}$, because the events $\mathcal{E}^x_{\mathrm{far}}$, $\mathcal{E}^x_{\mathrm{close}}$ depend on presence of elements of balls $B(x, 2^{m-1})$ and $B(x, 2^{m-2})$ in the subset W, the event $\mathcal{E}^y_{\mathrm{close}}$ depends on the presence of elements of the ball $B(y, \frac{3}{2} \cdot 2^{m-2})$ in the subset W, and the mentioned balls centered at x are disjoint with $B(y, \frac{3}{2} \cdot 2^{m-2})$ because of the assumption $d(x, y) \ge 2^m$.

Observe that the assumption $\log_2\left(\frac{|B(x,2^{m-1})|}{|B(x,2^{m-2})|}\right) < 1$ implies that $|B(x, 2^{m-2})| > 2^{s-1}$. Therefore (using the well-known fact that the sequence $\left\{\left(1-\frac{1}{n}\right)^n\right\}_n$ is increasing and tends to $\frac{1}{e}$)

$$\mu_s\left(\mathcal{E}^x_{\mathrm{close}}\right) \ge 1 - (1 - 2^{-s})^{2^{s-1}} > 1 - \sqrt{\frac{1}{e}},$$

$$\mu_s\left(\mathcal{E}^x_{\mathrm{far}}\right) \ge (1 - 2^{-s})^{2^{s+1}} \ge \frac{1}{16}.$$

We get

$$\begin{aligned}\|G(x)-G(y)\|^p &= \sum_t \int |g_t(x)-g_t(y)|^p d\mu_t \\ &\geq \int_{\mathcal{E}^y_{\text{close}}\cap \mathcal{E}^x_{\text{far}}} |g_s(x)-g_s(y)|^p d\mu_s \\ &\quad + \int_{\mathcal{E}^y_{\text{far}}\cap \mathcal{E}^x_{\text{close}}} |g_s(x)-g_s(y)|^p d\mu_s \\ &\geq (2^{m-3})^p \left(\mu_s(\mathcal{E}^y_{\text{close}})\mu_s(\mathcal{E}^x_{\text{far}}) + (1-\mu_s(\mathcal{E}^y_{\text{close}}))\mu_s(\mathcal{E}^x_{\text{close}})\right) \\ &\geq c(2^m)^p.\end{aligned}$$

□

Proof of Theorem 3.23. We apply Lemma 3.30 to two collections of maps: The collection $\{\varphi_m\}$ whose existence is assumed in Theorem 3.23 and the collection $\{f_\ell\}$ whose existence was proved in Corollary 3.20, see also Remark 3.21. We denote the obtained maps $\Phi : X \to \ell_p$ and $F : X \to \ell_p$, respectively. We concatenate the maps Φ, F, and G, and get a map

$$\Psi : X \to (\ell_p \oplus \ell_p \oplus \ell_p)$$

given by $\Psi(x) = \Phi(x) \oplus F(x) \oplus G(x)$. Using property (a) in Lemma 3.30 and property (a) in Lemma 3.31 we get $\text{Lip}(\Psi) \leq C(\log_2 n)^{\frac{1}{p}}$.

Now we turn to the lower bound. Fix $x, y \in X$, let m be such that $d(x,y) \in [2^m, 2^{m+1})$. Assume without loss of generality that $x = \gamma_{m-3}(x,y)$ and set

$$A = \log_2\left(\frac{|B(x,2^{m+1})|}{|B(x,2^{m-3})|}\right).$$

We have

$$\|\Psi(x)-\Psi(y)\|^p = \|F(x)-F(y)\|^p + \|\Phi(x)-\Phi(y)\|^p + \|G(x)-G(y)\|^p.$$

In the case where $A < 1$ we have, by Lemma 3.31, $\|G(x)-G(y)\| \geq c2^m$. In the case where $A \geq 1$ we use inequality (3.8) and get:

$$\|F(x)-F(y)\|^p \geq \frac{1}{C^p}\left(\frac{\lfloor A\rfloor}{(1+A)^p}\right)\cdot 2^m.$$

The constant C here is different from the constant in (3.7), because we use logarithms to different bases (here we recall our agreement in Remark 3.26).

$$\|\Phi(x)-\Phi(y)\|^p \geq \left(\frac{\lfloor A\rfloor}{K^p}\right)\cdot 2^m.$$

To get the desired estimate it remains to observe that for $A \geq 1$ we have $\frac{\lfloor A \rfloor}{(1+A)^p} + \frac{\lfloor A \rfloor}{K^p} \geq \frac{c}{K^{p-1}}$ for some absolute constant $c > 0$. (If $\lfloor A \rfloor \geq K$, the second term $\geq \frac{1}{K^{p-1}}$. If $1 \leq \lfloor A \rfloor \leq K$, the first term $\geq \min\left\{\frac{K}{(2+K)^p}, \frac{1}{(2+1)^p}\right\}$). Thus

$$\|\Psi(x) - \Psi(y)\|^p \geq \frac{c}{K^{p-1}} 2^m,$$

and the conclusion follows. □

Theorem 3.23 has the following important consequence, due to Bourgain. Historically this consequence preceded Theorem 3.23, which is an outgrowth of it.

Corollary 3.32. *$c_2(X) \leq C \log_2 n$ for each n-element metric space X.*

Proof. Corollary 3.20 implies that for each n-element metric space there exists and each $\ell \in \mathbb{Z}$ there exists a scale-2^ℓ embedding f_ℓ into ℓ_2 with deficiency $\leq C(1 + \log_2 n)$ satisfying $\|f_\ell(x)\| \leq 2^\ell$ for all $x \in X$. Combining this with Theorem 3.23 we get the desired result. □

We shall see in the next chapter that the estimate of Corollary 3.32 is sharp, up to a value of the constant.

Now we show how to prove the statement made in Remark 3.24. It follows immediately if we combine Theorem 3.23 with the next lemma.

Given a metric space M and $D > 0$, we denote by $M^{\leq D}$ the metric space M with a new distance $d_{M^{\leq D}}(x, y) = \min\{d_M(x, y), D\}$.

Lemma 3.33. *For every $D > 0$, $c_2(\ell_2^{\leq D}) \leq \sqrt{\frac{e}{e-1}}$. In fact, $\ell_2^{\leq D}$ $\sqrt{\frac{e}{e-1}}$-embeds into an ℓ_2-sphere.*

Here we sketch a proof which looks like a trick. In Notes and Remarks we describe how this lemma can be almost immediately derived from the theory of embeddability into a Hilbert space developed by Schoenberg.

Proof. Let $\{g_i\}_{i=1}^\infty$ be independent identically distributed standard Gaussian random variables defined on some probability space Ω. Let $H = L_2(\Omega)$ be the Hilbert space consisting of all complex valued square integrable functions on Ω. Let $K > 0$. Define $F : \ell_2 \to H$ by:

$$(F(x_1, x_2, \ldots))(\omega) = K \exp\left(\frac{i}{K} \sum_{j=1}^\infty x_j g_j(\omega)\right), \quad \omega \in \Omega.$$

Clearly $\|F(x)\|_2 = K$ for every $x \in \ell_2$. Observe that for every $x, y \in \ell_2$,

$$
\begin{aligned}
|(F(x))(\omega) - (F(y))(\omega)|^2 &= \\
&= K^2 \left| \exp\left(\frac{i}{K}\sum_{j=1}^{\infty} x_j g_j(\omega)\right) - \exp\left(\frac{i}{K}\sum_{j=1}^{\infty} y_j g_j(\omega)\right)\right|^2 \\
&= K^2 \left| \exp\left(\frac{i}{K}\sum_{j=1}^{\infty} y_j g_j(\omega)\right)\left[\exp\left(\frac{i}{K}\sum_{j=1}^{\infty}(x_j - y_j) g_j(\omega)\right) - 1\right]\right|^2 \\
&= K^2 \left| \exp\left(\frac{i}{K}\sum_{j=1}^{\infty}(x_j - y_j) g_j(\omega)\right) - 1\right|^2 \\
&= 2K^2\left[1 - \cos\left(\frac{1}{K}\sum_{j=1}^{\infty}(x_j - y_j) g_j(\omega)\right)\right].
\end{aligned}
$$

Now, $\sum_{j=1}^{\infty}(x_j - y_j)g_j$ has the same distribution as $g_1\sqrt{\sum_{j=1}^{\infty}(x_j - y_j)^2}$. Hence:

$$
\mathbb{E}|F(x) - F(y)|^2 = 2K^2\left[1 - \mathbb{E}\cos\left(\frac{g_1}{K}\|x - y\|_2\right)\right].
$$

Observe that by symmetry, $\mathbb{E}\sin\left(\frac{g_1}{K}\|x - y\|_2\right) = 0$, so that:

$$
\mathbb{E}\cos\left(\frac{g_1}{K}\|x - y\|_2\right) = \mathbb{E}\exp\left(i\frac{g_1}{K}\|x - y\|_2\right) = \exp\left(-\frac{\|x - y\|_2^2}{2K^2}\right),
$$

where we use the fact that $\mathbb{E}e^{iag_1} = e^{-a^2/2}$ (this fact can be derived using our computation on page 15).

Putting it all together, we have shown that:

$$
\|F(x) - F(y)\|_2 = \sqrt{2}K\sqrt{1 - e^{-\frac{\|x-y\|_2^2}{2K^2}}}.
$$

Using the elementary inequality:

$$
\frac{e-1}{e}\min\{1, a\} \le 1 - e^{-a} \le \min\{1, a\} \quad a > 0,
$$

we deduce that:

$$
\sqrt{\frac{e-1}{e}}\min\{\sqrt{2}K, \|x - y\|_2\} \le \|F(x) - F(y)\|_2 \le \min\{\sqrt{2}K, \|x - y\|_2\}.
$$

Letting $K = D/\sqrt{2}$ we get the desired result. □

3.5 Exercises

Exercise 3.34. Introduce graphs $R_{n,k}$ as graphs whose vertex sets are subsets in $\mathbb{R}^n$ with all n coordinates in the set $\{1, \dots, k\}$ and the edge set is determined by the condition: u and v are joined by an edge if and only if $\|u - v\|_1 = 1$ (where $\|\cdot\|_1$ is the norm of ℓ_1^n). So it is natural to call $R_{n,k}$ *n-dimensional cubic grids.*

Show that K_r is a minor of $R_{3,r}$ for each $r \in \mathbb{N}$ (and so 3-dimensional grids are not minor-excluded).

See the Hints to exercises section if needed.

An infinite graph is called *locally finite* if degrees of all vertices are finite.

Exercise 3.35. Use the techniques developed in Theorems 3.4 and 3.14 to prove the following result: Let $r \in \mathbb{N}$ and let G be an infinite locally finite unweighted connected graph which does not have K_r-minors, let d_G be the graph distance on G. Then (G, d_G) embeds coarsely into L_1.

See the Hints to exercises section if needed.

Exercise 3.36. If X is an unweighted graph with its graph distance and with n vertices. Show that the number of relevant scales in the sense of Remark 3.27 does not exceed $\lfloor \log_2 n \rfloor$.

Exercise 3.37 (Possibly the exact evaluation is an open problem). Let X be an n-element metric space. How large can the number of relevant scales be?

The general case of Exercise 3.37 could be difficult. Therefore I suggest to start by trying to get some partial results: (a) The number of relevant scales cannot reach the value $\frac{n(n-1)}{2}$ for nontrivially large values of n. (b) The number of relevant scales can be more than n. One of the ways of proving (a) and one of the ways of proving (b) are outlined in the Hints to exercises section.

Exercise 3.38. Prove the statement made in Remark 3.29 and give an example whose existence is claimed in the mentioned remark.

3.6 Notes and remarks

The embeddings of the type described at the beginning of this chapter are named after Fréchet, who introduced them in [155, pp. 161–162] to prove Proposition 1.17.

The first result on embeddability of general n-element metric space into a Hilbert space is due to Bourgain [75]. The upper estimates for distortion found in [75] turned out to be optimal, up to a multiplicative constant. Bourgain's ideas were further elaborated on by Matoušek in [311]. The book [314] (see also an updated version [315] of the chapter on metric spaces) contains a nice presentation of Bourgain's embeddability theorem. For this reason I decided not to present Bourgain's proof of $c_p(X) \leq O(\ln|X|)$ here. We derive Bourgain's estimate as a corollary of a theory which has grown out of [75]. (It it worth mentioning that there is a slight error in the proof of [314, Lemma 15.7.2]: r_j should be defined as the minimum of r_q and of the current definition. See [315] for a corrected proof.)

Padded decompositions. The concept of a padded decomposition first emerges in the paper of Rao [389]. Later this concept was used and studied by several other authors, see [143], [189], [254], [255], [275], [277]. The term *padded decomposition* was suggested in (the conference version (2003) of) the paper of Krauthgamer and Lee [254]. Theorem 3.4 goes back to Rao [389].

The theory of graph minors is a very important part of graph theory, an introduction to this theory can be found in [122].

Padded decompositions of minor-excluded graphs were discovered in [389]. They were based on the decomposition procedure suggested by Klein, Plotkin, and Rao [251]. Rao's estimates for parameters of padded decompositions of minor-excluded graphs were improved by Fakcharoenphol and Talwar [143]. Our presentation is very close to the proof in [143].

Padded decompositions in terms of ball growth are constructed using an approach originally suggested by Calinescu, Karloff, and Rabani [88], the analysis of which was improved on by Fakcharoenphol, Rao, and Talwar [142]. Our presentation follows the work of Krauthgamer, Lee, Mendel, and Naor [255], and the extension to general measures was observed in [277].

Section 3.4 is based on the paper of Lee [273]. The corresponding part of [273] is a development of ideas of [255]. See [94] and [25] for related developments. Lemma 3.31 is a piece of Bourgain's argument [75]. Lemma 3.33 and its proof are from [325, Lemma 5.2]. It is worth mentioning that Lee [273, Section 1.2] conjectured that one can achieve, in the Euclidean case, the estimate $(K + \sqrt{\log n})$ in Theorem 3.23 instead of $\sqrt{K \log n}$, but this conjecture was disproved by Jaffe, Lee, and Moharrami [210] who have proved that the bound $\sqrt{K \log n}$ is tight.

The proof that we mentioned before the proof of the truncation Lemma 3.33 is the following: The main step in the proof of Lemma 3.33 is to show that if we endow ℓ_2 with the metric $\sqrt{1 - e^{-\|x-y\|_2^2}}$, the obtained metric space embeds isometrically in Hilbert space. This follows from a classical characterization of Schoenberg [407] of isometric embeddability into a Hilbert space in terms of negative definite kernels, see [55, Chapter 8, Section 1] for a compact and accessible presentation of the corresponding results. Corollary 3.32 is from [75], see the beginning of these Notes and Remarks for related information.

3.7 Hints to exercises

To Exercise 3.34. Use the following description of minors. Let us introduce the *k-th level* in $R_{3,r}$ as the set of vertices whose third coordinate is k. Let us introduce the *k-th slice* in $R_{3,r}$ as the set of vertices whose second coordinate is k. We split the r-th slice into r vertical paths and introduce the branch sets $B_1, \dots, B_r$ as follows:

B_1 consists of vertices of first level, except vertices of the r-th slice, and of vertices of the first vertical path.

B_2 consists of vertices of second level, except vertices of the r-th slice, and of vertices of the second vertical path.

We continue in the obvious way: B_k, $3 \leq k \leq r$ consists of vertices of level k, except vertices of the r-th slice, and of vertices of the k-th vertical path.

Show that the the obtained sets $B_1, \dots, B_r$ are branch sets of a K_r-minor.

To Exercise 3.35. See the `arXiv` version of [355] if you have difficulties (the journal version does not contain the proof based on the suggested techniques).

To Exercise 3.37. (a) Assume that there are $\frac{n(n-1)}{2}$ different distances. Consider the metric space as a weighted graph K_n. Pick in this graph a subgraph isomorphic (as an unweighted graph) to $K_{1,n-1}$. Pick in the class containing $n-1$ vertices two vertices with the smallest distances to the only vertex in the other class. Analyze the possible values of the distances between these two vertices and the other vertices in the class containing $(n-1)$ vertices.

(b) Consider the graph $K_{1,n-1}$ with the following weights: $1, 2^2 - \varepsilon, 2^4 - \varepsilon, 2^6 - \varepsilon, \dots, 2^{2n-4} - \varepsilon$, where $\varepsilon \in (0,1)$.

Chapter 4

Obstacles for embeddability: Poincaré inequalities

The purpose of this chapter is to develop some techniques for estimates of distortion $c_Y(X)$ from below.

4.1 Definition of Poincaré inequalities for metric spaces

We start with a simple example: consider a 4-cycle C_4 and label its vertices in the cyclic order: v_1, v_2, v_3, v_4. We are going to show that the Euclidean distortion of C_4 can be estimated using the following inequality

$$\begin{aligned}&\|f(v_1) - f(v_3)\|^2 + \|f(v_2) - f(v_4)\|^2 \\ &\quad \le \|f(v_1) - f(v_2)\|^2 + \|f(v_2) - f(v_3)\|^2 \\ &\qquad + \|f(v_3) - f(v_4)\|^2 + \|f(v_4) - f(v_1)\|^2,\end{aligned} \tag{4.1}$$

which holds for an arbitrary collection $f(v_1), f(v_2), f(v_3), f(v_4)$ of elements of a Hilbert space.

Proof of (4.1). We use the identity $\|a-b\|^2 = \|a\|^2 - 2\langle a, b\rangle + \|b\|^2$ for each of the terms in (4.1). Then we move everything to the right-hand side and observe that the obtained inequality can be written in the form

$$0 \le \|f(v_1) - f(v_2) + f(v_3) - f(v_4)\|^2. \qquad \square$$

We postpone the computation of $c_2(C_4)$ slightly, introducing some terminology first. Inequality (4.1) can be considered as one of the simplest Poincaré inequalities for embeddings of metric spaces.

Definition 4.1. Let (X, d_X) and (Y, d_Y) be metric spaces, $\Psi : [0, \infty) \to [0, \infty)$ be a non-decreasing function, $a_{u,v}$, $b_{u,v}$ $(u, v \in X)$ be finitely nonzero arrays of non-negative real numbers (this condition definitely holds if X is finite). If for an arbitrary function $f : X \to Y$ the inequality

$$\sum_{u,v\in X} a_{u,v}\Psi(d_Y(f(u), f(v))) \ge \sum_{u,v\in X} b_{u,v}\Psi(d_Y(f(u), f(v))) \tag{4.2}$$

holds, we say that Y-valued functions on X satisfy the *Poincaré inequality* (4.2).

Observe that in this inequality the structure of X plays no role, we use X just as a set of labels for elements $f(u) \in Y$. On the other hand it is worth mentioning that the metric structure of X usually plays a significant role in our choice of the arrays $a_{u,v}, b_{u,v}$.

The inequality (4.2) is useful for the theory of embeddings only if a similar inequality does not hold for the identity map on X, that is, if

$$\sum_{u,v\in X} a_{u,v}\Psi(d_X(u,v)) < \sum_{u,v\in X} b_{u,v}\Psi(d_X(u,v)). \tag{4.3}$$

In such a case we get immediately that X is not isometric to a subset of Y.

Definition 4.2. We call the quotient

$$\frac{\sum_{u,v\in X} b_{u,v}\Psi(d_X(u,v))}{\sum_{u,v\in X} a_{u,v}\Psi(d_X(u,v))}$$

the *Poincaré ratio* of the metric space X corresponding to the Poincaré inequality (4.2) and denote it $P_{a,b,\Psi(t)}(X)$.

Having more information on the values of sides of (4.3) and on the function Ψ, we can get an estimate for the distortion $c_Y(X)$ and some information on the functions ρ_1 and ρ_2 from Definition 1.45 for any coarse embedding of X into Y. The corresponding estimate of $c_Y(X)$ is quite simple if $\Psi(t) = t^p$ for some $p > 0$.

Proposition 4.3. *If Y-valued functions on X satisfy the Poincaré inequality* (4.2) *with $\Psi(t) = t^p$, then*

$$c_Y(X) \geq \left(P_{a,b,t^p}(X)\right)^{1/p}. \tag{4.4}$$

Proof. In fact, assume the contrary. Then there exists $D < \left(P_{a,b,t^p}(X)\right)^{1/p}$, an embedding $f : X \to Y$, and $r > 0$ such that

$$\forall u,v \in X, \quad rd_X(u,v) \leq d_Y(f(u), f(v)) \leq rDd_X(u,v). \tag{4.5}$$

We get

$$\begin{aligned}\sum_{u,v\in X} b_{u,v}(d_X(u,v))^p &\overset{(4.5)}{\leq} \frac{1}{r^p}\sum_{u,v\in X} b_{u,v}(d_Y(f(u),f(v)))^p \\ &\overset{(4.2)}{\leq} \frac{1}{r^p}\sum_{u,v\in X} a_{u,v}(d_Y(f(u),f(v)))^p \\ &\overset{(4.5)}{\leq} D^p \sum_{u,v\in X} a_{u,v}(d_X(u,v))^p.\end{aligned}$$

Combining this inequality with $D < \left(P_{a,b,t^p}(X)\right)^{1/p}$ we get a contradiction with the definition of the Poincaré ratio. □

Example 4.4. Now we are ready to estimate $c_2(C_4)$. It is clear that (4.1) is a Poincaré inequality for ℓ_2-valued functions on C_4 (more precisely: for ℓ_2-valued functions on $V(C_4)$). The corresponding Poincaré ratio is:

$$\frac{d_{C_4}(v_1, v_3)^2 + d_{C_4}(v_2, v_4)^2}{d_{C_4}(v_1, v_2)^2 + d_{C_4}(v_2, v_3)^2 + d_{C_4}(v_3, v_4)^2 + d_{C_4}(v_4, v_1)^2} = 2.$$

By Proposition 4.3 we get $c_2(C_4) \geq \sqrt{2}$. This estimate is sharp, as it can be shown by considering an embedding whose image is the set

$$(0,0), (0,1), (1,1), (1,0) \in \ell_2^2.$$

An important example of a Poincaré inequality is given in the next section.

A very interesting fact is that the L_p-distortion can be characterized in terms of Poincaré ratios, see Section 4.3 in this chapter.

4.2 Poincaré inequalities for expanders

In this section we introduce a very important notion of a family of expander graphs (or expanders), establish a very useful Poincaré inequality for L_1-valued functions on expander graphs, and derive corollaries on the L_1-distortion and coarse embeddability of families of expander graphs.

Definition 4.5. For a graph G with vertex set V and a subset $F \subset V$ by ∂F we denote the set of edges connecting F and $V \backslash F$. The *expanding constant* (also known as *Cheeger constant* and as *conductance*) of G is

$$h(G) = \inf \left\{ \frac{|\partial F|}{\min\{|F|, |V \backslash F|\}} : F \subset V,\ 0 < |F| < +\infty \right\}$$

(where $|A|$ denotes the cardinality of a set A).

A sequence $\{G_n\}$ of graphs is called a *family of expanders* if all of G_n are finite, connected, *k-regular* for some $k \in \mathbf{N}$ (this means that each vertex is incident with exactly k edges), their expanding constants $h(G_n)$ are bounded away from 0 (that is, there exists $\varepsilon > 0$ such that $h(G_n) \geq \varepsilon$ for all n), and $|V(G_n)| \to \infty$ as $n \to \infty$.

Remark 4.6. It is far from being obvious that families of expanders exist. We will present several constructions of families of expanders in Chapter 5. In this chapter and in Chapter 7 we explain why families of expanders are important for metric geometry.

The following is a Poincaré inequality for L_1-valued functions on a vertex set of a graph. We denote the *adjacency matrix* of a graph $G = (V, E)$ by $\{a_{u,v}\}_{u,v \in V}$, that is

$$a_{u,v} = \begin{cases} 1 & \text{if } u \text{ and } v \text{ are adjacent} \\ 0 & \text{otherwise.} \end{cases}$$

Let h be the expanding constant of G.

Theorem 4.7. *The following Poincaré inequality holds for L_1-valued functions on V:*

$$\sum_{u,v\in V} a_{u,v}\|f(u)-f(v)\| \geq \sum_{u,v\in V} \frac{h}{|V|}\|f(u)-f(v)\|. \tag{4.6}$$

Lemma 4.8. *Let $G=(V,E)$ be a connected graph with the expanding constant h, and $f : V \to \mathbb{R}$ be a real-valued function on V. Then*

$$\sum_{v\in V} |f(v)-M| \leq \frac{1}{h}\sum_{uv\in E} |f(u)-f(v)|, \tag{4.7}$$

where M is a median of the set $\{f(v)\}_{v\in V}$.

Recall that a *median* of a finite set of real numbers is a number M such that at least half of the numbers are $\geq M$ and at least half the numbers are $\leq M$; note that the median may be non-unique in some cases.

Proof. Replacing f by $\tilde{f} = f - M$, we may assume that $M=0$. Also we assume (for simplicity) that the number of vertices is odd. (Only a slight modification is needed when the number of vertices is even.) Let

$$f_1 \leq f_2 \leq \cdots \leq f_k \leq 0 = f_{k+1} \leq f_{k+2} \leq \cdots \leq f_{2k+1}$$

be the values of the function. Then

$$\sum_{v\in V} |f(v)| = \sum_{i=1}^{2k+1} |f_i|.$$

We introduce *level sets* of the function f as

$$L_i^- := \{v : f(v) \leq f_i\}, \quad i = 1,\ldots,k,$$

and

$$L_i^+ = \{v : f(v) \geq f_i\}, \quad i = k+2,\ldots,2k+1.$$

We define also

$$f_i^\Delta := f_{i+1} - f_i,\ i = 1,\ldots,k$$

and

$$f_i^\nabla := f_i - f_{i-1},\ i = k+2,\ldots,2k+1.$$

We have

$$\sum_{i=1}^{2k+1} |f_i| = \sum_{i=1}^{k} |L_i^-| f_i^\Delta + \sum_{i=k+2}^{2k+1} |L_i^+| f_i^\nabla. \tag{4.8}$$

Cardinalities of the sets L_i^- and L_i^+ do not exceed k. Observe that the definition of the expanding constant implies $|\partial F| \geq h(G)|F|$ for each F with $|F| \leq |V|/2$. Hence we have

$$\begin{aligned}\sum_{i=1}^{k} |L_i^-| f_i^{\Delta} + \sum_{i=k+2}^{2k+1} |L_i^+| f_i^{\nabla} &\leq \sum_{i=1}^{k} \frac{1}{h} |\partial(L_i^-)| f_i^{\Delta} + \sum_{i=k+2}^{2k+1} \frac{1}{h} |\partial(L_i^+)| f_i^{\nabla} \\ &= \frac{1}{h}\Bigg(\sum_{i=1}^{k} |\partial(L_i^-)| f_i^{\Delta} + \sum_{i=k+2}^{2k+1} |\partial(L_i^+)| f_i^{\nabla} \Bigg) \qquad (4.9) \\ &= \frac{1}{h} \sum_{uv \in E(G)} |f(u) - f(v)|.\end{aligned}$$

To show the last equality, we need to observe that the contribution of the edge uv to the sum

$$\Bigg(\sum_{i=1}^{k} |\partial(L_i^-)| f_i^{\Delta} + \sum_{i=k+2}^{2k+1} |\partial(L_i^+)| f_i^{\nabla} \Bigg)$$

is equal to $|f(u) - f(v)|$. □

Proof of Theorem 4.7. Since continuous functions are dense in $L_1(0,1)$, it suffices to prove the inequality (4.6) in the case where the functions $f(u,t)$ are continuous as functions of t, and so $f(u,t)$ is well-defined for all $t \in [0,1]$. For each $t \in [0,1]$ we let $M(t)$ be a median of the set $\{f(u,t)\}_{u \in V}$. It is easy to show that the medians can be selected in such a way that $M(t)$ is a continuous function on $[0,1]$. Applying Lemma 4.8 for each value of t, we get

$$\sum_{uv \in E} |f(u,t) - f(v,t)| \geq h \sum_{v \in V} |f(v,t) - M(t)|.$$

Integrating this inequality over $[0,1]$ we get

$$\sum_{uv \in E} \|f(u) - f(v)\| \geq h \sum_{v \in V} \|f(v) - M\|. \qquad (4.10)$$

By the triangle inequality we have

$$\|f(u) - f(v)\| \leq \|f(u) - M\| + \|f(v) - M\|.$$

Therefore

$$\sum_{u,v \in V} \frac{h}{|V|} \|f(u) - f(v)\| \leq h \sum_{u \in V} \|f(u) - M\| + h \sum_{v \in V} \|f(v) - M\|.$$

Combining this inequality with (4.10) and the definition of the adjacency matrix we get (4.6). □

The Poincaré inequality (4.6) can be used to get an estimate for L_1-distortion of a k-regular graph with expanding constant h. In fact, to estimate such distortion from below we need to estimate from below the corresponding Poincaré ratio:

$$\frac{\sum_{u,v\in V} \frac{h}{|V|} d_G(u,v)}{\sum_{u,v\in V} a_{u,v} d_G(u,v)}. \tag{4.11}$$

The denominator of the ratio is equal to $2|E|$, where $|E|$ is the number of edges in G. Since the graph is k-regular, we have $2|E| = k|V|$.

On the other hand, the number of vertices at distance D to a given vertex in a k-regular graph is at most

$$1 + k + k(k-1) + \cdots + k(k-1)^{D-1} \le k^D + 1.$$

Let $D = \log_k\left(\frac{|V|}{2} - 1\right)$. Then there are at most $\frac{|V|}{2}$ vertices with distance $\le D$ to a given vertex. Therefore

$$\sum_{u,v\in V} \frac{h}{|V|} d_G(u,v) \ge \frac{h}{|V|} \cdot \frac{|V|^2}{2} \cdot \log_k\left(\frac{|V|}{2} - 1\right)$$

and the Poincaré quotient (4.11) is

$$\ge \frac{h}{2k} \log_k\left(\frac{|V|}{2} - 1\right) = \Omega(\ln|V|).$$

(Here and in some other places we use the standard for computer science notation: $f = \Omega(g)$ means $g = O(f)$.)

We get that distortions of members of a family of expanders grow as logarithms of their sizes. So the estimates for distortions obtained in Chapter 3 (see, for example, Corollary 3.32) have the best possible dependence on the size of the metric space.

The Poincaré inequality (4.6) can be used to prove the following result on coarse non-embeddability of expanders.

Theorem 4.9. *Families of expanders are not uniformly coarsely embeddable into L_1.*

Proof. Let $G = (V, E)$ be a finite graph and h be its expanding constant. Suppose that an embedding $f : V \to L_1$ satisfies

$$\forall u, v \in V \quad \rho_1(d_G(u,v)) \le \|f(u) - f(v)\| \le \rho_2(d_G(u,v))$$

Combining this inequality with (4.6) we get

$$\sum_{u,v\in V} \frac{h}{|V|} \rho_1(d_G(u,v)) \le \sum_{u,v\in V} a_{u,v}\|f(u) - f(v)\| \le k|V|\rho_2(1).$$

Now we recall that ρ_1 is non-decreasing and that we have already proved that for at least $\frac{|V|^2}{2}$ out of $|V|^2$ terms in the left-hand side of the last inequality we have $d_G(u,v) \geq \log_k\left(\frac{|V|}{2} - 1\right)$. We get

$$\frac{|V|^2}{2} \cdot \frac{h}{|V|} \cdot \rho_1(\log_k\left(\frac{|V|}{2} - 1\right)) \leq k|V|\rho_2(1)$$

or

$$\rho_1(\log_k\left(\frac{|V|}{2} - 1\right)) \leq \frac{2k\rho_2(1)}{h}. \tag{4.12}$$

Now we consider a family $\{G_n\}_{n=1}^{\infty}$ of expanders, $G_n = (V_n, E_n)$. It is clear that a function ρ_1 satisfying $\lim_{t\to\infty} \rho_1(t) = \infty$ cannot satisfy the inequality (4.12) for a sequence $\{|V_n|\}_{n=1}^{\infty}$ with $|V_n| \to \infty$ (if we plug each $|V_n|$ instead of $|V|$). □

Corollary 4.10. *A metric space containing isometric copies of all graphs of some family of expanders with their graph distances does not admit a coarse embedding into* L_1.

It is not difficult to construct spaces satisfying this condition provided we have a family of expanders. In fact, we can just build an infinite connected graph (V, E) out of $G_n = (V_n, E_n)$ by letting V to be the disjoint union of V_n and letting E to be the union of E_n plus a sequence $\{e_n\}_{n=1}^{\infty}$ of edges, where e_n is an edge joining a vertex of V_n with a vertex of V_{n+1}. An important property of this space is that it has bounded geometry in the sense of Definition 1.66 (see page 31).

Remark 4.11. Since (as we know from Proposition 1.21) finite subsets of ℓ_2 are isometric to finite subsets of L_1, Theorem 4.9 implies that families of expanders are not uniformly coarsely embeddable into ℓ_2.

We use repeatedly the following observation.

Observation 4.12. A separable metric space is coarsely embeddable into L_1 if and only if it is coarsely embeddable into a Hilbert space.

Proof. The "only if" part follows from the fact that L_1 is coarsely embeddable into ℓ_2 (see Example 1.40). The "if" part follows from the fact that ℓ_2 admits a linear bilipschitz (= isomorphic), and even a linear isometric, embedding into L_1.

The isometric version requires some knowledge of probability theory and measure theory. Its proof: we consider a sequence $\{g_i\}_{i=1}^{\infty}$ of independent normal (Gaussian) random variables with 0 mean and the same variance. They belong to L_1. Well-known properties of Gaussian random variables imply that the linear span of $\{g_i\}_{i=1}^{\infty}$ in L_1 is isometric to ℓ_2.

The existence of an isomorphic embedding of ℓ_2 into L_1 (and actually any L_p $(1 \le p < \infty)$) follows from the Khinchin inequality (Theorem 2.51). The Khinchin inequality immediately implies that the linear extension of the mapping $e_k \mapsto r_k$, where $\{e_k\}_{k=1}^{\infty}$ is the unit vector basis in ℓ_2 and $\{r_k\}$ are Rademacher functions considered as elements of L_1, to the linear span of unit vectors in ℓ_2 admits a continuous extension to ℓ_2. This extension is an isomorphic embedding of ℓ_2 to L_1. □

Later we shall need the following corollary which can be proved using the techniques of this section.

Corollary 4.13 (Of the proof of Theorem 4.7). *Let G be a k-regular graph with expanding constant h and $f : V(G) \to L_1$ be an embedding. Then*

$$\exists x \in L_1 \left|\left\{s \in V(G) : \ \|f(s) - x\| \le \frac{2k}{h}\mathrm{Lip}(f)\right\}\right| \ge \frac{3|V(G)|}{4}. \tag{4.13}$$

Proof. Observe that $\|f(u) - f(v)\| \le \mathrm{Lip}(f)$ if $uv \in E(G)$. Hence the left-hand side of (4.10) does not exceed $\frac{|V|k}{2}\mathrm{Lip}(f)$. By (4.10) we have

$$\sum_{v \in V} \|f(v) - M\| \le \frac{|V|k}{2h}\mathrm{Lip}(f).$$

Hence the amount of vertices v for which $\|f(v) - M\| > \frac{2k}{h}\mathrm{Lip}(f)$ is less than $\frac{|V|}{4}$. The conclusion follows with $x = M$. □

4.3 L_p-distortion in terms of constants in Poincaré inequalities

The purpose of this section is to show that the L_p-distortion $c_p(X)$ of a finite metric space can be characterized in terms of Poincaré inequalities of the form (4.2) with $\Psi(t) = t^p$.

Theorem 4.14. *Let (X, d_X) be a finite metric space. Then the L_p-distortion $c_p(X)$ is equal to the supremum of constants C for which there exist arrays $a_{u,v}$ and $b_{u,v}$ $(u, v \in X)$ of nonnegative and not-all-zero real numbers such that*

$$\sum_{u,v \in X} b_{u,v}(d_X(u,v))^p \ge C^p \sum_{u,v \in X} a_{u,v}(d_X(u,v))^p, \tag{4.14}$$

and for any map $f : X \to L_p$ the inequality

$$\sum_{u,v \in X} a_{u,v}\|f(u) - f(v)\|^p \ge \sum_{u,v \in X} b_{u,v}\|f(u) - f(v)\|^p \tag{4.15}$$

holds.

Proof of Theorem 4.14. We need to show that $c_p(X) \geq C$ for each C satisfying the condition above and that for each $C < c_p(X)$ we can find $a_{u,v}$ and $b_{u,v}$ satisfying (4.14) and (4.15).

The first part is completely straightforward: if $c_p(X) < C$, then there exist $\varepsilon > 0$ and a map $f : X \to L_p$ such that

$$\forall u, v \in X \quad d_X(u,v) \leq \|f(u) - f(v)\| \leq (C - \varepsilon) d_X(u,v)$$

Combining this inequality with (4.14) and (4.15), and using non-negativity of $a_{u,v}$ and $b_{u,v}$ we get

$$\begin{aligned} C^p \sum_{u,v \in X} a_{u,v}(d_X(u,v))^p &\leq \sum_{u,v \in X} b_{u,v}(d_X(u,v))^p \leq \sum_{u,v \in X} b_{u,v}\|f(u) - f(v)\|^p \\ &\leq \sum_{u,v \in X} a_{u,v}\|f(u) - f(v)\|^p \\ &\leq (C - \varepsilon)^p \sum_{u,v \in X} a_{u,v}(d_X(u,v))^p, \end{aligned}$$

a contradiction.

Now let $C < c_p(X)$, we have to find arrays $a_{u,v}$ and $b_{u,v}$ such that inequalities (4.14) and (4.15) hold. Let P be the set of all unordered pairs of distinct elements of X, we denote the linear space of all real-valued functions on P by $\mathbb{R}^P$. It is clear that each semimetric on X can be regarded as an element of $\mathbb{R}^P$. Let $\mathcal{L}_p \subset \mathbb{R}^P$ be the cone of the p-th powers of L_p-semimetrics on X, that is

$$\mathcal{L}_p = \{(\|f(u) - f(v)\|^p)_{uv \in P} : f : X \to L_p\} \subset \mathbb{R}^P.$$

Observe that $\mathcal{L}_p$ is a convex cone, it is closed with respect to sums because we can concatenate embeddings. We mean the following: if $f_1 : X \to L_p(\Omega_1)$ and $f_2 : X \to L_p(\Omega_2)$ are two embeddings, we define $f : X \to L_p(\Omega_1 \cup \Omega_2)$ (where Ω_1 and Ω_2 are assumed to be disjoint) by $f(u)|_{\Omega_1} = f_1(u)$ and $f(u)|_{\Omega_2} = f_2(u)$. It is clear that $\|f(u) - f(v)\|^p = \|f_1(u) - f_1(v)\|^p + \|f_2(u) - f_2(v)\|^p$. (Here it is worthwhile to recall that each separable space $L_p(\Omega, \Sigma, \mu)$ is isometric to a subspace of $L_p(0,1)$, see Fact 1.20.)

Since $C < c_p(X)$ the space (X, d_X) does not admit a C-bilipschitz embedding into L_p. We introduce the set

$$\mathcal{K} = \{(x_{uv})_{uv \in P} \in \mathbb{R}^P : \exists r > 0 \; \forall u, v \quad r(d_X(u,v))^p \leq x_{uv} \leq r C^p (d_X(u,v))^p\}.$$

The set $\mathcal{K}$ includes the p-th powers of all semimetrics arising by C-bilipschitz embeddings of (X, d_X). On the other hand, not all elements of $\mathcal{K}$ are semimetrics, since the triangle inequality may be violated. Since there is no C-bilipschitz embedding of (X, d_X) into L_p, we have $\mathcal{K} \cap \mathcal{L}_p = \emptyset$. Both $\mathcal{K}$ and $\mathcal{L}_p$ are convex sets in $\mathbb{R}^P$, and so they can be separated by a hyperplane. Moreover, since $\mathcal{L}_p$ and $\mathcal{K} \cup \{0\}$ are cones, the separating hyperplane should be of the form $\{x : \langle s, x\rangle = 0\}$ for some $s \in \mathbb{R}^P$, $s \neq 0$. So let $s \in \mathbb{R}^P$ be such that $\langle s, x\rangle \geq 0$ for all $x \in \mathcal{K}$, and $\langle s, x\rangle \leq 0$ for all $x \in \mathcal{L}_p$.

Using the vector s, we define the desired arrays $a_{u,v}$ and $b_{u,v}$ as follows

$$b_{u,v} = \begin{cases} s_{uv} & \text{if } s_{uv} > 0, \\ 0 & \text{otherwise;} \end{cases}$$

$$a_{u,v} = \begin{cases} -s_{uv} & \text{if } s_{uv} < 0, \\ 0 & \text{otherwise.} \end{cases}$$

It remains to show that these arrays satisfy the inequalities (4.14) and (4.15).

The condition $\langle s, x\rangle \leq 0$ for all $x \in \mathcal{L}_p$ means that

$$0 \geq \sum_{uv \in P} s_{uv} \|f(u) - f(v)\|^p \quad \text{for each } f : X \to L_p.$$

Moving those terms for which $s_{uv} < 0$ to the left-hand side, we get (4.15).

To get (4.14) we apply the condition "$\langle s, x\rangle \geq 0$ for all $x \in \mathcal{K}$" to the following sequence in $\mathcal{K}$:

$$x_{uv} = \begin{cases} (d_X(u,v))^p & \text{if } s_{uv} \geq 0, \\ C^p(d_X(u,v))^p & \text{if } s_{uv} < 0. \end{cases}$$

We get

$$\sum_{s_{uv}>0} s_{uv}(d_X(u,v))^p + C^p \sum_{s_{uv}<0} s_{uv}(d_X(u,v))^p \geq 0.$$

Moving the second sum in this inequality to the right-hand side we get (4.14). □

4.4 Euclidean distortion and positive semidefinite matrices

Analyzing the proof of Theorem 4.14 we see that the theorem can be restated in the following way:

$$c_p(X) = \sup \left(\frac{\sum_{u,v \in X} s_{uv}^+ (d_X(u,v))^p}{\sum_{u,v \in X} s_{uv}^- (d_X(u,v))^p} \right)^{1/p}, \tag{4.16}$$

where the supremum is over all collections $S = \{s_{uv}\}_{u,v \in X}$ satisfying $s_{uv} = s_{vu}$ and $s(\mathcal{L}_p) \leq 0$, that is, $\langle s, x\rangle \leq 0$ for all $x \in \mathcal{L}_p$, $s_{uv}^+ = \max\{s_{uv}, 0\}$, and $s_{uv}^- = -\min\{s_{uv}, 0\}$.

This restatement is particularly useful in the case $p = 2$, because in this case a useful characterization of s satisfying $s(\mathcal{L}_2) \leq 0$ is known. Recall that the inequality $s(\mathcal{L}_2) \leq 0$ means that

$$\sum_{u,v \in X} s_{uv} \|f(u) - f(v)\|^2 \leq 0 \tag{4.17}$$

for an arbitrary mapping $f : X \to \ell_2$.

Observe that s_{uu} is not really used in the inequalities and the formula (4.16), since $d_X(u,u) = \|f(u) - f(u)\| = 0$. Therefore we may pick the values s_{uu} arbitrarily. We pick them in such a way that

$$\sum_{u \in X} s_{uv} = \sum_{v \in X} s_{uv} = 0. \tag{4.18}$$

Definition 4.15. Let A be a real, symmetric $n \times n$ matrix. We say that A is *positive semidefinite* if

$$\forall x \in \mathbb{R}^n \quad x^T A x \geq 0. \tag{4.19}$$

(Here x is a column vector and x^T is its transpose, that is, the corresponding row vector. The product $x^T Ax$ is a 1×1 matrix which we regard as a real number.)

Lemma 4.16. *Any matrix $\{s_{uv}\}$ satisfying the conditions $s_{uv} = s_{vu}$,* (4.17)*, and* (4.18) *is positive semidefinite. Conversely, each positive semidefinite matrix $\{s_{uv}\}$ satisfying* (4.18) *satisfies* (4.17).

Proof. We can rewrite $\sum_{u,v} s_{uv}\|f(u) - f(v)\|^2$ as

$$\begin{aligned} &\sum_{u,v} s_{uv}\|f(u)\|^2 - \sum_{u,v} s_{uv}\langle f(u), f(v)\rangle + \sum_{u,v} s_{uv}\|f(v)\|^2 \\ &\overset{(4.18)}{=} -\sum_{u,v} s_{uv}\langle f(u), f(v)\rangle. \end{aligned} \tag{4.20}$$

Therefore (4.17) implies $\sum_{u,v} s_{uv}\langle f(u), f(v)\rangle \geq 0$ for all f. In particular, the condition should be satisfied for maps $f : X \to \mathbb{R}$ (that is, maps into a 1-dimensional Hilbert space). On the other hand, in the case of 1-dimensional maps f this condition is just a restatement of the definition (compare with (4.19)).

Conversely, assume that $\{s_{uv}\}$ is a positive semidefinite matrix satisfying the condition (4.18). Then it satisfies the inequality $\sum_{u,v} s_{uv}\langle f(u), f(v)\rangle \geq 0$ for all f, because such inequality can be written as a sum of inequalities corresponding to 1-dimensional functions f. Using equality (4.20) backward, we get that $\{s_{uv}\}$ satisfies (4.17). □

Lemma 4.16 and the discussion above imply

$$c_2(X) = \max \sqrt{\frac{\sum_{u,v \in X} s_{uv}^{+} d_X(u,v)^2}{\sum_{u,v \in X} s_{uv}^{-} d_X(u,v)^2}}, \tag{4.21}$$

where the maximum is over all (nonzero) positive semidefinite matrices S satisfying the condition $S\vec{1} = \vec{0}$ (this is shorthand for the condition (4.18)).

This corollary turned out to be a useful tool for estimates of the Euclidean distortion. Some examples are mentioned in Notes and Remarks, see also Exercises 4.33 and 4.34.

4.5 Fourier analytic method of getting Poincaré inequalities

We use $\mathbb{F}_2$ to denote the field with two elements which we (with some abuse of notation) denote 0 and 1, and with addition and multiplication given by Tables 4.1 and 4.2.

Table 4.1. Addition.

	0	1
0	0	1
1	1	0

Table 4.2. Multiplication.

	0	1
0	0	0
1	0	1

We call $\mathbb{F}_2$ a *field* because all axioms of a field are satisfied for these addition and multiplication. We consider the set $\mathbb{F}_2^n$ of all n-tuples of elements of $\mathbb{F}_2$ as a group with the addition given by

$$(u_1, \dots, u_n) + (v_1, \dots, v_n) = (u_1 + v_1, \dots, u_n + v_n).$$

We also define multiplication of elements of $\mathbb{F}_2^n$ by elements of $\mathbb{F}_2$ by

$$\alpha(u_1, \dots, u_n) = (\alpha u_1, \dots, \alpha u_n).$$

These operations satisfy all usual axioms of a vector space over the field $\mathbb{F}_2$. We are going to develop some Fourier analysis on the group $\mathbb{F}_2^n$.

Remark 4.17. Readers who are not familiar with Fourier analysis should not worry too much about this now, the presentation below can be understood without any prior knowledge of Fourier analysis.

Let μ be the uniform measure on $\mathbb{F}_2^n$, so $\mu(\{u\}) = 2^{-n}$ for each one-element set $\{u\}$. Given $A \subseteq \{1, \dots, n\}$ we define the *Walsh function* $W_A : \mathbb{F}_2^n \to \mathbb{R}$ by

$$W_A(x) = (-1)^{\sum_{j \in A} x_j}.$$

In this formula we abuse notation somewhat. In fact, the exponent in this formula is an element of $\mathbb{F}_2$, so we have to explain what we mean by (-1) to the power of an element of $\mathbb{F}_2$. We mean the most natural thing: $(-1)^0 = 1 \in \mathbb{R}$, where $0 \in \mathbb{F}_2$, and $(-1)^1 = (-1) \in \mathbb{R}$, where 1 in the exponent is an element of $\mathbb{F}_2$. Note that we also define W_A when A is empty: $W_\emptyset(x)$ is identically equal to 1.

Lemma 4.18. *The Walsh functions* $\{W_A : A \subseteq \{1, \ldots, d\}\}$ *form an orthonormal basis of* $L_2(\mathbb{F}_2^n, \mu)$. *In particular any* $f : \mathbb{F}_2^n \to \mathbb{R}$ *satisfies*

$$f = \sum_{A \subseteq \{1,\ldots,n\}} \hat{f}(A) W_A, \tag{4.22}$$

where

$$\hat{f}(A) = \int_{\mathbb{F}_2^n} f(x) W_A(x) d\mu(x) \in \mathbb{R}, \tag{4.23}$$

and

$$\int_{\mathbb{F}_2^n} (f(x))^2 d\mu(x) = \sum_{A \subseteq \{1,\ldots,n\}} (\hat{f}(A))^2. \tag{4.24}$$

Similar formulas hold for any $f : \mathbb{F}_2^n \to L_2$, *but in this case the coefficients* $\hat{f}(A)$ *given by* (4.23) *are elements of* L_2 *and the identity* (4.24) *is modified to*

$$\int_{\mathbb{F}_2^n} \|f(x)\|_2^2 d\mu(x) = \sum_{A \subseteq \{1,\ldots,n\}} \|\hat{f}(A)\|_2^2. \tag{4.25}$$

Definition 4.19. The coefficients $\hat{f}(A)$ given by (4.23) are called *Fourier coefficients* of f with respect to Walsh functions, the right-hand side of (4.22) is called the *Fourier expansion* of f. The identities (4.24) and (4.25) are called the *Parseval identities*.

Proof of Lemma 4.18. We start by proving the fact that for any two subsets A and B of $\{1, \ldots, n\}$ we have

$$\langle W_A, W_B \rangle = \begin{cases} 0 & \text{if } A \neq B \\ 1 & \text{if } A = B, \end{cases}$$

where the inner product is taken in $L_2(\mathbb{F}_2^n, \mu)$, that is $\langle W_A, W_B \rangle = \int_{\mathbb{F}_2^n} W_A W_B d\mu$.

In fact, it is easy to see that our explanation of the definition of W_A implies that

$$W_A(x) W_B(x) = (-1)^{\sum_{j \in A} x_j} \cdot (-1)^{\sum_{j \in B} x_j} = (-1)^{\sum_{j \in A \triangle B} x_j} = W_{A \triangle B}(x),$$

where $A \triangle B$ is the symmetric difference of A and B, that is the set of elements which are exactly in one of the sets. If $A \triangle B$ is empty, then $W_{A \triangle B}(x)$ is identically equal to 1, and $\int_{\mathbb{F}_2^n} W_A W_B d\mu = \int_{\mathbb{F}_2^n} W_{A \triangle B} d\mu = 1$.

It remains to show that $\int_{\mathbb{F}_2^n} W_A(x) d\mu(x) = 0$ if $A \neq \emptyset$. We can use the following symmetry argument to show this. Let $e_j \in \mathbb{F}_2^n$ be the element whose only nonzero coordinate is the j-th coordinate. Suppose that $j \in A$. Then, on one hand

$$\int_{\mathbb{F}_2^n} W_A(x + e_j) d\mu(x) = \int_{\mathbb{F}_2^n} W_A(x) d\mu(x), \tag{4.26}$$

because the measure μ is invariant with respect to shifts on the group $\mathbb{F}_2^n$ (actually it is invariant with respect to any bijections of the group $\mathbb{F}_2^n$).

On the other hand

$$W_A(x + e_j) = (-1)^{(\sum_{i \in A} x_i)+1} \text{ if } j \in A. \tag{4.27}$$

Therefore $W_A(x+e_j) = -W_A(x)$ and $\int_{\mathbb{F}_2^n} W_A(x+e_j)d\mu(x) = -\int_{\mathbb{F}_2^n} W_A(x)d\mu(x)$. Comparing this equation with (4.26) we get $\int_{\mathbb{F}_2^n} W_A(x)d\mu(x) = 0$.

The remaining part of the real-valued-case statements is standard. As for the proofs in the L_2-valued case: they can be done by considering an orthonormal basis $\{f_i\}$ in L_2, and by applying the real-valued-case statements to the functions $\langle f(x), f_i \rangle$. □

We write $e := e_1 + \cdots + e_n$ (addition in $\mathbb{F}_2^n$) for the vector $(1, \ldots, 1)$. We define the *partial differentiation operator* on the space of all functions $f : \mathbb{F}_2^n \to X$, where X is a Banach space, by

$$\partial_j f(x) := \frac{f(x + e_j) - f(x)}{2}.$$

First we consider the case where $X = \mathbb{R}$, and so $f \in L_2(\mathbb{F}_2^n, \mu)$. Using (4.27) we get

$$\partial_j W_A = \begin{cases} -W_A & j \in A \\ 0 & j \notin A, \end{cases}$$

for every $A \subseteq \{1, \ldots, n\}$. Using the Parseval identity for each of $\partial_j f$ we get that an arbitrary $f : \mathbb{F}_2^n \to \mathbb{R}$ satisfies

$$\sum_{j=1}^{n} \int_{\mathbb{F}_2^n} (\partial_j f(x))^2 d\mu(x) = \sum_{A \subseteq \{1,\ldots,n\}} |A| \hat{f}(A)^2. \tag{4.28}$$

Definition 4.20. We introduce the *Hamming cube* as $\mathbb{F}_2^n$ with its *Hamming metric*:

$$\rho(x, y) := |\{j \in \{1, \ldots, n\} : x_j \neq y_j\}| \text{ for } x, y \in \mathbb{F}_2^n.$$

Observe that

$$W_A(x + e) = \begin{cases} W_A(x) & \text{if } |A| \text{ is even} \\ -W_A(x) & \text{if } |A| \text{ is odd.} \end{cases} \tag{4.29}$$

Let $f : \mathbb{F}_2^n \to \mathbb{R}$ be an arbitrary function. Applying (4.29) and the Parseval identity (4.24) we get the following Poincaré inequality:

$$\begin{aligned}
&\int_{\mathbb{F}_2^n} |f(x) - f(x+e)|^2 d\mu(x) \\
&\quad = \int_{\mathbb{F}_2^n} \Big| \sum_{A\subseteq\{1,\dots,n\}} \widehat{f}(A)W_A(x) - \sum_{A\subseteq\{1,\dots,n\}} \widehat{f}(A)W_A(x+e)\Big|^2 d\mu(x) \\
&\quad = \int_{\mathbb{F}_2^n} \Big| \sum_{\substack{A\subseteq\{1,\dots,n\} \\ |A| \text{ is odd}}} 2\widehat{f}(A)W_A(x)\Big|^2 d\mu(x) \\
&\quad = \sum_{\substack{A\subseteq\{1,\dots,n\} \\ |A| \text{ is odd}}} 4\widehat{f}(A)^2 \\
&\quad \le 4 \sum_{A\subseteq\{1,\dots,n\}} |A|\widehat{f}(A)^2 \\
&\quad \overset{(4.28)}{=} 4\sum_{j=1}^{n} \int_{\mathbb{F}_2^n} |\partial_j f(x)|^2 d\mu(x).
\end{aligned} \tag{4.30}$$

Remark 4.21. We did not write this inequality in the standard form (4.2), but it obviously can be rewritten in such form.

The Poincaré inequality (4.30) implies the following Poincaré inequality for L_2-valued functions on $\mathbb{F}_2^n$:

$$\int_{\mathbb{F}_2^n} \|f(x) - f(x+e)\|_2^2 d\mu(x) \le 4\sum_{j=1}^{n} \int_{\mathbb{F}_2^n} \|\partial_j f(x)\|_2^2 d\mu(x). \tag{4.31}$$

Here we would like to show the standard way of getting L_2-valued Poincaré inequalities from $\mathbb{R}$-valued Poincaré inequalities in the case where the distances are squared (that is, in the case where $\Psi(t) = t^2$ in (4.2)). We mean getting an inequality of the form

$$\sum_{u,v\in X} a_{u,v}\|F(u) - F(v)\|_2^2 \ge \sum_{u,v\in X} b_{u,v}\|F(u) - F(v)\|_2^2 \tag{4.32}$$

for functions $F : X \to L_2$ from the inequality

$$\sum_{u,v\in X} a_{u,v}|f(u) - f(v)|^2 \ge \sum_{u,v\in X} b_{u,v}|f(u) - f(v)|^2 \tag{4.33}$$

for functions $f : X \to \mathbb{R}$. To get (4.32) from (4.33) we consider an orthonormal basis (g_α) in L_2 and let $f_\alpha(x) = \langle F(x), g_\alpha\rangle$. Applying (4.33) to each f_α and adding the obtained inequalities we get

$$\sum_{u,v\in X} a_{u,v} \sum_{\alpha} |f_\alpha(u) - f_\alpha(v)|^2 \geq \sum_{u,v\in X} b_{u,v} \sum_{\alpha} |f_\alpha(u) - f_\alpha(v)|^2.$$

It remains to observe that

$$\begin{aligned}\sum_{\alpha} |f_\alpha(u) - f_\alpha(v)|^2 &= \sum_{\alpha} |\langle F(u), g_\alpha\rangle - \langle F(v), g_\alpha\rangle|^2 \\ &= \sum_{\alpha} |\langle F(u) - F(v), g_\alpha\rangle|^2 \\ &\overset{\text{(Parseval identity)}}{=} \|F(u) - F(v)\|^2.\end{aligned}$$

Remark 4.22. Similar argument can be used to get ℓ_p-valued Poincaré inequalities out of $\mathbb{R}$-valued Poincaré inequalities with $\Psi(t) = t^p$. As for L_p-valued Poincaré inequalities, one has to use something similar to the argument preceding (4.10).

Now we are ready to prove

Theorem 4.23. $c_2(\mathbb{F}_2^n, \rho) = \sqrt{n}$.

Proof. By Proposition 4.3 and (4.31), to get an estimate for $c_2(\mathbb{F}_2^n, \rho)$ from below it suffices to compute the corresponding Poincaré ratio:

$$\begin{aligned}c_2(\mathbb{F}_2^n, \rho) &\geq \left(\frac{\int_{\mathbb{F}_2^n} (\rho(x, x+e))^2 d\mu(x)}{4\sum_{j=1}^n \int_{\mathbb{F}_2^n} \left(\frac{\rho(x+e_j, x)}{2}\right)^2 d\mu(x)}\right)^{1/2} \\ &= \left(\frac{\int_{\mathbb{F}_2^n} n^2 d\mu(x)}{4\sum_{j=1}^n \int_{\mathbb{F}_2^n} (1/2)^2 d\mu(x)}\right)^{1/2} = \sqrt{n}.\end{aligned}$$

The estimate $c_2(\mathbb{F}_2^n, \rho) \leq \sqrt{n}$ is obtained by embedding elements of $\mathbb{F}_2^n$ into ℓ_2^n as if they were sequences of real numbers 0 and 1. □

Using reasoning similar to that in the proof of (4.31) we prove the following Poincaré inequality which will be used later.

Lemma 4.24. *For every* $f : \mathbb{F}_2^n \to L_2$,

$$\begin{aligned}&\int_{\mathbb{F}_2^n\times\mathbb{F}_2^n} \|f(x) - f(y)\|_2^2 d\mu(x) d\mu(y) \\ &\qquad \leq \frac{2}{\min\{|A| : A \neq \emptyset,\ \widehat{f}(A) \neq 0\}} \sum_{j=1}^n \int_{\mathbb{F}_2^n} \|\partial_j f(x)\|_2^2 d\mu(x).\end{aligned}$$

Proof. Clearly

$$\int_{\mathbb{F}_2^n\times\mathbb{F}_2^n}\|f(x)-f(y)\|_2^2d\mu(x)d\mu(y)$$
$$=\int_{\mathbb{F}_2^n\times\mathbb{F}_2^n}\|f(x)-\widehat{f}(\emptyset)-(f(y)-\widehat{f}(\emptyset))\|_2^2d\mu(x)d\mu(y),$$

where $\widehat{f}(\emptyset)$ is the Fourier coefficient corresponding to the Walsh function $W_\emptyset\equiv 1$. Since we are in a vector-valued case, $\widehat{f}(\emptyset)\in L_2$. Direct computation shows that

$$\int_{\mathbb{F}_2^n\times\mathbb{F}_2^n}\|f(x)-\widehat{f}(\emptyset)-(f(y)-\widehat{f}(\emptyset))\|_2^2d\mu(x)d\mu(y)$$
$$=\int_{\mathbb{F}_2^n\times\mathbb{F}_2^n}(\|f(x)-\widehat{f}(\emptyset)\|_2^2+\|f(y)-\widehat{f}(\emptyset)\|_2^2)d\mu(x)d\mu(y).$$

We get

$$\int_{\mathbb{F}_2^n\times\mathbb{F}_2^n}\|f(x)-f(y)\|_2^2d\mu(x)d\mu(y)=2\int_{\mathbb{F}_2^n}\|f(x)-\widehat{f}(\emptyset)\|_2^2d\mu(x).$$

Applying the Parseval identity (4.25) we obtain

$$2\int_{\mathbb{F}_2^n}\|f(x)-\widehat{f}(\emptyset)\|_2^2d\mu(x)=2\sum_{\emptyset\neq A\subseteq\{1,\dots,n\}}\|\widehat{f}(A)\|_2^2. \tag{4.34}$$

Now we use the following L_2-valued version of (4.28):

$$\sum_{j=1}^n\int_{\mathbb{F}_2^n}\|\partial_j f(x)\|_2^2d\mu(x)=\sum_{A\subseteq\{1,\dots,n\}}|A|\,\|\widehat{f}(A)\|_2^2. \tag{4.35}$$

This version can be obtained from (4.28) in a standard way: we consider an orthonormal basis $\{g_\alpha\}$ in L_2 and apply (4.28) to the real-valued functions $\langle f(x),g_\alpha\rangle$.

Observe that

$$\sum_{A\subseteq\{1,\dots,n\}}|A|\,\|\widehat{f}(A)\|_2^2\geq\min\{|A|:\ A\neq\emptyset,\ \widehat{f}(A)\neq 0\}\sum_{\emptyset\neq A\subseteq\{1,\dots,n\}}\|\widehat{f}(A)\|_2^2.$$

Combining this inequality with (4.34) and (4.35), we get the desired inequality. □

In Section 4.2 we saw that for each family of expanders their L_1-distortions grow as logarithms of their sizes. Our next purpose is to construct another family of spaces sharing this property.

This family of metric spaces can be regarded as quotients of $\mathbb{F}_2^n$. To describe our construction we need to introduce some definitions and notation of coding theory.

Definition 4.25. A *linear code* C in $\mathbb{F}_2^n$ is defined as a linear subspace of $\mathbb{F}_2^n$. The *Hamming weight* $\mathrm{wt}(x)$ of a vector $x = \{x_i\}_{i=1}^n \in \mathbb{F}_2^n$ is the number of nonzero x_i. The *minimum distance* $d(C)$ of a code C is the minimum Hamming weight of nonzero elements of C.

The minimum distance is a very important parameter in coding theory because it indicates the number of errors which the code can correct.

We are going to use the $\mathbb{F}_2$-valued inner product on $\mathbb{F}_2^n$ given by

$$\langle x, y\rangle := \sum_{j=1}^{n} x_j y_j.$$

We set

$$C^{\perp} := \left\{x \in \mathbb{F}_2^n : \ \forall\, y \in C,\ \langle x, y\rangle = 0\right\}.$$

Observe that some properties of the $\mathbb{F}_2$-valued inner products are different from properties of conventional inner products. For example $\langle x, x\rangle = 0$ does not imply $x = 0$, it implies only that x has an even number of nonzero components. Nevertheless the following property remains true for $\mathbb{F}_2$-valued inner products.

Lemma 4.26. *If C is a subspace of $\mathbb{F}_2^n$, then $(C^{\perp})^{\perp} = C$.*

Proof. The fact that $C \subseteq (C^{\perp})^{\perp}$ is immediate from the definitions.

To show that $(C^{\perp})^{\perp} \subseteq C$ we use the notion of a *dimension* $\dim L$ of a subspace L of $\mathbb{F}_2^n$. It is defined in the same way as for real or complex spaces: as the number of elements in a set of vectors in L which is simultaneously linearly independent and spans L. A set of vectors with these properties is called a *basis* in L. The same argument as for real spaces shows that all bases in L have the same number of elements, therefore $\dim L$ is well-defined.

Also the same argument as for real spaces shows that if $\widetilde{L}$ is a proper subspace of L, then $\dim \widetilde{L} < \dim L$. Observe that $C^{\perp}$ is also a subspace of $\mathbb{F}_2^n$. Therefore to complete the proof it suffices to show that $\dim C + \dim(C^{\perp}) = n$.

Let $d = \dim C$. Using the standard manipulations with bases we can assume, possibly after renumbering coordinates in $\mathbb{F}_2^n$, that there is a basis in C whose elements are rows of a matrix of the form

$$\begin{pmatrix} 1 & 0 & \dots & 0 & x_{1,d+1} & \dots & x_{1,n} \\ 0 & 1 & \dots & 0 & x_{2,d+1} & \dots & x_{2,n} \\ \vdots & \vdots & \ddots & \vdots & \vdots & \ddots & \vdots \\ 0 & 0 & \dots & 1 & x_{d,d+1} & \dots & x_{d,n} \end{pmatrix}.$$

Now, it is easy to see that for each collection $y_{d+1}, \dots, y_n \in \mathbb{F}_2$ there is exactly one vector

$$(y_1, \dots, y_d, y_{d+1}, \dots, y_n)$$

which is in $C^{\perp}$. From here it is immediate that $\dim(C^{\perp}) = n - d$. $\square$

Now we give a new proof of the following result which was already proved in Section 4.2 of this chapter using expanders.

Theorem 4.27. *There exists an absolute constant $c > 0$ such that for each $N \in \mathbb{N}$ there exists a finite metric space X of cardinality at least N satisfying $c_1(X) \geq c \ln |X|$.*

Proof. The metric space will be constructed as the set of equivalence classes of $\mathbb{F}_2^n$, two elements are in the same equivalence class if and only if their difference belongs to $C^{\perp}$, where C is a "good code", whose construction we are going to describe now.

Theorem 4.28. *For each $n \in \mathbb{N}$ there exists a linear code $C \subset \mathbb{F}_2^n$ whose minimum distance and dimension satisfy $d(C) \geq \delta n$ and $\dim(C) \geq \frac{n}{4}$, respectively, where $\delta > 0$ is an absolute constant.*

Proof. Such codes can be obtained using the following greedy construction: fix $k \leq n/4$ and let V be a k-dimensional subspace of $\mathbb{F}_2^n$ with $d(V) > \delta n$ (it is clear that such subspaces exist for $k = 1$). Then V contains 2^k points. Let us count the number of vectors $x \in \mathbb{F}_2^n$ for which the linear span $\widetilde{V} = \mathrm{lin}(V \cup \{x\})$ does *not* satisfy $d(\widetilde{V}) \geq \delta n$. To prove the theorem it suffices to show that for small enough δ and for $k < \frac{n}{4}$ the number of such vectors is strictly smaller than the cardinality of $\mathbb{F}_2^n$.

The number of vectors satisfying the condition described in the previous paragraph can be estimated from above in the following way. We count, for each $v \in V$, the number of vectors x satisfying $\mathrm{wt}(v + x) \leq \delta n$ (observe that $x + v = x - v$ in $\mathbb{F}_2^n$). This number is $\sum_{\ell \leq \delta n} \binom{n}{l}$, the resulting upper estimate is

$$2^k \sum_{\ell \leq \delta n} \binom{n}{l}. \tag{4.36}$$

We assume that $\delta \leq \frac{1}{2}$. In such a case (as is easy to check and is well known) the last term in the sum in (4.36) is the largest one. Therefore (4.36) can be estimated from above by

$$2^k \lfloor \delta n \rfloor \cdot \frac{n!}{\lfloor \delta n \rfloor! \cdot (n - \lfloor \delta n \rfloor)!}. \tag{4.37}$$

It is clear that $\lfloor \delta n \rfloor = \widetilde{\delta} n$ for some $0 \leq \widetilde{\delta} \leq \delta \leq \frac{1}{2}$. Now we use the well-known Stirling formula:

$$n! = \sqrt{2\pi n}\, \frac{n^n}{e^n} \cdot e^{\alpha(n)}, \tag{4.38}$$

where $1/(12n + 1) < \alpha(n) < 1/(12n)$.

We get the following formula for (4.37)

$$
\begin{aligned}
2^k\widetilde{\delta}n & \frac{\sqrt{2\pi n}\, n^n\, e^{\alpha(n)}\, e^{\widetilde{\delta}n}\, e^{(1-\widetilde{\delta})n}}{e^n\, \sqrt{2\pi\widetilde{\delta}n}\, (\widetilde{\delta}n)^{\widetilde{\delta}n}\, e^{\alpha(\widetilde{\delta}n)} \sqrt{2\pi(1-\widetilde{\delta})n}\, ((1-\widetilde{\delta})n)^{(1-\widetilde{\delta})n}\, e^{\alpha((1-\widetilde{\delta})n)}} \\
&= 2^k\widetilde{\delta}n \frac{e^{\alpha(n)}}{\sqrt{\widetilde{\delta}n}\, (\widetilde{\delta})^{\widetilde{\delta}n}\, e^{\alpha(\widetilde{\delta}n)} \sqrt{2\pi(1-\widetilde{\delta})}\, (1-\widetilde{\delta})^{(1-\widetilde{\delta})n}\, e^{\alpha((1-\widetilde{\delta})n)}} \\
&= \left(\frac{e^{\alpha(n)}}{e^{\alpha(\widetilde{\delta}n)} \sqrt{2\pi(1-\widetilde{\delta})}\, e^{\alpha((1-\widetilde{\delta})n)}} \right) \cdot 2^k \sqrt{\widetilde{\delta}n} \cdot \left(\frac{1}{(\widetilde{\delta})^{\widetilde{\delta}} (1-\widetilde{\delta})^{1-\widetilde{\delta}}} \right)^n .
\end{aligned} \tag{4.39}
$$

The first term in the last line looks complicated, but it is easy to see that it is bounded from above by an absolute constant K. Let us recall that our purpose is to show that if we pick δ sufficiently small and if $k < n/4$, then the number in (4.39) is $< 2^n$.

We can estimate the number in (4.39) from above by

$$
2^{\log_2 K + \frac{n}{4} + \frac{1}{2}\log_2 \widetilde{\delta} + \frac{1}{2}\log_2 n - n(\widetilde{\delta}\log_2 \widetilde{\delta} + (1-\widetilde{\delta})\log_2(1-\widetilde{\delta}))}. \tag{4.40}
$$

Since $\lim_{\widetilde{\delta}\downarrow 0} \widetilde{\delta}\log_2 \widetilde{\delta} + (1-\widetilde{\delta})\log_2(1-\widetilde{\delta}) = 0$, it is clear that we can pick $\delta > 0$ such that the exponent in (4.40) is $< n$. □

We need the following Fourier analytic consequence of the fact that C is "good":

Lemma 4.29. *Assume that $f : \mathbb{F}_2^n \to L_2$ satisfies $f(x+y) = f(x)$ for every $x \in \mathbb{F}_2^n$ and $y \in C^\perp$. Then for every nonempty $A \subseteq \{1,\dots,n\}$ with $|A| < d(C)$, $\widehat{f}(A) = 0$.*

Proof. We use the notation $\mathbf{1}_A$ for the vector $\{x_i\} \in \mathbb{F}_2^n$ whose components x_i are equal to 1 if and only if $i \in A$. Since $\mathbf{1}_A \notin C$ and $(C^\perp)^\perp = C$ (by Lemma 4.26), we have $\mathbf{1}_A \notin (C^\perp)^\perp$, implying that there exists $v \in C^\perp$ such that $\langle \mathbf{1}_A, v\rangle = 1$. Now,

$$
\begin{aligned}
\widehat{f}(A) &= \int_{\mathbb{F}_2^n} f(x) W_A(x) d\mu(x) \\
&= \int_{\mathbb{F}_2^n} f(x+v) W_A(x) d\mu(x) \\
&= \int_{\mathbb{F}_2^n} f(x) W_A(x-v) d\mu(x) \\
&= (-1)^{\langle \mathbf{1}_A, v\rangle} \int_{\mathbb{F}_2^n} f(x) W_A(x) d\mu(x) \\
&= -\widehat{f}(A).
\end{aligned}
$$

So $\widehat{f}(A) = 0$. □

As we already mentioned, the metric space X which we are going to use for the proof of Theorem 4.27 is the space of equivalence classes of $\mathbb{F}_2^n$. Two elements are in the same class if their difference is in $C^\perp$. Our next purpose is to prove a Poincaré inequality for L_1-valued maps on $\mathbb{F}_2^n/C^\perp$.

Lemma 4.30. *Let C be a linear code, and $f : \mathbb{F}_2^n/C^\perp \to L_1$ be a map. We define $\tilde{f} : \mathbb{F}_2^n \to L_1$ by $\tilde{f}(x) = f(x + C^\perp)$. Then*

$$\int_{\mathbb{F}_2^n \times \mathbb{F}_2^n} \|\tilde{f}(x) - \tilde{f}(y)\|_1 d\mu(x) d\mu(y) \le \frac{2}{d(C)} \sum_{j=1}^n \int_{\mathbb{F}_2^n} \|\partial_j \tilde{f}(x)\|_1 d\mu(x) \quad (4.41)$$

Remark 4.31. The Poincaré inequality (4.41) is of somewhat unconventional form, because the integration (summation) is not over the metric space X, but over a "basic" set for which X is a quotient space. However, it is clear that one can write the inequality in terms of X by counting how many times each of the terms $\|\tilde{f}(x) - \tilde{f}(y)\|_1$ and $\|\partial_j \tilde{f}(x)\|_1$ is repeated. We prefer not to do that because it is more convenient to estimate the Poincaré ratio in terms of $\mathbb{F}_2^n$ (rather than $\mathbb{F}_2^n/C^\perp$), too.

Proof of Lemma 4.30. As was proved in Example 1.40 (see also Corollary 1.43), there exists a mapping $T : L_1 \to L_2$ such that for all $x, y \in L_1$,

$$\|T(x) - T(y)\|_2 = \sqrt{\|x - y\|_1}.$$

Define $h : \mathbb{F}_2^n \to L_2$ by $h = T \circ \tilde{f}$. We have

$$\begin{aligned}
&\int_{\mathbb{F}_2^n \times \mathbb{F}_2^n} \|\tilde{f}(x) - \tilde{f}(y)\|_1 d\mu(x) d\mu(y) \\
&\quad = \int_{\mathbb{F}_2^n \times \mathbb{F}_2^n} \|h(x) - h(y)\|_2^2 d\mu(x) d\mu(y) \\
&\quad \overset{\text{(Lemma 4.24)}}{\le} \frac{2}{\min\{|A| : A \neq \emptyset,\ \widehat{h}(A) \neq 0\}} \sum_{j=1}^n \int_{\mathbb{F}_2^n} \|\partial_j h(x)\|_2^2 d\mu(x) \\
&\quad \overset{\text{(Lemma 4.29)}}{\le} \frac{2}{d(C)} \sum_{j=1}^n \int_{\mathbb{F}_2^n} \|\partial_j h(x)\|_2^2 d\mu(x) \\
&\quad = \frac{2}{d(C)} \sum_{j=1}^n \int_{\mathbb{F}_2^n} \|\partial_j \tilde{f}(x)\|_1 d\mu(x).
\end{aligned}$$

□

Remark 4.32. It is worth mentioning that the way in which we used Example 1.40 in the proof of Lemma 4.30 has much wider applicability. It can be used in some other cases when we would like to get an L_p-valued Poincaré inequality from an L_q-valued Poincaré inequality.

Now we estimate from below the Poincaré ratio (see Definition 4.2) corresponding to the Poincaré inequality (4.41). That is, we estimate from below the ratio:

$$\frac{\int_{\mathbb{F}_2^n\times\mathbb{F}_2^n} \rho_{\mathbb{F}_2^n/C^\perp}(x,y)d\mu(x)d\mu(y)}{\frac{1}{d(C)}\sum_{j=1}^n \int_{\mathbb{F}_2^n} \rho_{\mathbb{F}_2^n/C^\perp}(x,x+e_j)d\mu(x)}. \tag{4.42}$$

Here, of course, we need to specify the definition of the metric $\rho_{\mathbb{F}_2^n/C^\perp}(x,y)$. The definition is: the distance between two equivalence classes X and Y is $\inf\{\rho(x,y) : x\in X, y\in Y\}$. One can check that because of the special form of equivalence classes this formula produces a metric. The reason is that for these classes one of the elements x and y could be fixed, that is,

$$\inf\{\rho(x,y) : x\in x_0+C^\perp, y\in y_0+C^\perp\} = \inf\{\rho(x_0,y) : y\in y_0+C^\perp\}. \tag{4.43}$$

To estimate the denominator from above we use the trivial estimate $\rho_{\mathbb{F}_2^n/C^\perp}(x,x+e_j)\le 1$, therefore

$$\frac{1}{d(C)}\sum_{j=1}^n \int_{\mathbb{F}_2^n} \rho_{\mathbb{F}_2^n/C^\perp}(x,x+e_j)d\mu(x) \le \frac{n}{d(C)}.$$

To estimate the numerator, we use almost the same computation as in the proof of Theorem 4.28. Namely, we show that there exists a positive constant $\alpha>0$ such that for each $x\in\mathbb{F}_2^n$ the number of elements $y\in\mathbb{F}_2^n$, for which $\rho_{\mathbb{F}_2^n/C^\perp}(x,y)\ge\alpha n$ is $\ge 2^n/2$.

We observe that we can estimate from above the number of elements y for which $\rho_{\mathbb{F}_2^n/C^\perp}(x,y)<\alpha n$ in the following way. First we count the number of elements v satisfying $\mathrm{wt}(v-x)<\alpha n$. Then we consider all equivalence classes containing all such v. The observation above (see (4.43)) implies that all elements y outside the union U of these equivalence classes satisfy $\rho_{\mathbb{F}_2^n/C^\perp}(x,y)\ge\alpha n$. So it remains to show that for sufficiently small α the number of elements in all the equivalence classes $<2^{n-1}$.

To achieve this goal we observe that the number of elements in U does not exceed $2^{3n/4}\sum_{\ell\le\alpha n}\binom{n}{\ell}$. Observe that the sum is exactly of the same form as in (4.36) with $3n/4$ instead of k and α instead of δ. We proceed in the same way as in the proof of Theorem 4.28 and get the existence of the desired α.

From here we get the following estimate for the numerator of (4.42):

$$\int_{\mathbb{F}_2^n\times\mathbb{F}_2^n} \rho_{\mathbb{F}_2^n/C^\perp}(x,y)d\mu(x)d\mu(y)\ge\alpha n/2.$$

Therefore the value of (4.42) is $\ge\frac{\alpha d(C)}{2}\ge\frac{\alpha\delta n}{2}$. As we know (see Proposition 4.3), this implies the estimate $c_1(\mathbb{F}_2^n/C^\perp)\ge\frac{\alpha\delta n}{2}\ge\frac{\alpha\delta\log_2|\mathbb{F}_2^n/C^\perp|}{2}$. □

4.6 Exercises

Exercise 4.33. Prove the inequality $c_2(\mathbb{F}_2^n) \geq \sqrt{n}$ using (4.21) with the matrix T whose entries are

$$t_{u,v} = \begin{cases} n-1 & \text{if } u = v, \\ -1 & \text{if } \rho(u,v) = 1, \\ 1 & \text{if } u = v + e, \end{cases}$$

where $u, v \in \mathbb{F}_2^n$.

In the Hints to exercises section we suggest an approach to Exercise 4.33.

Exercise 4.34 (Optimal embedding of C_n). Let us label the vertices of the cycle C_n using $\{1, \dots, n\}$ (in their natural order on the cycle). Show that the (most natural) embedding $f : \{1, \dots, n\} \to \ell_2^2$ given by

$$f(j) = \begin{bmatrix} \cos\left(\frac{2\pi j}{n}\right) \\ \sin\left(\frac{2\pi j}{n}\right) \end{bmatrix}$$

is optimal, in the sense that it has the least distortions among all embeddings of C_n into ℓ_2.

In the Hints to exercises section we suggest an approach to Exercise 4.34.

Exercise 4.35. Read the definition of the diamond graphs given at the beginning of Section 9.3.2. In this exercise it is convenient for us to assume that the lengths of all edges of D_n are equal to 1 (so we multiply the length of edges of D_i, as they are defined on page 297, by 2^i).

Use inequality (4.1) and iteration to get a Poincaré inequality for D_n. Derive from this inequality an estimate for the Euclidean distortion of D_n of the form $c_2(D_n) \geq c\sqrt{\log |V(D_n)|}$ for some absolute constant $c > 0$.

Exercise 4.36 (Poincaré inequalities for L_p-valued functions on expanders). Let $1 \leq p < \infty$, $x \in \mathbb{R}$. We use the notation x^p for $|x|^p \cdot \operatorname{sign} x$.

Use the inequality (4.7) for $(f(v))^p$ and apply the following lemma (you are welcome to prove it if you wish):

Lemma 4.37. *For any real numbers a, b, and any $p \geq 1$, we have*

$$|a^p - b^p| \leq p|a-b|(|a|^{p-1} + |b|^{p-1})$$

to get the following version of (4.7):

$$\left(\sum_{v \in V} |f(v) - M|^p\right)^{\frac{1}{p}} \leq cp\left(\sum_{uv \in E} |f(u) - f(v)|^p\right)^{\frac{1}{p}}, \tag{4.44}$$

where c is a constant which depends only on the expanding constant and the degree of the graph, but not on p.

Use inequality (4.44) to get the fact that the distortion of embeddings of a family $\{G_n\}$ of expanders into L_p satisfies $c_p(G_n) \geq C\frac{1}{p}\ln|V(G_n)|$, where C depends only on the expanding constants and the degree of the family $\{G_n\}$.

Exercise 4.38. If a k-regular graph G with n vertices is embedded into ℓ_∞^d with distortion α, then $d \geq n^{c/\alpha}$, where the constant $c > 0$ depends on the expanding constant of the graph and k.

In the Hints to exercises section we suggest an approach to Exercise 4.38.

4.7 Notes and remarks

The method of Poincaré inequalities for lower distortion estimates was implicit in the pioneering paper [133] where the first estimate of this type was proved. Inequalities which could be called Poincaré inequalities appear in the papers [290] and [312] dealing with lower estimates of distortion for embeddings of expanders. The term "Poincaré inequality" in this context was systematically used in [292] and later papers.

As previously stated, in Chapter 5 we present some of the known constructions of expanders.

The Poincaré inequality for ℓ_2-valued functions on expanders can be obtained using the spectral characterization of expanders (due to [10] and [125], we present the spectral characterization in Section 5.2). The use of spectral characterization for getting a Poincaré inequality for expanders is demonstrated, for example, in [314, Section 15.5]. It is well known that the L_1-Poincaré inequality can be obtained using a direct argument, see, for example, the 1993 edition of [297, Problem 11.30] (however this argument is absent in the 1979 edition).

The general L_p Poincaré inequality (Theorem 4.14) is from [314]; see Proposition 15.5.2 (where the proof contains some errors, corrected in [315]) and Exercise 4 on p. 383. The idea goes back to [290], Matoušek writes [314, p. 380] that the result of Theorem 4.14 was communicated to him by Rabinovich.

The formula (4.21) for Euclidean distortion first appears in [290, Corollary 3.5]. It was applied in many different situations in [291] and [292].

The material covered in Section 4.5 appears in [249]. The paper [249] contains many other interesting applications of the Fourier-analytic approach developed there. The famous Poincaré inequality (4.31) was first proven by Enflo in [133] via a geometric argument, he also proved Theorem 4.23. Using Matoušek's extrapolation lemma for Poincaré inequalities [312], [42] (see Exercise 4.36 in this chapter and the corresponding references below), it is possible to prove that for a code C as in Theorem 4.27, for every $p \geq 1$, $c_p(\mathbb{F}_2^n/C^\perp) \geq c(p)d$.

The coding theory specific terminology used in this book corresponds to that which appears in [307], however the definition of a *linear code* presented in the latter is more general than that which we present. Theorem 4.28 is well known in coding theory, see [307, Chapter 1].

A proof of the Stirling formula can be found in [146, Chapter II, §9].

The quotient metrics were studied in [44],[82],[177],[325].

Newman and Rabinovich [339] have shown that the metric spaces $\mathbb{F}_2^n/C^\perp$ can be obtained as Cayley graphs of finite commutative groups with respect to sets of generators whose size is the logarithm of the size of the group. These Cayley graphs can, in turn, be regarded as unbounded degree expanders.

Exercises 4.33 and 4.34 are taken from [291]. Exercise 4.35 is a result of [338]. Exercise 4.36 is a result of [312]. Exercise 4.38 is a result of [274, Proposition 4.1].

In connection with Exercise 4.35 it is worth mentioning that combining the Kuratowski theorem (mentioned in Remark 3.13), Theorem 3.14 on padded decompositions of minor excluded graphs, Theorem 3.4, and Exercise 3.36 we get that the Euclidean distortion of any unweighted planar graph with n vertices is $O(\sqrt{\log n})$. This result is originally due to Rao [389]. Exercise 4.35 shows that the estimate of Rao is sharp up to a value of the constant.

In connection with Exercise 4.36 it is worth mentioning that Matoušek's extrapolation argument (which is described in Exercise 4.36) was later used in [42, p. 691] to show that even a reasonably large part of an expander does not admit a "good" embedding into L_p. The extrapolation argument was also used in [351] to get graph-theoretical versions of some Sobolev inequalities.

4.8 A bit of history of coarse embeddability

Gromov [175, Remark (b), p. 218] wrote: "There is no known geometric obstruction for uniform embeddings into infinite dimensional Banach spaces". (By *uniform embeddings* here Gromov means what we call *coarse embeddings.*) At the time Gromov was unaware of Enflo's work, [134], in which it was shown that there are no uniform (in the sense: uniformly continuous with the uniformly continuous inverse) embeddings of the Banach space c_0 into a Hilbert space. Dranishnikov, Gong, Lafforgue, and Yu observed [127, Section 6] that the construction of [134] can be used to prove that there exist locally finite metric spaces which are not coarsely embeddable into Hilbert spaces. (Later the examples of [134] and [127] were combined into one example in [247].)

But the spaces constructed in [127] do not have bounded geometry. Spaces with bounded geometry which are not coarsely embeddable into a Hilbert space were suggested by Gromov after he became acquainted with the results of [127]. Gromov's observation was published in Gromov [178, p. 158]. His example is based on the notion of the expander. I would like to mention that the paper of Matoušek [312] contains inequalities which imply immediately that expanders are not coarsely embeddable into ℓ_p for $1 \le p < \infty$.

In [179] Gromov wrote the following in reference to the aforementioned citation: "There is no known ...", should have been replaced with "There is a well-known example due to P. Enflo ...", who constructed a sequence of finite graphs (of growing degrees) admitting no uniform embedding into the Hilbert space (see [134]).

4.9 Hints to exercises

To Exercise 4.33. Show that Walsh functions form a set of eigenfunctions of T with nonnegative eigenvalues. (The positive semi-definiteness of the matrix in question can also be derived using the standard Enflo-type argument about the diagonals of the cube.)

To Exercise 4.34. One of the ways of doing this is to use the formula (4.21) and the matrix $\{s_{ij}\}$ given by

$$s_{i,j} = \begin{cases} 2\cos^2\left(\frac{\pi}{n}\right) & \text{if } i = j \\ -1 & \text{if } d_{C_n}(i,j) = 1 \\ 2\sin^2\left(\frac{\pi}{n}\right) & \text{if } d_{C_n}(i,j) = \frac{n}{2} \\ 0 & \text{otherwise.} \end{cases}$$

To Exercise 4.38. Suggested tools: To compose the embedding of G into ℓ_∞^d with the canonical embedding of ℓ_∞^d into ℓ_p^d with $p = \log_2 d$. Use (or prove, if you wish) the fact that the distortion of the canonical embedding of ℓ_∞^d into ℓ_p^d is $d^{\frac{1}{p}} = d^{\frac{1}{\log_2 d}} = 2$. Combine this computation with the result of Exercise 4.36.

Chapter 5

Families of expanders and of graphs with large girth

5.1 Introduction

In Section 4.2 of Chapter 4 we introduced the notion of a family of expanders and proved that families of expanders are, in a certain sense, the worst for embeddings into such spaces as L_1 and L_2. We plan to say more on their poor embeddability in Chapters 6 and 8. The purpose of the first part of this chapter is to provide several constructions of expanders. First we recall the definitions.

Definition 5.1. For a graph G with vertex set V and a subset $F \subset V$ by ∂F we denote the set of edges connecting F and $V \backslash F$. The *expanding constant* (also known as *Cheeger constant* and as *conductance*) of G is

$$h(G) = \inf\left\{\frac{|\partial F|}{\min\{|F|, |V\backslash F|\}} : F \subset V,\ 0 < |F| < +\infty\right\} \tag{5.1}$$

(where $|A|$ denotes the cardinality of a set A).

A sequence $\{G_n\}$ of graphs is called a *family of expanders* if all of G_n are finite, connected, k-*regular* for some $k \in \mathbf{N}$ (this means that each vertex is incident with exactly k edges), their expanding constants $h(G_n)$ are bounded away from 0 (that is, there exists $\varepsilon > 0$ such that $h(G_n) \geq \varepsilon$ for all n), and $|V(G_n)| \to \infty$ as $n \to \infty$.

Remark 5.2. Observe that for finite graphs the formula (5.1) can be written as

$$h(G) = \inf\left\{\frac{|\partial F|}{|F|} : F \subset V,\ 0 < |F| \leq |V|/2\right\}. \tag{5.2}$$

So far the simplest known proofs of existence of expanders are probabilistic. We present one such proof in Section 5.8 and describe some other probabilistic proofs in Notes and Remarks. Probabilistic proofs of existence of expanders are based directly on the definition of expanding constants. For some of the other proofs it is more convenient to use the spectral characterization of expanders which we present in Section 5.2. It is worth mentioning that there exist polynomial algorithms for computing the spectral characteristics of expanders, this makes it possible to find expanders in the following way: first we pick an expander at random using any of the probabilistic constructions producing an expander with high probability, then we check whether what we get is really an expander using the spectral characterization.

A substantial portion of this chapter is devoted to another series of graphs which are very important in metric geometry. We mean graphs with large girth, that is, graphs which do not have short cycles. As we shall see in Chapter 8 large girth is also an obstruction for embeddability into certain Banach spaces. We start our study of graphs with large girth in Section 5.6.

5.2 Spectral characterization of expanders

Everywhere in this section G is a finite simple graph. The *adjacency matrix* A of G is a matrix whose rows and columns are indexed by vertices of the graph, so the matrix is $(A_{u,v})$ and the u,v-entry $A_{u,v}$ of the graph is given by

$$A_{u,v} = \begin{cases} 1 & \text{if } u \text{ and } v \text{ are adjacent} \\ 0 & \text{otherwise.} \end{cases}$$

It is clear that A is a symmetric matrix. Therefore, by the general spectral theory it has real eigenvalues and has an orthogonal basis of eigenvectors. These eigenvalues are called *eigenvalues of the graph* G.

We consider the matrix A as an operator acting on $\ell_2(V)$, which is the space of real-valued (sometimes complex-valued, when it is specifically mentioned) functions on V with the norm

$$\|f\|_2 = \Big(\sum_{v\in V(G)} |f(v)|^2 \Big)^{1/2}.$$

For simplicity we denote $V(G)$ by V if the graph is clear from context.

We shall focus mostly on the case where the graph is k-regular. It is clear that in this case the indicator function $\mathbf{1}_V$ is an eigenvector of A with eigenvalue k.

Our next purpose is to prove some facts on the eigenvalues of the adjacency matrix A, which we denote $\lambda_1, \dots, \lambda_n$, where $n = |V|$ and

$$\lambda_1 \ge \lambda_2 \ge \dots \ge \lambda_{n-1} \ge \lambda_n.$$

We use the notation $\lambda_1(G), \dots, \lambda_n(G)$ if we need to specify the graph.

Definition 5.3. Recall that a graph G is called *bipartite* if its vertex set V can be represented as a disjoint union $V = X \cup Y$ in such a way that each edge has one end in X and one end in Y. The representation $V = X \cup Y$ satisfying the mentioned conditions is called a *bipartition.*

Proposition 5.4. *If G is k-regular, then*

(a) $|\lambda_i| \le k$ *for all* $i = 1, \dots, n$ *and* $\lambda_1 = k$.

(b) $\lambda_2 = k$ *if and only if the graph G is disconnected.*

(c) $\lambda_n = -k$ *if and only if the graph G has a bipartite connected component.*

Proof. (a) Since the matrix A has k ones in each row and all other entries are zeros, we get

$$|(Af)(v)| \leq k \max_{v \in V} |f(v)|.$$

This immediately implies that no eigenvalue of A can have absolute value $> k$. Since $A\mathbf{1}_V = k\mathbf{1}_V$, we have $\lambda_1 = k$.

(b) If G is disconnected and U is a vertex set of a connected component, then the function $\mathbf{1}_U \in \ell_2(V)$ satisfies $A\mathbf{1}_U = k\mathbf{1}_U$. Thus A has two linearly independent eigenvectors corresponding to eigenvalue k, and $\lambda_2 = k$.

Conversely, suppose that G is connected. It suffices to show that each eigenvector corresponding to the eigenvalue k is a multiple of $\mathbf{1}_V$. So let $f \in \ell_2(V)$ be such that $Af = k \cdot f$. Let v be such that $f(v)$ is maximal. The condition $Af = k \cdot f$ means

$$\sum_{u \in V, uv \in E(G)} f(u) = kf(v).$$

Since the summation is over the k-element set, our assumption that $f(v)$ is maximal implies that $f(u) = f(v)$ for all vertices u adjacent to v. Now we can apply the same reasoning to $f(u)$. Thus for all vertices u contained in the same connected component as v we have $f(v) = f(u)$. Since G is assumed to be connected, this implies that f is a multiple of $\mathbf{1}_V$.

(c) Suppose that G contains a connected component H which is a bipartite graph. Let $X \cup Y$ be the bipartition of the vertex set of H. Consider the function $f = \mathbf{1}_X - \mathbf{1}_Y \in \ell_2(V)$. We claim that $Af = -k \cdot f$. In fact, let $v \in X$, then all k neighbors of v are in Y, therefore

$$(Af)(v) = \sum_{u \in X,\ uv \in E(G)} f(u) = -k = -k \cdot f(v).$$

Similar verification works for $v \in Y$. Finally, it is clear that for $v \in V \backslash (X \cup Y)$ we have $f(v) = (Af)(v) = 0$.

Conversely, suppose that $-k$ is an eigenvalue of G and f is such that $Af = -k \cdot f$. Let v be such that $|f(v)|$ is maximal. Then

$$-kf(v) = (Af)(v) = \sum_{u \in V,\ uv \in E(G)} f(u).$$

Since $|f(v)|$ is maximal, this equality implies that $f(u) = -f(v)$ for each u adjacent to v. Thus $|f(u)|$ is also maximal for each such u and we can use the same argument for u. Therefore we get that for each u in the connected component containing v we have either $f(u) = f(v)$ or $f(u) = -f(v)$, and for adjacent vertices the signs are the opposite. This clearly implies that the set X of vertices u in this connected component with positive $f(u)$ and the set Y of vertices with negative $f(u)$ form a bipartition of the connected component containing v. □

The following theorem is called the *spectral characterization of expanders.*

Theorem 5.5. *A family $\{G_n\}$ of k-regular graphs with $|V(G_n)| \to \infty$ is a family of expanders if and only if the differences $k - \lambda_2(G_n)$ are uniformly bounded away from zero.*

Definition 5.6. If G is a k-regular graph, the difference $k - \lambda_2$, where λ_2 is the second (from above) eigenvalue of the adjacency matrix of G, is called the *spectral gap* of G.

Theorem 5.5 is an obvious consequence of the following more precise result.

Theorem 5.7. *For each finite k-regular graph G the inequality*

$$\frac{k-\lambda_2}{2} \le h(G) \le \sqrt{2k(k-\lambda_2)} \tag{5.3}$$

holds.

Proof. We start with the left-hand side inequality which is easier. The main ideas here are the following:

(a) If U and W are two subsets of V, then $\langle A\mathbf{1}_U, \mathbf{1}_W\rangle$ is the number of edges joining U and W.

(b) Thus $\langle A\mathbf{1}_U, \mathbf{1}_V\rangle = k|U|$, and in an expander, $\langle A\mathbf{1}_U, \mathbf{1}_{V\setminus U}\rangle \ge h(G)|U|$ if $|U| \le \frac{|V|}{2}$; and therefore $\langle A\mathbf{1}_U, \mathbf{1}_U\rangle \le (k - h(G))|U|$. The converse is also true: If $\langle A\mathbf{1}_U, \mathbf{1}_U\rangle \le (k-s)|U|$ for all $|U| \le \frac{|V|}{2}$, then $h(G) \ge s$.

(c) To prove the left-hand side of (5.3) we use the fact that for each function $f \in \ell_2(V)$ with zero mean the inequality $\frac{\langle Af,f\rangle}{\|f\|^2} \le \lambda_2(G) \le k$ holds, and show that a similar inequality holds for the quotient $\frac{\langle A\mathbf{1}_F, \mathbf{1}_F\rangle}{\|\mathbf{1}_F\|^2}$ in the case where $|F| \le \frac{|V|}{2}$.

(d) This leads to the following idea: we decompose $\mathbf{1}_F$ into an orthogonal sum of a multiple of $\mathbf{1}_V$ and a vector with zero average, and use this decomposition to estimate $\langle A\mathbf{1}_F, \mathbf{1}_F\rangle$ from above. This approach has a chance to lead to the desired conclusion since in the case where $|F| \le \frac{|V|}{2}$ the first term of the decomposition (that is, multiple of $\mathbf{1}_V$) is not dominating. To get the stated formula we just compute.

We have

$$\mathbf{1}_F = \frac{|F|}{|V|}\mathbf{1}_V + f,$$

where f has zero mean. For the norms we have

$$\|\mathbf{1}_F\|_2^2 = |F|, \quad \left\|\frac{|F|}{|V|}\mathbf{1}_V\right\|_2^2 = \frac{|F|^2}{|V|},$$

therefore

$$\|f\|_2^2 = |F| - \frac{|F|^2}{|V|}.$$

Since $\mathbf{1}_V$ is an eigenvector of A corresponding to the eigenvalue k and f is an orthogonal sum of eigenvectors corresponding to eigenvalues $\leq \lambda_2(G)$, we get

$$\begin{aligned}
\langle A\mathbf{1}_F, \mathbf{1}_F\rangle &= \left\langle A\frac{|F|}{|V|}\mathbf{1}_V, \frac{|F|}{|V|}\mathbf{1}_V\right\rangle + \langle Af, f\rangle \\
&\leq k\frac{|F|^2}{|V|} + \lambda_2(G)\left(|F| - \frac{|F|^2}{|V|}\right) \\
&= (k - \lambda_2(G))\frac{|F|^2}{|V|} + \lambda_2(G)|F|.
\end{aligned}$$

Of course, somewhere in the proof we have to use $|F| \leq \frac{|V|}{2}$. It is time to do so now. We continue the chain of inequalities above in the following way

$$\leq (k - \lambda_2(G))\frac{|F|}{2} + \lambda_2(G)|F| = \frac{k + \lambda_2(G)}{2}|F|.$$

By the comment in item (b), we get $h(G) \geq k - \frac{k+\lambda_2(G)}{2} = \frac{k-\lambda_2(G)}{2}$.

Proof of the right-hand side inequality in (5.3) is more difficult. Let $g \in \ell_2(V)$ be a mean zero (real-valued) eigenvector of A associated with the eigenvalue $\lambda_2(G)$. Let $V^+ = \{x \in V : g(x) > 0\}$ and let g^+ and g^- be functions defined by

$$g^+(x) = \max\{g(x), 0\} \text{ and } g^-(x) = \min\{g(x), 0\}.$$

We may assume without loss of generality that $|V^+| \leq |V|/2$: we replace g by $-g$ otherwise. We claim that

$$\langle Ag^+, g^+\rangle \geq \lambda_2\langle g^+, g^+\rangle. \tag{5.4}$$

In fact, $Ag = \lambda_2 g$, therefore

$$Ag^+ + Ag^- = \lambda_2 g^+ + \lambda_2 g^-$$

and

$$\langle Ag^+, g^+\rangle + \langle Ag^-, g^+\rangle = \lambda_2\langle g^+, g^+\rangle.$$

It remains to observe that $\langle Ag^-, g^+\rangle \leq 0$.

The inequality (5.4) can be rewritten as

$$\langle (kI - A)g^+, g^+\rangle \leq (k - \lambda_2)\langle g^+, g^+\rangle. \tag{5.5}$$

This rewriting is useful because it turns out that both sides of (5.5) admit suitable estimates in terms of the following discrete analogue of the gradient.

We introduce the space $\ell_2(E)$ of real-valued functions on the edge set $E = E(G)$ with the norm $\|\psi\| = (\sum_{e \in E} |\psi(e)|^2)^{1/2}$. So $\ell_2(E)$ is a Hilbert space with the inner product

$$\langle \psi_1, \psi_2 \rangle = \sum_{e \in E} \psi_1(e)\overline{\psi_2(e)}.$$

We also pick an *orientation of the graph* G, that is, for each edge e assign one of its ends to be a *head* of e and the other to be a *tail* of e. We denote the head of e by e^+ and the tail of e by e^-.

Now, for each function $f \in \ell_2(V)$ we define its *gradient* as a function $\nabla f \in \ell_2(E)$ given by

$$\nabla f(e) = f(e^+) - f(e^-). \tag{5.6}$$

The following formula is very useful (it can be obtained by writing detailed representations of both sides):

$$\langle (kI - A)f, f \rangle = \|\nabla f\|^2. \tag{5.7}$$

This relation is similar to the relation of the Laplace operator and the gradient in the classical analysis, and some of the results below were derived by using this analogy with the classical case.

To get an estimate of the right-hand side of (5.5) in terms of the gradient we use the following inequality proved in Chapter 4.

Lemma (Lemma 4.8). *Let $G = (V, E)$ be a connected graph with the expanding constant h, and $f : V \to \mathbb{R}$ be a real-valued function on V. Then*

$$\sum_{v \in V} |f(v) - M| \leq \frac{1}{h} \sum_{uv \in E} |f(u) - f(v)|, \tag{5.8}$$

where M is a median of the set $\{f(v)\}_{v \in V}$.

We apply (5.8) to the function $f(v) = (g^+(v))^2$. Observe that the assumption $|V^+| \leq |V|/2$ implies that 0 is a median of this function. We get

$$\|g^+\|^2 \leq \frac{1}{h(G)} \sum_{uv \in E(G)} |(g^+(u))^2 - (g^+(v))^2|. \tag{5.9}$$

Now we estimate the right-hand side of (5.9) in terms of $\nabla(g^+)$. We have

$$\begin{aligned}\sum_{uv\in E(G)} |(g^+(u))^2-(g^+(v))^2| &= \sum_{e\in E} |g^+(e^+)-g^+(e^-)|\cdot|g^+(e^+)+g^+(e_-)|\\ &\le \Big(\sum_{e\in E} |g^+(e^+)-g^+(e^-)|^2\Big)^{\frac12}\cdot\Big(\sum_{e\in E} |g^+(e^+)+g^+(e^-)|^2\Big)^{\frac12}\\ &\le \|\nabla(g^+)\|\cdot\Big(\sum_{e\in E} 2(|g^+(e^+)|^2+|g^+(e^-)|^2)\Big)^{\frac12}\\ &\le \|\nabla(g^+)\|\cdot\sqrt{2k}\cdot\|g^+\|.\end{aligned}$$

Combining this inequality with (5.9) we get

$$h(G)\|g^+\| \le \sqrt{2k}\,\|\nabla(g^+)\|. \tag{5.10}$$

On the other hand, combining (5.5) and (5.7) we get

$$\|\nabla(g^+)\| \le \sqrt{k-\lambda_2}\|g^+\|. \tag{5.11}$$

Combining (5.10) and (5.11) we get the right-hand side inequality of (5.3). □

Remark 5.8. 1. The lower bound in (5.3) is tight. One of the examples is the Hamming cube $\mathbb{F}_2^k$. Its eigenvalues are $k, k-2, \ldots, -k$ (see Exercise 5.82 at the end of this chapter), so $k-\lambda_2 = 2$. On the other hand it is not difficult to show that $h(\mathbb{F}_2^k) = 1$, and the bound is achieved on the set A of all vertices for which the first coordinate is 0.

2. The upper bound in (5.3) is close to being tight. Consider the cycle C_n with even n. It is easy to check that $h(C_n) = \frac{4}{n}$. On the other hand $k-\lambda_2$ in this case is $2-2\cos\left(\frac{2\pi}{n}\right)$ (see Exercise 5.81 at the end of this chapter). Therefore

$$\sqrt{2k(k-\lambda_2)} \approx \frac{4\pi}{n}.$$

5.3 Kazhdan's property (T) and expanders

In this section we are going to use some results on group representations to construct some classes of expanders. The expanders which we construct here are Cayley graphs of finite groups. We already introduced the corresponding definition (see page 3). However, it will be convenient to introduce it again in a slightly different form.

Definition 5.9. Let F be a group, and let S be a finite subset of F. We assume that S satisfies the condition $(s \in S) \Leftrightarrow (s^{-1} \in S)$ and does not contain the identity. We call such a subset *symmetric*. The *right-invariant Cayley graph* $\mathrm{Cay}(F, S)$ is defined to be the graph with vertex set F and edge set defined by the condition: uv is an edge if and only if $u = sv$ for some $s \in S$. The *left-invariant Cayley graph* (we do not introduce a special notation for it, reasons are explained below) is defined to be the graph with vertex set F and edge set defined by the condition: uv is an edge if and only if $u = vs$ for some $s \in S$.

Remark 5.10. The term *right-invariant* Cayley graph is chosen because the multiplication of all group elements by a fixed element g on the right is an isomorphism of $\mathrm{Cay}(F, S)$. (The reason for the choice of the term *left-invariant Cayley graph* is similar.) The right-invariant Cayley graph and the left-invariant Cayley graph of the same group with the same set S are isomorphic, the map $f \mapsto f^{-1}$ is one of the isomorphisms. This is the reason why we are going to study only right-invariant Cayley graphs and call them just *Cayley graphs*. It is also clear that for commutative groups there is no difference between the right-invariant Cayley graph and the left-invariant Cayley graph.

Observation 5.11. 1. It is easy to see that the graph $\mathrm{Cay}(F, S)$ is $|S|$-regular.

2. The graph $\mathrm{Cay}(F, S)$ is connected if and only if S is a generating set (that is, if and only if each element of F can be written as a finite product of elements of S).

Example 5.12. The Hamming cube (see pages 118 and 181) can be regarded as the Cayley graph of the group $\mathbb{F}_2^n$ with respect to the set

$$S = \{(1,0,\dots,0), (0,1,\dots,0), \dots, (0,0,\dots,1)\}.$$

Observe that each element of S is an inverse of itself in the group $\mathbb{F}_2^n$.

Expansion of Cayley graphs can be characterized in terms of set-products in groups. If A and B are two subsets in a group F with operation $\cdot$, their *product* is defined by

$$A \cdot B = \{a \cdot b : a \in A, \ b \in B\}.$$

Proposition 5.13. *Let $\{F_n\}_{n=1}^{\infty}$ be a family of groups of growing sizes, and $\{S_n\}$ be symmetric subsets $S_n \subset F_n$ such that the cardinality $|S_n|$ is the same for each n. The graphs $\{\mathrm{Cay}(F_n, S_n)\}_{n=1}^{\infty}$ form a family of expanders if and only if there is a constant $c > 0$ independent of n such that*

$$|A_n \cup (S_n \cdot A_n)| \geq (1 + c)|A_n| \tag{5.12}$$

for all subsets $A_n \subset F_n$ with $|A_n| \leq |F_n|/2$.

Proof. Observe that $S_n \cdot A_n$ is the set of vertices which are adjacent to vertices of A_n in $\mathrm{Cay}(F_n, A_n)$. The inequality (5.12) implies that $|(S_n \cdot A_n) \backslash A_n| \geq c|A_n|$. Therefore $|\partial A_n| \geq c|A_n|$, and the expanding constant of $\mathrm{Cay}(F_n, A_n)$ is at least c.

In the other direction. Let k be the cardinality of S_n (for each n). Suppose that $|\partial A_n| \geq h|A_n|$ for some $h > 0$ and each subset $A_n \subset F_n$ with $|A_n| \leq |F_n|/2$.

Observe that this implies that the number of elements in F_n which are adjacent to elements of A_n, but are not in A_n is at least $\frac{|\partial A_n|}{k}$. The set of all such elements and the set A_n are disjoint and both of them are contained in the union $A_n \cup (S_n \cdot A_n)$. Hence

$$|A_n \cup (S_n \cdot A_n)| \geq |A_n| + \frac{|\partial A_n|}{k} \geq \left(1 + \frac{h}{k}\right)|A_n|. \qquad \square$$

Examples of Cayley graphs which are expanders can be easily constructed from any infinite finitely generated group satisfying the two conditions, one of which we are going to discuss now (we discuss the second condition and complete the construction in Section 5.4). We start by recalling some standard definitions.

Definition 5.14. Let $\mathcal{H}$ be a complex Hilbert space with inner product $\langle \cdot, \cdot \rangle$ and T be a bounded linear operator in $\mathcal{H}$. The set $L(\mathcal{H})$ of all such operators forms an algebra. It has an identity, which is the identity operator I of $\mathcal{H}$. We define an *involution* $T \mapsto T^*$ on $L(\mathcal{H})$ by

$$\forall x, y \in \mathcal{H}\ \langle T^* x, y \rangle = \langle x, Ty \rangle.$$

An operator $U : \mathcal{H} \to \mathcal{H}$ is called *unitary* if $UU^* = U^*U = I$ or, equivalently, if $\langle Ux, Uy \rangle = \langle x, y \rangle$ for all $x, y \in \mathcal{H}$ and if U is onto. All unitary operators in $L(\mathcal{H})$ form a group (with respect to the product of operators) which is called the *unitary group*, denoted $U(\mathcal{H})$. A *unitary representation* of a group G on a Hilbert space $\mathcal{H}$ is a group homomorphism of G into $U(\mathcal{H})$.

Remark 5.15. Studying unitary representations of *topological* groups it is natural and customary to introduce certain continuity conditions (strong continuity) as a part of the definition of unitary representation.

Now we introduce one of the most important definitions of the theory of infinite-dimensional representations of groups, restricted to the case of finitely generated groups because it is the only case which we are going to consider.

Definition 5.16. Let G be a finitely generated group and $\pi : G \to U(\mathcal{H})$ be a unitary representation of G on a Hilbert space $\mathcal{H}$. If $\varepsilon > 0$ and K is a finite subset of G, then a norm-one vector x in $\mathcal{H}$ is called an *(ε, K)-invariant vector* if $\|\pi(g)x - x\| < \varepsilon$ for all $g \in K$. An *invariant vector* of the representation π is a vector $x \in \mathcal{H}$ such that $\pi(g)x = x$ for each $g \in G$. We say that G has *Kazhdan's property* (T), or just *property* (T) if (and only if) there exists an $\varepsilon > 0$ and a finite subset K of G such that every unitary representation having an (ε, K)-invariant vector has a nonzero invariant vector.

We also introduce the following numerical parameter which is defined for all groups (and can be introduced for infinite subsets K as well).

Definition 5.17. Let $\pi : G \to U(\mathcal{H})$ be a unitary representation of a finitely generated group G, and let K be a finite subset of G. The *Kazhdan constant* $\mathcal{K}(G, K, \pi)$ is defined as the supremum of all $\varepsilon > 0$ for which

$$\sup_{g \in K} \|\pi(g)x - x\| \geq \varepsilon \|x\|$$

for all $x \in \mathcal{H}$. The *Kazhdan constant* $\mathcal{K}(G, K)$ is defined as $\inf_\pi \mathcal{K}(G, K, \pi)$, where π ranges over all unitary representations of G with no nontrivial invariant vectors.

Remark 5.18. 1. It is easy to see that a finitely generated group has Kazhdan's property (T) if and only if $\mathcal{K}(G, K) > 0$ for at least one finite set $K \subset G$.

2. The constant $\mathcal{K}(G, K, \pi)$ vanishes whenever the representation contains nontrivial invariant vectors (such representations exist for all groups, for example the representation which maps all elements of the group onto the identity operator).

3. It is easy to show that each finitely generated group has a unitary representation with no nontrivial invariant vectors: for finite groups see Observation 5.21, for infinite groups one of the examples is the analogue of the left regular representation introduced in Definition 5.20.

Remark 5.19. If $\|x\| = 1$, the identity

$$\|\pi(g)(x) - x\|^2 = 2 - 2\mathrm{Re}\langle \pi(g)(x), x\rangle$$

holds. Using this identity we can restate the definition of the Kazhdan constant in the following way:

$$\mathcal{K}(G, K) = \inf_{\pi} \inf_{x \in \mathcal{H}, \|x\|=1} \sup_{g \in K} \sqrt{2} \cdot \sqrt{1 - \mathrm{Re}\langle \pi(g)(x), x\rangle}, \tag{5.13}$$

where the infimum is over unitary representations of G with no nontrivial invariant vectors.

At first glance it could be unclear how property (T) of groups and the Kazhdan constant are related to expanders. But if we recall the spectral characterization of expanders (Theorem 5.5) and compare it with (5.13), we see some connection because $\mathcal{K}(G, K) > 0$ implies that there is $\alpha = \alpha(\mathcal{K}(G, K)) > 0$ such that for each unitary representation π without nontrivial invariant vectors and for each normalized vector $x \in \mathcal{H}$ there is $g \in K$ such that

$$\mathrm{Re}\langle \pi(g)(x), x\rangle \leq 1 - \alpha. \tag{5.14}$$

Inequality (5.14) is already somewhat similar to the inequality which we would like to get:

$$|\langle Ay, y\rangle| \leq k - \varepsilon,$$

where A is the adjacency matrix and k is the vertex degree, and y is a normalized vector with zero mean. Soon we shall see (Observation 5.21) that the subspace of vectors with zero mean has the representation-theoretical property which we need.

It remains to see how $\pi(g)$ can be related to A. For this we need to recall some basic notions of finite-dimensional representations of finite groups. At this point we need one of the most straightforward constructions of a representation.

Definition 5.20. Let F be a finite group. The *right regular representation* of F is the representation of F on the space $\ell_2(F)$ (in this context considered as a space over $\mathbb{C}$) given by $\pi(g)(f(x)) = f(xg)$. The *left regular representation* of F is the representation of F on the space $\ell_2(F)$ given by $\pi(g)(f(x)) = f(g^{-1}x)$.

Observation 5.21. The restriction of a (right or left) regular representation to the subspace consisting of functions with zero mean does not have invariant vectors.

In fact, one can see that all of the coordinates of an invariant vector of a left (right) regular representation have to be equal, and a vector with equal coordinates has zero mean if and only if it is 0.

Now we are ready to prove the following crucial observation:

Theorem 5.22. *The spectral gap of the graph* $\mathrm{Cay}(F, S)$ *can be estimated from below in terms of a function of* $\mathcal{K}(F, S)$*, and this function is strictly positive for strictly positive* $\mathcal{K}(F, S)$.

Proof. We start with the following observation: the adjacency matrix $(A_{u,v})_{u,v \in F}$ of $\mathrm{Cay}(F, S)$ can be written as a sum of matrices of operators of the left regular representation. In fact, the adjacency matrix of $\mathrm{Cay}(F, S)$ is the sum of matrices $M(g) = (M_{u,v}(g))_{u,v \in F}$ corresponding to $g \in S$, where $M(g)$ (defined for any $g \in F$) is the matrix given by

$$M_{u,v}(g) = \begin{cases} 1 & \text{if } uv^{-1} = g \\ 0 & \text{otherwise.} \end{cases}$$

It is also easy to check that $M(g)$ is the matrix of $\pi(g^{-1})$. Therefore the quantity $\langle Ax, x\rangle$, where A is the adjacency matrix of $\mathrm{Cay}(F, S)$ can be written as

$$\Big\langle \sum_{g \in S} M(g)x, x \Big\rangle.$$

We need to estimate $\sup_x |\langle Ax, x\rangle|$ where the supremum is over $x \in \ell_2(F)$ with norm one and zero average. Since A is self-adjoint, we have

$$\sup_x |\langle Ax, x\rangle| = \sup_x \mathrm{Re}\langle Ax, x\rangle.$$

This leads to the desired connection with the Kazhdan constant:

$$\sup_x \mathrm{Re}\Big\langle \sum_{g\in S} M(g)x, x\Big\rangle \le \sum_{g\in S} \sup_x \mathrm{Re}\langle M(g)x, x\rangle \le |S| - \alpha(\mathcal{K}(F,S)). \qquad \square$$

This theorem shows that in order to construct a sequence of expanders we need to find a family of groups F_n with $|F_n| \to \infty$ and generating subsets $S_n \subset F_n$ with $|S_n| = const$ such that the Kazhdan constants $\mathcal{K}(F_n, S_n)$ are bounded away from zero (that is, there exists $c > 0$ such that $\mathcal{K}(F_n, S_n) \ge c$ for all n).

Remark 5.23. Proof of Theorem 5.22 gives a bit more, namely it shows that the sequence $\{\mathrm{Cay}(F_n, S_n)\}_{n=1}^{\infty}$ is a family of expanders, if the family of groups $\{F_n\}_{n=1}^{\infty}$ with $|F_n| \to \infty$ and generating subsets $S_n \subset F_n$ with $|S_n| = const$ are such that the Kazhdan constants $\mathcal{K}(F_n, S_n, \rho_n)$ are bounded away from zero (that is, there exists $c > 0$ such that $\mathcal{K}(F_n, S_n, \rho_n) \ge c$ for all n), where ρ_n is the restriction of the left regular representation of F_n to the space of vectors with zero average.

By now many examples of such sequences $\{F_n, S_n\}$ are known, for example, it is known that the Kazhdan constants are bounded away from zero if F_n is the alternating group on n elements and S_n is a suitably chosen subset in F_n (with $|S_n| = const$).

Historically the first examples of this type were obtained by considering infinite finitely generated groups with property (T) which have finite quotients of growing sizes providing the desired examples. Our next section is devoted to this construction.

5.4 Groups with property (T)

One of the ways to complete the construction of expanders presented in the previous section is to find an infinite finitely generated group G with property (T) having finite quotient groups $\{G_i\}_{i=1}^{\infty}$ with indefinitely increasing cardinalities $|G_i|$ and such that the restrictions of the quotient maps $\varphi_i : G \to G_i$ to one of the finite sets $K \subset G$ satisfying $\mathcal{K}(G, K) > 0$ are injective.

Proposition 5.24. *The graphs* $\{\mathrm{Cay}(G_i, \varphi_i(K))\}_{i=1}^{\infty}$ *form a family of expanders.*

Proof. Since the restrictions of φ_i to K are injective, all of the obtained graphs are $|K|$-regular.

To estimate the spectral gap of the obtained graphs we denote by ρ_i the restriction of the left regular representation of G_i to the space of vectors with zero average and observe that Definition 5.17 implies

$$\mathcal{K}(G, K) \le \mathcal{K}(G, K, \rho_i \circ \varphi_i) = \mathcal{K}(G_i, \varphi_i(K), \rho_i).$$

Application of Remark 5.23 completes the proof. $\square$

Remark 5.25. The injectivity of the restrictions of φ_i to K can be dropped if we agree to consider expanders with loops and multiple edges.

Some of the well-known groups have property (T). Possibly the most well-known groups which have property (T) are the groups $SL_n(\mathbb{Z})$, $n \geq 3$.

Definition 5.26. The *special linear group* $SL_n(\mathbb{Z})$ is the group of all $n \times n$ matrices with integer coefficients and determinant equal to 1.

Remark 5.27. The same definition is used to define $SL_n(R)$ for any commutative ring R with unit.

5.4.1 Finite generation of $SL_n(\mathbb{Z})$

Proposition 5.28. *The group $SL_n(\mathbb{Z})$ is finitely generated for each n and is an infinite group if $n \geq 2$.*

Proof. The second statement is trivial.

The fact that $SL_n(\mathbb{Z})$ is finitely generated with rather weak and depending on n bound on the number of generators is an easy consequence of the basic techniques of linear algebra known as *Gaussian elimination*. This bound can be obtained in the following way.

Let $A \in SL_n(\mathbb{Z})$. We start by observing that the greatest common divisor of entries of the first column of A is equal to 1 (otherwise the determinant cannot be equal to one). Therefore, by Euclid's algorithm for getting the greatest common divisor, subtracting rows one from another, and adding rows, we can get a matrix A_1 whose first column has exactly one nonzero entry, and the entry is ± 1.

Observe that the addition of row $\#j$ to row $\#i$ in a matrix A is equivalent to multiplication of the matrix A from the left by one of the so-called *elementary matrices* of a special form.

Definition 5.29. An $n \times n$ matrix is called *elementary* if all of its diagonal entries are equal to 1, and it has exactly one nonzero non-diagonal entry equal to t. We denote such a matrix with (i, j)-entry equal to t by $E_{ij}(t)$.

The addition of a row mentioned before Definition 5.29 is equivalent to multiplication on the left by $E_{ij}(1)$, the subtraction of row $\#j$ from row $\#i$ in a matrix A is equivalent to multiplication of the matrix A from the left by $E_{ij}(-1)$.

Suppose that the ± 1 in the first column of A_1 is the $(i_1, 1)$-entry. It is easy to see that the greatest common divisor of all entries of the second column except $(i_1, 2)$-entry is equal to 1. Repeating the procedure of multiplication on the left by matrices $E_{ij}(1)$ and $E_{ij}(-1)$, we get a matrix A_2 whose first column is the same as the first column of A_1 and whose second column has at most two nonzero entries. One of them is the

$(i_1, 2)$-entry, which can be zero. The other is an $(i_2, 2)$-entry, $i_2 \neq i_1$, which is equal to ± 1.

Now we can subtract/add (repeatedly, if needed) the i_2^{th} row from/to the i_1^{th} row and make the $(i_1, 2)$-entry equal to 0.

We continue in an obvious manner (in the third step we use the fact that the greatest common divisor of all entries of the third column except the $(i_1, 3)$-entry and $(i_2, 3)$-entry is equal to 1).

At the end of this computation we get a *signed permutation matrix*.

Definition 5.30. A one-to-one mapping of the set $\{1, \dots, n\}$ onto itself is called a *permutation*. A (n by n) *permutation matrix* is a matrix of the form $\sum_{i=1}^{n} e_{i,\pi(i)}$, where $\pi : \{1, \dots, n\} \to \{1, \dots, n\}$ is a permutation and e_{ij} is the matrix with only one nonzero entry, which is an (i, j)-entry, and which is equal to 1.

By a *signed permutation matrix* we mean a matrix which is obtained from a permutation matrix if we replace some (may be all) of its nonzero entries (which are equal to 1) by -1.

This completes the proof of the theorem. We have proved that each $A \in SL_n(\mathbb{Z})$ is a product of elementary matrices of the form $E_{ij}(\pm 1)$ (it is important that all of these matrices are in $SL_n(\mathbb{Z})$) and some signed permutation matrix (which should be $SL_n(\mathbb{Z})$ since $A \in SL_n(\mathbb{Z})$ and all elementary matrices are in $SL_n(\mathbb{Z})$). Therefore the set consisting of (1) all matrices of the form $E_{ij}(\pm 1)$, and (2) all signed permutation matrices with determinant 1; is a finite generating set of $SL_n(\mathbb{Z})$. □

Remark 5.31. The groups $SL_n(\mathbb{Z})$ have much smaller generating sets, see references in Notes and Remarks. See also Exercise 5.91.

5.4.2 Finite quotients of $SL_n(\mathbb{Z})$

Finite quotients of $SL_n(\mathbb{Z})$ satisfying the conditions mentioned at the beginning of Section 5.4 are obtained by reduction modulo p. This means the following. We consider the fields $\mathbb{Z}_p$ of residues modulo p, where p is a prime number (the field $\mathbb{Z}_2$ can be identified with the field $\mathbb{F}_2$ which we introduced at the beginning of Section 4.5; in some papers notation $\mathbb{F}_p$ is used for the fields $\mathbb{Z}_p$). After that we introduce groups $SL_n(\mathbb{Z}_p)$ as groups of matrices with entries in $\mathbb{Z}_p$ with determinants equal to 1 (considered as an element of $\mathbb{Z}_p$). It is easy to see that the natural map $SL_n(\mathbb{Z}) \mapsto SL_n(\mathbb{Z}_p)$, which maps a matrix $\{a_{ij}\}$ to the matrix $\{a_{ij} \pmod p\}$ is a surjective homomorphism. This homomorphism is injective on a finite generating set unless the set contains two matrices which are the same if considered $\mod p$. It never happens if $p \geq 3$ and the entries of the matrices are 0 and ± 1 (observe that it is the case for all sets generating $SL_n(\mathbb{Z})$, which we have considered so far).

5.4.3 Property (T) for groups $SL_n(\mathbb{Z})$

Theorem 5.32 (without proof). *$SL_n(\mathbb{Z})$, $n \geq 3$, has property (T).*

The group $SL_3(\mathbb{Z})$ is historically the first example of a group with property (T). Property (T) for this and many other groups is proved in the following way. The definition of property (T) is generalized to the case of locally compact groups. The difference between the definition of property (T) for locally compact groups and Definition 5.16 is that we consider a *compact* subset K instead of finite subset K. In the next step property (T) is established for $SL_n(\mathbb{R})$ (or continuous analogues of some other groups). And finally there is a general result showing that in many cases the property (T) passes from locally compact groups to their discrete subgroups. Detailed presentation of this approach requires substantial amount of material from representation theory, in Notes and Remarks we provide references to sources containing the proof of the fact that $SL_n(\mathbb{Z})$, $n \geq 3$, has property (T) on these lines. It is worth mentioning that $SL_2(\mathbb{Z})$ *does not* have property (T).

There is another, in a certain sense, more algebraic and direct proof of the fact that $SL_n(\mathbb{Z})$ has property (T). In this approach the property (T) is established for $SL_n(\mathbb{Z})$ directly, without using $SL_n(\mathbb{R})$. In Notes and Remarks we refer to a source containing a nice account of this approach.

5.4.4 Criterion for property (T)

All proofs mentioned in the previous section are rather complicated. Their complexity contrasts, to a certain extent, with the existence of a relatively simple sufficient criterion for property (T), which we are going to present now.

Let G be a group generated by a finite symmetric set S (we remind that according to our definition of a symmetric set, S does not contain the identity of G). We define a finite graph Z whose vertex set is S and whose edge set $E(Z)$ is the set of all pairs uv satisfying the conditions $u, v, u^{-1}v \in S$. We assume that S is chosen in such a way that the graph Z is connected. It is easy to see that a symmetric generating set S for which Z is connected can be chosen in each finitely generated group. For a vertex u in Z we denote by $\deg u$ its degree in Z. Let $\ell_2(S, \deg)$ be a weighted Hilbert space of real-valued functions on S. The Hilbert norm of a function $f : S \to \mathbb{R}$ in this space is given by

$$\|f\| = \sum_{u \in S} (f(u))^2 \deg u.$$

Let Δ be a *discrete Laplace operator* on $\ell_2(S, \deg)$ given by

$$(\Delta f)(u) = f(u) - \frac{1}{\deg u} \sum_{v,\, uv \in E(Z)} f(v)$$

The operator Δ is a nonnegative, self-adjoint operator on $\ell_2(S, \deg)$. If Z is connected then zero is a simple eigenvalue of Δ. Let $\zeta_1(Z)$ be the smallest nonzero eigenvalue of Δ considered as an operator on $\ell_2(S, \deg)$.

Remark 5.33. In the case where Z is a k-regular graph, the smallest eigenvalue of the discrete Laplace operator is closely related to the spectral gap (introduced in Definition 5.6). In fact, one can check that in such a case

$$\Delta = I - \frac{1}{k}A, \quad \zeta_1(Z) = \frac{k - \lambda_2(Z)}{k}.$$

Theorem 5.34 (Żuk's criterion). *Let G be a group generated by a finite symmetric subset S. If the graph Z is connected and $\zeta_1(Z) \geq \frac{1}{2}$ then G has Kazhdan's property (T). Moreover*

$$\mathcal{K}(G, S) \geq \frac{2}{\sqrt{3}}\left(2 - \frac{1}{\zeta_1(Z)}\right).$$

This criterion leads to many new examples of groups with property (T). Unfortunately, all explicit examples known to me of groups to which this criterion is directly applicable are rather complicated (even to describe). The criterion can be used to show that some classes of random groups have property (T), but such examples are not interesting in the context of explicit constructions of expander families. (The proof of Theorem 5.34 is relatively simple, we do not present it because we are not going to apply Theorem 5.34.) There exist more complicated criteria which generalize and strengthen Theorem 5.34 (improve the value of Kazhdan's constant), and which are applicable to relatively simple examples of groups, but their study is beyond the scope of this book.

5.5 Zigzag products

Our purpose in this section is to present a combinatorial construction of families of expanders based on the so-called *zigzag product* of graphs. The construction starts with a slightly different definition of a spectral gap than the one introduced in Definition 5.6 (page 134).

Recall that a nonnegative matrix $A = (a_{ij})_{i,j=1}^n$ is called *stochastic* if $\sum_{j=1}^n a_{ij} = 1$ for each $i \in \{1, \dots, n\}$. We are going to study symmetric (that is, satisfying $a_{ij} = a_{ji}$) stochastic matrices. A typical example which is of interest for us in the present context is the normalized adjacency matrix of a k-regular graph, where "normalized" means that each entry is divided by k.

We are going to introduce (Definitions 5.35 and 5.38) and study quantities closely related to the spectral gap introduced in Definition 5.6. After that we show (Theorem 5.42) that these quantities behave nicely with respect to the combinatorial construction called *zigzag product* and thus lead to a construction of families of expanders.

Definition 5.35. Let A be an $n \times n$ symmetric stochastic matrix, and let $1 = \lambda_1(A) \geq \lambda_2(A) \cdots \geq \lambda_n(A) \geq -1$ be its eigenvalues. Let $\lambda(A) := \max_{2 \leq i \leq n} |\lambda_i(A)|$. The quantity $1 - \lambda(A)$ is called the *absolute spectral gap* of A.

Remark 5.36. Comparing Definitions 5.35 and 5.6 and recalling the spectral characterization of expanders (Theorem 5.5) we see that any family of k-regular graphs with indefinitely growing numbers of vertices and with absolute spectral gaps of their normalized adjacency matrices bounded away from zero is a family of expanders.

Proposition 5.37. *The reciprocal of the absolute spectral gap of A is the smallest constant $\gamma_+ > 0$ such that for all $x_1, \ldots, x_n, y_1, \ldots, y_n \in L_2$ we have*

$$\frac{1}{n^2} \sum_{i=1}^{n} \sum_{j=1}^{n} \|x_i - y_j\|_2^2 \leq \frac{\gamma_+}{n} \sum_{i=1}^{n} \sum_{j=1}^{n} a_{ij} \|x_i - y_j\|_2^2. \tag{5.15}$$

Proof. It is clear that it suffices to prove (5.15) in the case where

$$x_1, \ldots, x_n, \quad y_1, \ldots, y_n$$

are scalars. We consider $\{x_i\}_{i=1}^n$ and $\{y_i\}_{i=1}^n$ as vectors of ℓ_2^n.

Subtracting from each of y_j their average, and adding this average to each of x_i we may assume that $\sum_{j=1}^n y_j = 0$. With this assumption we have:

$$\begin{aligned}
\frac{1}{n^2} \sum_{i=1}^{n} \sum_{j=1}^{n} (x_i - y_j)^2 &= \frac{1}{n^2} \sum_{i=1}^{n} \sum_{j=1}^{n} (x_i^2 + y_j^2 - 2x_i y_j) \\
&= \frac{1}{n^2} \left(n\|x\|^2 + n\|y\|^2 - 2 \sum_{i=1}^{n} \sum_{j=1}^{n} x_i y_j \right) \\
&= \frac{1}{n} (\|x\|^2 + \|y\|^2).
\end{aligned}$$

As for the right-hand side, we get

$$\frac{\gamma_+}{n} \sum_{i=1}^{n} \sum_{j=1}^{n} \left(a_{ij} x_i^2 + a_{ij} y_j^2 - 2a_{ij} x_i y_j \right) = \frac{\gamma_+}{n} (\|x\|^2 + \|y\|^2 - 2\langle Ay, x\rangle).$$

After multiplying by n the inequality (5.15) becomes

$$(\|x\|^2 + \|y\|^2) \leq \gamma_+ (\|x\|^2 + \|y\|^2 - 2\langle Ay, x\rangle),$$

or

$$\langle Ay, x\rangle \leq \left(1 - \frac{1}{\gamma_+}\right) \frac{\|x\|^2 + \|y\|^2}{2}.$$

Now it remains to observe that the smallest constant C for which

$$\forall x, y \in \ell_2^n \quad \langle Ay, x\rangle \leq C\,\frac{\|x\|^2 + \|y\|^2}{2}$$

holds, is $\lambda(A)$. In fact, since A is a symmetric stochastic matrix, the eigenvalue 1 corresponds to the vector with all coordinates equal to 1, the condition $\sum_{j=1}^n y_j = 0$ implies that y is orthogonal to this vector. Therefore $|\langle Ay, x\rangle| \leq \lambda(A)\|y\|\|x\| \leq \lambda(A)(\|x\|^2 + \|y\|^2)/2$. The equality is attained because we may choose $\{x_i\}_{i=1}^n$ and $\{y_i\}_{i=1}^n$ in such a way that $\|x\| = \|y\|$ and y is an eigenvector of A whose respective eigenvalue has absolute value $\lambda(A)$. □

Definition 5.38. Let X be an arbitrary set. A *kernel* on X is a symmetric function $K : X \times X \to [0, \infty)$. The *reciprocal absolute spectral gap of A with respect to the kernel K*, denoted $\gamma_+(A, K)$, is defined as the infimum of numbers $C \geq 0$ satisfying the condition: for all $x_1, \dots, x_n, y_1, \dots, y_n \in X$ the inequality

$$\frac{1}{n^2}\sum_{i=1}^n\sum_{j=1}^n K(x_i, y_j) \leq \frac{C}{n}\sum_{i=1}^n\sum_{j=1}^n a_{ij}K(x_i, y_j) \tag{5.16}$$

holds.

Remark 5.39. Observe that the quantity γ_+ defined in Proposition 5.37 is the reciprocal absolute spectral gap of A with respect to the kernel $K : L_2 \times L_2 \to [0, \infty)$ given by $K(x, y) = \|x - y\|_2^2$. We introduce the very general Definition 5.38 because one of the important combinatorial results which we are going to prove (Theorem 5.42) holds in this generality.

In what follows we deal with regular graphs, which in this section are allowed to have loops and multiple edges. (Recall that two or more edges are called *parallel edges* or *multiple edges* if they have the same pair of end vertices.) We allow loops and parallel edges because in the constructions presented below it is too difficult to get rid of them, on the other hand loops and multiple edges do not cause any problems for the most important applications of expanders.

We use the convention that each loop contributes 1 to the degree of a vertex (this convention agrees with the algebraic results on adjacency matrices which we are going to use, but does not agree with the well-known identity: the sum of all degrees = twice the number of edges).

Definition 5.40. The *normalized adjacency matrix* of a (not necessarily simple) k-regular graph $G = (V, E)$, denoted A_G, is the matrix whose rows and columns are labeled by elements of V, and whose uv-entry is the number of edges joining u and v ($u, v \in V$) divided by k. When discussing reciprocal absolute spectral gaps we will interchangeably identify G with A_G. Thus, for example, we define $\gamma_+(G, K)$ as $\gamma_+(A_G, K)$.

The quantities $\gamma_+(G, K)$ behave nicely under the zigzag product of graphs. This product is defined only for pairs of graphs satisfying a certain matching condition.

Definition 5.41. Let $G_1 = (V_1, E_1)$ be an n_1-vertex graph which is d_1-regular and let $G_2 = (V_2, E_2)$ be a d_1-vertex graph which is d_2-regular. Since the number of vertices in G_2 is the same as the degree in G_1, we can identify V_2 with the edges emanating from a given vertex $u \in V_1$. Formally, we fix for every $u \in V_1$ a bijection $\pi_u : \{e \in E_1 : u \text{ is incident with } e\} \to V_2$. We also fix for every $a \in V_2$ a bijection between $[d_2] := \{1, \ldots, d_2\}$ and the multiset of the vertices adjacent to a in G_2 (*multiset* is a set in which elements are allowed to have positive integer multiplicities, we have to consider multisets here because we allow multiple edges). We denote this bijection by $\kappa_a : [d_2] \to \{b \in V_2 : \{a, b\} \in E_2\}$. The *zigzag product* $G_1 ⓩ G_2$ is the graph whose vertex set is $V_1 \times V_2$ and two vertices $(u, a), (v, b) \in V_1 \times V_2$ are joined by an edge if and only if there exist $i, j \in [d_2]$ such that:

$$uv \in E_1, \quad a = \kappa_{\pi_u(uv)}(i), \text{ and } b = \kappa_{\pi_v(uv)}(j). \tag{5.17}$$

We can view this construction as follows: think of the vertex set of $G_1 ⓩ G_2$ as a disjoint union of "clouds" which are copies of $V_2 = \{1, \ldots, d_1\}$ indexed by V_1. Thus (u, a) is the point indexed by a in the cloud labeled by u. Every edge $\{(u, a), (v, b)\}$ of $G_1 ⓩ G_2$ can be regarded as the result of a three-step walk:

- A "zig" step in G_2 from a vertex a belonging to the cloud of the vertex $u \in V_1$ to $\pi_u(\{u, v\})$ in the same cloud (for this to be defined a and $\pi_u(\{u, v\})$ should be adjacent in G_2).
- A "zag" step in G_1 from u's cloud to v's cloud along the edge $\{u, v\}$.
- A final "zig" step in G_2, in the cloud of the vertex v, from $\pi_v(\{u, v\})$ to b.

The graph $G_1 ⓩ G_2$ is a d_2^2-regular graph with $n_1 d_1$ vertices. The zigzag product depends on the choice of the bijections $\{\pi_u\}_{u \in V_1}$, and in fact different bijections (which are also called labelings) of the same graphs can produce non-isomorphic zigzag products. (The choice of bijections (labelings) $\{\kappa_a\}_{a \in V_2}$ does not affect the structure of the zigzag product but these bijections are useful in the subsequent analysis.) However, all of our results below will be independent of the actual choice of the labelings, so while our notation should formally depend on the labelings, for the sake of simplicity, we do not mention the choice of labelings in our notation.

Let us now examine how the constant $\gamma_+(G_1 ⓩ G_2, K)$ is related to $\gamma_+(G_1, K)$ and $\gamma_+(G_2, K)$, where $K : X \times X \to [0, \infty)$ is an arbitrary kernel.

Theorem 5.42 (Sub-multiplicativity of the reciprocal absolute spectral gap). *Let G_1, $G_1 = (V_1, E_1)$, be an n_1-vertex graph which is d_1-regular and let $G_2 = (V_2, E_2)$ be a d_1-vertex graph which is d_2-regular. Then for every kernel $K : X \times X \to [0, \infty)$,*

$$\gamma_+(G_1 ⓩ G_2, K) \le \gamma_+(G_1, K) \cdot \gamma_+(G_2, K)^2. \tag{5.18}$$

Proof. Take any $f, g : V_1 \times V_2 \to X$ and note that the definition of $\gamma_+(G_1, K)$ implies that for any pair $a, b \in V_2$ we have:

$$\frac{1}{n_1^2} \sum_{u,v \in V_1} K(f(u,a), g(v,b)) \le \frac{\gamma_+(G_1, K)}{n_1 d_1} \sum_{(u,v) \in E_1} K(f(u,a), g(v,b)). \quad (5.19)$$

We would like to explain the notation used in the right-hand side. It is adjusted to the matrix form of the definition of the reciprocal absolute spectral gap. In this notation we mean that for each edge $\{u, v\}$ which is not a loop there are two elements in E_1: (u, v) and (v, u), and so the sum in the right-hand side contains two terms: $K(f(u,a), g(v,b))$ and $K(f(v,a), g(u,b))$. For a loop $\{u, u\}$ there is one element (u, u) in E_1. Division by d_1 makes the sum $\frac{1}{d_1} \sum_{(u,v) \in E_1} K(f(u,a), g(v,b))$ the same as the sum $\sum_{u,v \in V_1} a_{u,v} K(f(u,a), g(v,b))$, where $a_{u,v}$ are entries of the normalized adjacency matrix of G_1.

Using (5.19) we get the following inequality

$$\begin{aligned} \frac{1}{|V_1 \times V_2|^2} & \sum_{(u,a),(v,b) \in V_1 \times V_2} K(f(u,a), g(v,b)) \\ &= \frac{1}{d_1^2} \sum_{a,b \in V_2} \frac{1}{n_1^2} \sum_{u,v \in V_1} K(f(u,a), g(v,b)) \\ &\le \frac{\gamma_+(G_1, K)}{n_1 d_1^3} \sum_{a,b \in V_2} \sum_{(u,v) \in E_1} K(f(u,a), g(v,b)). \end{aligned} \quad (5.20)$$

Now we show that the definition of $\gamma_+(G_2, K)$ implies that for all $u \in V_1$ and $b \in V_2$ we have

$$\begin{aligned} \frac{1}{d_1^2} & \sum_{a \in V_2} \sum_{\substack{v \in V_1: \\ (u,v) \in E_1}} K(f(u,a), g(v,b)) \\ &\le \frac{\gamma_+(G_2, K)}{d_1 d_2} \sum_{\substack{v \in V_1: \\ (u,v) \in E_1}} \sum_{i \in [d_2]} K\left(f(u, \kappa_{\pi_u(\{u,v\})}(i)), g(v,b)\right). \end{aligned} \quad (5.21)$$

To see that this inequality follows from the definition of $\gamma_+(G_2, K)$ we observe the following:

- For fixed u, $f(u,a)$ is an X-valued function of a.
- For fixed u and b, and for v satisfying $(u, v) \in E_1$, we may regard $g(v,b)$ as a function of $\pi_u(\{u, v\})$, which is a vertex in the u's cloud; and each vertex in the u's cloud is of this form.
- Hence the sum in the left-hand side can be regarded as a sum over all pairs of points of the u's cloud.

- Therefore we may use the definition of $\gamma_+(G_2, K)$ to estimate from above the sum in the left-hand side.

- After using the definition of $\gamma_+(G_2, K)$ we get the sum shown in the right-hand side because (as we have already mentioned) the term $g(v,b)$ can be regarded as a value of an X-valued function at $\pi_u(\{u,v\})$, and thus the sum

$$\sum_{\substack{v\in V_1:\\ (u,v)\in E_1}} \sum_{i\in[d_2]} K\left(f(u,\kappa_{\pi_u(\{u,v\})}(i)), g(v,b)\right)$$

is actually the sum over all pairs $(\pi_u(\{u,v\}), \text{neighbor of } \pi_u(\{u,v\}))$, that is, over the set of edges of G_2.

Summing (5.21) over $u \in V_1$ and $b \in V_2$ and plugging it into (5.20), yields the bound:

$$\begin{aligned}
&\frac{1}{|V_1 \times V_2|^2} \sum_{(u,a),(v,b)\in V_1\times V_2} K(f(u,a), g(v,b)) \\
&\quad \le \frac{\gamma_+(G_1,K)}{n_1 d_1} \cdot \frac{1}{d_1^2} \sum_{b\in V_2} \sum_{u\in V_1} \sum_{a\in V_2} \sum_{\substack{v\in V_1:\\ (u,v)\in E_1}} K(f(u,a), g(v,b)) \\
&\quad \le \frac{\gamma_+(G_1,K)}{n_1 d_1} \sum_{b\in V_2} \sum_{u\in V_1} \frac{\gamma_+(G_2,K)}{d_1 d_2} \sum_{\substack{v\in V_1:\\ (u,v)\in E_1}} \sum_{i\in[d_2]} K\left(f(u,\kappa_{\pi_u(\{u,v\})}(i)), g(v,b)\right) \\
&\quad = \frac{\gamma_+(G_1,K)\gamma_+(G_2,K)}{n_1 d_1^2 d_2} \sum_{v\in V_1} \sum_{i\in[d_2]} \sum_{\substack{u\in V_1:\\ (u,v)\in E_1}} \sum_{b\in V_2} K\left(f(u,\kappa_{\pi_u(\{u,v\})}(i)), g(v,b)\right).
\end{aligned} \tag{5.22}$$

Another application of the definition of $\gamma_+(G_2, K)$ implies that for all $v \in V_1$ and $i \in [d_2]$ we have:

$$\begin{aligned}
&\frac{1}{d_1^2} \sum_{\substack{u\in V_1:\\ (u,v)\in E_1}} \sum_{b\in V_2} K\left(f(u,\kappa_{\pi_u(\{u,v\})}(i)), g(v,b)\right) \\
&\quad \le \frac{\gamma_+(G_2,K)}{d_1 d_2} \sum_{\substack{u\in V_1:\\ (u,v)\in E_1}} \sum_{j\in[d_2]} K\left(f(u,\kappa_{\pi_u(\{u,v\})}(i)), g(v,\kappa_{\pi_v(\{u,v\})}(j))\right).
\end{aligned} \tag{5.23}$$

Here we use the following reasoning: for fixed v, $g(v,b)$ might be considered as a function of b. On the other hand, for fixed i the function $f(u,\kappa_{\pi_u(\{u,v\})}(i))$ might be considered as a function of $\pi_v(\{u,v\})$. Hence we get a function on the Cartesian product of v's cloud with itself. Now we use the fact that $\{\kappa_{\pi_v(\{u,v\})}(j)\}_{j\in[d_2]}$ is just the list of neighbors of $\pi_v(\{u,v\})$.

Summing (5.23) over $v \in V_1$ and $i \in [d_2]$, and combining the resulting inequality with (5.22) yields the desired inequality:

$$\begin{aligned}
&\frac{1}{|V_1 \times V_2|^2} \sum_{(u,a),(v,b) \in V_1 \times V_2} K(f(u,a), g(v,b)) \\
&\leq \frac{\gamma_+(G_1,K)\gamma_+(G_2,K)}{n_1 d_1^2 d_2} \sum_{v \in V_1} \sum_{i \in [d_2]} \sum_{\substack{u \in V_1: \\ (u,v) \in E_1}} \sum_{b \in V_2} K\left(f(u, \kappa_{\pi_u(\{u,v\})}(i)), g(v,b)\right) \\
&\leq \frac{\gamma_+(G_1,K)\gamma_+(G_2,K)^2}{n_1 d_1 d_2^2} \\
&\quad \times \sum_{v \in V_1} \sum_{i \in [d_2]} \sum_{\substack{u \in V_1: \\ (u,v) \in E_1}} \sum_{j \in [d_2]} K\left(f(u, \kappa_{\pi_u(\{u,v\})}(i)), g(v, \kappa_{\pi_v(\{u,v\})}(j))\right) \\
&= \frac{\gamma_+(G_1,K)\gamma_+(G_2,K)^2}{n_1 d_1 d_2^2} \sum_{\{(u,a),(v,b)\} \in E(G_1 ⓩ G_2)} K\left(f(u,a), g(v,b)\right)). \qquad \square
\end{aligned}$$

Combining Theorem 5.42 and Proposition 5.37 we get

Corollary 5.43. *In the special case $X = \mathbb{R}$ and $K(x,y) = (x-y)^2$, the inequality* (5.18) *becomes:*

$$\frac{1}{1 - \lambda(G_1 ⓩ G_2)} \leq \frac{1}{1 - \lambda(G_1)} \cdot \frac{1}{(1 - \lambda(G_2))^2}. \tag{5.24}$$

Now we use Corollary 5.43 to construct expander families.

Definition 5.44. For a graph $G = (V, E)$ and for $t \in \mathbb{N}$, let G^t be the graph with the vertex set V in which an edge between $u, v \in V$ is drawn for every walk in G of length t which starts at u and terminates at v. We call G^t the t-th *power of the graph* G.

For our next observation it is important that loops contribute 1 to the degree of a vertex.

Observation 5.45. It is easy to see that $A_{G^t} = (A_G)^t$, therefore $\lambda(G^t) = (\lambda(G))^t$. Also it is easy to see that G^t is d^t-regular if G is d-regular.

Theorem 5.46. *The zigzag product construction leads to a construction of families of expanders if there is a regular graph G satisfying the following conditions:*

(a) *The number n_0 of its vertices and its degree d_0 satisfy $n_0 = d_0^{2t_0}$, where $t_0 \in \mathbb{N}$ is a number satisfying the following condition.*

(b) *There exists $a \in (0,1)$ satisfying $\left(1 - (1-a)^{t_0}\right)(1 - \lambda(G))^2 \geq a$.*

Proof. In fact, these conditions allow to construct expanders inductively as follows:

(a) We start with $G_1 = G^2$. (We square G because Definition 5.41 implies that all further graphs in our construction are d_0^2-regular, so we need G_1 also to be d_0^2-regular.)

(b) We let $G_2 = G_1^{t_0} ⓩ G$. This graph is well-defined because $G_i^{t_0}$ is $d_0^{2t_0}$-regular (=n_0-regular) and G has n_0 vertices. By Definition 5.41, G_2 is d_0^2-regular and has n_0^2-vertices.

(c) We repeat this construction in all of the further steps, namely, we let $G_{i+1} = G_i^{t_0} ⓩ G$ for $i \geq 2$. It is easy to verify that each G_n has n_0^i vertices and is d_0^2-regular.

As for the absolute spectral gaps, combining inequality (5.24) and Observation 5.45 we get

$$1 - \lambda(G_{i+1}) \geq (1 - (\lambda(G_i))^{t_0})(1 - \lambda(G))^2. \tag{5.25}$$

Now we use induction to prove that absolute spectral gaps of all graphs G_i are at least a.

The spectral gap of G_1 is $1 - \lambda(G^2) = 1 - (\lambda(G))^2 \geq (1 - \lambda(G))^2 > a$.

If we assume that the spectral gap of G_i is at least a, for the spectral gap of G_{i+1} we get

$$1 - \lambda(G_{i+1}) \geq (1 - (\lambda(G_i))^{t_0})(1 - \lambda(G))^2 \geq (1 - (1 - a)^{t_0})(1 - \lambda(G))^2 \geq a. \quad \square$$

Definition 5.47. A graph G satisfying the assumptions of Theorem 5.46 is called a *base graph.*

Our next purpose is to present one of the known constructions of a base graph. Let $r, s \in \mathbb{N}$ and $q = 2^s$. Let $\mathbb{F}_q$ be the field with q elements and let $\mathbb{F}_q^{r+1}$ be the $(r + 1)$-dimensional vector space over this field. We introduce a q^2-regular graph $G(q, r)$ whose vertex set is $\mathbb{F}_q^{r+1}$. The edge set of this graph is described as follows. For each $a = (a_0, a_1, \dots, a_r) \in \mathbb{F}_q^{r+1}$ the set of all vertices adjacent to a is the set of vertices of the form

$$a + y \cdot (1, x, x^2, \dots, x^r),$$

where (x, y) is any pair of elements of $\mathbb{F}_q$.

Theorem 5.48. *For any $q = 2^s$ and any $r < q$ the obtained graph satisfies*

$$\lambda_2(G(q, r)) \leq qr.$$

Proof. The vector space $\mathbb{F}_q^{r+1}$ has $n = 2^{s(r+1)}$ elements. Let M be the $n \times n$ adjacency matrix of the graph $G(q, r)$.

We need to recall the easy fact that $\mathbb{F}_q$ might be considered as a vector space over the field $\mathbb{F}_2$, and so there exists a surjective and linear over $\mathbb{F}_2$ map $L : \mathbb{F}_q \to \mathbb{F}_2$.

Let us describe eigenvectors of M considered as an operator on $\mathbb{R}^n$. Both the eigenvectors and their entries will be labeled by elements of $\mathbb{F}_q^{r+1}$. So, for every $a = (a_0, \dots, a_r) \in \mathbb{F}^{r+1}$, let v_a be the vector whose b-th entry, where $b = (b_0, \dots, b_r) \in \mathbb{F}^{r+1}$, satisfies $v_a(b) = (-1)^{L(\sum_{i=0}^r a_i b_i)}$. Here $(-1)^\ell$, where $\ell \in \mathbb{F}_2$ is understood as $(-1)^0 = 1$, $(-1)^1 = -1$.

Let us show that the vectors $\{v_a\}_{a\in\mathbb{F}_q^{r+1}}$ are pairwise orthogonal. In fact,

$$\begin{aligned}\langle v_a, v_c\rangle &= \sum_{b\in\mathbb{F}_q^{r+1}} (-1)^{L(\sum_{i=0}^r a_i b_i)}(-1)^{L(\sum_{i=0}^r c_i b_i)}\\ &= \sum_{b\in\mathbb{F}_q^{r+1}} (-1)^{L(\sum_{i=0}^r (a_i+c_i) b_i)}.\end{aligned} \tag{5.26}$$

It is clear that if $a_i = c_i$ for all i, then $L(\sum_{i=0}^r (a_i + c_i)b_i) = L(\sum_{i=0}^r a_i b_i + \sum_{i=0}^r a_i b_i) = 0$ since L is linear over $\mathbb{F}_2$. Therefore $\langle v_a, v_a\rangle = 2^{s(r+1)}$.

Now we consider the case where there is $j \in \{0, \dots, r\}$ such that $a_j \neq c_j$. We can rewrite the last line in (5.26) as

$$\sum_{b_j\in\mathbb{F}_q} (-1)^{L((a_j+c_j)b_j)} \sum_{b\in\mathbb{F}_q^r} (-1)^{L(\sum_{i=0,i\neq j}^r (a_i+c_i)b_i)}.$$

The first sum in this product is equal to 0 because the mapping L is surjective and thus is equal to 0 on half of the elements of $\mathbb{F}_q$ and is equal to 1 on the other half. On the other hand, since $c_j + a_j \neq 0$ each element of $\mathbb{F}_q$ has a unique representation in the form $(a_j + c_j)b_j$.

Let us show that $\{v_a\}_{a\in\mathbb{F}_q^{r+1}}$ are eigenvectors of M. Observe that entries of M are naturally labeled by pairs of elements of $\mathbb{F}_q^{r+1}$. We have

$$\begin{aligned}(Mv_a)(b) = \sum_{c\in\mathbb{F}_q^{r+1}} M_{bc} v_a(c) &= \sum_{x,y\in\mathbb{F}_q} v_a(b + y(1, x, \dots, x^r))\\ &= \left(\sum_{x,y\in\mathbb{F}_q} v_a(y, yx, \dots, yx^r)\right)\cdot v_a(b),\end{aligned}$$

where we used the observation: $v_a(b + c) = v_a(b)\cdot v_a(c)$ for any $b, c \in \mathbb{F}_q^{r+1}$. Therefore v_a is an eigenvector, and the corresponding eigenvalue is equal to

$$\lambda_a = \sum_{x,y\in\mathbb{F}_q} v_a(y, yx, \dots, yx^r).$$

We have

$$\lambda_a = \sum_{x,y\in\mathbb{F}} (-1)^{L(a_0 y + a_1 yx + \dots + a_r yx^r)} = \sum_{x,y\in\mathbb{F}} (-1)^{L(y\cdot p_a(x))},$$

where $p_a(x) = \sum_{i=0}^{r} a_i x^i$. Therefore

$$\lambda_a = \sum_{x,y\in\mathbb{F}_q,\ p_a(x)=0} (-1)^{L(y\cdot p_a(x))} + \sum_{x,y\in\mathbb{F}_q,\ p_a(x)\neq 0} (-1)^{L(y\cdot p_a(x))}.$$

From here it is clear that if $a = (0, 0, \dots, 0)$, then $\lambda_a = q^2$.

Now consider the case where $a \neq (0, 0, \dots, 0)$. In this case there are at most r elements $x \in \mathbb{F}_q$ such that $p_a(x) = 0$ and therefore the first sum is a positive integer $\leq qr$. As for the second sum, it is equal to zero because $\sum_{y\in\mathbb{F}_q} (-1)^{L(y\cdot z)} = 0$ for each $z \neq 0$. (This can be proved using the same argument which we used above: Since L is surjective it takes value 0 on exactly half of the elements of $\mathbb{F}_q$; since $z \neq 0$, $y \mapsto y \cdot z$ is a bijection of $\mathbb{F}_q$.) This completes the proof of the theorem. □

It remains to check that the family of graphs whose spectral gaps are estimated in Theorem 5.48 contains a suitable base graph.

The graphs constructed in Theorem 5.48 have q^{r+1} vertices and are q^2-regular. So the relation $n_0 = d_0^{2t_0}$ of Theorem 5.46(a) is satisfied for some t_0 if and only if $r + 1$ is divisible by 4, that is, $r + 1 = 4t_0$, so $q^{r+1} = q^{4t_0} = (q^2)^{2t_0}$.

Observe that by Theorem 5.48, $\lambda(A_G)$, where A_G is the normalized adjacency matrix of $G(q, r)$ (see Definitions 5.35 and 5.40), is equal $\frac{r}{q}$. Therefore the condition (b) of Theorem 5.46 becomes: there exists $a \in (0, 1)$ satisfying

$$\left(1 - (1-a)^{\frac{r+1}{4}}\right)\left(1 - \frac{r}{q}\right)^2 \geq a.$$

This condition is satisfied for an arbitrary $a \in (0, 1)$ if we pick q and r in a suitable way. One of the ways is the following:

- First we pick r to be large enough so that

$$\left(1 - (1-a)^{\frac{r+1}{4}}\right) \geq \sqrt{a}.$$

- Then we pick q large enough so that

$$\left(1 - \frac{r}{q}\right)^2 \geq \sqrt{a}.$$

5.6 Graphs with large girth: basic definitions

Definition 5.49. The *girth* of a graph is the length of a shortest cycle in it. (We define girth for graphs having some cycles only.) We denote the girth of a graph G by $g(G)$.

We would like to mention that a loop (an edge joining a vertex to itself) is considered as a cycle of length 1, and that a pair of parallel edges (that is, edges with the same pair of end vertices) are considered as forming a cycle of length 2. Thus, a graph has girth at least 3 if and only if it is simple.

As with expanders, talking about graphs with large girth we usually mean families $\{G_n\}_{n=1}^{\infty}$ of graphs with $g(G_n) \to \infty$. The simplest such family is a family of cycles with growing length. For metric theory this family is not very interesting because of its simplicity (only precise isometric results could be difficult).

Our main interest is the following types of families of graphs with large girth:

(a) k-regular, $k \geq 3$, families $\{G_n\}_{n=1}^{\infty}$ of graphs with $g(G_n) \to \infty$.

(b) Families $\{G_n\}_{n=1}^{\infty}$ of graphs with $g(G_n) \to \infty$ and all degrees ≥ 3.

(c) Families $\{G_n\}_{n=1}^{\infty}$ of graphs with $g(G_n) \to \infty$ and average degrees of G_n at least $2 + \varepsilon$ for some $\varepsilon > 0$.

As we shall see in Chapter 8 such families do not admit uniformly bilipschitz embeddings into a Hilbert space and even into any uniformly convex Banach space. However, as we shall see in this chapter, some such families admit uniformly bilipschitz embeddings into ℓ_1. There are still several important intriguing open problems about families of graphs with growing girth, see Section 11.3 and Problem 11.4 in Chapter 11.

The described types of families of graphs with large girth are such that it is far from being clear that such families exist, especially families of type (a).

However, it is easy to find k-regular graph of girths 3 and 4: The complete graph K_{k+1} is a k-regular graph of girth 3 and the Hamming cube $\mathbb{F}_2^k$ (see the definition in Exercise 5.82 on page 181) is a k-regular graph of girth 4. As we shall see these simple examples could serve as a basis for construction of k-regular graphs with arbitrarily large girth.

In further sections of this chapter we present some of the known constructions of families of graphs with growing girth. Important parameters of such constructions are: *order* (defined as the number of vertices) and diameter of the graphs. One of the goals of some of the constructions is to minimize the order (or diameter) for a given girth.

5.7 Graph lift constructions and ℓ_1-embeddable graphs with large girth

Very simple constructions of k-regular graphs with arbitrarily large girth are the constructions based on the notion of a lift of a graph, also known as *graph covering*. We prefer the term *graph lift* because the term *graph covering* is also used for many other purposes.

Definition 5.50. Let L be a finite set. A *lift* $\widetilde{G}$ of a graph $G = (V(G), E(G))$ is a graph with vertex set $V(\widetilde{G}) = V(G) \times L$. The edge set $\widetilde{G}$ is the union of matchings corresponding to edges of $E(G)$. The matching corresponding to an edge uv matches all vertices of $\{u\} \times L$ with vertices of $\{v\} \times L$.

Observe that to be more formal we need to consider edges uv as directed edges and to specify to each of such edges a permutation of L, and the permutation corresponding to the direction for which v is the head and u is the tail is the inverse of the permutation corresponding to the direction for which u is the head and v is the tail.

Definition 5.50 immediately implies that there are well-defined projections

$$E(\widetilde{G}) \to E(G) \text{ and } V(\widetilde{G}) \to V(G):$$

edges of the matching corresponding to uv are projected onto uv and vertices of $\{u\} \times L$ are projected onto u. We denote both of the projections by π. It is clear from the definition that the degrees of all vertices in $\widetilde{G}$ whose projection in G is u are the same as the degree of u. In particular, any lift of a k-regular graph is k-regular.

Remark 5.51. It is easy to see that for each walk $\{e_i\}_{i=1}^n$ in G and each vertex $\widetilde{u} \in V(\widetilde{G})$ of the form $\widetilde{u} = (u, \ell)$ with $\ell \in L$ and u being the initial vertex of the walk $\{e_i\}_{i=1}^n$; there is a uniquely determined *lifted walk* $\{\widetilde{e}_i\}_{i=1}^n$ in $\widetilde{G}$ for which $\pi(\widetilde{e}_i) = e_i$ and $\widetilde{u}$ is the initial vertex. It is important for us that if G is connected, this remark remains true if instead of $\widetilde{G}$ we consider a connected component of it.

Remark 5.52. It is clear that if a walk in G has an edge e which is *backtracked* (that is, the walk contains two consecutive edges e), then the corresponding edge in the lifted walk is also backtracked.

Remark 5.52 implies that the projection of a cycle in $\widetilde{G}$ to G cannot be such that its edges induce in G a subgraph having vertices of degree 1. In particular, the graph induced by edges of the projection of a cycle in $\widetilde{G}$ contains cycles in G. This immediately implies $g(\widetilde{G}) \geq g(G)$.

Remark 5.52 can be used to find relatively simple graph-lift-based constructions of k-regular graphs $k \geq 3$ with arbitrarily large girth.

Example 5.53. We start with an arbitrary simple k-regular graph G (example of such a graph: K_{k+1}). We consider $L = \{0,1\}^{E(G)}$, that is, L is the set of all $\{0,1\}$-valued functions on the edge set of G. The permutation of L corresponding to $e \in E(G)$ maps each function f on $E(G)$ to the function h, which has the same values as f everywhere except the edge e, and on the edge e its value is the other one. (Recall that we consider $\{0,1\}$-valued functions. Observe that in this construction we do not have a need to consider directed edges: the change of "level" in L is the same in both directions.)

Claim 5.54. The girth of the obtained graph lift $\widetilde{G}$ is $\geq 2g(G)$.

Remark 5.55. (1) The obtained graph lift $\widetilde{G}$ does not have to be connected (and usually is disconnected). This is not a problem: if we need a connected k-regular graph of girth $\geq 2g(G)$, we consider a connected component of $\widetilde{G}$.
(2) One can check that actually $g(\widetilde{G}) = 2g(G)$.

Proof of Claim 5.54. Consider a shortest cycle C in $\widetilde{G}$. The comment after Remark 5.52 implies that its projection $\pi(C)$ to G contains a cycle C_0 in G. This cycle has length $\geq g(G)$. We consider C as a closed walk: we initialize and terminate it at an arbitrary vertex of C. Then $\pi(C)$ is a closed walk in G containing each edge of C_0. It is clear that the lift of the walk $\pi(C)$ is C. Since the lift of $\pi(C)$ to $\widetilde{G}$ is also a closed walk, each of the edges of C_0 should be contained in $\pi(C)$ an even number of times (otherwise, the walk would end on a different "level", see the description of Example 5.53), so at least twice. Hence the length of C is $\geq 2g(G)$. □

A similar graph lift construction can be used to find an example of a sequence $\{G_n\}$ of connected k-regular graphs with $g(G_n) \to \infty$ and uniformly bounded ℓ_1-distortions. Before proceeding to this construction we need to mention the following observation on minimum-diameter k-regular graphs of girth $\geq g$ (we know, by Example 5.53, that k-regular graphs of girth $\geq g$ exist for all values of k and g).

Proposition 5.56. *Let H be a k-regular graph with girth $\geq g$ and such that H has the minimal amount of vertices among all such graphs. Then the diameter of H is $\leq g$.*

Proof. Suppose the diameter of H is $> g$, and let u and v be two vertices in H with $d_H(u,v) \geq g+1$. Let $U = \{u_1,\dots,u_k\}$ and $V = \{v_1,\dots,v_k\}$ be the sets of all vertices adjacent to u and v, respectively. We delete u and v as well as all edges incident to u and v from H, and denote the obtained graph D. Next we add edges $u_1v_1,\dots u_kv_k$ to D and denote the obtained graph $\widehat{H}$.

Observe that

$$d_H(u_i,v_j) \geq g-1 \quad \forall i,j \in \{1,\dots,k\}. \tag{5.27}$$

Therefore $\widehat{H}$ is a simple k-regular graph. It is clear that $\widehat{H}$ has fewer vertices than H. Thus if we prove that $g(\widehat{H}) \geq g$, we get a contradiction and complete the proof of the proposition.

To achieve this goal we need to show that

$$d_{\widehat{H}\backslash u_iv_i}(u_i,v_i) \geq g-1,$$

where $\widehat{H}\backslash u_iv_i$ is the graph $\widehat{H}$ from which we deleted the edge u_iv_i. To show this we assume the contrary. Then, by (5.27), the shortest u_iv_i-path in $\widehat{H}\backslash u_iv_i$ should contain one of the edges u_jv_j $(i \neq j)$.

Let a u_iw_1-path be the beginning of this path until the first edge of it which is not in D and let a w_2v_i-path be the end of this path after the last edge of it which is not in D. Now we observe that for $i \neq j$ both $d_D(u_i,u_j) \geq g-2$ and $d_D(v_i,v_j) \geq g-2$ (we use $g(H) \geq g$ and the fact that we deleted paths of length 2 joining u_i with u_j, and v_i with v_j). Together with (5.27) this implies that the length of the $u_i v_i$-path which we consider is at least $2(g-2)+1 = 2g-3$ which is at least g if $g \geq 3$. □

Now we turn to a construction of a family of graphs with indefinitely growing girths and uniformly bounded ℓ_1-distortions. It is worth mentioning that it will be proved in Chapter 8 that families of k-regular, $k \geq 3$ graphs with indefinitely growing girths do not admit uniformly bilipschitz embeddings into a uniformly convex Banach space.

Theorem 5.57. *For each $k \geq 3$ there exists a family $\{\widetilde{G}_n\}_{n=1}^{\infty}$ of k-regular graphs with $g(\widetilde{G}_n) \to \infty$ and uniformly bounded ℓ_1-distortions.*

(Our notation hints that the desired graphs will be obtained using the graph lifts.)

Remark 5.58. There exist many parallels between families of graphs of two types: families of expanders and families of k-regular graphs of indefinitely growing girth. We shall see in Chapter 8 that both types of families do not admit uniformly bilipschitz embeddings into any (fixed) uniformly convex Banach space. There are many other parallels which one can see in many parts of this book. These parallels even led to a hope (which seems to never be mentioned in print) that each family of k-regular graphs with indefinitely growing girths contain some family of expanders in some weak but meaningful sense, and so this parallel extends forever. Comparing Theorem 5.57 with Theorem 4.9 we see that such hopes are not justified.

Proof. We start with a sequence of k-regular connected graphs $\{G_n\}$ with indefinitely increasing girths $g(G_n)$, such that

$$g(G_n) \geq \alpha \operatorname{diam}(G_n) \tag{5.28}$$

for some constant $\alpha > 0$. We know, by Example 5.53 and Proposition 5.56, that such families exist for $\alpha = 1$.

We apply the graph lift construction to the graphs $\{G_n\}$. The fact that we get k-regular graphs with indefinitely increasing girths follows immediately from the observations which we made immediately after the definition of a graph lift (see page 156). It remains to specify lifts for which there are suitable estimates for ℓ_1 distortions of connected components of the obtained graphs. The bounds for the distortions which we get are in terms of the constant α in (5.28).

For each $G \in \{G_n\}_{n=1}^{\infty}$ we do the following. Let L be the set $\{0,1\}^{E(G)}$, so each element of L can be regarded as a $\{0,1\}$-valued function on $E(G)$. For each $uv \in E(G)$ we need to specify a perfect matching of $\{u\} \times L$ and $\{v\} \times L$. To specify the perfect matching it suffices, for each edge in E, to pick a bijection of the set L. We do this as follows. The bijection corresponding to $e \in E(G)$ maps each function f on $E(G)$ to the function h, which has the same values as f everywhere except the edge e, and on the edge e its value is the other one. (Recall that we consider $\{0,1\}$-valued functions. Observe that in this construction we do not have a need to consider directed edges: the change of "level" in L is the same in both directions.)

We denote the graphs obtained from $\{G_n\}$ using such lifts by $\{\widehat{G}_n\}$. Observe that the graphs $\{\widehat{G}_n\}$ are disconnected. To see this consider two vertices: (v, f_1) and (v, f_2), where $v \in V(G_n)$, $f_1, f_2 \in \{0,1\}^{E(G_n)}$, $f_1 \neq f_2$. If (v, f_1) and (v, f_2) are connected in $\widehat{G}_n$ by a path P, it is clear that the projection $\pi(P)$ of this path is a closed walk in G which contains each of the edges corresponding to different values of f_1 and f_2 an odd number of times. Each of the other edges contained in $\pi(P)$ is contained there an even number of times. This implies that degrees of all vertices in the subgraph of G_n spanned by edges at which f_1 and f_2 have different values are even (we give a detailed proof of a more general statement in Lemma 5.61 below). Therefore (v, f_1) and (v, f_2) are not connected in $\widehat{G}_n$ if the condition of the last sentence does not hold.

For this reason (ℓ_1-distortions are well-defined for connected graphs only) we pick in each of $\widehat{G}_n$ a connected component which we denote $\widetilde{G}_n$. It is easy to see that a connected component of a k-regular graph with girth g is a k-regular graph with girth $\geq g$. We are going to show that the graphs $\{\widetilde{G}_n\}$ satisfy the conditions of Theorem 5.57.

The main steps in our proof are presented as lemmas, where G is one of the $\{G_n\}$. Recall that an *edge cut* in a graph is any set of all edges joining two complementary nonempty sets of vertices in the graph.

Lemma 5.59. *For each edge $e \in E(G)$ the set of all edges $\widetilde{e} \in E(\widetilde{G})$ for which $\pi(\widetilde{e}) = e$ forms an edge cut in $\widetilde{G}$.*

Proof. Recall that G is assumed to be connected. Combining this with Remark 5.51 we get that for each $e \in E(G)$ and each $v \in V(G)$ the sets $\{\widetilde{e} \in E(\widetilde{G}) : \pi(\widetilde{e}) = e\}$ and $\{\widetilde{v} \in V(\widetilde{G}) : \pi(\widetilde{v}) = v\}$ are nonempty.

Now let $e \in E(G)$. We can just describe the sets separated by the set of edges $\{\widetilde{e} \in E(\widetilde{G}) : \pi(\widetilde{e}) = e\}$: they are the sets $(V(G) \times A_{e,0}) \cap V(\widetilde{G})$ and $(V(G) \times A_{e,1}) \cap V(\widetilde{G})$ where $A_{e,0}$ and $A_{e,1}$ are the sets of functions in $\{0,1\}^{E(G)}$ whose values on e are equal to 0 and 1, respectively.

The only thing which is not obvious is that the sets $(V(G) \times A_{e,0}) \cap V(\widetilde{G})$ and $(V(G) \times A_{e,1}) \cap V(\widetilde{G})$ are nonempty. Let $e = uv$ and let $\widetilde{v} \in V(\widetilde{G})$ be such that $\pi(\widetilde{v}) = v$. Then $\widetilde{v}$ is in one of the sets $(V(G) \times A_{e,0}) \cap V(\widetilde{G})$ or $(V(G) \times A_{e,1}) \cap V(\widetilde{G})$. We lift the one-edge-walk e to $\widetilde{G}$ starting at $\widetilde{v}$. The other end of this walk is in the other set of the pair $(V(G) \times A_{e,0}) \cap V(\widetilde{G})$ and $(V(G) \times A_{e,1}) \cap V(\widetilde{G})$. □

Let $\ell_1(E(G))$ be the space of real-valued functions on $E(G)$ with its ℓ_1-norm. By the *ℓ_1-norm* of a function $H : E(G) \to \mathbb{R}$ we mean the norm

$$\|H\|_1 = \sum_{e \in E(G)} |H(e)|.$$

It is easy to see that the space $\ell_1(E(G))$ is isometric to a subspace of ℓ_1 and therefore to prove Theorem 5.57 it suffices to find an embedding F of $V(\widetilde{G})$ into $\ell_1(E(G))$ with distortion bounded from above by a universal constant.

For each edge cut $R(e)$ defined by the set of edges $\widetilde{e}$ in $\widetilde{G}$ satisfying $\pi(\widetilde{e}) = e$, we call one of the sides of the cut $R(e)$ the 0-*side*, and the other side the 1-*side*, and define a function $F : V(\widetilde{G}) \to \ell_1(E(G))$ by

$$(F(x))(e) = \begin{cases} 1 & \text{if } x \text{ is in the 1-side of } R(e) \\ 0 & \text{if } x \text{ is in the 0-side of } R(e). \end{cases}$$

The Lipschitz constant of this embedding is 1. In fact, the cuts $R(e)$ are disjoint and each edge of $\widetilde{G}$ is in exactly one of the cuts. Therefore $\|F(x) - F(y)\|_1 = 1$ if x and y are adjacent vertices of $\widetilde{G}$.

To estimate the Lipschitz constant of F^{-1} we consider $x, y \in V(\widetilde{G})$, denote by $d_{\widetilde{G}}(x, y)$ their distance in $\widetilde{G}$, and observe the following:

Observation 5.60. If P is an xy-walk in $\widetilde{G}$, then $d_{\widetilde{G}}(x, y) \le \text{length}(P)$ and $\|F(x) - F(y)\|_1$ is the number of edges in the walk $\pi(P)$ which are repeated in the walk an odd number of times.

Let us denote by $D(P)$ the number of edges repeated in $\pi(P)$ an odd number of times. Observation 5.60 shows that to complete the proof of Theorem 5.57 it suffices, for each $x, y \in V(\widetilde{G})$, to find an xy-walk P in $\widetilde{G}$ for which

$$\text{length}(P) \le \beta D(P) \tag{5.29}$$

for some absolute constant β. This is our next goal.

Let $x = (u, f)$ and $y = (v, g)$ be two vertices of $\widetilde{G}$, where $u, v \in V(G)$ and $f, g \in \{0, 1\}^{E(G)}$. Let $S \subset E(G)$ be the subset on which the functions f and g differ (recall that we consider f and g as $\{0, 1\}$-valued functions on $E(G)$). Denote by H the subgraph of G induced by edges of S.

Lemma 5.61. *If $u \neq v$, then degrees of all vertices of H, except u and v, are even. If $u = v$, then degrees of all vertices of H are even.*

Proof. We use the assumption that $\widetilde{G}$ is connected, so there is an xy-walk Q in $\widetilde{G}$. We claim that the walk $\pi(Q)$ has to contain each of the edges of S an odd number of times and each of the other edges an even number of times (possibly 0).

To see this we recall our construction of the lift of G and our definition of a lifted walk (see Remark 5.51). It is obvious that Q is a lifted walk of $\pi(Q)$. Our definitions are such that the change in the L-coordinate in each step (when we walk along the lifted walk) is made in exactly one value of the corresponding $\{0, 1\}$-valued function on $E(G)$, the choice of this coordinate depends only on the π-projection of the edge which we are passing, and not on the direction in which we pass it, or on the L-coordinate of the vertex we are at (this is a very important property of the graph lift which we consider). Also, we need an obvious observation that if we change some value of a $\{0, 1\}$-valued function twice, it returns to its original value.

We denote the subgraph of G induced by edges of $\pi(Q)$ by $I(Q)$ ($\pi(Q)$ and $I(Q)$ are slightly different objects: $\pi(Q)$ is a sequence of edges in which some edges can be repeated, $I(Q)$ is the subgraph of G induced by edges which are listed in $\pi(Q)$ at least once). Now we introduce a non-simple graph $N(Q)$ having $I(Q)$ as its underlying simple graph and having as many parallel edges for each edge of $I(Q)$, as many times the edge is repeated in $\pi(Q)$.

It is clear that the graph $N(Q)$ contains an Euler trail which is a uv-walk (to get it, we can just follow the walk $\pi(Q)$, each time using a different parallel edge for edges repeated in $\pi(Q)$). Therefore, if $u \neq v$ degrees of all vertices of $N(Q)$, except u and v, are even. If $u = v$ degrees of all vertices of $N(Q)$ are even. It is clear that if we delete from $N(Q)$ an even number of parallel edges, this property continues to hold. In particular, it continues to hold if we delete all edges parallel to edges repeated in $\pi(Q)$ an even number of times, and leave one copy of each edge repeated in $\pi(Q)$ an odd number of times. It is clear that what we get after this deletion is the graph H. □

To complete the proof of Theorem 5.57 it remains to construct an xy-walk P satisfying (5.29). We use the graph H introduced above. Lemma 5.61 implies that all components of H, except possibly the one that contains u and v, contain cycles. Therefore each of them has at least $g(G)$ edges. It is also clear that the component containing u and v (if it exists) has an Euler trail whose initial vertex is u and whose terminal vertex is v; and all other components have Euler tours (that is, closed Euler trails).

Let $H_1, \dots, H_t$ be the components of H. We assume that H_1 contains u and v. Observe that in the case where $u = v$ such a component does not have to exist. In this case we introduce H_1 as a trivial component containing one vertex $u = v$. Observe that we may assume that this trivial component contains an Euler trail joining u and v, it is just a trail with no edges.

Observe that any two components of H can be joined by a path in G of length $\leq \operatorname{diam}(G)$. Let $M_1, \dots, M_{t-1}$ be paths of length $\leq \operatorname{diam}(G)$ each, such that

(a) M_i joins H_i and H_{i+1}.

(b) The terminal vertex of M_i coincides with the initial vertex of M_{i+1}.

Now we form the following uv-walk M in G:

- It starts at u and follows an Euler trail of H_1 to the initial vertex of M_1.
- It follows M_1 to H_2.
- It follows the Euler tour of H_2.

- It follows M_2 to H_3.
- It continues in an obvious way to H_t.
- It follows the Euler tour of H_t.
- It follows M_{t-1} back to H_{t-1}.
- It follows M_{t-2} back to H_{t-2}.
- It continues in an obvious way until it reaches H_1.
- It follows the final part of the Euler trail of H_1 (the initial part of that Euler trail was followed in the first step) and completes it at v.

We lift the walk M taking $x = (u, f)$ as the initial vertex of the lifted walk. Denote the obtained walk by P. It is clear that all edges of S are used in M an odd number of times and that all other edges are used in M an even number of times. Therefore the lifted walk P has $y = (v, g)$ as its terminal vertex (to see this we use observations made in the proof of Lemma 5.61). So P is an xy-walk. Also it is clear that $D(P) = |S|$.

Counting the number of edges in P, we get that its length does not exceed

$$2(t-1)\text{diam}(G) + |E(H_1)| + |E(H_2)| + \cdots + |E(H_t)| \\ = 2(t-1)\text{diam}(G) + |S|. \quad (5.30)$$

In the case where $t = 1$, we use the right-hand side of (5.30) and get

$$\text{length}(P) \le D(P),$$

so (5.29) holds with $\beta = 1$. In the case where $t > 1$ we use the left-hand side of (5.30) to get

$$\text{length}(P) \le |E(H_1)| + \sum_{i=2}^{t}(|E(H_i)| + 2\,\text{diam}(G)).$$

After that we combine the fact that $|E(H_i)| \ge g(G)$ for $i \ge 2$ with the assumption that $g(G) \ge \alpha\,\text{diam}(G)$, and get

$$\begin{aligned}\text{length}(P) &\le |E(H_1)| + \sum_{i=2}^{t}|E(H_i)|\left(1 + \frac{2}{\alpha}\right) \\ &\le \left(1 + \frac{2}{\alpha}\right)|S| \\ &= \left(1 + \frac{2}{\alpha}\right)D(P).\end{aligned}$$

Thus (5.29) holds with $\beta = \left(\frac{2}{\alpha} + 1\right)$. This completes the proof of Theorem 5.57. □

5.8 Probabilistic proof of existence of expanders

Historically the first proof of existence of expanders was probabilistic. Until now the simplest known proofs of existence of expanders are probabilistic. We present a version of such a proof which uses graph lifts as a tool for achieving k-regularity. For simplicity, we prove the result only for sufficiently large k, although a similar proof (with more complicated details) works for each $k \geq 3$. In Notes and Remarks we describe other probabilistic proofs and provide references.

The expander family which we construct is a sequence of random lifts of the graph K_{k+1} (the complete graph with $k+1$ vertices). The number of edges in K_{k+1} is $\frac{k(k+1)}{2}$, we denote this number by d.

For convenience of the reader we give a detailed definition of a random lift. Let $L_n = \{1, 2, 3, \ldots, n\}$, we use the standard notation S_n for the set of all permutations of L_n. We consider the following probability space. Let

$$\Omega_{n,k} = \underbrace{S_n \times S_n \times \cdots \times S_n \times S_n}_{d \text{ times}},$$

with the probability measure $\mathbb{P}_{n,k}$ defined by $\mathbb{P}_{n,k}(\omega) = \frac{1}{(n!)^d}$ for each d-tuple ω of permutations, so $\omega = \{\tau_e\}_{e \in E(K_{k+1})}$, where $\tau_e \in S_n$ (we label permutation by edges of K_{k+1}).

For each $\omega \in \Omega_{n,k}$ we introduce the corresponding graph $G_{n,k}(\omega)$ in the following way. The vertex set of $G_{n,k}(\omega)$ is the Cartesian product $V(K_{k+1}) \times L_n$ (it is the same for all ω). The edge set of $G_{n,k}(\omega)$ is defined as follows.

We pick an orientation of the graph K_{k+1}, that is, for each edge e assign one of its ends to be a *head* of e and the other to be a *tail* of e. We denote the head of e by e^+ and the tail of e by e^-. The edge set $E(G_{n,k}(\omega))$ contains n edges per each edge of K_{k+1}. They are the edges joining (e^-, m) with $(e^+, \tau_e(m))$ for each $m \in \{1, \ldots, n\}$.

It is easy to check that for each d-tuple $\omega \in \Omega_{n,k}$ of permutations we get a k-regular graph $G_{n,k}(\omega)$, just because K_{k+1} is k-regular. The following theorem immediately implies that the constructed collection of graphs contains families of expanders.

Theorem 5.62. *There exist $k \in \mathbb{N}$ and $\eta > 0$ such that*

$$\lim_{n \to \infty} \mathbb{P}_{n,k}\{\omega : h(G_{n,k}(\omega)) \geq \eta\} = 1,$$

where $h(G_{n,k}(\omega))$ is the expanding constant.

Proof. We need to show that for sufficiently large k and n the graphs $G_{n,k}(\omega)$ have expanding constants $\geq \eta$ for the vast majority of $\omega \in \Omega_{n,k}$. This means that for such ω each $A \subset V(G_{n,k}(\omega))$ such that $|A| \leq \frac{|V(G_{n,k}(\omega))|}{2}$ satisfies

$$|\partial_{G_{n,k}(\omega)}(A)| \geq \eta |A|, \tag{5.31}$$

where we indicate the graph with respect to which we take the boundary in the subscript.

To achieve this goal we estimate the probability of the complement of this event, that is the probability of the set of those ω for which there is $A \subset V(G_{n,k}(\omega))$ satisfying $|A| \le \frac{|V(G_{n,k}(\omega))|}{2}$ which violates condition (5.31).

Following the notation introduced in Section 5.7 we denote, for each $v \in V(K_{k+1})$, the set $\{v\} \times L_n \subset V(G_{n,k}(\omega))$ by $\pi^{-1}(v)$. Among the intersections $A \cap \pi^{-1}(v)$ there is one of maximal cardinality, we denote it by S and let $s \in V(K_{k+1})$ be the vertex for which $S = A \cap \pi^{-1}(s)$. There is also an intersection of the least cardinality, we use the notation $T = A \cap \pi^{-1}(t)$ for it. It is clear that

$$|S| \ge \frac{|A|}{k+1} \ge |T|.$$

We estimate $|\partial_{G_{n,k}(\omega)}(A)|$ from below in the following way. Observe that there are k edge-disjoint paths joining s and t in K_{k+1}. According to Remark 5.51 (page 157), for each ω and each vertex $(s,m) \in S$, each such path can be lifted to the graph $G_{n,k}(\omega)$ starting from (s,m). If such a lifted path ends outside T, then it contains an edge in $\partial_{G_{n,k}(\omega)}(A)$. Note also that the lifted paths are edge-disjoint, because the original paths are edge-disjoint. Hence the cardinality of $\partial_{G_{n,k}(\omega)}(A)$ can be estimated from below by the number of lifted paths which, originating in S, terminate outside T. If the inequality (5.31) does not hold, then the number of such paths is $< \eta|A|$ and thus there exists a set $T' \in \pi^{-1}(t)$ of cardinality $< \eta|A|$ such that all paths obtained after lifting the k edge-disjoint paths joining s and t originating at vertices of S terminate in $\widetilde{T} = T \cup T'$. So for each set A and graph $G_{n,k}(\omega)$ satisfying $|\partial_{G_{n,k}(\omega)}(A)| < \eta|A|$ there exist sets S and $\widetilde{T}$ such that all lifted to $G_{n,k}(\omega)$ paths mentioned above which originate at S terminate in $\widetilde{T}$ and

$$|S| \ge \frac{|A|}{k+1}, \quad |\widetilde{T}| < \frac{|A|}{k+1} + \eta|A| = \frac{|A|}{k+1}(1 + \eta(k+1)).$$

Observe that we consider A satisfying $|A| \le \frac{(k+1)n}{2}$ and that the condition above will still be satisfied if we delete some elements of the set S. Also, for simplicity, let us restrict our attention to even n. It is easy to see that to complete the proof of the theorem it suffices to show that for sufficiently large k we can choose η in such a way that the probability $\mathbb{P}_{n,k}$ of the following event goes to 0 as $n \to \infty$.

(a) There exist $s, t \in V(K_{k+1})$, $s \neq t$, and $S \subset \pi^{-1}(s)$ and $\widetilde{T} \subset \pi^{-1}(t)$ satisfying $|S| \le \frac{n}{2}$, $|\widetilde{T}| \le (1 + \eta(k+1))|S|$ and such that lifts to $G_{n,k}(\omega)$ starting at vertices of S of all k edge-disjoint paths from s to t in K_{k+1} terminate in $\widetilde{T}$.

This observation allows us to proceed in the following way. We make the following choice: after the selection of k, the value of η is determined by

$$\eta = \frac{1}{2} \cdot \frac{1}{k+1}$$

(the number $\frac{1}{2}$ could be replaced by any fixed number in the interval $(0, 1)$, but we stick with the choice of $\frac{1}{2}$). We consider all possible pairs of subsets S and $\widetilde{T}$, satisfying

$S \subset \pi^{-1}(s)$, $\widetilde{T} \subset \pi^{-1}(t)$, $s, t \in V(K_{k+1})$, $s \neq t$, satisfying $|S| \leq \frac{n}{2}$, $|\widetilde{T}| = \lfloor \frac{3}{2}|S| \rfloor$ and estimate from above the probability $\mathbb{P}_{n,k}$ of the event that lifts to $G_{n,k}(\omega)$ starting at vertices of S of all k edge-disjoint paths from s to t in K_{k+1} terminate in $\widetilde{T}$. If we add the probabilities of all such events, over all such choices of S and $\widetilde{T}$, we get an upper estimate for the probability of the event described above.

Our next observation is crucial. It is the fact that the probability that one of the paths originating at s and ending at t will have all of its lifts starting at S terminate in $\widetilde{T}$ is the same as the probability of a permutation of L_n to map the set of L_n-coordinates of S into set of L_n-coordinates of $\widetilde{T}$. Using this and the fact that the choices of such permutations for different paths are independent, we get the following upper estimate for the probability associated with S and $\widetilde{T}$:

$$\left(\frac{\binom{R}{r}}{\binom{n}{r}}\right)^k,$$

where $r = |S|$ and $R = |\widetilde{T}|$.

Therefore the probability of the event described in (a) can be estimated from above by

$$k(k+1) \sum_{1 \leq r \leq \frac{n}{2}} \binom{n}{r}\binom{n}{R}\left(\frac{\binom{R}{r}}{\binom{n}{r}}\right)^k, \tag{5.32}$$

where $R = \lfloor \frac{3}{2} r \rfloor$. Now we observe that $R \leq n$ implies

$$\frac{\binom{R}{r}}{\binom{n}{r}} = \left(\frac{R}{n}\right) \cdot \dots \cdot \left(\frac{R-r+1}{n-r+1}\right) \leq \left(\frac{R}{n}\right)^r.$$

As for the remaining part of the product, we use the Stirling formula (which we reminded the reader of on page 123) and the inequality $R \leq 2r$, to get an estimate

$$\binom{n}{r}\binom{n}{R} \leq \left(\frac{n^r}{r!}\right)\left(\frac{n^R}{R!}\right) \leq \frac{n^{R+r}}{\frac{r^r \cdot R^R}{e^r \cdot e^R}} \leq \frac{n^{R+r}e^{R+r}2^r}{R^{R+r}}.$$

Now we estimate the probability in (5.32) by

$$\begin{aligned} k(k+1) &\sum_{1 \leq r \leq \frac{n}{2}} \frac{n^{R+r}e^{R+r}2^r}{R^{R+r}}\left(\frac{R}{n}\right)^{rk} \\ &\leq k(k+1) \sum_{1 \leq r \leq \frac{n}{2}} e^{4r}\left(\frac{R}{n}\right)^{rk-R-r} \qquad (5.33) \\ &\leq k(k+1) \sum_{1 \leq r \leq \frac{n}{2}} \left(e^4 \cdot \left(\frac{R}{n}\right)^{k-3}\right)^r. \end{aligned}$$

To get the inequality in the last line we used the inequalities $\frac{R}{n} \le 1$ and $R \le 2r$. Now we pick k in such a way that

$$e^4 \cdot \left(\frac{3}{4}\right)^{k-3} < 1. \tag{5.34}$$

We fix such a value of k, denote the corresponding number in the left-hand side of (5.34) by α, and let $\eta = \frac{1}{2(k+1)}$ (as we agreed).

Observe that each of the terms in the sum in the last line of (5.33) does not exceed α^r because $R \le \lfloor \frac{3}{2} \cdot \frac{n}{2} \rfloor \le \frac{3}{4} n$.

It remains only to show that now, if we let $n \to \infty$, the sum in (5.33) goes to 0. This is a standard argument: since the r-th term in the sum (5.33) does not exceed $k(k+1)\alpha^r$, the series $k(k+1)\sum_{r=1}^{\infty} \alpha^r$ is convergent, and the r-th term in the sum (5.33) approaches 0 as $n \to \infty$, by the Lebesgue dominated convergence theorem, the sum in (5.33) approaches 0 as $n \to \infty$. □

5.9 Size and diameter of graphs with large girth: basic facts

In this section we consider not necessarily regular graphs with large girth. For a graph G we denote the *minimum degree* of vertices of G by $\delta(G)$. Denote by $n(g, \delta)$ the minimum number of vertices in a graph of girth $\ge g$ and minimum degree $\ge \delta$. As we have already mentioned, we are interested in the case where $\delta > 2$.

Theorem 5.63.

$$n(g, \delta) \ge \begin{cases} 1 + \delta \dfrac{(\delta-1)^{(g-1)/2} - 1}{\delta - 2} & \textit{if } g \textit{ is odd} \\ 2 \cdot \dfrac{(\delta-1)^{g/2} - 1}{\delta - 2} & \textit{if } g \textit{ is even.} \end{cases}$$

Proof. Let G be a graph with $g(G) \ge g$ and $\delta(G) \ge \delta$. First we consider the case where g is odd, $g = 2m + 1$. Let v be a vertex in G and let $N(m, v)$ be the set of vertices at distance $\le m$ to v. It is clear that $N(m, v) = \cup_{k=0}^{m} D(k, v)$ where

$$D(k, v) = \{u \in V(G) : d_G(u, v) = k\},$$

and that the sets $D(k, v)$ are disjoint. Obviously

$$|D(0, v)| = 1 \quad \text{and} \quad |D(1, v)| \ge \delta.$$

As for all further sets, we claim that

$$|D(k, v)| \ge (\delta - 1)|D(k-1, v)|, \quad \text{for } k \in \{2, \dots, m\}. \tag{5.35}$$

This inequality holds because the following two conditions are satisfied.

- Each vertex in $D(k, v)$, $k \in \{1, \ldots, m\}$ has exactly one edge joining it to a vertex in $D(k-1, v)$, since otherwise the graph would contain a cycle of length $\leq 2m$.
- There are no edges between vertices in $D(k, v)$, $k \in \{1, \ldots, m-1\}$, such edges would create a cycle of length $\leq 2(m-1) + 1 = 2m - 1$.

Therefore for each vertex in $D(k, v)$, $k \in \{1, \ldots, m-1\}$, there are at least $\delta - 1$ edges which join it to vertices in $D(k+1, v)$, and each vertex in $D(k+1, v)$ is incident to exactly one such edge. Inequality (5.35) follows. We get

$$|V(G)| \geq \sum_{k=0}^{m} |D(k, v)| \geq 1 + \delta \sum_{t=0}^{m-1} (\delta - 1)^t,$$

which is the desired estimate.

In the case where g is even, $g = 2m$, we consider a pair v, w of adjacent vertices in G and consider the set $N(m-1, v, w)$ of all vertices at distance $\leq (m-1)$ to v or to w. It is clear that $N(m-1, v, w) = \cup_{k=0}^{m-1} D(k, v, w)$ where

$$D(k, v, w) = \{u \in V(G) : \min\{d_G(u, v), d_G(u, w)\} = k\},$$

and that the sets $D(k, v, w)$ are disjoint. Similarly to the first part of the proof we get

$$|D(0, v, w)| = 2,$$

$$|D(k, v, w)| \geq (\delta - 1) \cdot |D(k-1, v, w)|, \text{ for } k \in \{1, \ldots, m-2\}.$$

Therefore

$$|V(G)| \geq \sum_{k=0}^{m-1} |D(k, v, w)| \geq \sum_{k=0}^{m-1} 2(\delta - 1)^k,$$

which is the desired estimate. □

The following simple estimate is also useful.

Theorem 5.64. *If g is odd then*

$$n(g+1, \delta) \leq 2n(g, \delta).$$

Proof. Take a graph H of order $n(g, \delta)$ such that $\delta(H) \geq \delta$, $g(H) \geq g$. We consider the following lift $\widetilde{H}$ of H: $L = \{0, 1\}$ and the matching corresponding to an arbitrary edge of H is induced by a transposition of 0 and 1. It is clear that the obtained graph is bipartite with the bipartition

$$X = \{(v, 0) : v \in V(H)\} \quad Y = \{(v, 1) : v \in V(H)\}.$$

On the other hand, by the comments made after Definition 5.50 and Remark 5.52, we have $\delta(\widetilde{H}) = \delta(H)$ and $g(\widetilde{H}) \geq g(H)$. Furthermore, since $\widetilde{H}$ is bipartite, it is easy to see that $\widetilde{H}$ does not contain *odd cycles* (that is, cycles of odd length), therefore $g(\widetilde{H}) \geq g(H) + 1 \geq g + 1$. □

Remark 5.65. It is well known that absence of odd cycles characterizes bipartite graphs.

5.10 Random constructions of graphs with large girth

It turns out that if we consider a random graph with N vertices and δN edges, then the average number of cycles contained in such a graph is rather small and, after removal of a relatively small number of edges, we get a graph with large girth. Using standard techniques this leads to a construction of a graph with large girth and a given bound on the minimum degree.

Theorem 5.66. *Given $\delta \geq 3$ and $g \geq 3$ there exists a graph with $n \leq (2\delta)^g$ vertices with minimum degree at least δ and girth at least g.*

Proof. Let $N = (2\delta)^g$. Consider all graphs with vertex set $\{1, \dots, N\}$ having δN edges. Note that there are

$$\binom{\binom{N}{2}}{\delta N}$$

such graphs. One can form

$$Y(\ell) := \frac{1}{2}(\ell - 1)!\binom{N}{\ell} = \frac{N \cdot (N-1) \cdot \dots \cdot (N - \ell + 1)}{2\ell}$$

cycles of length ℓ on N vertices. On the other hand, the number of graphs containing a given cycle of length ℓ is

$$\binom{\binom{N}{2} - \ell}{\delta N - \ell}.$$

Consequently the average number of cycles of length at most $g - 1$ contained in the considered set of graphs does not exceed

$$\begin{aligned}
\sum_{\ell=3}^{g-1} Y(\ell) \cdot \binom{\binom{N}{2} - \ell}{\delta N - \ell}\binom{\binom{N}{2}}{\delta N}^{-1} &= \sum_{\ell=3}^{g-1} Y(\ell) \cdot \frac{\delta N(\delta N - 1)\dots(\delta N - \ell + 1)}{\binom{N}{2}\left(\binom{N}{2} - 1\right)\dots\left(\binom{N}{2} - \ell + 1\right)} \\
&= \sum_{\ell=3}^{g-1} (2\delta)^\ell \frac{1}{2\ell} \frac{N^2\left(N - \frac{1}{\delta}\right)(N-1)\dots\left(N - \frac{(\ell-1)}{\delta}\right)(N - (\ell - 1))}{(N^2 - N)(N^2 - N - 2)(N^2 - N - 4)\dots(N^2 - N - 2(\ell - 1))} \\
&< \sum_{\ell=3}^{g-1} (2\delta)^\ell < N.
\end{aligned}$$

The first inequality in the last line can be shown by using

$$\frac{N^2}{2(N^2 - N)} < 1 \quad \text{and} \quad \frac{\left(N - \frac{k}{\delta}\right)(N - k)}{N^2 - N - 2k} \leq 1 \quad \forall k \in \{1, \dots, \ell - 1\}.$$

Thus there is a graph G_0 with N vertices and δN edges which contains at most $N-1$ cycles whose length does not exceed $g-1$. We delete an edge from each such cycle and get a graph G_1 with N vertices and M edges, $M > (\delta-1)N$, and girth at least g.

Therefore the average degree of this graph is $> 2(\delta-1)$. A useful observation is: If we delete from G_1 all vertices whose degree is $\le (\delta-1)$, the average degree of the obtained graph will still be $> 2(\delta-1)$. In fact, denote the number of deleted vertices by D, so the number of deleted edges is $\le (\delta-1)D$. The average degree of the obtained graph is

$$\ge 2\,\frac{M-(\delta-1)D}{N-D} > 2\,\frac{(\delta-1)N-(\delta-1)D}{N-D} = 2(\delta-1).$$

Of course, after this deletion new vertices with degrees $\le (\delta-1)$ may appear. We repeat the deletion of all such vertices if needed. This procedure will terminate eventually because we consider finite graphs and it will terminate when the remaining graph is nonempty because of the bound on average degree. Thus the result will be a nonempty graph G with minimum degree $\ge \delta$ girth $\ge g$ and at most $(2\delta)^g$ vertices. □

5.11 Graphs with large girth using variational techniques

The purpose of this section is to prove the existence of graphs with girth g and minimum degree δ and to estimate from above the number of vertices in such graphs. Our first result (Theorem 5.67) proves the existence of even δ-regular graphs satisfying the mentioned conditions. Our second result (Theorem 5.69) is somewhat weaker, but is algorithmic.

Theorem 5.67. *Let $\delta, g \in \mathbb{N}$, $g \ge 3$, $\delta \ge 2$, and let*

$$m \ge \sum_{i=0}^{g-2} (\delta-1)^i = (\textit{if}\ \delta \neq 2) = \frac{(\delta-1)^{g-1}-1}{\delta-2} \tag{5.36}$$

be an integer. Then there exists a δ-regular graph of order $2m$ and girth at least g.

Remark 5.68. A useful consequence of this is that for even values of n there exist δ-regular graphs with n vertices and girth $\ge c \log_\delta n$.

It is also worthwhile to mention that the main difference between (5.36) and the inequalities in Theorem 5.63 is that the power in (5.36) is twice larger: $g-1$ vs. $(g-1)/2$.

Proof of Theorem 5.67. We fix $m = m_0$ satisfying (5.36) for some $\delta = \delta_0$. We use induction on $\delta \in \{2, \dots, \delta_0\}$ to show that for each $\delta \in \{2, \dots, \delta_0\}$ there is a δ-regular graph with $2m_0$ vertices and girth at least g.

If $\delta = 2$ the result is trivial because a cycle of length $2m_0$ is the desired graph, so suppose $\delta > 2$. Induction hypothesis: assume that for $k \in \{2, \dots, \delta_0 - 1\}$ we have proved the existence of a k-regular graph with $2m_0$ vertices and girth at least g. In remains to show that under this hypothesis we can find a $(k+1)$-regular graph with $2m_0$ vertices and girth at least g.

We are going to use $\Delta(G)$ to denote the *maximum degree* of vertices of a graph G.

Let G be a graph with $2m_0$ vertices, girth $\geq g$, $k \leq \delta(G) \leq \Delta(G) \leq k+1$ (at least one graph exists by the induction hypothesis) and the maximal possible number of edges. To prove the theorem it suffices to show that it is $(k+1)$-regular.

Assume the contrary, then G has at least two vertices with degree k. Let u and v have degree k. If $d_G(u,v) \geq g-1$, we add the edge uv to the graph, and get a contradiction with the assumption that G has the maximal possible number of edges. So we assume $d_G(u,v) \leq g-2$ and that any pair of vertices of degree k in G is at distance $\leq g-2$.

Let us show first that if we find an adjacent pair of vertices, $y_1, y_2 \in V(G)$, such that $d_G(y_1,u) \geq g-1$ and $d_G(y_2,v) \geq g-1$, we can complete the argument. In fact, in such a case we delete the edge y_1, y_2 and add edges y_1u and y_2v. It is clear that doing so we increase the total number of edges and do not create cycles of length $\leq g-1$. (Here we need to proceed with some care because by adding two edges for pairs at distance $\geq g-1$ we can create short cycles. So we add them one by one. After the addition of the edge y_1u the distance between y_2 and v is still $\geq g-1$ because a y_2v-path of length $< g-1$ cannot go through y_1. In fact, the distance between y_1 and y_2 after the deletion of y_1y_2 is at least $g-1$.) Thus we get a contradiction with the assumption that G has the maximal possible number of edges.

To find a pair y_1, y_2 we consider the balls of radii $g-2$ centered at u and v:

$$B_1 := \{w : d_G(w,u) \leq g-2\}, \qquad B_2 := \{w : d_G(w,v) \leq g-2\}.$$

Since the degrees of u and v are equal to k, we have

$$|B_1| \leq \sum_{i=0}^{g-2} k^i \leq \sum_{i=0}^{g-2} (\delta_0-1)^i \leq m_0, \quad |B_2| \leq \sum_{i=0}^{g-2} k^i \leq \sum_{i=0}^{g-2} (\delta_0-1)^i \leq m_0.$$

The inequality $d_G(u,v) \leq g-2$ implies that B_1 and B_2 intersect. Since $|V(G)| = 2m_0$, the complement of the union of the balls contains at least one vertex $y \in V(G)\backslash(B_1 \cup B_2)$. Now we analyze: how can it happen that a vertex $y \in V(G)\backslash(B_1 \cup B_2)$ does not have a neighbor which is at distance $\geq g-1$ to u or to v? (If y has such a neighbor z, then y and z form the desired pair y_1, y_2.)

An easy analysis shows that this can happen only if all neighbors of y are in the intersection $S_1 \cap S_2$, where

$$S_1 := \{w : d_G(w,u) = g-2\}, \qquad S_2 := \{w : d_G(w,v) = g-2\}.$$

So it remains to show that each element of $V(G)\backslash(B_1 \cup B_2)$ cannot have all of its neighbors in $S_1 \cap S_2$.

Observe that the degree of each vertex in $V(G)\backslash(B_1 \cup B_2)$ is $k+1$ (otherwise such a vertex y together with u or v would form a pair of vertices of degree k with distance $\geq g-1$ between them, contrary to our assumption).

Our next observation is:

$$|V(G)\backslash(B_1 \cup B_2)| > |S_1 \cap S_2|. \tag{5.37}$$

In fact, we have $|V(G)\backslash(B_1 \cup B_2)| = |V(G)| - |B_1 \cup B_2| = |V(G)| - |B_1| - |B_2| + |B_1 \cap B_2|$. On the other hand, $|B_1 \cap B_2| \geq |S_1 \cap S_2| + 2$, because, in addition to $S_1 \cap S_2$, the set $B_1 \cap B_2$ contains at least two vertices u and v.

Since degrees of all vertices of $V(G)\backslash(B_1 \cup B_2)$ are $k+1$ and degrees of all vertices of $S_1 \cap S_2$ are at most $k+1$, inequality (5.37) implies that it cannot happen that all neighbors of $V(G)\backslash(B_1 \cup B_2)$ are in $S_1 \cap S_2$. □

Our next purpose is to show that if we do not insist on having a δ-regular graph and are content with a graph G satisfying $\delta - 1 \leq \delta(G) \leq \Delta(G) \leq \delta + 1$, we can make the construction of such graph algorithmic. For simplicity we consider the case where the number of vertices is even.

Theorem 5.69. *Let $n \in \mathbb{N}$ be even, $n = 2k$, and let $V = \{1, \ldots, n\}$. For each $\delta \geq 2$ there is a graph $G(\delta)$ with vertex set V for which the average degree is δ, the minimum degree $\geq \delta - 1$, the maximum degree $\leq \delta + 1$ and the girth is $> \log_\delta\left(\frac{n}{2}\right) + 1$.*

Proof. We construct such graphs algorithmically and inductively. In the case $\delta = 2$ we let $G(\delta) = G(2)$ be a cycle with n vertices. It is clear that all of the conditions are satisfied.

We produce $G(3)$ using the following simple algorithm. We add $k = \frac{n}{2}$ edges to $G(2)$, one by one. Each time we pick one of the ends of the edge which we are adding to be a vertex u of the least degree among the vertices of the (intermediate) graph $\mathcal{G}$ which we modify, and the second vertex to be the vertex which is the farthest in $\mathcal{G}$ from u among vertices whose degree does not exceed 3.

It is clear that in this way we get a graph whose average degree is 3 (average degree increases by 1 just because we add $\frac{n}{2}$ edges), the maximum degree is ≤ 4 and the minimum degree is ≥ 2. It remains to estimate the girth.

To do this we need to estimate from below distances between vertices which we join by edges. When we add an edge uv to $\mathcal{G}$, it means that all vertices w satisfying

$$d_{\mathcal{G}}(u, w) > d_{\mathcal{G}}(u, v) \tag{5.38}$$

have degree 4 (otherwise we would add a different edge). Since the average degree of $\mathcal{G}$ is < 3 (we reach average degree 3 only at the end) and the minimum degree is 2, we get that the number of vertices w satisfying (5.38) is $< \frac{n}{2}$.

Since the maximum degree of $\mathcal{G}$ is 4 and the degree of u is 2, estimating the number of vertices w which *do not* satisfy (5.38) we get

$$1 + 2 + 2 \cdot 3 + 2 \cdot 3^2 + \cdots + 2 \cdot 3^{d_{\mathcal{G}}(u,v)-1} > \frac{n}{2}.$$

This inequality can be rewritten as

$$3^{d_{\mathcal{G}}(u,v)} > \frac{n}{2}.$$

Therefore

$$d_{\mathcal{G}}(u,v) > \log_3 \left(\frac{n}{2}\right)$$

and the girth of the obtained graph is

$$> \log_3 \left(\frac{n}{2}\right) + 1,$$

which is the inequality claimed in Theorem 5.69.

Now we turn to an induction step. At the beginning we have a graph $G(\delta)$ with average degree δ, the minimum degree $\geq \delta - 1$, the maximum degree $\leq \delta + 1$ and girth $\geq \log_\delta \left(\frac{n}{2}\right) + 1$. We add $\frac{n}{2}$ edges to $G(\delta)$, one by one. Each time we pick one of the ends of the edge which we are adding to be a vertex u of the least degree among the vertices of the (intermediate) graph $\mathcal{G}$ which we modify, and the second vertex to be the vertex which is the farthest in $\mathcal{G}$ from u among vertices whose degree does not exceed $\delta + 1$.

It is clear that in this way we get a graph $G(\delta + 1)$ whose average degree is $\delta + 1$ (average degree increases by 1 just because we add $\frac{n}{2}$ edges), the maximum degree is $\leq \delta + 2$.

We need to estimate the minimum degree and the girth. The fact that the minimum degree becomes $\geq \delta$ follows from the following argument. It is clear that $G(\delta)$ cannot have more than $\frac{n}{2}$ vertices of degree $\delta - 1$. Since we add $\frac{n}{2}$ edges and each time pick one of the ends to be a vertex with minimum degree of the intermediate graph $\mathcal{G}$, all of the vertices of degree $\delta - 1$ in $G(\delta)$ will increase their degree when we complete our construction of $G(\delta + 1)$.

To estimate the girth we need to estimate from below distances between vertices which we join by edges. When we add an edge uv to $\mathcal{G}$, it means that all vertices w satisfying $d_{\mathcal{G}}(u,w) > d_{\mathcal{G}}(u,v)$ have degree $\delta + 2$ (otherwise we would add a different edge). The number of such vertices is $< \frac{n}{2}$ because initially (when we started to add edges to $G(\delta)$) there were no such vertices, and in each step we could increase the number of such vertices at most by 1.

Since the maximum degree of $\mathcal{G}$ is $\delta + 2$ and the degree of u is $\leq \delta$, we get that the number of vertices w which *do not* satisfy (5.38) is

$$1 + \delta + \delta \cdot (\delta + 1) + \delta \cdot (\delta + 1)^2 + \cdots + \delta \cdot (\delta + 1)^{d_{\mathcal{G}}(u,v)-1} > \frac{n}{2}.$$

This inequality can be rewritten as

$$(\delta + 1)^{d_{\mathcal{G}}(u,v)} > \frac{n}{2}.$$

Therefore

$$d_G(u, v) > \log_{\delta+1}\binom{n}{2}$$

and the girth of the obtained graph is

$$> \log_{\delta+1}\binom{n}{2} + 1.$$

This proves Theorem 5.69. □

5.12 Inequalities for the spectral gap of graphs with large girth

There exists a very elegant argument imposing restrictions on the spectrum of adjacency matrices of graphs with large girth whose number of vertices is very close to the lower bound of Theorem 5.63.

Let A be the adjacency matrix of a k-regular graph G with n vertices of girth $g = 2r + 1$. Let $n(g, k)$ be the lower bound of Theorem 5.63 on the number of vertices of this graph, that is

$$n(g, k) = 1 + \frac{k}{k-2}((k-1)^r - 1).$$

The number $n - n(g, k)$ is called the *excess* of G.

Let us denote by A_j the following generalization of the adjacency matrix: the u, v-entry of this matrix is 1 if $d_G(u, v) = j$ and is 0 otherwise. In this notation the usual adjacency matrix is A_1, and A_0 is the identity matrix.

It is easy to find a recursive polynomial expression for the matrix $F_r(A) = A_0 + \cdots + A_r$ with rows and columns indexed by vertices of G and such that the entry is 1 if the distance between the vertices is $\leq r$. More precisely we have:

$$A_0 = I, \quad A_1 = A, \quad A_2 = A^2 - kI,$$

$$A_{k+1} = A \cdot A_k - (k-1)A_{k-1} \text{ if } 3 \leq k \leq r - 1.$$

In fact, the u, v-entry of the matrix $A \cdot A_k$ is the number of uv-walks of length $k + 1$ satisfying the following condition: the walk goes to a vertex which is at distance k from u, and then reaches v using just one edge. Now we use the fact that in a graph of girth $2r + 1$ there is just one shortest wz-walk for any two vertices w and z with $d_G(w, z) \leq r$. This implies that the u, v-entry of the matrix $A \cdot A_k$ is 1 if $d_G(u, v) = k + 1$ and is $k - 1$ if $d_G(u, v) = k - 1$.

Theorem 5.70. *Let λ_2 be the second eigenvalue of the adjacency matrix A. Then*

$$|F_r(\lambda_2)| \leq n - n(g, k). \tag{5.39}$$

Proof. Observe that the number of ones in the column (or row) of the matrix $F_r(A)$ corresponding to a vertex u is the number of vertices in G which are at distance $\leq r$ from u in G. It is easy to see that this number is $n(g,k)$.

Therefore the eigenvalue of the matrix $F_r(A)$ corresponding to all-ones vector $\mathbf{1}_{V(G)}$ is $n(g,k)$. On the other hand, the matrix J (all-ones matrix) is equal to $A_0 + \cdots + A_D$, where D is the diameter of the graph. Observe that the matrix $J - F_r(A)$ has the same amount of ones in each column and each row, and this amount is $n - n(g,k)$, the excess of the graph G.

Observe that both $F_r(A)$ and J are symmetric real matrices and both have $\mathbf{1}_{V(G)}$ as an eigenvector. Also the restriction of J to the orthogonal complement of $\mathbf{1}_{V(G)}$ is a zero-operator. Therefore $F_r(A)$ and J have a common basis of eigenvectors.

Observe that all eigenvalues of the difference $J - F_r(A)$ do not exceed $n - n(g,k)$ (the argument is the same as in Proposition 5.4) and that J has one eigenvalue equal to n and all other eigenvalues equal to 0. Thus $|F_r(\lambda)| \leq n - n(g,k)$ for each eigenvalue of A except the first one. □

5.13 Biggs's construction of graphs with large girth

Now we turn to more algebraic constructions of families of graphs with fixed degree and growing girths.

Definition 5.71. We say that a group G *acts on a set* X if there is a map $G \times X \to X$, $(g,x) \mapsto g(x)$ such that

- $e(x) = x$ for all $x \in X$, where e is the identity of G.
- $f(g(x)) = (fg)(x)$ for all $f, g \in G$ and all $x \in X$.

Definition 5.72. Let G be a finite group which acts on a finite set X. Let S be a symmetric subset of G (this means that $e \notin S$ and $(s \in S) \Leftrightarrow (s^{-1} \in S)$) which *acts freely* on X in the sense that $s(x) \neq x$ for all $x \in X$ and $s \in S$, and that $s_1(x) \neq s_2(x)$ for all $s_1, s_2 \in S$ with $s_1 \neq s_2$ and all $x \in X$. Then the *Schreier graph* $\mathrm{Sch}(X,S)$ is defined as the graph with vertex set X and edge set defined by the condition uv is an edge if and only if $u = s(v)$ for some $s \in S$.

Biggs's construction produces Schreier graphs of subgroups of symmetric groups.

First we produce the corresponding set X. It consists of labelings of vertices of a d-regular rooted tree T of depth r. This means the following: T has

$$1 + d + d(d-1) + \cdots + d(d-1)^{r-1}$$

vertices which are partitioned into generations having $1, d, d(d-1), \ldots, d(d-1)^{r-1}$ vertices, respectively. The generation with $d(d-1)^{j-1}$ vertices, $j \in \{2, \ldots, r\}$, is called the *generation* j; the generation with d vertices is called the *generation* 1; the generation with 1 vertex is called the *generation* 0. The only vertex in the generation 0 is called the *root*. The edge set of T is (up to an isomorphism) determined by the following rules: Each vertex of generation j, $1 \le j \le r$, is adjacent to one vertex of the generation $j-1$, each vertex of generation j, $1 \le j \le r-1$, is adjacent to $d-1$ vertices of generation $j+1$, and the root is adjacent to d vertices of generation 1.

It is easy to see that all vertices in T except vertices of generation r have degree d. Vertices of generation r have degree 1.

We denote by V the set of vertices of T, and let L be some set of labels satisfying $|L| = |V|$. We consider some initial labeling of V with elements of L. The set X consists of all labelings of V obtained from this initial labeling in the way described below.

We consider a proper d-coloring of edges of T. We mean that we color edges using d different colors in such a way that edges having the same color do not have a common end. (It is easy to construct such coloring starting from edges incident with the root.)

For each of the d colors we introduce the corresponding permutation of labels: The permutation σ_i corresponding to the color i transposes pairs of labels which are at the ends of all edges having color i (more details: if an edge has color i and its ends have labels ℓ_1 and ℓ_2 before the action of σ_i, after the action they have labels ℓ_2 and ℓ_1, respectively).

Let X be the set of all labelings obtained from the original labeling using all possible combinations of permutations σ_i, $i = 1, \dots, d$. It is clear that $\sigma_i^{-1} = \sigma_i$. Therefore $S = \{\sigma_i : i = 1, \dots, d\}$ is a symmetric set. Let $G = \mathrm{Sch}(X, S)$ be the corresponding Schreier graph.

It is clear that the graph G is d-regular. Now we estimate the girth of G from below. We claim that $g(G) \ge 2r + 1$. It is not difficult to see that to prove this claim we need to show that for $k < 2r + 1$ the composition $\sigma_{n_k}\sigma_{n_{k-1}} \cdots \sigma_{n_2}\sigma_{n_1}$ cannot satisfy the following two conditions simultaneously:

(a) Its product (in the given order) is the identity permutation.

(b) There is no i such that $\sigma_{n_i} = \sigma_{n_{i+1}}$.

To show this we apply $\sigma_{n_k}\sigma_{n_{k-1}} \cdots \sigma_{n_2}\sigma_{n_1}$ to some labeling x in X. We see that the condition (b) implies that the label of the root in x will be moved away (in the graph distance of T) from its original position by at least first r permutations: in order to move the label in the direction of the root a permutation of the set $\sigma_{n_1}, \sigma_{n_2}, \dots, \sigma_{n_r}$ has to coincide with the immediately preceding one. This shows that the girth is at least $2r + 1$.

5.14 Margulis's 1982 construction of graphs with large girth

Recall that the *special linear group* $SL_2(\mathbb{Z})$ is the group of all 2×2 matrices with integer coefficients and determinant equal to 1. Let a and b be two elements in a group G. We say that a and b generate a *free subgroup* in G if none of the nonempty words in a and b which do not contain any of the subwords from the list

$$aa^{-1}, a^{-1}a, bb^{-1}, b^{-1}b \tag{5.40}$$

represent the identity of G.

Theorem 5.73. *The matrices*

$$a = \begin{pmatrix} 1 & 2 \\ 0 & 1 \end{pmatrix} \quad \textit{and} \quad b = \begin{pmatrix} 1 & 0 \\ 2 & 1 \end{pmatrix} \tag{5.41}$$

generate a free subgroup of $SL_2(\mathbb{Z})$.

Proof. These matrices act in a natural way on $\mathbb{R}^2$. We prove the theorem by studying the action of these matrices on the regions $X = \{(x, y) \in \mathbb{R}^2 : |x| > |y|\}$ and $Y = \{(x, y) \in \mathbb{R}^2 : |y| > |x|\}$. Straightforward verification shows that any nonzero power of a (positive or negative) maps Y into X and that any nonzero power of b (positive or negative) maps X into Y.

Observe that any word in a and b which does not contain subwords from the list (5.40) can be written in one of the forms:

(a) $a^{n_1}b^{n_2}a^{n_3}\dots a^{n_{k-2}}b^{n_{k-1}}a^{n_k}$ (k is odd).

(b) $b^{n_1}a^{n_2}b^{n_3}\dots b^{n_{k-2}}a^{n_{k-1}}b^{n_k}$ (k is odd).

(c) $a^{n_1}b^{n_2}a^{n_3}\dots a^{n_{k-1}}b^{n_k}$ (k is even).

(d) $b^{n_1}a^{n_2}b^{n_3}\dots b^{n_{k-1}}a^{n_k}$ (k is even).

The elements of the form (a) cannot represent the identity because they map Y into X. The elements of the form (b) cannot represent the identity because they map X into Y.

To show that words of the forms (c) and (d) cannot represent the identity we use the following trick to show that if a product of powers of even length represents the identity, then there is a product of powers of odd length word representing the identity. We do this for words of the form (c) (the (d) case is similar).

Observation: If $a^{n_1}b^{n_2}a^{n_3}\dots a^{n_{k-1}}b^{n_k}$ is equal to the identity and $n_1 > 0$, then $a^{n_1+1}b^{n_2}a^{n_3}\dots a^{n_{k-1}}b^{n_k}a^{-1}$ is also equal to the identity. If $n_1 < 0$, we consider $a^{n_1-1}b^{n_2}a^{n_3}\dots a^{n_{k-1}}b^{n_k}a$. In both cases we get a product of powers of odd length representing the identity. □

We consider the natural quotient mapping of $SL_2(\mathbb{Z})$ onto $SL_2(\mathbb{Z}_p)$, where p is a prime number (see Section 5.4.2 for more details on this quotient) and the Cayley graph of the group $SL_2(\mathbb{Z}_p)$ with respect to the set S of generators which are the natural images of the matrices (5.41) and their inverses. The main result of this section is that these Cayley graphs have large girth.

Theorem 5.74. $g(\mathrm{Cay}(SL_2(\mathbb{Z}_p), S)) \geq \log_{\sqrt{2}+1}(p-1)$.

Proof. We estimate the cycle length from below in the following way: the length of the shortest cycle is the length of the shortest word of one of the forms (a)–(d) which is equal to the identity of $SL_2(\mathbb{Z}_p)$. We consider the corresponding word in $SL_2(\mathbb{Z})$. Since the matrices (5.41) generate a free subgroup in $SL_2(\mathbb{Z})$, the word cannot be the identity in $SL_2(\mathbb{Z})$, so it contains entries whose differences with the respective entries of the identity matrix are multiples of p. Thus it contains at least one entry whose absolute value is $\geq p-1$.

To estimate the number of terms in the product of matrices (5.41) in which at least one of the entries has absolute value $\geq p-1$ we use the submultiplicativity of matrix norms.

This means the following: we consider each of the matrices as an operator on ℓ_2^2 and estimate its operator norms. It is easy to see that the norm of a matrix containing an entry with absolute value $\geq p-1$ is $\geq p-1$. On the other hand, it is easy to compute that $\|a\| = \|b\| = \sqrt{2}+1$ (using either Lagrange multipliers or some linear algebra). Hence each cycle in $\mathrm{Cay}(SL_2(\mathbb{Z}_p), S)$ should contain at least $\log_{\sqrt{2}+1}(p-1)$ edges. □

5.15 Families of expanders which are not coarsely embeddable one into another

In this section we allow expanders to have loops and assume that each loop contributes 1 to the degree of the vertex it is attached to.

Definition 5.75. Saying that a family $\mathcal{F}_1$ of metric spaces is *uniformly coarsely embeddable* into a family $\mathcal{F}_2$ of metric spaces we mean that there exist ρ_1 and ρ_2 such that each metric space in $\mathcal{F}_1$ is coarsely embeddable into some metric space of the family $\mathcal{F}_2$, and the corresponding inequality (see Definition 1.45) holds for each of the embeddings with the same ρ_1 and ρ_2.

Theorem 5.76. *There exist families $\mathcal{F}_1$, $\mathcal{F}_2$ of expanders such that $\mathcal{F}_1$ is not uniformly coarsely embeddable into $\mathcal{F}_2$.*

We are going to prove this theorem picking the family $\mathcal{F}_2$ consisting of expanders with indefinitely growing girth, and the family $\mathcal{F}_1$ consisting of expanders containing plenty of cycles of different lengths (satisfying an additional condition). Existence of families of the first type is well known (see Notes and Remarks). Families of the second type can be easily constructed as is shown in the next proposition. The main idea of this construction is: if we consider a union of an expander and a regular graph with the same set of vertices, we get an expander. Recall that a subset A in a metric space (X,d) is called *δ-separated*, $\delta > 0$, if $d(u,v) \geq \delta$ for any $u, v \in A$, $u \neq v$.

Proposition 5.77. *Let $\{P_n\}_{n=1}^{\infty}$ be a d_1-regular family of expanders. It is clear that we can find non-decreasing sequences of positive integers $\{a_n\}_{n=1}^{\infty}$ and $\{b_n\}_{n=1}^{\infty}$ with* $\lim_{n\to\infty} a_n = \lim_{n\to\infty} b_n = \infty$, *and such that P_n contains an a_n-separated set of vertices of cardinality b_n. Let $\{S_n\}_{n=1}^{\infty}$ be graphs satisfying $\Delta(S_n) \leq d_2$ ($\Delta(S_n)$ is the maximum degree of S_n), $|V(S_n)| \leq b_n$ and* $\operatorname{diam} S_n \leq a_n$. *Then there is a $(d_1 + d_2)$-regular family $\{G_n\}$ of expanders such that G_n contains an isometric copy of S_n.*

Proof. We just add to these a_n-separated sets of vertices of cardinality b_n the edge structure of S_n. The condition $\operatorname{diam} S_n \leq a_n$ implies that it is an isometric embedding of S_n into P_n. The maximum degree of the obtained graph is $\leq d_1 + d_2$. We add the necessary amount of loops to get a $(d_1 + d_2)$-regular graph. It is clear that the expanding constant of the graph obtained in this way from P_n cannot be less that the expanding constant of P_n. □

Proof of Theorem 5.76. Let S_n be a *square grid graph* defined as the graph with the vertex set $\{1,2,3,\dots,n\} \times \{1,2,3,\dots,n\}$; vertices (k_1,m_1) and (k_2,m_2) in S_n are adjacent if and only if $|k_1 - k_2| + |m_1 - m_2| = 1$. Applying Proposition 5.77 to an arbitrary family of expanders with $\{S_n\}_{n=1}^{\infty}$ being square grid graphs we get a family of expanders which we denote $\{G_n\}_{n=1}^{\infty}$. We may omit some graphs in this sequence and assume without loss of generality that that G_n contains a subgraph isometric to S_n. We claim that the family $\{G_n\}_{n=1}^{\infty}$ does not admit uniformly coarse embeddings into a family $\{H_n\}$ of expanders with indefinitely growing girths.

In fact, assume the contrary, let $f_n : G_n \to H_{m(n)}$ be a family of uniformly coarse embeddings, ρ_1 and ρ_2 be the corresponding functions. Consider restrictions of $\{f_n\}$ to $\{S_n\}$. It is clear that the graph S_n contains cycles of lengths $4, 8, \dots, 4n$. Furthermore, these cycles can be selected in such a way that vertices in a cycle of length $4k$, which are at distance γ in the (graph distance of the) cycle are at distance $\geq \min\{k,\gamma\}$ in S_n (we can just pick the cycles to be naturally defined "squares" in S_n).

The vertex set of such a cycle of length $4k$ is mapped onto a sequence $\{w_i\}_{i=1}^{4k}$ in $H_{m(n)}$ satisfying $d_{H_{m(n)}}(w_i, w_{i+1}) \leq \rho_2(1)$ (we let $i + 1 = 1$ for $i = 4k$). We join w_i and w_{i+1} with a shortest path in $H_{m(n)}$ and get closed walks of lengths $\leq 4\rho_2(1), \dots, 4n\rho_2(1)$ in $H_{m(n)}$. It is clear that for k satisfying $4k\rho_2(1) < g(H_{m(n)})$, where $g(H_{m(n)})$ is the girth of $H_{m(n)}$, the subgraph R_k of $H_{m(n)}$ induced by the introduced above walk of length $\leq 4k\rho_2(1)$ is a tree. Let k_0 be the largest k satisfying $4k\rho_2(1) < g(H_{m(n)})$. Now we use:

Proposition 5.78. *Let T be a thickening of a graph-theoretical tree, that is, a metric space obtained from a graph-theoretical tree if we identify each of its edges with a line segment of length 1, with the metric defined as the length of the shortest curve joining the points. Then, for any continuous map $f : \mathbb{S} \to T$, where $\mathbb{S}$ is the unit circle in the plane with its geodesic distance $d_{\mathbb{S}}$, there exist points $x, y \in \mathbb{S}$ such that $d_{\mathbb{S}}(x, y) \geq 2\pi/3$ and $f(x) = f(y)$.*

Proof. Divide $\mathbb{S}$ into three equal parts of length $2\pi/3$ each, let a, b, c be the ends of the parts. Then the images of the parts contain continuous images of line segments joining each of the pairs $(f(a), f(b))$, $(f(a), f(c))$ $(f(b), f(c))$. By a well-known simple fact such images of line segments in a thickening of a tree contain a common point p. Considering pre-images of p in all three parts of the circle $\mathbb{S}$ we get a triple of points with distance $\geq 2\pi/3$ at least between one of the pairs. □

We apply this proposition to the map constructed in the following way. As is clear from our construction, a cycle of length $4k_0\lceil\rho_2(1)\rceil$ admits a 1-Lipschitz map onto R_{k_0} which is an extension of the restriction of f_n to the $4k_0$-cycle: we replace each edge of the cycle by a path of length $\lceil\rho_2(1)\rceil$ and map this path in a 1-Lipschitz way onto the shortest path joining the corresponding w_i and w_{i+1} in $H_{m(n)}$ (we do not require this map to be injective or a coarse embedding). We use this map to construct a $\frac{4k_0\lceil\rho_2(1)\rceil}{2\pi}$-Lipschitz map of $\mathbb{S}$ onto the thickening T_{k_0} of R_{k_0} in the most natural way: we identify vertices of the $4k_0\lceil\rho_2(1)\rceil$-cycle with the points of $\mathbb{S}$ forming a regular convex polygon with $4k_0\lceil\rho_2(1)\rceil$ vertices and extend the map to the intervals between these vertices by mapping them onto the corresponding shortest paths in the thickening.

By Proposition 5.78, there are two points $x, y \in \mathbb{S}$ with $d_{\mathbb{S}}(x, y) \geq 2\pi/3$ with the same image. We consider the nearest to x and y points x_S and y_S in $\mathbb{S}$ corresponding to images of vertices of the $4k_0$-cycle. Then $d_{\mathbb{S}}(x, x_S) \leq \frac{2\pi}{4k_0}$ and $d_{\mathbb{S}}(y, y_S) \leq \frac{2\pi}{4k_0}$. Therefore

$$d_{H_{m(n)}}(f_n(x_S), f_n(y_S)) \leq d_{T_{k_0}}(f_n(x_S), f_n(y_S)) \leq \frac{4\pi}{4k_0} \cdot \frac{4k_0\lceil\rho_2(1)\rceil}{2\pi} = 2\lceil\rho_2(1)\rceil.$$

It is also easy to estimate that the distance between x_S and y_S in the $4k_0$-cycle is at least $\lfloor\frac{4}{3}k_0 - 2\rfloor$. Recalling the way in which the distance on a $4k_0$-cycle compares with the distance in square grid graph S_n, as well as the definition of a coarse embedding, we get

$$\rho_1\left(\min\left\{k_0, \left\lfloor\frac{4}{3}k_0 - 2\right\rfloor\right\}\right) \leq 2\lceil\rho_2(1)\rceil,$$

where k_0 is the largest integer satisfying $4k\rho_2(1) < g(H_{m(n)})$. This shows that $\lim_{n\to\infty} g(H_{m(n)}) = \infty$ contradicts $\lim_{t\to\infty} \rho_1(t) = \infty$. □

5.16 Exercises

Exercise 5.79. The complete bipartite graph $K_{n,n}$ is n-regular. Show that its eigenvalues are $\lambda_1 = n$, $\lambda_2 = \cdots = \lambda_{2n-1} = 0$ and $\lambda_{2n} = -n$.

Exercise 5.80. The complete graph K_n is $(n-1)$-regular. Show that its eigenvalues are $\lambda_1 = n-1$, $\lambda_2 = \cdots = \lambda_n = -1$.

Exercise 5.81. For each n, let C_n be the cycle of length n, that is, the graph whose vertex set is $\{1, \ldots, n\}$ for which two vertices are adjacent if and only if either the difference between the corresponding number is ± 1, or one of the numbers is 1 and the other is n. Show that the eigenvalues of C_n are equal to $2\cos\left(\frac{2\pi j}{n}\right)$, $j = 1, \ldots, n$.

If needed, see a hint in the Hints to exercises section below.

Exercise 5.82. The Hamming cube defined on page 118 can be introduced as a graph with its graph distance. In fact, one can check that $\mathbb{F}_2^n$ is naturally isometric to the graph whose vertex set is the set of all n-element sequences consisting of 0 and 1 and whose edge set consists of edges joining sequences with exactly one distinct entry.

Show that the eigenvalues of the adjacency matrix of this graph are $n, n-2, n-4, \ldots, -n+2, -n$.

If needed, see a hint in the Hints to exercises section below.

Exercise 5.83. Let G be a k-regular graph on n vertices with expanding constant h. Show that there is a constant $\alpha > 1$ depending only on k and h such that for every vertex $v \in V(G)$ and each radius $r \in \mathbb{N}$, we have the following estimate for the cardinality of the ball

$$|B(v,r)| \geq \min\{\alpha^r,\ |V(G)|\}.$$

(The conclusion is that the diameter of G does not exceed $\log_\alpha |V(G)|$.)

If needed, see a hint in the Hints to exercises section below.

Exercise 5.84. Let G be a k-regular graph on n vertices with expanding constant h, and let $f : V(G) \to \mathbb{R}$ (since the graph is finite the function is necessarily Lipschitz). Let M be a median of f. Show that

$$\{v \in V(G) :\ |f(v) - M| \geq t\,\mathrm{Lip}(f)\}| \leq Cne^{-ct} \tag{5.42}$$

for all $t > 0$ and some constants $0 < c, C < \infty$ depending only on h.

The next two exercises are related to Proposition 5.13.

Exercise 5.85. Assuming that you already know that there exist a family $\{F_n\}_{n=1}^{\infty}$ of finite groups with $\lim_{n\to\infty}|F_n| = \infty$ and symmetric subsets $S_n \subset F_n$ such that $|S_n| = k$ for all k and $\{\mathrm{Cay}(F_n, S_n)\}_{n=1}^{\infty}$ is a family of expanders, show that the condition (5.12) cannot be replaced by

$$|S_n \cdot A_n| \geq (1+c)|A_n|. \tag{5.43}$$

If needed, see a hint in the Hints to exercises section below.

Exercise 5.86. Let $\{F_n\}$ be a sequence of finite commutative groups with

$$\lim_{n\to\infty} |F_n| = \infty$$

and $S_n \subset F_n$ be symmetric subsets of F_n such that the cardinality $|S_n|$ does not depend on n. Then the sequence $\{\mathrm{Cay}(F_n, S_n)\}_{n=1}^{\infty}$ cannot be a sequence of expanders.

If needed, see a hint in the Hints to exercises section below.

Exercise 5.87. Let G be a 2^n-regular graph, $n \in \mathbb{N}$. Then the edge set of G can be represented as a union of edge sets of 2-regular spanning subgraphs of G.

If needed, see a hint in the Hints to exercises section below.

Exercise 5.88. (This exercise if substantially more difficult than Exercise 5.87. This result is due to Petersen.) Let G be a $2k$-regular graph ($k \in \mathbb{N}$). Then the edge set of G can be represented as a union of edge sets of 2-regular spanning subgraphs of G.

If needed, see a hint in the Hints to exercises section below.

Exercise 5.89. (The purpose of this exercise is to show a very interesting application of Exercise 5.88.) Let G be a $2k$-regular graph ($k \in \mathbb{N}$). Then G is a Schreier graph.

If needed, see a hint in the Hints to exercises section below.

Exercise 5.90. Show that there exist families of k-regular graphs for which the absolute spectral gaps (see Definitions 5.35 and 5.40) are not bounded away from 0, but the usual spectral gaps (see Definition 5.6) are bounded away from zero.

If needed, see a hint in the Hints to exercises section below.

Exercise 5.91. Show that the following three matrices, together with their inverses, form a symmetric generating set of $SL_n(\mathbb{Z})$, $n \geq 3$.

$$\begin{bmatrix} 0 & 1 & 0 & \dots & 0 & 0 \\ -1 & 0 & 0 & \dots & 0 & 0 \\ 0 & 0 & 1 & \dots & 0 & 0 \\ \vdots & \vdots & \vdots & \ddots & \vdots & \vdots \\ 0 & 0 & 0 & \dots & 1 & 0 \\ 0 & 0 & 0 & \dots & 0 & 1 \end{bmatrix}, \begin{bmatrix} 0 & 1 & 0 & \dots & 0 & 0 \\ 0 & 0 & 1 & \dots & 0 & 0 \\ 0 & 0 & 0 & \dots & 0 & 0 \\ \vdots & \vdots & \vdots & \ddots & \vdots & \vdots \\ 0 & 0 & 0 & \dots & 0 & 1 \\ (-1)^{n-1} & 0 & 0 & \dots & 0 & 0 \end{bmatrix}, \begin{bmatrix} 1 & 0 & 0 & \dots & 0 & 0 \\ 1 & 1 & 0 & \dots & 0 & 0 \\ 0 & 0 & 1 & \dots & 0 & 0 \\ \vdots & \vdots & \vdots & \ddots & \vdots & \vdots \\ 0 & 0 & 0 & \dots & 1 & 0 \\ 0 & 0 & 0 & \dots & 0 & 1 \end{bmatrix}.$$

If needed, see a hint in the Hints to exercises section below.

Exercise 5.92. Let G be a finitely generated group. Show that one can find in G a symmetric generating set (see Definition 5.9), such that the corresponding graph Z (see the beginning of Section 5.4.4) is connected.

Exercise 5.93. Give an example of graphs G_1 and G_2 such that the zigzag product $G_1 ⓩ G_2$ is defined, and for different choices of bijections π_u (see Definition 5.41) we get non-isomorphic zigzag products.

Exercise 5.94. Prove the statement (2) in Remark 5.55.

Exercise 5.95. Let $k \geq 3$, $t \geq 3$, $k, t \in \mathbb{N}$, and let $c_3, c_4, \ldots, c_m \in (\mathbb{N} \cup \{0\})$. Prove that there exists a k-regular graph having exactly c_j cycles of length j, $3 \leq j \leq m$. (There are no restrictions on longer cycles.)

If needed, see a hint in the Hints to exercises section below.

5.17 Notes and remarks

General references: Recommended sources for readers willing to learn more about expanders and their applications in different areas are [110], [205], [299], [301], [423], see also references therein.

The existence of expanders was first proven by Kolmogorov–Barzdin in [253] and then later by Pinsker in [374]. Both proofs are probabilistic and while Pinkser was the first to refer to the objects as expanders, his definition deviates slightly from the one presented in Definition 5.1. Gromov and Guth recently published a paper, [180], containing multidimensional generalizations of the Kolmogorov–Barzdin results.

A probabilistic construction of expanders which is very close to Pinsker's is given in Sarnak [404, Section 3.1.2] (it was also reproduced in [299, p. 6]). It is the following:

We consider two sets of cardinality n, denote them I and O, we consider five permutations $\pi_1, \pi_2, \pi_3, \pi_4, \pi_5$ in the symmetric group S_n. We connect each element i of I with $\pi_1(i), \ldots, \pi_5(i)$ in O (with some abuse of notation we assume that both I and O are identified with $\{1, \ldots, n\}$).

Theorem 5.96. *There exists a collection $\pi_1, \pi_2, \pi_3, \pi_4, \pi_5$ such that for each subset $A \subset I$ with $|A| \leq n/2$ the number of neighbors of elements of A in O is at least $\frac{3}{2}|A|$.*

Remark 5.97. If we consider such graphs for all n and "paste" sets I and O according to one of the possible bijections, we get an expanding family of graphs, but such graphs may have parallel edges and loops, so we have to widen our notion of an expander family to include graphs which are not simple.

Other probabilistic constructions were suggested and it was later discovered that such constructions could be used to determine expander families consistent with the definition provided in Definition 5.1. For more information see [70] and [297]. We are going to provide more details on the construction contained in the problem book of Lovasz [297, Problem 11.33], where 3-regular expanders, without loops and parallel edges were constructed. Here is the description of this construction.

Definition 5.98. Let G be a graph. A 1-*factor* in G is a collection F of edges satisfying the condition: each vertex of G is incident to exactly one edge in F.

Let G be a union of three random 1-factors on the set $\{1, \ldots, 2n\}$. The proof goes in two steps:

Theorem 5.99. (a) *If n is large enough, the probability that G is a* 3*-regular simple graph is between* 0.1 *and* 0.9.

(b) *The probability that G has expanding constant $\geq \frac{1}{1000}$ is at least* 95%.

To prove the statement (a) one is expected to use inclusion-exclusion formulas developed in [297].

A similar construction with less technical estimates, but also with a weaker result (only k-regular expanders for sufficiently large k) is presented in the lecture course [423].

Very strong results on probabilistic expanders were obtained by Friedman [156]. He proved that for an even $d \geq 4$, a random d-regular graph model formed from $d/2$ uniform, independent permutations on $\{1, \ldots, n\}$ has a very large spectral gap $k - \lambda_2$. Namely, he proved that for any $\varepsilon > 0$ all eigenvalues aside from $\lambda_1 = d$ are bounded by $2\sqrt{d-1} + \varepsilon$ with probability $1 - O(n^{-\tau})$, where $\tau = \lceil(\sqrt{d-1}+1)/2\rceil$. The work [156] contains results for other models of random graphs and is indispensable for everyone interested in random expanders.

Amit and Linial introduced another approach to the construction of random expanders in [21]. The notion of graph lift, which we introduce in Section 5.7, serves as the cornerstone of their method. A version of their construction can be found in Section 5.8 and for more general constructions, which include the 3-regular case, refer to [21].

Section 5.2 contains a very short introduction to spectral graph theory, which is a well-developed area with various applications in other areas of mathematics. Interested readers are referred to [61], [100], [106], [170], and [297, §11] for additional information and references. We strongly recommend the elegant exposition in Biggs [61] and the rather small collection of problems in Lovasz [297, §11], which nevertheless develops many useful and important techniques.

The right-hand side inequality in the spectral characterizations of expanders (Theorem 5.7) is due to Alon [10] and Dodziuk [125]. The left-hand side inequality is due to Alon [10], a close inequality was proved by Tanner [421]. Alon's proof of the spectral characterization is based on an inequality proved in [15]. These results can be considered as discrete versions of results of Cheeger [95]. See [402] for an interesting and accessible introduction to Poincaré and Sobolev inequalities in a continuous setting.

The method, first mentioned in the introduction, in which the expanding properties of randomly-generated expanders are checked using spectral characterization, was developed by Alon [10]. Results on algorithms for computing eigenvalues can be found in [388, Chapter 10]. The Cheeger constant itself is computationally hard. More precisely, [64] proved that its computation is co-NP-complete (see the paper for explanation of the term). They proved the statement in somewhat different setting, see [10] for relations between the notions used in [64] and the Definition 5.1.

The property (T) was introduced by Kazhdan [246] and actively studied since then. See [53] for a recent accessible account. Property (T) was used by Margulis [308] to find the first deterministic construction of expanders, Section 5.3 is mostly based on his ideas. These ideas were further developed by Alon and Milman in [15], the source which we also use, as well as the lecture notes [423].

The estimates of Kazhdan constants of alternating groups which were mentioned in Section 5.3 are due to Kassabov [244].

Proposition 5.28 is a classical result whose history I did not investigate. We refer interested readers to [306, Section 22], [104, Section 7.2], [340, Section 5 in Chapter VII], [439], [195, Chapter III], and references therein both for the history of the study of finite generation of $SL_n(\mathbb{Z})$ and related groups, and the smallest known symmetric sets generating $SL_n(\mathbb{Z})$. Currently the smallest symmetric set known to generate $SL_n(\mathbb{Z})$ has cardinality 4, see [340].

The first proof of the fact that $SL_n(\mathbb{Z})$ has property (T) mentioned in Section 5.4.3 is the original proof of Kazhdan [246]. The second proof is due to Shalom [410]. Both proofs are presented in [53]. The first proof and some related results and approaches are presented in [423].

Theorem 5.34 is due to Żuk [446]. Mentioned generalizations and improvements of Theorem 5.34, as well as examples showing applicability of the criteria, can be found in [53, Chapter 5].

Construction of families of expanders based on zigzag products was suggested by Reingold, Vadhan, and Wigderson [392]. (A variant of this construction was analyzed earlier by Gromov [174].) This construction was developed in many different directions, see, for example, [14], [397]. Mendel and Naor [329] suggested a new approach to the proof of the main step of the construction of [392]. In our presentation we use the approach suggested in [329]. It is worth mentioning that inequality 5.24 implies the main result of [392]. Actually it leads to the later bound of Reingold–Trevisan–Vadhan [391]. Our construction of a base graph follows the paper [17], this construction goes back to the paper [16]. We refer to [307, Chapter 4] for the construction and properties of the fields $\mathbb{F}_q$, where q is a power of a prime number. The existence and uniqueness of $\mathbb{F}_q$, where q is a power of prime is shown in [307, Chapter 4, §3]; the fact which we use is shown in the introductory remarks of [307, Chapter 4, §2].

Graph lifts also known as *covering graphs* and *graph covers* are classical objects in topology and topological graph theory. They were studied from different perspectives in [20], [21], [22], [23], [27], [43], [128], [140], [183], [184], [185], [205, Section 6],

[279], [337], [359], [417]. In some works these notions are studied using such terms as *voltage graph* and *voltage assignment*.

Example 5.53 is a version of [68, Exercise 2, p. 161], it can also be found in [140], where an interesting strengthening of it is also proved. Proposition 5.56 was proved in [139, p. 254]. Theorem 5.57 was proved in [359], the construction was inspired by [27]. Theorem 5.57 solves the problem posed by Linial, Magen, and Naor [292], and repeated by Linial in [289, Open Problem 7] and [317, Problem 2.3].

See, for example, [71, Section 3.3] for well-known results on Euler trails and tours which we use in the proof of Theorem 5.57.

Theorem 5.62 is a weakened and simplified version of a result of Amit and Linial [21, Theorem 2.1]. Some pieces of the argument are borrowed from the notes of Tao [423].

Theorems 5.63 and 5.64 are from [139]. Theorem 5.66 is from [68, Theorem 1.1, p. 104], its proof is close in spirit to the argument of Erdős in [138].

Theorem 5.67 goes back to Erdős and Sachs [139], it was refined by Walther [433, 434], and by Sauer [405], whose bounds are the strongest, see [68, Chapter III, §1] and [69, §3] in this connection. Theorem 5.69 is due to Chandran [92], where the readers can also find the modification of the proof needed in the case where the number of vertices is odd.

The main result of Section 5.12 is due to Biggs [58]. See [19] for related results.

The construction presented in Section 5.13 is due to Biggs [60, p. 57], who mentions the fact that it was inspired by the group-theoretical results of [49]. The Biggs's construction was presented in a more general and explicit form in [11, p. 1752]. Biggs also mentions [60, p. 57] that is seems that the actual girth of the constructed graphs is much larger than $2r + 1$ and that this was verified for some special cases in [203]. See [59] for related results and discussions.

The construction presented in Section 5.14 is due to Margulis [309]. Our presentation is influenced by the presentation in [110, Appendix] (where one can find more detailed arguments and ideas which can be used to improve the estimates). The better girth parameters were achieved by a modification due to Imrich [206]. Lubotzky observed (see [300, Proposition 2.2.2, p. 164]) that Margulis's graphs [309] presented in Section 5.14 are expanders.

It is worth mentioning that if we consider an analogue of Margulis's construction with matrices

$$\begin{pmatrix} 1 & 3 \\ 0 & 1 \end{pmatrix}, \quad \begin{pmatrix} 1 & 0 \\ 3 & 1 \end{pmatrix} \tag{5.44}$$

instead of (5.41), the expansion of the obtained graphs does not follow from the argument of [300] and Lubotzky suggested the corresponding open problem [300, Open Problem 2.2.3, p. 165] and [304, Problem 5.18, p. 90]. This problem was solved in the positive by Bourgain and Gamburd [78, Theorem 3].

The main result of Section 5.15 is due to Mendel and Naor [329, Section 9.1]. The topological part of the argument (Proposition 5.78) goes back to Rabinovich and Raz

[387]. As for expanders with large girth, which we need for Theorem 5.75, one of the examples is provided by Margulis's graphs presented in Section 5.14 (see the previous paragraph). A simpler proof of existence of such families can be obtained by doing some more work on the examples of random graphs presented in Section 5.10. See [175, Section 1.B] for the general notion of *thickening of a graph*.

The result of Exercise 5.83 was proved in [15]. Results of the type of Exercise 5.84 establish the so-called *concentration of measure phenomenon*. The first phenomenon of this type goes back to the result of Levy, the importance of such phenomena was recognized by V. Milman and developed by him and many other authors, see [177, Chapter $3\frac{1}{2}$], [271], and [332] for accounts on the obtained theory. Exercise 5.84 can be easily derived from results of [15] and basic results of this theory.

It is difficult to find out who first observed that commutative groups do not lead to expander families (Exercise 5.86). Lubotzky and Weiss [303] discovered several classes of families of finite groups which are not expanders with respect to any families of generators (of constant size). For example they proved that an infinite family of quotient groups of a finitely generated amenable group is not a family of expanders.

The result of Exercise 5.88 was proved by Petersen [370]. See also [63, Chapter 10]. The hint to Exercise 5.88 is borrowed from [185, pp. 41–42].

The result stated in Exercise 5.89 is due to Gross [182]. The result of Exercise 5.95 is due to Sachs [398],[399].

5.17.1 Bounds for spectral gaps

We would like to finish this section by mentioning some more important constructions of expanders and graphs with large girth.

The following theorem bounds from above the spectral gap of a k-regular graph.

Theorem 5.100 (Alon-Boppana). *If $\{G_n\}_{n=1}^{\infty}$ is a family of k-regular graphs with $|V(G_n)| \to \infty$, then*

$$\liminf_{n\to\infty} \lambda_2(G_n) \geq 2\sqrt{k-1}. \tag{5.45}$$

This result is mentioned in [10, p. 96]. To the best of my knowledge this result was never published by its authors. Nilli [341] obtained (and published) a more precise estimate. See [110, Sections 1.3–1.4] for a presentation of Theorem 5.100 and related estimates.

5.17.2 Graphs with very large spectral gaps

Graphs attaining the bound (5.45) in a precise (non-asymptotic) way and for absolute values of all eigenvalues except the first one have a special name:

Definition 5.101. A finite, connected, k-regular graph G is called a *Ramanujan graph* if for every eigenvalue $\lambda_j(G)$, except $\lambda_1(G)$, the inequality

$$|\lambda_j(G)| \leq 2\sqrt{k-1}$$

holds.

The existence of families of Ramanujan graphs with indefinitely growing orders (numbers of vertices) was proved independently and almost simultaneously by Lubotzky–Phillips–Sarnak [302] and Margulis [310]. The families of expanders constructed in these papers are Cayley graphs and have very large girth. More precisely, their girths satisfy

$$g(G_n) \geq \left(\frac{4}{3} + o(1)\right) \log_{k-1} |V(G_n)|.$$

These constructions of Ramanujan graphs are explained and partially presented in an accessible text [110, Section 4.3], see also [299].

All the known families of Ramanujan graphs of indefinitely growing orders have *essentially large girth*, that is, for all $L \in \mathbb{N}$, we have

$$\lim_{n\to\infty} \frac{c_L(G_n)}{|V(G_n)|} = 0,$$

where $c_L(G_n)$ denotes the number of cycles of length L in G_n. Recently, it was proved [1] that this is not a coincidence.

Theorem 5.102 ([1]). *Let Γ be a group generated by a finite symmetric set S and let $\{H_n\}_{n=1}^{\infty}$ be a sequence of subgroups of finite index with indices $\to \infty$ as $n \to \infty$, and such that the Schreier graphs $\mathrm{Sch}(\Gamma/H_n, S)$ are Ramanujan for all n. Then the sequence $\{\mathrm{Sch}(\Gamma/H_n, S)\}_{n=1}^{\infty}$ has essentially large girth.*

5.17.3 Some more results and constructions

- There are many other recent constructions of expander families and families of graphs with growing girth: [78], [81], [157], [245]. One can find an overview of some of these constructions in [301] and [423]. The notes of Tao [423] contain a nice presentation of some of them.

- Ajtai [7] presented an algorithm which in $n^3(\log n)^3$ constructs a 3-regular graph with "large" expanding constant on n vertices. In each step of the algorithm one pair of edges in the graph is replaced by another pair of edges so that the total number of cycles of length $\lfloor c \log n \rfloor$ decreases (for some fixed absolute constant c). The graph has a "large" expanding constant when a local minimum in the number of cycles of length $\lfloor c \log n \rfloor$ is reached.

- The papers [158] and [413] contain results on connections between girth and expansion.
- Lubotzky [298] constructed a sequence of graphs with indefinitely growing girths using the theory of Coxeter groups.
- The paper [419] is a relatively recent contribution to the rather old study of the set of possible cycle lengths in a graph of average degree d and girth g.

5.18 Hints to exercises

To Exercise 5.81. The corresponding matrix can be written as a sum of two permutation matrices. If we consider the matrices in the corresponding complex space $\mathbb{C}^n$, each of them is diagonalizable in the basis consisting of the following vectors

$$\{e^{\frac{2\pi ikm}{n}}\}_{m=1}^{n}, \quad k = 1, \ldots, n.$$

See [297, Problem 11.1] for a detailed solution and related problems.

To Exercise 5.82. Consider the basis consisting of the Walsh functions.

See [297, Problem 11.9] for a detailed solution.

To Exercise 5.83. First use the definition of the Cheeger constant to show that there is a number $\beta > 1$ such that

$$|B(v,r)| \geq \min\{\beta^r, \frac{|V(G)|}{2}\}.$$

Then use this fact to estimate the diameter of $V(G)$. Finally, combine the steps to get the desired estimate, possibly, with α somewhat smaller than β from the first step.

To Exercise 5.84. We may assume that $\mathrm{Lip}(f) = 1$ and $M = 0$. The set of vertices satisfying (5.42) with $t > 1$ consists of two parts: where $f(v) \geq t$ and where $f(v) \leq t$. Use the definition of the Cheeger constant to show that cardinalities of both sequences of sets are exponentially decaying as $t = 1, 2, 3, \ldots$.

To Exercise 5.85. Consider products of the groups F_n with the group $\mathbb{F}_2$ consisting of two elements.

To Exercise 5.86. Combine Proposition 5.13 with the result of Exercise 5.83 and the study of the growth in cardinality of products $S_n \cdot S_n \cdot \cdots \cdot S_n$.

To Exercise 5.87. Use Euler circuits.

To Exercise 5.88. Allow graphs to have loops (assuming that each loop contributes 2 to the degree of the vertex it is incident to) and use the backward induction of the number of loops in a $2k$-regular graph with n vertices. To prove the result for graphs with fewer loops do the following: let uv and vw be two adjacent edges in G. Consider the graph G' obtained from G by deletion of uv and vw and addition of the edge uw and the loop at v. Use the result of Exercise 5.87 for 4-regular graphs.

To Exercise 5.89. 2-regular graphs are unions of cycles. Consider the permutations of $V(G)$ corresponding to these unions of cycles.

To Exercise 5.90. You are expected to use Proposition 5.4(c), the existence of expander families, their spectral characterization and the fact that they can be chosen to be bipartite.

To Exercise 5.91. Use the fact that the permutation of $\{1,\dots,n\}$, which transposes the first two elements, and the cyclic permutation, which acts as $1 \mapsto 2$, $2 \mapsto 3, \dots,$ $(n-1) \mapsto n$, $n \mapsto 1$, generate the group of all permutations of $\{1,\dots,n\}$. It is also useful to compute the square of the first matrix.

To Exercise 5.95. Try to follow the argument of Theorem 5.67, starting with a family of disjoint cycles with prescribed lengths and a large cycle.

Chapter 6

Banach spaces which do not admit uniformly coarse embeddings of expanders

Since each n-element metric space is isometric to a subset of ℓ_∞^n (Proposition 1.17) and each Banach space without nontrivial cotype admits, by the Maurey–Pisier theorem (Theorem 2.56), $(1+\varepsilon)$-bilipschitz embeddings of $\{\ell_\infty^n\}_{n=1}^\infty$, we get that any family of finite metric spaces and in particular, any family of expanders, admits uniformly bilipschitz embeddings into an arbitrary Banach space without nontrivial cotype. It turns out that families of expanders are resistant to uniformly coarse and uniformly bilipschitz embeddings into all other Banach spaces ("resistant" in the sense that no such embeddings are known so far). So far there are no examples of a Banach space X with nontrivial cotype for which there is a family $\{G_n\}_{n=1}^\infty$ of expanders admitting uniformly coarse embeddings into X, so it is still possible that such Banach spaces X do not exist. Such a negative answer is not only known for uniformly coarse, but also for uniformly bilipschitz embeddings. In Chapter 11 we list some more specific problems of this type.

The purpose of this chapter is to present known non-embeddability results for families of expanders as well as to describe some classes of Banach spaces and some particular Banach spaces for which the problem of embeddability of families of expanders into them is open.

It is known that wide classes of Banach spaces admit no uniformly coarse embeddings of families of expanders into them. For even wider classes of Banach spaces it is known that there are no uniformly bilipschitz embeddings of expanders into them, we discuss such results for uniformly convex Banach spaces in Chapter 8. There is also a wide class of Banach spaces for which it is known that *there are* families of expanders which are not uniformly coarsely embeddable into them. For bilipschitz embeddings the last version of the problem is relatively easy and a complete answer is known: there exist families $\{G_n\}$ of expanders which admit uniformly bilipschitz embeddings into a Banach space X if and only if X has no nontrivial cotype. The idea of constructions of such families of expanders is the following: to construct expanders containing uniformly bilipschitz images of nets of balls of ℓ_∞^n (See Exercise 10.21 in Chapter 10).

6.1 Banach spaces whose balls admit uniform embeddings into L_1

It turns out that Theorem 4.9, stating that families of expanders do not admit uniformly coarse embeddings into L_1, can be transferred to any Banach space whose unit ball admits a uniform embedding into L_1. The *unit ball* of a Banach space X is the set $B_X = \{x \in X : \|x\| \le 1\}$. We use the term "uniform embedding" in the way in which it was introduced in Definition 1.36.

Theorem 6.1. *If a Banach space X is such that B_X admits a uniform embedding into L_1 and $\{G_n\}$ is a family of expanders with their graph distances, then $\{G_n\}$ are not uniformly coarsely embeddable into X.*

Proof. Let G be a k-regular graph and h be its Cheeger constant (saying *h is a Cheeger constant of G* we mean $h(G) \ge h$). We are going to prove Theorem 6.1 on the same lines as the similar result for L_1 (Theorem 4.9). We are going to present our proof in such a way that it can be read independently, using two versions of Poincaré type inequalities for L_1 proved in Section 4.2.

The main step is to prove that for each (finite) k-regular graph with Cheeger constant h and each mapping $f : V(G) \to X$ there is a constant $D = D(k, h, X)$ such that for some $x \in X$

$$|\{s \in V(G) : \|f(s) - x\|_X \le D \operatorname{Lip}(f)\}| \ge \frac{3|V(G)|}{4}. \tag{6.1}$$

Let $D > 0$ be the least number satisfying

$$\exists x \in X \; |\{s \in V(G) : \|f(s) - x\| \le D \operatorname{Lip}(f)\}| \ge \frac{3|V(G)|}{4}. \tag{6.2}$$

At this point we allow the number D to depend on the graph G and on the embedding f. (We need to modify the argument slightly if the infimum is not attained.)

We are going to find an inequality for D in terms of k, h, ω_φ and $\omega_{\varphi^{-1}}$; where $\varphi : B_X \to L_1$ is a uniform embedding, and ω_φ and $\omega_{\varphi^{-1}}$ are the moduli of continuity of φ and φ^{-1}. The inequality which we get leads to an estimate of the desired type.

Lemma 6.2. *Let G be a k-regular graph with Cheeger constant h. Let f be a map of the vertex set of G into a Banach space X, $x \in X$. Then*

$$|\{s \in V(G) : \|f(s) - x\| \le D_1 \operatorname{Lip}(f)\}| \ge \frac{|V(G)|}{2}, \tag{6.3}$$

implies

$$|\{s \in V(G) : \|f(s) - x\| \le \left(\frac{2k}{h} + D_1\right) \operatorname{Lip}(f)\}| \ge \frac{3|V(G)|}{4}. \tag{6.4}$$

Proof. It is clearly enough to consider functions f having Lipschitz constant 1. Consider the function $r(s) = \|f(s) - x\|$. It is also 1-Lipschitz. The inequality (6.3) implies that its median m satisfies $m \le D_1$. Applying Lemma 4.8 we get

$$\sum_{s \in V(G)} \big| \|f(s) - x\| - m \big| \le \frac{k|V(G)|}{2h}.$$

Therefore

$$\left|\left\{s \in V(G) : \big| \|f(s) - x\| - m \big| \le \frac{2k}{h}\right\}\right| \ge \frac{3|V(G)|}{4},$$

and we get (6.4). □

In the proof of the theorem it also suffices to consider the case where $\mathrm{Lip}(f) = 1$. We map the smallest ball $B(x, R)$ (we define balls by $B(x, R) = \{z \in X : \|z - x\| \le R\}$) containing $f(V(G))$ into L_1 in the following way:

- First we project all points of $f(V(G))$ which are not in $B(x, D)$ onto the sphere $S(x, D)$ (where $S(x, D) = \{z \in X : \|z - x\| = D\}$). We mean the following projection P:

$$P(y) = \begin{cases} y & \text{if } y \in B(x, D) \\ x + D\dfrac{y - x}{\|y - x\|} & \text{if } y \in B(x, R) \setminus B(x, D). \end{cases}$$

- Then we translate the ball $B(x, D)$ to $B(0, D)$.
- Then we map the ball $B(0, D)$ homothetically onto B_X.
- Then we apply φ.

We denote the obtained composite mapping by g, so $g : f(V(G)) \to L_1$. Let us estimate $\mathrm{Lip}(g)$. It is easy to check that the Lipschitz constant P is ≤ 2 (see hint to Exercise 6.27). The Lipschitz constant of the translation is 1, of the homothety is $\frac{1}{D}$. Therefore each edge of G is mapped by the composition of P, the translation, and the homothety to a pair of vectors at distance $\le \frac{2}{D}$. Combining this with the definition of ω_φ we get $\mathrm{Lip}(g) \le \omega_\varphi(2/D)$.

Since $g(B(x, R))$ is in L_1, by Corollary 4.13 (which proves a version of (6.1) for L_1), we can find a subset $A \subset V(G)$ of size $\ge \frac{3}{4}|V(G)|$ and a vector $m(g) \in L_1$ such that

$$\|g(f(s)) - m(g)\| \le \frac{2k}{h}\omega_\varphi(2/D) \tag{6.5}$$

for any $s \in A$. Decreasing the subset A by at most $\frac{|V(G)|}{4}$ elements, we get a subset of size at least $\frac{|V(G)|}{2}$ of points in A which were not mapped by f into $B(x, R) \setminus B(x, D)$ and therefore were mapped according to the first option when we applied P. Also, doubling the constant in (6.5), we may assume that $m(g)$ was picked in $g(f(V(G)))$. Therefore there is $s_0 \in V(G)$ with $f(s_0) \in B(x, D)$ such that

$$|\{s : \|g(f(s)) - g(f(s_0))\| \leq \frac{4k}{h}\,\omega_\varphi(2/D)\}| \geq \frac{|V(G)|}{2}.$$

We consider pre-images (with respect g) of these points $g(f(s))$ in $B(x, D)$. We get

$$|\{s : \|f(s) - f(s_0)\| \leq D\omega_{\varphi^{-1}}\Big(\frac{4k}{h}\,\omega_\varphi(2/D)\Big)\}| \geq \frac{|V(G)|}{2}.$$

We recall that $\mathrm{Lip}(f) = 1$ and apply Lemma 6.2. We get

$$|\{s : \|f(s) - f(s_0)\| \leq \Big(\frac{2k}{h} + D\omega_{\varphi^{-1}}\Big(\frac{4k}{h}\omega_\varphi(2/D)\Big)\Big)\}| \geq \frac{3|V(G)|}{4}.$$

By the minimality of D we get

$$D \leq \Big(\frac{2k}{h} + D\omega_{\varphi^{-1}}\Big(\frac{4k}{h}\omega_\varphi(2/D)\Big)\Big).$$

Hence at least one of the following is true

$$D \leq \frac{4k}{h}$$

or

$$D \leq 2D\omega_{\varphi^{-1}}\Big(\frac{4k}{h}\,\omega_\varphi(2/D)\Big). \tag{6.6}$$

It is clear that both equations lead to estimates of D from above in terms of k, h and the moduli of continuity.

The inequality (6.1) is an obstruction to a uniformly coarse embeddability of $\{G_n\}$ into X. In fact, it is easy to see that uniformly coarse embeddings of unweighted graphs with their graph distances are Lipschitz maps, and their Lipschitz constants do not exceed $\rho_2(1)$ (ρ_1, ρ_2 are functions used in the definition of a coarse embedding). So we get that there is a sequence of balls in X of uniformly bounded radiuses $D(k, h, X)\cdot\rho_2(1)$ which contain an indefinitely increasing, as $n \to \infty$, numbers of images of vertices of $\{G_n\}_{n=1}^{\infty}$. Combining this with the definition of uniformly coarse embeddings and the fact that in a k-regular graph the number of vertices at distance M to a given vertex does not exceed

$$1 + k + k(k-1) + \cdots + k(k-1)^{M-1} \leq k^M + 1$$

(the main point is that this number does not depend on n), we get a contradiction. □

Banach spaces having nontrivial cotype and satisfying certain additional conditions, like having unconditional basis or being a Banach lattice, satisfy the assumption of

Theorem 6.1. As we know from the introduction to this chapter, Banach spaces with no nontrivial cotype do not produce any obstructions to embeddability of finite metric spaces into them (and Theorem 2.63 shows that this statement can be generalized to locally finite metric spaces). However, there are important and interesting Banach spaces which have nontrivial cotype, and even nontrivial type, but do not satisfy the assumptions of Theorem 6.1. We study such spaces in Sections 6.4 and 6.5. To study them we need a characterization of reflexivity presented in Section 6.3.

6.2 Banach spaces not admitting coarse embeddings of expander families, using interpolation

In this section we describe one more way of proving that some Banach spaces do not admit uniformly coarse embeddings of any expander families.

Let $\{G_n\}_{n=1}^{\infty}$, $G_n = (V_n, E_n)$, be a sequence of k-regular graphs satisfying

$$\lim_{n\to\infty} |V_n| = \infty,$$

$\{A_n\}_{n=1}^{\infty}$ be their normalized adjacency matrices (see page 148), and let X be a Banach space. We introduce the spaces $\ell_p(V_n, X)$ as spaces of X-valued functions on V_n with the norm

$$\|f\|_p = \left(\sum_{v\in V_n} \|f(v)\|_X^p\right)^{\frac{1}{p}}.$$

We denote the operator of averaging in $\ell_p(V_n, X)$ by R, that is,

$$\forall u \in V_n \quad Rf(u) = \frac{1}{|V_n|}\sum_{v\in V_n} f(v).$$

We consider $\{A_n\}$ as operators on $\ell_p(V_n, X)$ (sometimes called *Markov operators*), we mean:

$$(A_n f)(u) = \sum_{v\in V_n} a_{v,u} f(v) = \frac{1}{k}\sum_{v,\ uv\in E_n} f(v).$$

We are going to use the following criterion for coarse non-embeddability of the family $\{G_n\}$ into X.

Lemma 6.3. *Suppose that there exists $m \in \mathbb{N}$ such that*

$$\delta(\{G_n\}, X, m) \stackrel{\text{def}}{=} \sup_n \|A_n^m - R\|_2 < 1 \tag{6.7}$$

for all $n \in \mathbb{N}$, where the operator norm is taken in $\ell_2(V_n, X)$. Then the graphs $\{G_n\}$ do not admit uniformly coarse embeddings into the Banach space X.

Proof. Assume the contrary, let $f_n : V_n \to X$ be uniformly coarse embeddings, $\rho_1, \rho_2 : [0,\infty) \to [0,\infty)$ be the corresponding functions.

Since V_n is finite we may regard f_n as an element of $\ell_2(V_n, X)$. Observe that $A_n R = RA_n = R$ and $R^2 = R$. Therefore

$$\begin{aligned}\|f_n - Rf_n\|_2 &\le \|f_n - A_n^m f_n\|_2 + \|A_n^m f_n - Rf_n\|_2\\ &= \|f_n - A_n^m f_n\|_2 + \|(A_n^m - R)(f_n - Rf_n)\|_2\\ &\le \|f_n - A_n^m f_n\|_2 + \delta(\{G_n\}, X, m)\|f_n - Rf_n\|_2\end{aligned}$$

Hence

$$(1 - \delta(\{G_n\}, X, m))\|f_n - Rf_n\|_2 \le \|f_n - A_n^m f_n\|_2. \tag{6.8}$$

We need the following standard inequality (in the proof we use the fact that Rf_n is a constant function on V_n)

$$\begin{aligned}&\Big(|V_n|^{-2} \sum_{(x,y)\in V_n\times V_n} \|f_n(x) - f_n(y)\|^2\Big)^{1/2}\\ &\quad\le \Big(|V_n|^{-2} \sum_{(x,y)\in V_n\times V_n} \|f_n(x) - Rf_n\|^2\Big)^{1/2}\\ &\quad\quad + \Big(|V_n|^{-2} \sum_{(x,y)\in V_n\times V_n} \|Rf_n - f_n(y)\|^2\Big)^{1/2}\\ &\quad= 2\Big(|V_n|^{-1} \sum_{x\in V_n} \|f_n(x) - Rf_n\|^2\Big)^{1/2}.\end{aligned} \tag{6.9}$$

Combining (6.8) and (6.9) we get

$$\frac{1}{|V_n|^2} \sum_{(x,y)\in V_n\times V_n} \|f_n(x) - f_n(y)\|^2 \le \frac{4}{|V_n|(1-\delta)^2}\|f_n - A_n^m f_n\|_2^2. \tag{6.10}$$

Now we recall that A_n^m is the adjacency matrix of the m-th power of the graph G_n (see page 152), which we denote $H = (V_n, E(H))$. Therefore $g = f_n - A_n^m f_n$ can be written as

$$g(u) = \frac{1}{k^m} \sum_{v,\ uv\in E(H)} (f_n(u) - f_n(v))$$

and

$$\|f_n - A_n^m f_n\|_2^2 = \sum_{u\in V_n} \left\| \frac{1}{k^m} \sum_{v,\ uv\in E(H)} (f_n(u) - f_n(v)) \right\|_X^2.$$

Observe that $uv \in E(H)$ implies that $d_{G_n}(u, v) \le m$, and hence

$$\|f_n(u) - f_n(v)\|_X \le \rho_2(1)m$$

(since G_n is an unweighted graph, the map f_n is $\rho_2(1)$-Lipschitz). Therefore

$$\left\| \frac{1}{k^m} \sum_{v,\, uv \in E(H)} (f_n(u) - f_n(v)) \right\|_X^2 \le (\rho_2(1)m)^2.$$

Thus (6.10) implies

$$\frac{1}{|V_n|^2} \sum_{(u,v)\in V_n\times V_n} \| f_n(u) - f_n(v)\|_X^2 \le \left(\frac{2\rho_2(1)m}{(1-\delta)^2}\right)^2. \tag{6.11}$$

Hence

$$\frac{1}{|V_n|^2} \sum_{(u,v)\in V_n\times V_n} (\rho_1(d_{G_n}(u,v)))^2 \le \left(\frac{2\rho_2(1)m}{(1-\delta)^2}\right)^2. \tag{6.12}$$

Now we use the same argument as in the estimate for L_1-distortion of expanders (see page 110). Namely, we observe the number of vertices at distance D to a given vertex in a k-regular graph is at most

$$1 + k + k(k-1) + \cdots + k(k-1)^{D-1} \le k^D + 1.$$

Let $D_n = \log_k \left(\frac{|V_n|}{2} - 1\right)$. Then there are at most $\frac{|V_n|}{2}$ vertices with distance $\le D_n$ to a given vertex in G_n. Therefore

$$\sum_{(u,v)\in V_n\times V_n} (\rho_1(d_{G_n}(u,v)))^2 \ge \frac{|V_n|^2}{2} \cdot \left(\rho_1\left(\log_k\left(\frac{|V_n|}{2} - 1\right)\right)\right)^2.$$

Combining with (6.12) we get

$$\frac{1}{2} \cdot \left(\rho_1\left(\log_k\left(\frac{|V_n|}{2} - 1\right)\right)\right)^2 \le \left(\frac{2\rho_2(1)m}{(1-\delta)^2}\right)^2.$$

Letting $n \to \infty$ we get a contradiction because the left-hand side $\to \infty$ and the right-hand side does not depend on n. □

Thus the main topic of this chapter makes us interested in the problem:

Problem 6.4. Characterize Banach spaces X such that for each family $\{G_n\}$ of expanders there exists $m \in \mathbb{N}$ satisfying $\delta(\{G_n\}, X, m) < 1$.

First we make some *observations* related to the problem.

(a) The definition of the absolute spectral gap (see page 147) implies that the family $\{G_n\}$ of expanders satisfies $\delta(\{G_n\}, \mathbb{R}, 1) < 1$ if and only if their absolute spectral gaps are uniformly bounded away from 0. In fact, one can easily check that the operator norm of $A_n - R$ in $\ell_2(V_n, \mathbb{R})$ coincides with the norm of the restriction of A_n to the space of vectors in $\ell_2(V_n, \mathbb{R})$ with zero average.

(b) If we would like to prove that any family of expanders is not uniformly coarsely embeddable into X it suffices to do so for families with *absolute* spectral gaps uniformly bounded away from 0. To see this we do the following: for a family $\{G_n\}$ of expanders we consider the family $\{H_n\}_{n=1}^{\infty}$ of graphs defined as follows. The vertex set of H_n is the same as the vertex set of G_n. The edge set of H_n is the union of the edge set of G_n and the edge set of G_n^2 (the second power of G_n, see the definition on page 152). Simple linear algebra implies that if $\{G_n\}_{n=1}^{\infty}$ is a family of expanders, the absolute spectral graphs of $\{H_n\}_{n=1}^{\infty}$ are uniformly bounded away from 0. On the other hand, using the definition of the power of the graph and the definition of $\{H_n\}_{n=1}^{\infty}$ we get that the identity mapping on V_n considered as a map from G_n to H_n is 1-Lipschitz, and considered as a map from H_n to G_n is 2-Lipschitz.

(c) Operator norms of R and A_n (and thus of $A_n^m, m \in \mathbb{N}$), considered as operators on the spaces $\ell_1(V_n, X)$ and $\ell_\infty(V_n, X)$ are equal to 1 for any X, and so the norms of $A_n^m - R$ do not exceed 2.

In connection with Lemma 6.3 and the observations made above the following terminology was introduced.

Definition 6.5. Let X be a Banach space, $\varepsilon > 0$. Let $\varepsilon \mapsto \Delta_X(\varepsilon)$ be a function given by

$$\Delta_X(\varepsilon) = \sup \|T_X\|,$$

where the supremum is over all measure spaces (Ω, Σ, μ) and (Ω', Σ', μ'), all operators $T : L_2(\mu) \to L_2(\mu')$ satisfying $\|T_{L_2(\mu)\to L_2(\mu')}\| \le \varepsilon$, $\|T_{L_1(\mu)\to L_1(\mu')}\| \le 1$, $\|T_{L_\infty(\mu)\to L_\infty(\mu')}\| \le 1$, and T_X is the operator mapping the space $L_2(\mu, X)$ of X-valued functions into $L_2(\mu', X)$, which in tensor-product notation is denoted $T \otimes I_X$, and which is defined as the operator mapping each product $\mathbf{1}_A x$ of an indicator function of $A \in \Sigma$ and $x \in X$ onto the product $T(\mathbf{1}_A)x$.

The Banach space X is called *curved* if $\Delta_X(\varepsilon_0) < 1$ for some $\varepsilon_0 < 1$; X is called *fully curved* if $\Delta_X(\varepsilon) < 1$ for any $\varepsilon < 1$; X is called *uniformly curved* if $\Delta_X(\varepsilon) \to 0$ when $\varepsilon \to 0$.

Definition 6.5 is relevant to our discussion because Lemma 6.3 implies the following corollary.

Corollary 6.6. *If X is a uniformly curved Banach space, X does not admit uniformly coarse embeddings of any family of expanders.*

Proof. Observation (b) on page 198 implies that it suffices to consider an arbitrary family $\{G_n\}$ of expanders with absolute spectral gaps uniformly bounded away from 0. Therefore $\delta = \delta(\{G_n\}, \mathbb{R}, 1) < 1$ (Observation (a)). We pick large enough $m \in \mathbb{N}$ so that $2\Delta_X(\delta^m/2) < 1$. This is possible since X is uniformly curved. Consider the operators $\{\frac{1}{2}(A_n^m - R)\}_{n=1}^{\infty}$. By Observation (c), their operator norms in $\ell_1(V_n, \mathbb{R})$ and $\ell_\infty(V_n, \mathbb{R})$ do not exceed 1. The easily verified equality $(A_n - R)^m(I - R) = A_n^m - R$ implies that the operator norms of $\{\frac{1}{2}(A_n^m - R)\}_{n=1}^{\infty}$ in $\ell_2(V_n, \mathbb{R})$ do not exceed $\frac{1}{2}\delta^m$. By the definition of Δ_X, we get that norms of $\left(\frac{1}{2}(A_n^m - R)\right)_X$ in $\ell_2(V_n, X)$ do not exceed $\Delta_X(\delta^m/2)$, and the norms of $(A_n^m - R)_X$ do not exceed $2\Delta_X(\delta^m/2)$. Applying Lemma 6.3 we get that X does not admit uniformly coarse embeddings of $\{G_n\}$. □

Remark 6.7. It seems to be an open problem whether there are any uniformly curved spaces which do not satisfy the assumption of Theorem 6.1.

It is quite natural to look for a characterization of uniformly curved Banach spaces in terms of interpolation spaces. An important step in this direction is the description of Banach spaces X satisfying $\Delta_X(\varepsilon) \in O(\varepsilon^\alpha)$ for some $\alpha > 0$, which is given in Theorem 6.10.

Definition 6.8. Let $D = \{z \in \mathbb{C} \mid |z| < 1\}$ and ∂D be its boundary. Consider a measurable family $\{\| \ \|_z\}$ of norms on $\mathbb{C}^n$ indexed by $z \in \partial D$. (This means that $z \to \|x\|_z$ is measurable for any x in $\mathbb{C}^n$.) Assume that there are positive functions $k_1 > 0$ and $k_2 > 0$ such that their logarithms are in $L_1(m)$ on ∂D equipped with its normalized Lebesgue measure m, and

$$\forall z \in \partial D \quad \forall x \in \mathbb{C}^n \quad k_1(z)\|x\| \le \|x\|_z \le k_2(z)\|x\|, \tag{6.13}$$

where $\|x\|$ denotes the Euclidean norm of x (here any other fixed norm would do just as well).

The class N^+ is defined as the set of all analytic functions $f\colon D \to \mathbb{C}$ such that the family $\{\log^+|f(rz)| : 0 < r < 1\}$ is uniformly integrable on the unit circle. The vector-valued class $N^+(D, \mathbb{C}^n)$ is defined as that of all f of the form $f(z) = \sum_{j=1}^k f_j x_j$ with $x_j \in \mathbb{C}^n$, $f_j \in N^+$ $(k \ge 1)$. One can show that functions of the class $N^+(D, \mathbb{C}^n)$ have non-tangential limit on ∂D (*non-tangential* limit at $z \in \partial D$ means limit along any set of directions approaching z from inside the circle with some neighborhoods of directions close to tangential directions to the circle excluded).

Finally we introduce the space $X(0)$ as the space $\mathbb{C}^n$ endowed with the norm

$$\|x\|_{X(0)} = \inf\{\operatorname{ess\,sup} \|f(z)\|_z\} \tag{6.14}$$

where the infimum runs over all functions f in the class $N^+(D, \mathbb{C}^n)$ such that $f(0) = x$.

Definition 6.9. A finite-dimensional Banach space X is called *θ-Euclidean* if X is isometric to the complex interpolation space $X(0)$ associated to a family $\{X(z) \mid z \in \partial D\}$ of Banach spaces satisfying (6.13) and such that $X(z)$ is Hilbertian for all z in a subset of ∂D of (normalized) Lebesgue measure $\ge \theta$.

A Banach space is called *θ-Hilbertian* if it is isometric to an ultraproduct of a family of θ-Euclidean finite-dimensional spaces.

Theorem 6.10 (Without proof). *A Banach space X satisfies $\Delta_X(\varepsilon) \in O(\varepsilon^\alpha)$ for some $\alpha > 0$ if and only if X is isomorphic to a quotient of a subspace of an ultraproduct of θ-Hilbertian spaces for some $\theta > 0$.*

This is a deep result of the theory of interpolation spaces, it is beyond the scope of this book, see references in the Notes and Remarks section. It is known that uniformly curved spaces are isomorphic to uniformly convex spaces, and therefore the approach based on Corollary 6.6 cannot be used outside this class of spaces. However, it is unknown (and would be good to know) whether each uniformly convex space is uniformly curved. To the best of my knowledge it is an open problem: whether uniformly curved space or θ-Hilbertian spaces satisfy the assumption of Theorem 6.1.

6.3 Banach space theory: a characterization of reflexivity

Recall that a Banach space X is called *reflexive* if its canonical image in X^{**} coincides with X^{**}. Saying *canonical image* we mean the following: by definition X^{**} is the space of continuous linear functionals on X^*, on the other hand each element $x \in X$ can be considered as a continuous linear functional on X^* given by $x^* \mapsto x^*(x)$. Thus each element of X can be identified with a uniquely determined element of X^{**}. It is easy to see that the canonical embedding of X into X^{**} is isometric if we endow X^{**} with its standard norm (of the dual space). We denote the canonical embedding of X into X^{**} by π_X (or just π if there is no danger of confusion).

This section is devoted to a characterization of non-reflexive Banach spaces in terms of sequences.

Theorem 6.11. *The following five conditions for a Banach space X are equivalent:*

(a) *The Banach space X is nonreflexive.*

(b) *For each $0 < \theta < 1$, there are sequences $\{x_i\}_{i=1}^\infty \subset B_X$ and $\{f_i\}_{i=1}^\infty \subset B_{X^*}$ such that*
$$f_n(x_k) = \begin{cases} \theta & \text{if } n \le k \\ 0 & \text{if } n > k. \end{cases}$$

(c) *For some $0 < \theta < 1$, there are sequences $\{x_i\}_{i=1}^\infty \subset B_X$ and $\{f_i\}_{i=1}^\infty \subset B_{X^*}$ such that*
$$f_n(x_k) = \begin{cases} \theta & \text{if } n \le k \\ 0 & \text{if } n > k. \end{cases}$$

(d) *For each $0 < \theta < 1$ there is a sequence $\{x_i\}_{i=1}^\infty \subset B_X$ such that for each $n \in \mathbb{N}$*
$$d(\operatorname{conv}(\{x_i\}_{i=1}^n), \operatorname{conv}(\{x_i\}_{i=n+1}^\infty)) \ge \theta. \tag{6.15}$$

(e) *For some* $0 < \theta < 1$ *there is a sequence* $\{x_i\}_{i=1}^{\infty} \subset B_X$ *such that inequality* (6.15) *holds for each* $n \in \mathbb{N}$.

In this theorem we use the standard notation: $\operatorname{conv}(A)$ denotes the convex hull of a set A and $d(A, B)$ (in a metric space with metric ρ) is $\inf\{\rho(a,b) : a \in A, b \in B\}$, so in a Banach space $d(A, B) := \inf\{\|a - b\| : a \in A, b \in B\}$.

We need the following lemma:

Lemma 6.12 (Helly's condition). *Let* Y *be an infinite-dimensional normed linear space,* $g_1, \dots, g_n$ *be bounded linear functionals on* Y, $M > 0$ *and* $c_1, \dots, c_n$ *be scalars. Then the following two conditions are equivalent:*

(M) *For every* $\varepsilon > 0$, *there exists* $y \in Y$ *satisfying* $\|y\| = M + \varepsilon$ $g_k(y) = c_k$ *for* $k = 1, \dots, n$.

(H) *For every sequence* $a_1, \dots, a_n$ *of scalars the inequality*

$$\left|\sum_{i=1}^{n} a_i c_i\right| \le M \left\|\sum_{i=1}^{n} a_i g_i\right\| \tag{6.16}$$

holds

We call the conditions **(M)** and **(H)** because sometimes the existence of y satisfying **(M)** is called the *moment problem* and the condition (6.16) is called *Helly's condition*

Proof of Lemma 6.12. First we prove **(M)**$\Rightarrow$**(H)**. If there is $y \in Y$ satisfying **(M)**, then, for every sequence $a_1, \dots, a_n$ of scalars we have

$$\left|\sum_{i=1}^{n} a_i c_i\right| = \left|\sum_{i=1}^{n} a_i g_i(y)\right| \le \|y\| \cdot \left\|\sum_{i=1}^{n} a_i g_i\right\| = (M + \varepsilon)\left\|\sum_{i=1}^{n} a_i g_i\right\|.$$

Since $\varepsilon > 0$ is arbitrary, the condition **(H)** follows.

Now we turn to the proof of **(H)**$\Rightarrow$**(M)**. We may assume without loss of generality that some of $\{g_i\}$ are nonzero (otherwise the condition **(H)** implies that all of $\{c_i\}$ are zeros and any $y \in Y$ satisfying $\|y\| = M + \varepsilon$ satisfies **(M)**). Let k be the maximum cardinality of a linearly independent subset of $\{g_i\}$, so $1 \le k \le n$. We may assume without loss of generality that $g_1, \dots, g_k$ are linearly independent.

It is easy to verify that condition **(H)** implies the following: if $y \in Y$ satisfies $g_i(y) = c_i$ for $i = 1, \dots, k$, then $g_i(y) = c_i$ for $i = 1, \dots, n$. Therefore it suffices to restrict our attention to $\{g_i\}_{i=1}^{k}$. Let $H_i = \{y \in Y : g_i(y) = c\}$. Then $I = \cap_{i=1}^{k} H_i$ is an affine subspace of codimension k. We need to show that

$$\inf_{y \in I} \|y\| \le M.$$

Let us denote $\inf_{y \in I} \|y\|$ by N. Then I does not intersect the open ball

$$\{y \in Y : \|y\| < N\}.$$

Let $\widetilde{g}$ be a functional on the linear span of I given by $\widetilde{g}(y) = N$ for every $y \in I$. It is easy to check that this functional is uniquely determined and that $\|\widetilde{g}\| = 1$. By the Hahn–Banach theorem there is a bounded linear functional $g \in Y^*$ such that $\|g\| = 1$ and g is an extension of $\widetilde{g}$, in particular, $g(y) = N$ for each $y \in I$.

By the definition of I, the conditions $g_i(y) = c_i$, $i = 1, \ldots, k$, imply $g(y) = N$. Therefore $\cap_{i=1}^k \ker(g_i) \subset \ker(g)$ and $g = \sum_{i=1}^k a_i g_i$ for some scalars a_i, $i = 1, \ldots, k$. It is clear that for the same scalars we also have $N = \sum_{i=1}^k a_i c_i$. Applying **(H)** we get

$$N = \frac{|N|}{\|g\|} = \frac{\left|\sum_{i=1}^k a_i c_i\right|}{\left\|\sum_{i=1}^k a_i g_i\right\|} \le M.$$

Therefore for each $\varepsilon > 0$ there is $\widetilde{y} \in Y$ satisfying $\widetilde{y} \in I$ and $\|\widetilde{y}\| \le M + \varepsilon$. To get the equality in the last inequality, we need to use the infinite-dimensionality of Y. It implies that there is $\widehat{y} \in Y$ such that $g_i(\widehat{y}) = 0$ for all $i = 1, \ldots, n$. We consider the line $\widetilde{y} + \alpha\widehat{y}$, $\alpha \in \mathbb{R}$. It is clear that $g_i(\widetilde{y} + \alpha\widehat{y}) = g_i(\widetilde{y})$ and that $\lim_{\alpha\to\infty} \|\widetilde{y} + \alpha\widehat{y}\| = \infty$. Therefore this line contains a vector y satisfying $g_i(y) = g_i(\widetilde{y})$ and $\|y\| = M + \varepsilon$. □

Proof of Theorem 6.11. (a)⇒(b). We assume that X is nonreflexive. As we know $\pi_X : X \to X^{**}$ is an isometric embedding. Since a Banach space is, by definition, complete, it is easy to see that X is a closed subspace of X^{**}. Therefore we can find a bounded linear functional $T \in X^{***}$ such that $T|_{\pi_X(X)} = 0$ and $\|T\| = 1$. Let θ be a real number satisfying $0 < \theta < 1$. We choose $D \in X^{**}$ such that $\|D\| < 1$ and $T(D) > \theta$.

We are going to use induction to show the existence of sequences $\{x_i\}_{i=1}^\infty \subset B_X$ and $\{f_i\}_{i=1}^\infty \subset B_{X^*}$ satisfying the condition (b) of Theorem 6.3 as well as the condition $D(f_n) = \theta$. The main idea of the construction is first to find vectors with the desired properties in dual spaces of higher orders (X^{**}, X^{***}) and then to "push" them back to the desired spaces using the fact that Helly's condition is satisfied for suitable finite sets.

The first step is so simple that we do not really need Helly's condition. Everything follows from definitions: Since $T(D) > \theta$ and $\|T\| = 1$, we get that $\|D\| > \theta$. Therefore there is $f_1 \in X^*$ satisfying $\|f_1\| = 1$ and $D(f_1) = \theta$. Next we choose $x_1 \in X$ satisfying $\|x_1\| = 1$ and $f_1(x_1) = \theta$ (to see the existence of f_1 and x_1 it suffices to use the definitions and the argument with the line used at the end of the proof of Lemma 6.12).

The second step in the construction is similar to all further steps, so we provide a generic description.

Suppose that we have already constructed $f_1, \ldots, f_k$ and $x_1, \ldots, x_k$ such that condition (b) of Theorem 6.11 is satisfied and $D(f_i) = \theta$ for $i = 1, \ldots, k$. Now we describe how to construct f_{k+1} and x_{k+1}.

Observe that $T(x_i) = 0$ for $i = 1, \ldots, k$ and $T(D) = \lambda > \theta$. This implies that the vectors $x_1, \ldots, x_k, D$, considered as functionals on X^*, and the numbers $0, \ldots, 0, \theta$ satisfy the Helly condition **(H)** with $M = \frac{\theta}{\lambda}$. In fact, inequality (6.16) in this case looks like

$$|a_{k+1}\theta| \leq \frac{\theta}{\lambda}\left\| \sum_{i=1}^{k} a_i x_i + a_{k+1} D \right\|.$$

This inequality immediately follows from $\|T\| = 1$ and

$$T\left(\sum_{i=1}^{k} a_i x_i + a_{k+1} D\right) = a_{k+1}\lambda.$$

Since $0 < \frac{\theta}{\lambda} < 1$, applying Lemma 6.12 we get that there exists $f_{k+1} \in B_{X^*}$ such that $f_{k+1}(x_i) = 0$ for $i = 1, \ldots, k$ and $D(f_{k+1}) = \theta$.

Now we observe that the functionals $f_1, \ldots, f_k, f_{k+1}$ and the numbers $\theta, \ldots, \theta, \theta$ satisfy the Helly condition **(H)** with $M = \|D\|$. In fact, inequality (6.16) in this case looks like

$$\left| \sum_{i=1}^{k+1} a_i \theta \right| \leq \|D\| \left\| \sum_{i=1}^{k+1} a_i f_i \right\|.$$

This inequality immediately follows from

$$D\left(\sum_{i=1}^{k+1} a_i f_i\right) = \sum_{i=1}^{k+1} a_i \theta.$$

Applying Lemma 6.12 we get a vector $x_{k+1} \in B_X$ such that $f_i(x_{k+1}) = \theta$ for $i \leq k+1$. By induction, this completes the proof of the existence of sequences satisfying the condition (b) of Theorem 6.11.

The implication (b)$\Rightarrow$(c) is obvious.

(b)$\Rightarrow$(d) and (c)$\Rightarrow$(e) We show that a sequence $\{x_i\}_{i=1}^{\infty}$ such that the condition in (b) (or (c)) is satisfied for some sequence $\{f_i\}_{i=1}^{\infty}$, satisfies also the condition in (d) (or (e)).

In fact, let $\sum_{i=1}^{n} \alpha_i x_i \in \operatorname{conv}(\{x_i\}_{i=1}^{n})$ and $\sum_{i=n+1}^{\infty} \alpha_i x_i \in \operatorname{conv}(\{x_i\}_{i=n+1}^{\infty})$ be convex combinations, so $\sum_{i=1}^{n} \alpha_i = \sum_{i=n+1}^{\infty} \alpha_i = 1$. Then

$$\left\| \sum_{i=1}^{n} \alpha_i x_i - \sum_{i=n+1}^{\infty} \alpha_i x_i \right\| \geq f_{n+1}\left(\sum_{i=1}^{n} \alpha_i x_i - \sum_{i=n+1}^{\infty} \alpha_i x_i\right) = \sum_{i=n+1}^{\infty} \alpha_i \theta = \theta.$$

The implication (d)$\Rightarrow$(e) is obvious.

It remains to prove (e)⇒(a). Assume the contrary, that is, suppose that there exists a reflexive Banach space X containing a sequence $\{x_i\}_{i=1}^{\infty}$ satisfying the condition (6.15) for some $0 < \theta < 1$. Let us introduce the sets

$$C_n = \overline{\operatorname{conv}(\{x_i\}_{i=n}^{\infty})},$$

where bar denotes the closure in the strong topology. Then $\{C_n\}_{n=1}^{\infty}$ are bounded closed convex sets in X. The reflexivity of X implies that they are compact in the weak topology. Since

$$C_1 \supset C_2 \supset \cdots \supset C_n \supset \ldots,$$

this implies that $\{C_n\}_{n=1}^{\infty}$ have a nonempty intersection.

Let u be a point in the intersection. Then there is $v \in \operatorname{conv}(\{x_i\}_{i=1}^{\infty})$ such that $\|u - v\| < \frac{\theta}{2}$. Each element of $\operatorname{conv}(\{x_i\}_{i=1}^{\infty})$ is a convex combination of finitely many elements of the sequence $\{x_i\}_{i=1}^{\infty}$. Let n be such that $v \in \operatorname{conv}(\{x_i\}_{i=1}^{n})$.

There is also $w \in \operatorname{conv}(\{x_i\}_{i=n+1}^{\infty})$ such that $\|u - w\| < \frac{\theta}{2}$. Then $\|v - w\| < \theta$ and we get a contradiction with (6.15). □

6.4 Some classes of spaces whose balls are not uniformly embeddable into L_1

One of the important problems in the theory of metric embeddings can be stated as: characterize the class of Banach spaces which admit uniformly coarse embeddings of some families of expanders. See Section 11.1 in Chapter 11 for further discussion of this problem.

The main result of Section 6.1 states that Banach spaces whose unit balls are uniformly homeomorphic to subsets of L_1 (or L_2) are not in this class. On the other hand, comparing the characterization of Banach spaces with no nontrivial cotype (Theorem 2.56) and the Fréchet proposition on embeddings of finite metric spaces into ℓ_∞^n (Proposition 1.17) we get that spaces with no nontrivial cotype are in the class.

This observation implies that unit balls of Banach spaces with no nontrivial cotype are not uniformly homeomorphic to subsets of L_1. Also, it is clear that to answer the problem posed at the beginning of this section, it remains to investigate classes of Banach spaces which have nontrivial cotype, but whose balls are not uniformly homeomorphic to subsets of L_1. The purpose of this section is to single out a class of Banach spaces having these properties.

6.4.1 Stable metric spaces and iterated limits

For problems of embeddability into a Hilbert space it turns out to be important that Hilbert spaces have the following property.

Definition 6.13. A metric space (X,d) is called *stable* if for any two bounded sequences $\{x_n\}$ and $\{y_m\}$ in X and for any two free ultrafilters $\mathcal{U}$ and $\mathcal{V}$ on $\mathbb{N}$

$$\lim_{n,\mathcal{U}} \lim_{m,\mathcal{V}} d(x_n, y_m) = \lim_{m,\mathcal{V}} \lim_{n,\mathcal{U}} d(x_n, y_m).$$

Example 6.14. A Hilbert space is stable. Indeed, let $x = w - \lim_{n,\mathcal{U}} x_n$ and $y = w - \lim_{m,\mathcal{V}} y_m$ (the weak limits exist by the reflexivity). Then

$$\lim_{n,\mathcal{U}} \lim_{m,\mathcal{V}} \|x_n - y_m\| = \Big(\lim_{n,\mathcal{U}} \|x_n\|^2 + \lim_{m,\mathcal{V}} \|y_m\|^2 - 2\langle x, y\rangle\Big)^{1/2} = \lim_{m,\mathcal{V}} \lim_{n,\mathcal{U}} \|x_n - y_m\|.$$

(In this computation we assumed that the space is over the real numbers.)

Example 6.15. The space $L_1(0,1)$ is stable. The stability of L_1 follows from the previous example and the observation that L_1 endowed with the metric $\|x-y\|_{L_1}^{1/2}$ is isometric to a subset of L_2 (see Corollary 1.43). In general, it is easy to check that for a metric space (X,d), the function $(x,y) \mapsto (d(x,y))^\varepsilon$ is also a metric for $0 < \varepsilon < 1$, it is the so-called *snowflaked* version of d. It is also easy to see that a metric space (X,d) is stable if and only if (X,d^ε) is stable.

In the remainder of this section we omit ultrafilters from the notation of the limit meaning that $\lim_i$ where $i \in \mathbb{N}$ is a limit over some free ultrafilter on $\mathbb{N}$.

Example 6.16. The space c_0 is not stable. In fact, let $x_n = e_n$, $y_m = -\sum_{i=1}^m e_i$. Then

$$\lim_n \lim_m \|x_n - y_m\| = 2, \quad \lim_m \lim_n \|x_n - y_m\| = 1.$$

We need to consider limits of multi-indexed sequences (= functions of many natural variables).

Lemma 6.17. *Let (X,d) be a stable metric space, $\{x_{i_1,\dots,i_k}\}$ and $\{y_{j_1,\dots,j_\ell}\}$ be bounded multi-indexed sequences in X. Then*

$$\lim_{i_1} \dots \lim_{i_k} \lim_{j_1} \dots \lim_{j_\ell} d(x_{i_1,\dots,i_k}, y_{j_1,\dots,j_\ell}) = \lim \dots \lim d(x_{i_1,\dots,i_k}, y_{j_1,\dots,j_\ell}), \quad (6.17)$$

where the right-hand side limits are taken over all indices $(i_1,\dots,i_k,j_1,\dots,j_\ell)$, in a different, but restricted in the following way, order: the restrictions of the orders of limits to the orders of $\lim_{i_1} \dots \lim_{i_k}$ and $\lim_{j_1} \dots \lim_{j_\ell}$ in the right-hand side are the same as in the left-hand side.

Proof. We call orders satisfying the restriction *admissible*. Let us say that two admissible orders are *equivalent* if they produce the same limits for an arbitrary choice of a bounded multi-indexed sequence in any stable metric space X. It is clear that it is really an equivalence relation. We need to show that each admissible order is equivalent to the order in the left-hand side of (6.17), we call it *original*.

We do this by induction on $k + \ell$, for $k, \ell \geq 1$ and arbitrary families of ultrafilters. For $k + \ell = 2$ the statement follows immediately from the definition of a stable metric space.

Suppose that we have proved the statement for $k + \ell = n$ and consider families with $k + \ell = n + 1$.

Observation 6.18. If we consider an admissible order for which there is a pair of neighbors of the form (j_p, j_{p+1}) or (i_p, i_{p+1}), then this order is equivalent to the original order by the induction hypothesis.

Proof. Assume that two of consecutive limits are $\lim_{j_p}$ and $\lim_{j_{p+1}}$. Then the limit, both in the original iterated limit and in the limit in the admissible order which we consider, does not change if we replace the pair of limits $\lim_{j_p} \lim_{j_{p+1}}$ over ultrafilters $\mathcal{V}_p$ and $\mathcal{V}_{p+1}$ by a limit over one ultrafilter which can be called a "product ultrafilter". We mean the following construction: we consider any bijection $\beta : \mathbb{N} \times \mathbb{N} \to \mathbb{N}$. We consider the filter $\mathcal{U}$ on $\mathbb{N}$ described as the set of images under the bijection β of sets $A \in \mathbb{N} \times \mathbb{N}$ for which the sets $\{k \in \mathbb{N} : (A \cap (\{k\} \times \mathbb{N})) \in \mathcal{V}_{p+1}\}$ are in $\mathcal{V}_p$. Using Observation 2.14 one can easily check that $\mathcal{U}$ is an ultrafilter. It is also clear that $\lim_{j_p, \mathcal{V}_p} \lim_{j_{p+1}, \mathcal{V}_{p+1}}$ is equivalent to $\lim_{\beta(j_p . j_{p+1}), \mathcal{U}}$.

After making this replacement we get into the situation where $k + \ell = n$ and so we may apply the induction hypothesis. □

If we consider an admissible order which does not contain any consecutive pairs of the form (j_p, j_{p+1}) or (i_p, i_{p+1}), then this order ends either with $j_{\ell-1}, i_k, j_\ell$ or with i_{k-1}, j_ℓ, i_k. In both cases we can interchange the last two limits just because the space is stable. In both cases we get an admissible order which satisfies the condition of Observation 6.18, and thus is equivalent to the original order. □

6.4.2 Non-embeddability result

Definition 6.19. A sequence of linear bilipschitz embeddings $T_i : X_i \to Y_i$, where $\{X_i\}_{i=1}^{\infty}$ and $\{Y_i\}_{i=1}^{\infty}$ are normed spaces, are called *uniformly isomorphic* if

$$\sup_i \|T_i\| \cdot \|(T_i|_{T_i(X_i)})^{-1}\| < \infty.$$

If $T_i : X_i \to Y$ are uniformly isomorphic embeddings, we say that Y contains the spaces $\{X_i\}_{i=1}^{\infty}$ *uniformly.*

Theorem 6.20. *If E is a nonreflexive Banach space and B_E is uniformly homeomorphic to a subset of a stable metric space (X, d), then E contains the spaces $\{\ell_1^n\}_{n=1}^{\infty}$ uniformly.*

Corollary 6.21. *The unit ball of a nonreflexive space which does not contain the spaces $\{\ell_1^n\}_{n=1}^{\infty}$ uniformly is not uniformly homeomorphic to a subset of L_2.*

Remark 6.22. Spaces satisfying the assumptions of Corollary 6.21 are difficult to construct, see Notes and Remarks for some history. We describe one of the constructions in Section 6.5 of this chapter.

Proof of Theorem 6.20. We prove Theorem 6.20 in the following way. We assume that E is a nonreflexive Banach space whose unit ball admits a uniform embedding into a stable metric space and show that in such a case E contains $\{\ell_1^n\}_{n=1}^\infty$ uniformly.

We use the characterization of nonreflexive spaces proved in Theorem 6.11. It implies that for each $0 < \theta < 1$, there are sequences $\{x_i\}_{i=1}^\infty \subset B_E$ and $\{f_i\}_{i=1}^\infty \subset B_{E^*}$ such that

$$f_n(x_k) = \begin{cases} \theta & \text{if } n \le k \\ 0 & \text{if } n > k. \end{cases} \tag{6.18}$$

We fix some $\theta \in (0,1)$ and let $\{x_i\}_{i=1}^\infty \subset B_E$ and $\{f_i\}_{i=1}^\infty \subset B_{E^*}$ be some sequences satisfying (6.18).

We use the same notation for vectors in B_E and for their images of the uniform embedding (whose existence we assumed). By the definition of uniform continuity, there exists $\omega > 0$ such that $d(x,y) < \omega$ implies $\|x-y\| < \theta/2$ and there exists $r > 0$ such that $\|x-y\| < r$ implies $d(x,y) < \omega/2$.

Now we show that for each sequence $\{b_i\} \in S_1^n$ (where S_1^n is the unit sphere of ℓ_1^n) we have

$$\lim_{k_1,\mathcal{U}_1} \dots \lim_{k_n,\mathcal{U}_n} \left\| \sum_{i=1}^n b_i x_{k_i} \right\| \ge r = r \sum_{i=1}^n |b_i|, \tag{6.19}$$

where $\mathcal{U}_1, \dots, \mathcal{U}_n$ is any collection of free ultrafilters on $\mathbb{N}$.

We write the sum $\sum_{i=1}^n b_i x_{k_i}$ as $\sum_{i\in A} b_i x_{k_i} + \sum_{i\in B} b_i x_{k_i}$ where i is in A if b_i is positive and i is in B if b_i is negative. Changing the sign of the whole sum, if necessary, we may assume without loss of generality that $\sum_{i\in B} |b_i| \ge 1/2$.

Assume that (6.19) does not hold. Observe that this assumption can be written as

$$\lim_{k_1,\mathcal{U}_1} \dots \lim_{k_n,\mathcal{U}_n} \left\| \sum_{i\in A} b_i x_{k_i} - \Big(- \sum_{i\in B} b_i x_{k_i} \Big) \right\| < r.$$

With our choice of numbers r and ω, this assumption implies that

$$\lim_{k_1,\mathcal{U}_1} \dots \lim_{k_n,\mathcal{U}_n} d\Big(\sum_{i\in A} b_i x_{k_i}, -\sum_{i\in B} b_i x_{k_i} \Big) \le \omega/2.$$

Since (X,d) is a stable metric space we may apply Lemma 6.17 and change the order of limits, considering first (means: closer to $d(\cdot,\cdot)$) the limits over indices from B and then over indices from A; and keeping the relative order of limits for indices in A (and the same for B). Since the limit does not change, it is still $\le \omega/2$. Hence we can find a pair $\sum_{i\in A} b_i x_{k_i}, -\sum_{i\in B} b_i x_{k_i}$ satisfying

$$d\Big(\sum_{i\in A} b_i x_{k_i}, -\sum_{i\in B} b_i x_{k_i} \Big) < \omega$$

and $\max\{k_i : i \in A\} < \min\{k_i : i \in B\}$. By the choice of ω we have

$$\left\| \sum_{i \in A} b_i x_{k_i} + \sum_{i \in B} b_i x_{k_i} \right\| < \theta/2.$$

Let $j = \min_{i \in B} k_i$. Then

$$\left| f_j \left(\sum_{i \in A} b_i x_{k_i} + \sum_{i \in B} b_i x_{k_i} \right) \right| = \left| \sum_{i \in B} \theta b_i \right| = \theta \sum_{i \in B} |b_i| \geq \theta/2,$$

and we get a contradiction.

Inequality (6.19) immediately implies the following conclusion: for each finite collection $\{\{b_i(j)\}_{i=1}^n\}_{j=1}^m \subset S_1^n$ there is a sequence $\{e_i\}_{i=1}^n \subset E$ such that

$$\frac{r}{2} \sum_{i=1}^n |b_i(j)| \leq \left\| \sum_{i=1}^n b_i(j) e_i \right\| \leq \sum_{i=1}^n |b_i(j)|$$

To complete the proof of the theorem we use this conclusion for a finite $\frac{r}{4}$-net in S_1^n. Let us show that in such a case the embedding $T : \ell_1^n \to E$ given by $T(\{d_i\}_{i=1}^n) = \sum_{i=1}^n d_i e_i$ satisfies

$$\frac{r}{4} \sum_{i=1}^n |d_i| \leq \left\| T(\{d_i\}_{i=1}^n) \right\| \leq \sum_{i=1}^n |d_i|. \tag{6.20}$$

It suffices to prove the left-hand side inequality in the case where $\{d_i\}_{i=1}^n \in S_1^n$. Let $\{a_i\}_{i=1}^n \in S_1^n$ be an element of the $\frac{r}{4}$-net satisfying $\sum_{i=1}^n |a_i - d_i| \leq \frac{r}{4}$. We have

$$\left\| T(\{d_i\}_{i=1}^n) \right\| = \left\| \sum_{i=1}^n d_i e_i \right\| \geq \left\| \sum_{i=1}^n a_i e_i \right\| - \left\| \sum_{i=1}^n (a_i - d_i) e_i \right\| \geq \frac{r}{2} - \frac{r}{4} = \frac{r}{4}.$$

Since r does not depend on n, inequality (6.20) implies that E contains $\{\ell_1^n\}_{n=1}^\infty$ uniformly. □

6.5 Examples of non-reflexive spaces with nontrivial type

Corollary 6.21 states that the unit ball of a nonreflexive Banach space which does not contain $\{\ell_1^n\}_{n=1}^\infty$ uniformly is not uniformly homeomorphic to a subset of L_2 (or L_1). Such a space also cannot be isomorphic to a uniformly convex space (in Theorem 8.26 we prove that uniformly convex spaces are reflexive). Therefore none of the known general results (we mean Theorem 6.1 and Corollary 6.6) on lack of uniformly coarse embeddability of expanders is applicable to such spaces. This makes it very interesting and important to find out, for each such space X, whether there exists a family of expanders which is uniformly coarsely embeddable into X? (See Section 11.1 for more open problems of the same spirit.)

Remark 6.23. Combining Theorem 2.56 and an easy fact that ℓ_1 does not have nontrivial type (can be immediately derived from the definition), we see that a Banach space contains the spaces $\{\ell_1^n\}_{n=1}^{\infty}$ uniformly if and only if it does not have nontrivial type.

The purpose of this section is to describe one of the known constructions of a non-reflexive space with a nontrivial type. It should be mentioned that all of the known constructions of such spaces are complicated. We present the construction for which the formula for the norm is the simplest (in my opinion). This (relative) simplicity increases chances to succeed in determining whether the space admits uniformly coarse embeddings of a family of expanders.

The spaces which we are going to describe are the real interpolation spaces usually denoted $(v_1, \ell_\infty)_{\theta,q}$, where $0 < \theta < 1$, $1 < q < \infty$. Let us define all so far undefined parts of this formula.

We use v_1 to denote the Banach space of all sequences of real numbers $(\alpha_n)_{n\geq 0}$ of bounded variation, that is, such that

$$|\alpha_0| + |\alpha_1 - \alpha_0| + \cdots + |\alpha_n - \alpha_{n-1}| + \cdots < \infty.$$

We define the norm in v_1 as the left-hand side of this inequality. Clearly v_1 (as a set of sequences) is a subset of ℓ_∞, and the natural embedding of v_1 into ℓ_∞ is a linear operator of norm 1.

Let A_0 and A_1 be two Banach spaces which are continuously embedded into a Hausdorff topological vector space $\mathcal{C}$. In such a case we call the Banach spaces A_0 and A_1 *compatible* and call the pair (A_0, A_1) a *Banach couple*. It is convenient to introduce some notation for a Banach couple. We let $\bar{A} = (A_0, A_1)$. The described compatibility is needed for us in order to be able to claim that the following two conditions are satisfied:

(a) The (set-theoretical) intersection $A_0 \cap A_1$ is a Banach space with respect to the norm

$$\|x\|_{A_0\cap A_1} := \max\{\|x\|_{A_0}, \|x\|_{A_1}\}.$$

This Banach space is denoted $\Delta(\bar{A})$.

(b) The set of all linear combinations of the form x_0+x_1, where $x_0 \in A_0$ and $x_1 \in A_1$, is a Banach space with respect to the norm

$$\|x\|_{A_0+A_1} := \inf\{\|x\|_{A_0} + \|x\|_{A_1} : x = x_0 + x_1,\ x_0 \in A_0,\ x_1 \in A_1\}.$$

This Banach space is denoted $\Sigma(\bar{A})$.

It is not difficult to check that the described spaces are indeed Banach spaces for an arbitrary compatible pair of Banach spaces. We do not do this because we use these constructions in the case where this verification is completely trivial.

Let $\bar{A} = (A_0, A_1)$ be a Banach couple, $t > 0$. We define the *K-functional* on $\Sigma(\bar{A})$ by

$$\underbrace{K_t(a) = K(t,a) = K(t,a,\bar{A})}_{\text{three forms of notation}} = \inf_{a=a_0+a_1} (\|a_0\|_{A_0} + t\|a_1\|_{A_1}). \tag{6.21}$$

For $0 \le \theta \le 1$ and $1 \le q \le \infty$ we define the functional $\Phi_{\theta,q}$ on nonnegative functions $\varphi : (0,\infty) \to [0,\infty)$ by

$$\Phi_{\theta,q}(\varphi(t)) = \left(\int_0^\infty (t^{-\theta}\varphi(t))^q \frac{dt}{t}\right)^{1/q}. \tag{6.22}$$

The interpolation space $(A_0, A_1)_{\theta,q} = \bar{A}_{\theta,q}$ (sometimes also denoted $\bar{A}_{\theta,q;K}$ and $K_{\theta,q}(\bar{A})$) is defined as the space consisting of all elements in $\Sigma(\bar{A})$ for which the condition

$$\Phi_{\theta,q}(K(t,a)) < \infty \tag{6.23}$$

holds. For such vectors we define the norm by

$$\|a\|_{(A_0,A_1)_{\theta,q}} = \|a\|_{\theta,q} = \Phi_{\theta,q}(K(t,a)). \tag{6.24}$$

We get a Banach space. We say that the space $(A_0, A_1)_{\theta,q} = \bar{A}_{\theta,q}$ is obtained using the *K-method of interpolation*.

Remarks above imply that (v_1, ℓ_∞) form a Banach couple in a natural way (both of the spaces are continuously embedded into ℓ_∞).

Theorem 6.24. *The spaces $(v_1, \ell_\infty)_{\theta,q}$, where $0 < \theta < 1$, $1 < q < \infty$, are nonreflexive spaces of nontrivial type.*

We are going to prove only the fact that $(v_1, \ell_\infty)_{\theta,q}$ is nonreflexive. In Notes and Remarks we mention sources where one can find a (much more complicated) proof of the nontrivial type of $(v_1, \ell_\infty)_{\theta,q}$.

Proof of non-reflexivity. Observe that for $x \in v_1 \cap \ell_\infty$ and any representation $x = x_v + x_\ell$ with $x_v \in v_1$ and $x_\ell \in \ell_\infty$ we have $\|x_v\|_v + t\|x_\ell\|_\infty \ge \|x_v\|_\infty + t\|x_\ell\|_\infty$, and therefore

$$\|x\|_\infty \le K(t,x) \le \|x\|_v \quad \text{if } t \ge 1$$

and

$$t\|x\|_\infty = K(t,x) \le t\|x\|_v \quad \text{if } t \le 1.$$

Hence

$$\|x\|_{(v_1,\ell_\infty)_{\theta,q}} \ge \|x\|_\infty \cdot \left(\int_0^\infty (t^{-\theta}\min\{1,t\})^q \frac{dt}{t}\right)^{\frac{1}{q}} = \frac{\|x\|_\infty}{((1-\theta)\theta q)^{\frac{1}{q}}}. \tag{6.25}$$

and

$$\|x\|_{(v_1,\ell_\infty)_{\theta,q}} \le \frac{\|x\|_v}{((1-\theta)\theta q)^{\frac{1}{q}}}. \tag{6.26}$$

To prove non-reflexivity of $(v_1, \ell_\infty)_{\theta,q}$ we are going to use condition (e) of Theorem 6.11. Let $\{e_i\}_{i=1}^\infty$ be the unit vector sequence in v_1 and ℓ_∞. Let $x_i = \sum_{j=1}^i e_j$. The sequence $\{x_i\}_{i=1}^\infty$ satisfies the following conditions:

$$\|x_i\|_v = 2, \quad \|x_i\|_\infty = 1,$$

and for any $n \in \mathbb{N}$ and any sequence $\{\alpha_i\}_{i=1}^\infty$ satisfying $\alpha_i > 0$, $\sum_{i=1}^n \alpha_i = 1$, and $\sum_{i=n+1}^\infty \alpha_i = 1$ we have

$$\left\| \sum_{i=1}^n \alpha_i x_i - \sum_{i=n+1}^\infty \alpha_i x_i \right\|_\infty = 1$$

and

$$\left\| \sum_{i=1}^n \alpha_i x_i - \sum_{i=n+1}^\infty \alpha_i x_i \right\|_v = 2.$$

Hence, by (6.25) and (6.26) we get

$$\frac{1}{((1-\theta)\theta q)^{\frac{1}{q}}} \le \|x_i\|_{(v_1,\ell_\infty)_{\theta,q}} \le \frac{2}{((1-\theta)\theta q)^{\frac{1}{q}}}$$

and

$$\frac{1}{((1-\theta)\theta q)^{\frac{1}{q}}} \le \left\| \sum_{i=1}^n \alpha_i x_i - \sum_{i=n+1}^\infty \alpha_i x_i \right\|_{(v_1,\ell_\infty)_{\theta,q}} \le \frac{2}{((1-\theta)\theta q)^{\frac{1}{q}}}.$$

Therefore $\left\{\frac{((1-\theta)\theta q)^{\frac{1}{q}}}{2} x_i\right\}_{i=1}^\infty$ satisfies the condition (e) in $(v_1, \ell_\infty)_{\theta,q}$ with θ (parameter in the condition (e), but not in $(v_1, \ell_\infty)_{\theta,q}$) equal to $\frac{1}{2}$. By the implication (e)⇒(a) of Theorem 6.11, the space $(v_1, \ell_\infty)_{\theta,q}$ is nonreflexive. □

Our next goal is to get a more explicit formula for the norm of the space $(v_1, \ell_\infty)_{\theta,q}$. Now our purpose is to look for quantities which are equivalent to $K_t(x)$ and are more convenient computationally.

For $x \in \ell_\infty$ let

$$\mathrm{V}_k(x) = \sup \sum_{j=1}^k |x_{n_j} - x_{n_{j-1}}|,$$

where the supremum runs over all increasing sequences $n_0 \le n_1 \le \cdots \le n_k$ of integers. As we shall see this functional can be used to get a good approximation of $K_t(x)$ for $t = k$.

Let

$$\begin{aligned}
n_0 &= 0 \\
n_1 &= \inf\{n : n > n_0 \;\&\; |x_n - x_{n_0}| > \mathrm{V}_k(x)/k\} \\
n_2 &= \inf\{n : n > n_1 \;\&\; |x_n - x_{n_1}| > \mathrm{V}_k(x)/k\} \\
&\dots \\
n_k &= \inf\{n : n > n_{k-1} \;\&\; |x_n - x_{n_{k-1}}| > \mathrm{V}_k(x)/k\}.
\end{aligned} \tag{6.27}$$

Then $n_k = \infty$. In fact, otherwise we get

$$\sum_{j=1}^{k} |x_{n_j} - x_{n_{j-1}}| > \mathrm{V}_k(x) = \sup_{\substack{n_0 \le n_1 \le \dots \le n_k \\ n_j \in \mathbb{N}}} \sum_{j=1}^{k} |x_{n_j} - x_{n_{j-1}}|,$$

a contradiction.

We consider $y \in v_1$ (actually, an eventually constant sequence):

$$y_i = \begin{cases} x_{n_0} & \text{if } i < n_1 \\ x_{n_1} & \text{if } n_1 \le i < n_2 \\ \dots & \\ x_{n_{k-2}} & \text{if } n_{k-2} \le i < n_{k-1} \\ x_{n_{k-1}} & \text{if } n_{k-1} \le i < \infty. \end{cases} \tag{6.28}$$

Equalities (6.27) imply that $\|x-y\|_\infty \le \mathrm{V}_k(x)/k$. It is also clear that $\mathrm{V}_k(y) \le \mathrm{V}_k(x)$ ("level changes" of y are present in x also).

Lemma 6.25. *Let $K_t(x)$ be the K-functional of an element $x \in \ell_\infty$ corresponding to the couple (v_1, ℓ_∞). Then*

$$\max\left\{\frac{\mathrm{V}_k(x)}{2}, |x_0|\right\} \le K_k(x) \le 2\mathrm{V}_k(x) + |x_0| \tag{6.29}$$

for each $k \in \mathbb{N}$.

Proof. Recall that

$$K_k(x) = \inf_{x = x_v + x_l} (\|x_v\|_v + k\|x_\ell\|_\infty). \tag{6.30}$$

To estimate $K_k(x)$ from above we let $x_v = y$, where y is defined by (6.28), and let $x_\ell = x - y$.

The vector y has $k-1$ level changes, therefore $\|y\|_v = \mathrm{V}_{k-1}(y) + |x_0|$, the same level changes are present in x, therefore $\|y\|_v \le \mathrm{V}_{k-1}(x) + |x_0| \le \mathrm{V}_k(x) + |x_0|$.

The definition (6.28) implies immediately that $\|x-y\|_\infty \le \mathrm{V}_k(x)/k$. These estimates imply

$$K_k(x) \le (\mathrm{V}_k(x) + |x_0|) + k \cdot \frac{\mathrm{V}_k(x)}{k} = 2\mathrm{V}_k(x) + |x_0|,$$

that is, the right-hand side inequality in (6.29).

To get the left-hand side inequality of (6.29) we consider an arbitrary decomposition $x = x_v + x_\ell$ and let $m_0 \le m_1 \le \cdots \le m_k$ be such that

$$\sum_{j=1}^{k} |x_{m_j} - x_{m_{j-1}}| \ge (1-\varepsilon)\mathrm{V}_k(x).$$

We have

$$\begin{aligned}\sum_{j=1}^{k} |x_{m_j} - x_{m_{j-1}}| &\le \sum_{j=1}^{k} |(x_v)_{m_j} - (x_v)_{m_{j-1}}| + \sum_{j=1}^{k} |(x_\ell)_{m_j} - (x_\ell)_{m_{j-1}}| \\ &\le \|x_v\|_v + 2k\|x_\ell\|_\infty \le 2(\|x_v\|_v + k\|x_\ell\|_\infty).\end{aligned}$$

Since $\varepsilon > 0$ and the decomposition $x = x_v + x_\ell$ is arbitrary, we get the inequality $K_k(x) \ge \mathrm{V}_k(x)/2$ from here. The inequality $K_k(x) \ge |x_0|$ follows immediately from the definitions ($k \ge 1$). We get the left-hand side inequality in (6.29). □

Now we are going to present an equivalent formula for the norm in $(v_1, \ell_\infty)_{\theta,q}$ which does not use integrals. In order to do this we need to recall the *discrete version of the K-method of real interpolation*. Denote by $\lambda^{\theta,q}$ the space of sequences $\{\tau_j\}_{j=-\infty}^{\infty}$ satisfying

$$\sum_{j=-\infty}^{\infty} (2^{-j\theta}|\tau_j|)^q < \infty$$

with the norm

$$\|\{\tau_j\}\|_{\lambda^{\theta,q}} = \left(\sum_{j=-\infty}^{\infty} (2^{-j\theta}|\tau_j|)^q\right)^{1/q} < \infty.$$

(The space $\lambda^{\theta,q}$ is just a weighted ℓ_q-space.)

Lemma 6.26. *For $x \in \Sigma(\bar{A})$ let $\alpha_j = K(2^j, x, \bar{A})$. The vector x belongs to $\bar{A}_{\theta,q}$ if and only if the sequence $\{\alpha_j\}_{j=-\infty}^{\infty}$ is in $\lambda^{\theta,q}$. Furthermore, the inequality*

$$2^{-\theta}(\ln 2)^{\frac{1}{q}}\|\{\alpha_j\}\|_{\lambda^{\theta,q}} \le \|x\|_{\theta,q} \le 2(\ln 2)^{\frac{1}{q}}\|\{\alpha_j\}\|_{\lambda^{\theta,q}} \tag{6.31}$$

holds.

Proof. We have

$$\|x\|_{\theta,q} = \Big(\sum_{j=-\infty}^{\infty} \int_{2^j}^{2^{j+1}} (t^{-\theta}K(t,x))^q \frac{dt}{t}\Big)^{1/q}.$$

Let $t \in [2^j, 2^{j+1}]$. The definition of the K-functional immediately implies

$$K(2^j,x) \le K(t,x) \le K(2^{j+1},x) \le 2K(2^j,x),$$

and hence

$$2^{-\theta}2^{-j\theta}\alpha_j \le t^{-\theta}K(t,x) \le 2\cdot 2^{-j\theta}\alpha_j. \tag{6.32}$$

Since $\int_{2^j}^{2^{j+1}} \frac{dt}{t} = \ln 2$, integrating and summing the inequality (6.32), we get (6.31). □

By Lemma 6.26 we get that $x \in \ell_\infty$ is in the space $(v_1,\ell_\infty)_{\theta,q}$ if and only if the sequence $\{K(2^j,x)\}_{j=-\infty}^{\infty}$ is in the space $\lambda^{\theta,q}$ and

$$\|x\|_{(v_1,\ell_\infty)_{\theta,q}} \approx \Big\|\Big\{K(2^j,x)\Big\}_{j=-\infty}^{\infty}\Big\|_{\lambda^{\theta,q}}. \tag{6.33}$$

We need the following observation on the case $j \le 0$. By the definition of the K-functional and $\|x\|_\infty \le \|x\|_v$, we get that in this case $K(2^j,x) = 2^j\|x\|_\infty$. Therefore

$$\begin{aligned}\Big(\sum_{j=-\infty}^{0}(2^{-j\theta}K(2^j,x))^q\Big)^{1/q} &= \Big(\sum_{j=-\infty}^{0}(2^{-j\theta}2^j\|x\|_\infty)^q\Big)^{1/q}\\ &= \frac{\|x\|_\infty}{(1-2^{q(\theta-1)})^{1/q}}.\end{aligned} \tag{6.34}$$

For $j \ge 1$, by (6.29), we have

$$\begin{aligned}\max&\Big\{\frac{1}{2}\Big(\sum_{j=1}^{\infty}(2^{-j\theta}\mathrm{V}_{2^j}(x))^q\Big)^{1/q}, |x_0|\Big(\sum_{j=1}^{\infty}2^{-j\theta q}\Big)^{1/q}\Big\}\\ &\le \Big(\sum_{j=1}^{\infty}(2^{-j\theta}K(2^j,x))^q\Big)^{1/q}\\ &\le 2\Big(\sum_{j=1}^{\infty}(2^{-j\theta}\mathrm{V}_{2^j}(x))^q\Big)^{1/q} + |x_0|\Big(\sum_{j=1}^{\infty}2^{-j\theta q}\Big)^{1/q}.\end{aligned} \tag{6.35}$$

We can combine (6.34) and (6.35), and write

$$\begin{aligned}c(\|\{\mathrm{V}_{2^j}(x)\}_{j=1}^{\infty}\|_+ + \|x\|_\infty) &\le \|x\|_{(v_1,\ell_\infty)_{\theta,q}}\\ &\le C(\|\{\mathrm{V}_{2^j}(x)\}_{j=1}^{\infty}\|_+ + \|x\|_\infty),\end{aligned} \tag{6.36}$$

where $\|\cdot\|_+$ is the "positive" part of $\|\cdot\|_\lambda$, that is,

$$\|\{\tau_j\}_{j=1}^{\infty}\|_+ = \left(\sum_{j=1}^{\infty}(2^{-j\theta}|\tau_j|)^q\right)^{1/q}.$$

The formula for the norm in (6.36) is reasonably simple, but it still resists the analysis for the expander embeddability problem.

6.6 Exercises

Exercise 6.27. Let X be a Banach space and B_X be its unit ball. Consider the following map $P : X \to B_X$:

$$P(x) = \begin{cases} x & \text{if } x \in B_X \\ \frac{x}{\|x\|} & \text{if } x \notin B_X. \end{cases}$$

Show that $\text{Lip}(P) \le 2$.

There is a hint to this exercise.

Exercise 6.28. Verify the statement made in Observation (b) on page 198.

Exercise 6.29. A Banach space X is called *uniformly non-square* if there is a number $\delta < 1$ such that for any pair x, y in the unit ball we have either $\|2^{-1}(x+y)\| \le \delta$ or $\|2^{-1}(x-y)\| \le \delta$. Consider the operator τ: $\ell_2^2 \to \ell_2^2$ given by $\tau(x,y) = (2^{-1}(x+y), 2^{-1}(x-y))$. Evaluate the norm of this operator. Use the result to show that a fully curved Banach space (see Definition 6.5 on page 198) is uniformly non-square.

Exercise 6.30. Show that a curved Banach space (see Definition 6.5 on page 198) has nontrivial type.

There is a hint to this exercise.

Exercise 6.31. Show that the spaces ℓ_p, $1 \le p < \infty$, are stable (see Definition 6.13 on page 204).

6.7 Notes and remarks

The main result of Section 6.1 is due to Ozawa [363]. There are surprisingly many Banach spaces satisfying the condition of Theorem 6.1. Mazur [320] proved that unit balls of L_p, $1 \le p < \infty$ and ℓ_p, $1 \le p < \infty$ are uniformly homeomorphic to each other. A very important achievement in this direction is due to Odell and Schlumprecht [346], who proved that unit balls of spaces with unconditional bases and nontrivial cotype are uniformly homeomorphic to the unit ball of a separable Hilbert space (see also [55, Chapter 9]). Chaatit [91] proved a generalization of this result for Banach lattices. Thus we conclude that Banach spaces with unconditional bases and nontrivial cotype, or, more generally, separable Banach lattices with nontrivial cotype do not admit uniformly coarse embeddings of families of expanders. The paper [91] contains also nonseparable generalizations of the Odell–Schlumprecht result.

Section 6.2 is based on [382], where Lemma 6.3 is attributed to V. Lafforgue. Complex interpolation for families of Banach spaces, which is introduced in Definition 6.8, is due to Coifman, Cwikel, Rochberg, Sagher, and Weiss [102] (in an equivalent form). It is worth mentioning that Pisier [382, §9] gives a characterization of uniformly curved spaces in terms similar to the ones used in the real method of interpolation (see [56, Chapter 3] for real method of interpolation). This characterization is rather technical and we do not state it here. The results of Exercises 6.29 and 6.30 are taken from [382]. We refer to [382, Chapter 2] for more discussion on the notions of (uniformly/fully) curved Banach spaces and related open problems. Here we would like to present the proof (given in [382, p. 15]) of the fact that uniformly curved Banach spaces are isomorphic to uniformly convex Banach spaces.

Fix $n \geq 1$. Let $\Gamma(n) = \{\Gamma(n)_{i,j}\}_{i,j=1}^{n}$ be the Hilbert matrix defined by

$$\Gamma(n)_{i,j} = \begin{cases} (n-(i+j))^{-1} & \text{if } n-(i+j) \neq 0, \\ 0 & \text{otherwise.} \end{cases}$$

It is well known that the operator $\Gamma(n)$ satisfies $\|\Gamma(n)_{\ell_2^n \to \ell_2^n}\| \leq C$, $\|\Gamma(n)_{\ell_1^n \to \ell_1^n}\| \leq C\log(n+1)$ and $\|\Gamma(n)_{\ell_\infty^n \to \ell_\infty^n}\| \leq C\log(n+1)$ for some constant C (independent of n), see [427]. Thus, for any Banach space X, we have

$$\|\Gamma(n)_X\| \leq C\log(n+1)\Delta_X((\log(n+1))^{-1}).$$

Therefore, if X is uniformly curved, then $\|\Gamma(n)_X\| = o(\log(n))$. By an observation made in [377], this implies that X is isomorphic to a uniformly convex space.

Section 6.3 presents a characterization of reflexivity which goes back to [386]. It was further developed by [412], [368], [212, see Theorem 8 (35)], and [331]. It was presented also in some texts, see [50, Theorem 6, p. 51]. Helly's condition is from [201, p. 73].

The notion of a stable space is due to Krivine and Maurey [256], they also proved that L_p is stable for each $p < \infty$. Useful expositions of results on stable Banach spaces can be found in [51], [159], and [187].

Results of Section 6.4.2 are due to Raynaud [390]. It is worth mentioning that up to 1969 it was an open problem, suggested by Yu. M. Smirnov: does there exist a separable metric space which is not uniformly homeomorphic to a subset of L_2? Smirnov's problem was solved in the positive by Enflo [134] who proved that the Banach space c_0 cannot be embedded uniformly into a Hilbert space.

Examples of nonreflexive Banach spaces which do not contain $\{\ell_1^n\}_{n=1}^{\infty}$ uniformly are difficult to construct, for some time (1964–1974) their existence was an open problem and there was a series of publications devoted to it. The first example was constructed by James [215]. See [217], [216], and [384] for more constructions of such spaces. See [144] for an exposition of the James–Lindenstrauss construction and [383] for a detailed exposition of the Pisier–Xu construction.

As we have already mentioned, Theorem 2.56 (in combination with Exercise 2.69) implies that a Banach space contains $\{\ell_1^n\}_{n=1}^\infty$ uniformly if and only if it has no nontrivial type. The construction of a nonreflexive space with nontrivial type presented in Section 6.5 is due to Pisier and Xu [384]. A detailed presentation of the construction and the proof of nontrivial type of $(v_1, \ell_\infty)_{\theta,q}$ are given in [383]. The simplified formula for the equivalent norm is derived using the ideas which are present in [384], some of them go back to [57].

Papers [112], [113], [145] contain related studies on degrees of non-reflexivity.

The standard reference on the real interpolation method is [56, Chapters 2 and 3], which in its presentation of the method borrows from [364]. [383] features an introduction to the real interpolation method while [86] is the most comprehensive advanced monograph on the real interpolation method.

6.8 Hints to exercises

To Exercise 6.27. To estimate $\|Px_1 - Px_2\|$ in terms of $\|x_1 - x_2\|$ first consider the trivial case $\|x_1\| = \|x_2\|$. Then, assume that $\|x_1\| \le \|x_2\|$ and $\|x_2\| > 1$ (other cases are either similar or trivial). Let

$$x_3 = \max\left\{\frac{\|x_1\|}{\|x_2\|}, \frac{1}{\|x_2\|}\right\} \cdot x_2.$$

Complete the estimate using the observations:

- $Px_3 = Px_2$.
- $\|Px_1 - Px_3\|$ is easy to estimate in terms of $\|x_3 - x_1\|$, because it is reduced to one of the trivial cases.
- $\|x_3 - x_1\| \le 2\|x_2 - x_1\|$.

To Exercise 6.30. Since X is curved, there is a value of $n \in \mathbb{N}$ satisfying

$$\Delta_X(2^{-n/2}) < 1.$$

Consider the operator

$$\mathrm{T}_n := \underbrace{\tau \otimes \tau \otimes \cdots \otimes \tau}_{n \text{ times}}$$

on $\ell_2^{2^n} = \ell_2^2 \otimes \cdots \otimes \ell_2^2$, where τ is the operator introduced in Exercise 6.29. Use the fact that $\|(\mathrm{T}_n)_X\| \le \Delta_X(2^{-n/2}) < 1$ to show that ℓ_1 (as a Banach space over $\mathbb{R}$) is not finitely representable in X. Use Theorem 2.56.

Chapter 7

Structure properties of spaces which are not coarsely embeddable into a Hilbert space

7.1 Expander-like structures implying coarse non-embeddability into L_1

There are several known obstructions for coarse embeddability of one metric space into another. But only one such obstruction is known for embeddings of spaces with bounded geometry into a Hilbert space (or any other infinite-dimensional Banach space, see Theorem 2.7). We mean Theorem 4.9 and its corollary, that is:

Families of expanders are not uniformly coarsely embeddable into L_1 or into a Hilbert space.

An easy consequence of this result and the fact that a composition of two coarse embeddings is a coarse embedding is:

Theorem 7.1. *If a metric space X admits uniformly coarse embeddings of a family of expanders into it, then X is not coarsely embeddable into a Hilbert space.*

As we have already mentioned, in the case where X is a space of bounded geometry no other obstructions for coarse embeddability into a Hilbert space are known (see Problem 11.9 and discussion after it in this connection).

The main results of this chapter were proved in attempts to find expander-like structures in metric spaces which do not admit coarse embeddings into a Hilbert space.

The following property of metric spaces with bounded geometry is the main candidate for being equivalent to coarse non-embeddability into ℓ_2.

Definition 7.2. Let X be a space with bounded geometry and $\{Y_n\}_{n=1}^{\infty}$ be a family of expanders. We say that X *weakly contains* $\{Y_n\}$ if there are maps $f_n : Y_n \to X$ satisfying (with some abuse of notation we use Y_n to denote the vertex set of Y_n)

(a) Lipschitz constants $\mathrm{Lip}(f_n)$ are uniformly bounded

(b) $\displaystyle\lim_{n\to\infty} \max_{y\in Y_n} \frac{|f_n^{-1}(f_n(y))|}{|Y_n|} = 0.$

The images of Y_n in X are called *weak expanders*.

Observation 7.3. If a sequence $\{Y_n\}_{n=1}^{\infty}$ is uniformly coarsely embeddable into a metric space X with bounded geometry, then X weakly contains $\{Y_n\}$.

Proof. Let $f_n : Y_n \to X$ be uniformly coarse embeddings. We are going to check that $\{f_n\}$ satisfy the conditions of Definition 7.2.

It is an easy observation (see Remark 1.50 on page 21) that uniformly coarse embeddings of (unweighted) graphs with their graph distances are Lipschitz maps with uniformly bounded Lipschitz constants.

To show that the second condition is satisfied we prove that the cardinalities

$$|f_n^{-1}(f_n(y))|$$

are uniformly bounded. To this end we observe that otherwise for each $r > 0$ there are $n \in \mathbb{N}$ and $y \in Y_n$ such that $|f_n^{-1}(f_n(y))| > M(r)$ (the function $M(r)$ is from the definition of bounded geometry, see Definition 1.66 on page 31). Therefore for each $r > 0$ there are $n = n(r)$ and $u = u(r,n), v = v(r,n) \in Y_n$ satisfying $f_n(u) = f_n(v)$ and $d_{Y_n}(u,v) > r$. This contradicts the assumption that $\{f_n\}$ are uniformly coarse embeddings. □

Example 7.4. For an arbitrary family $\{Y_n\}$ of expanders there is a metric space X with bounded geometry and Lipschitz maps $f_n : Y_n \to X$ which embed $\{Y_n\}$ weakly into X, but the embeddings $\{f_n\}$ are not uniformly coarse.

Proof. Let $\{Y_n\}_{n=1}^{\infty}$ be an arbitrary family of expanders. Let $u_n, v_n \in V(Y_n)$ be such that $d_{Y_n}(u_n, v_n) = \operatorname{diam} Y_n$. It is not difficult to see that $\lim_{n\to\infty} \operatorname{diam} Y_n = \infty$. (In fact, since Y_n are k-regular, for each $u \in V(Y_n)$ and $r \in \mathbb{N}$ the number of vertices at distance $\le r$ from u does not exceed $1 + k + (k-1)k + \cdots + (k-1)^{r-1}k$. As $\lim_{n\to\infty} |V(Y_n)| = \infty$, the conclusion follows.) Let Z_n be the graph obtained from Y_n if we identify u_n and v_n. We mean the standard modification of a graph consisting in replacing u_n and v_n by a single vertex incident to all the edges which were incident in Y_n to either u_n or v_n.

It is clear that $\{Z_n\}_n$ are graphs with uniformly bounded degrees and therefore it is easy to form a bounded geometry metric space X containing isometric copies of all of $\{Z_n\}$ (one of the ways of constructing such X is described after Corollary 4.10).

It is clear that the natural embeddings $f_n : Y_n \to X$, formed by identifying u_n and v_n first, and then applying an isometric embedding into X, weakly embed $\{Y_n\}$ into X. On the other hand, these embeddings are not uniformly coarse because $d_{Y_n}(u_n, v_n) \to \infty$ but $d_X(f(u_n), f(v_n)) = 0$. □

Observation 7.5. If X is a metric space with bounded geometry weakly containing a family $\{Y_n\}$ of expanders, then X is not coarsely embeddable into L_1.

Proof. This proof is similar to the proof of Corollary 4.10. Let $f_n : Y_n \to X$ be maps showing that X weakly contains $\{Y_n\}$ (see Definition 7.2). Assume that there is a coarse embedding $f : X \to L_1$. Then there are functions ρ_1 and ρ_2 such that

$$\begin{aligned} \forall n \in \mathbb{N}\ \forall u, v \in Y_n \quad \rho_1(d_X(f_n(u), f_n(v))) &\le \|f(f_n(u)) - f(f_n(v))\| \\ &\le \rho_2(d_X(f_n(u), f_n(v))). \end{aligned} \tag{7.1}$$

and $\lim_{t\to\infty} \rho_1(t) = \infty$.

We apply the inequality (4.6) to $\varphi_n(u) = f(f_n(u))$ and get

$$\sum_{u,v \in V(Y_n)} a_{u,v} \|\varphi_n(u) - \varphi_n(v)\| \geq \sum_{u,v \in V(Y_n)} \frac{h}{|V(Y_n)|} \|\varphi_n(u) - \varphi_n(v)\|. \tag{7.2}$$

Since Y_n are k-regular, we use (7.1) and get that the left-hand side is $\leq k|V(Y_n)| \cdot \rho_2(\sup_n \operatorname{Lip}(f_n))$. The right-hand side can be estimated from below by

$$\sum_{u,v \in V(Y_n)} \frac{h}{|V(Y_n)|} \rho_1(d_X(f_n(u), f_n(v))).$$

We estimate this sum from below using the following argument. We claim that for each $0 < C < \infty$ we can find $n \in \mathbb{N}$ such that for each $u \in V(Y_n)$ the number of vertices $v \in V(Y_n)$ satisfying $d_X(f_n(u), f_n(v)) \geq C$ is at least $|V(Y_n)|/2$. In fact, the number of elements $x \in X$ satisfying $d_X(f_n(u), x) \leq C$ does not exceed $M_X(C)$ (where $M_X(\cdot)$ is the function from the definition of bounded geometry, see Definition 1.66). Therefore the number of elements $v \in V(Y_n)$ satisfying $d_X(f_n(u), f_n(v)) \leq C$ does not exceed $M_X(C) \cdot \max_{x \in f_n(Y_n)} |f_n^{-1}(x)|$. Condition (b) of Definition 7.2 (of weak containment of expanders) implies that

$$\lim_{n \to \infty} \frac{M_X(C) \cdot \max_{x \in f_n(Y_n)} |f_n^{-1}(x)|}{|V(Y_n)|} = 0.$$

Therefore the desired n exists.

For such C and n we get, by inequalities (7.1), (7.2), and the argument immediately after them, that:

$$\frac{|V(Y_n)|^2}{2} \cdot \frac{h}{|V(Y_n)|} \cdot \rho_1(C) \leq k|V(Y_n)|\rho_2(\sup_n \operatorname{Lip}(f_n)).$$

Thus

$$\rho_1(C) \leq \frac{2k}{h} \rho_2(\sup_n \operatorname{Lip}(f_n)).$$

Since C can be chosen to be arbitrarily large, this leads to a contradiction. □

7.2 On the structure of locally finite spaces which do not admit coarse embeddings into a Hilbert space

At the moment it is unknown whether each metric space with bounded geometry which does not admit a coarse embedding into a Hilbert space contains weak expanders. In an attempt to prove this the following result was obtained.

Recall (see Observation 4.12) that coarse embeddability into a (separable infinite-dimensional) Hilbert space is equivalent to coarse embeddability into $L_1(0, 1)$.

Theorem 7.6. *Let (M, d_M) be a locally finite metric space which is not coarsely embeddable into L_1. Then there exists a constant D, depending on M only, such that for each $n \in \mathbb{N}$ there exists a finite set $B_n \subset M \times M$ and a probability measure μ on B_n such that*

- $d_M(u, v) \geq n$ *for each* $(u, v) \in B_n$.
- *For each Lipschitz function* $f : M \to L_1$ *we have*

$$\int_{B_n} \|f(u) - f(v)\|_{L_1} d\mu(u, v) \leq D\,\mathrm{Lip}(f). \tag{7.3}$$

Lemma 7.7. *Let (M, d_M) be a locally finite metric space which is not coarsely embeddable into L_1. Then there exists a constant C depending on M only such that for each Lipschitz function $f : M \to L_1$ there exists a subset $B_f \subset M \times M$ such that $\sup_{(x,y)\in B_f} d_M(x, y) = \infty$, but $\sup_{(x,y)\in B_f} \|f(x) - f(y)\|_{L_1} \leq C\,\mathrm{Lip}(f)$.*

Proof. Assume the contrary. Then, for each $n \in \mathbb{N}$, the number n^3 cannot serve as C. This means, that for each $n \in \mathbb{N}$ there exists a Lipschitz map $f_n : M \to L_1$ such that for each subset $U \subset M \times M$ with

$$\sup_{(x,y)\in U} d_M(x, y) = \infty,$$

we have

$$\sup_{(x,y)\in U} \|f_n(x) - f_n(y)\| > n^3 \mathrm{Lip}(f_n).$$

We choose a point in M and denote it by O. Without loss of generality we may assume that $f_n(O) = 0$. Consider the mapping $f : M \to (\oplus_{k=1}^{\infty} L_1)_1$ given by

$$f(x) = \oplus_{k=1}^{\infty} \left(\frac{1}{Kk^2} \cdot \frac{f_k(x)}{\mathrm{Lip}(f_k)} \right),$$

where $K = \sum_{k=1}^{\infty} \frac{1}{k^2}$ (ℓ_1 direct sums were defined on p. 82). It is clear that the series converges and $\mathrm{Lip}(f) \leq 1$. Considering $L_1(a, b)$ for infinitely many disjoint subintervals of [0, 1] it is easy to show that $\left(\oplus_{k=1}^{\infty} L_1\right)_1$ admits an isometric embedding into $L_1(0, 1)$. Thus f may be regarded as a mapping into $L_1(0, 1)$.

Let us show that f is a coarse embedding. We need an estimate from below only (the estimate from above is satisfied because f is Lipschitz).

The assumption implies that for each $n \in \mathbb{N}$ there is $N \in \mathbb{N}$ such that

$$d_M(x, y) \geq N \Rightarrow \|f_n(x) - f_n(y)\| > n^3 \mathrm{Lip}(f_n).$$

On the other hand, $\|f_n(x) - f_n(y)\| > n^3 \mathrm{Lip}(f_n)$ implies

$$\|f(x) - f(y)\| = \sum_{k=1}^{\infty} \frac{1}{Kk^2} \cdot \frac{\|f_k(x) - f_k(y)\|}{\mathrm{Lip}(f_k)} \geq \frac{1}{Kn^2} \cdot \frac{\|f_n(x) - f_n(y)\|}{\mathrm{Lip}(f_n)} > \frac{n}{K}.$$

Hence $f : M \to L_1$ is a coarse embedding and we get a contradiction. □

Lemma 7.8. *Let C be the constant whose existence is proved in Lemma 7.7, and let $\varepsilon > 0$ be arbitrary. Then for each $n \in \mathbb{N}$ there is a finite subset $M_n \subset M$ such that for each Lipschitz mapping $f : M \to L_1$ there is a pair $(u_{f,n}, v_{f,n}) \in M_n \times M_n$ such that*

- $d_M(u_{f,n}, v_{f,n}) \geq n.$
- $\|f(u_{f,n}) - f(v_{f,n})\| \leq (C + \varepsilon)\mathrm{Lip}(f).$

Proof. The ball in the space M of radius R centered at O will be denoted by $B(R)$. It is clear that it suffices to prove the result for 1-Lipschitz mappings satisfying $f(O) = 0$.

Assume the contrary. Since M is locally finite, this implies that for each $R \in \mathbb{N}$ there is a 1-Lipschitz mapping $f_R : M \to L_1$ such that $f_R(O) = 0$ and, for $u, v \in B(R)$, the inequality $d_M(u, v) \geq n$ implies $\|f_R(u) - f_R(v)\|_{L_1} > C + \varepsilon$.

We form an ultraproduct of the mappings $\{f_R\}_{R=1}^{\infty}$, that is, a mapping $f : M \to (\prod L_1)_{\mathcal{U}}$, given by $f(m) = \{f_R(m)\}_{R=1}^{\infty}$, where $\mathcal{U}$ is a free ultrafilter on $\mathbb{N}$ and $(\prod L_1)_{\mathcal{U}}$ is the corresponding ultraproduct.

It is well known that an ultraproduct of L_1 spaces is isometric to an L_1 space on some measure space (see Notes and Remarks for references). It is known (see Fact 1.20) that separable subspaces of an L_1 space on any measure space are isometric to subspaces of $L_1(0,1)$. Therefore we may consider f as a mapping into $L_1(0,1)$. It is easy to verify that $\mathrm{Lip}(f) \leq 1$ and that f satisfies the condition

$$d_M(u, v) \geq n \Rightarrow \|f(u) - f(v)\|_{L_1} \geq (C + \varepsilon).$$

We get a contradiction with the definition of C. □

Proof of the Theorem 7.6. Let D be a number satisfying $D > C$, and let B be a number satisfying $C < B < D$.

By Lemma 7.8, there is a finite subset $M_n \subset M$ such that for each 1-Lipschitz function f on M there is a pair (u, v) in M_n such that $d_M(u, v) \geq n$ and $\|f(u) - f(v)\| \leq B$.

We choose a point in M_n and denote it by O (it should be different from the point O chosen above if that point is not in M_n, otherwise we can choose the same point). Proving the theorem it is enough to consider 1-Lipschitz functions $f : M_n \to L_1$ satisfying $f(O) = 0$. As we know (Lemma 1.24) each subset of L_1 of cardinality s is isometric to a subset of ℓ_1^k for $k \leq s!$. (Actually much better estimates for k are known, see Notes and Remarks to Chapter 1, but we do not need them here.) Therefore it suffices to prove the result for 1-Lipschitz functions $M_n \mapsto \ell_1^k$ satisfying $f(O) = 0$. It is easy to see that this set of functions is compact with respect to the metric

$$\tau(f, g) = \max_{m \in M_n} \|f(m) - g(m)\|.$$

Clearly it suffices to prove the inequality

$$\int_{B_n} \|f(u) - f(v)\| d\mu(u, v) \leq B$$

for a $\left(\frac{D-B}{2}\right)$-net in the set of all functions satisfying the conditions mentioned above, endowed with the metric τ. By compactness there exists a finite net satisfying the condition. Let N be such a net.

We are going to use the von Neumann minimax theorem.

Let A be the matrix whose columns are labeled by functions belonging to N, whose rows are labeled by pairs (u, v) of elements of M_n satisfying $d_M(u, v) \geq n$, and whose entry on the intersection of the column corresponding to f, and the row corresponding to (u, v) is $\|f(u) - f(v)\|$.

Then, for each column vector $x = \{x_f\}_{f \in N}$ with $x_f \geq 0$ and $\sum_{f \in N} x_f = 1$, the entries of the product Ax are the differences $\|F(u) - F(v)\|$, where

$$F : M \to \left(\oplus_{f \in N} \ell_1^k\right)_1$$

is given by $F(m) = \oplus_{f \in N} x_f f(m)$. It is easy to show that the space $(\oplus_{f \in N} \ell_1^k)_1$ is isometric to $\ell_1^{k|N|}$ and thus admits a linear isometric embedding into L_1. Therefore the function F can be considered as a function into L_1. It satisfies $\mathrm{Lip}(F) \leq 1$. Hence there is a pair (u, v) in M_n satisfying $d_M(u, v) \geq n$ and $\|F(u) - F(v)\| \leq B$. Therefore we have

$$\max_x \min_\mu \mu A x \leq B,$$

where the minimum is taken over all vectors $\mu = \{\mu(u, v)\}$, indexed by $u, v \in M_n$, $d_M(u, v) \geq n$, and satisfying the conditions $\mu(u, v) \geq 0$ and $\sum \mu(u, v) = 1$.

By the von Neumann minimax theorem we have

$$\min_\mu \max_x \mu A x \leq B,$$

which is exactly the inequality we need to prove because μ can be regarded as a probability measure on the set of pairs from M_n with distance $\geq n$. □

7.3 Expansion properties of metric spaces not admitting a coarse embedding into a Hilbert space

Our next purpose is to find some expansion properties of the sets M_n whose existence is proved in Theorem 7.6.

Let s be a positive integer. We consider graphs $G(n, s) = (M_n, E(M_n, s))$, where the edge set $E(M_n, s)$ is obtained by joining those pairs of vertices of M_n which are at distance $\leq s$. The graphs $\{G(n, s)\}_{n=1}^{\infty}$ have uniformly bounded degrees if the metric space M has bounded geometry.

We need to recall some more definitions of graph theory. A *component* of a graph G is a maximal subset $D \subset V(G)$ satisfying the condition: for each u, v in D there is a uv-walk in G. A *vertex cut* C in G is a set of vertices of G whose deletion from G

(together with all edges incident with vertices in C) increases the number of components. If two vertices are in the same component of G, but in different components of $G - C$ (the graph obtained after the deletion of C), we say that they are *separated* by the vertex cut C.

Observation 7.9. If vertices x and y in $G(n,s)$ can be separated by a vertex cut, then $d_M(x,y) > s$. Informally: each vertex cut of $G(n,s)$ separates it into pieces with d_M-distance between them at least s.

It is clear that for fixed s the identical embeddings on the graphs $G(n,s)$ with their graph metrics into X have uniformly bounded Lipschitz constants and are bijective. For this reason if we could prove that for X having bounded geometry, the graphs $G(n,s)$ contain weak expanders, this would provide a characterization of spaces with bounded geometry which do not admit coarse embeddings into a Hilbert space in terms of weak containment of expanders. (Since $G(n,s)$ is a family of metric spaces rather than a metric space, we need to extend our definitions slightly: saying $G(n,s)$ *contain weak expanders*, we mean that there is a family of expanders $\{Y_m\}_{m=1}^{\infty}$, a sequence $\{n(m)\}_{m=1}^{\infty}$ of positive integers, and embeddings $f_m : Y_m \to G(n(m),s)$ satisfying the conditions of Definition 7.2.)

At the moment we are able to prove only the following weaker expansion property of the graphs $G(n,s)$. We denote the measure μ whose existence was proved in Theorem 7.6 by μ_n. We introduce the measure ν_n on M_n by $\nu_n(A) = \mu_n(A \times M_n)$. Let F be a subgraph of $G(n,s)$ and A be a set of vertices in F. The set of all vertices of F which are adjacent to vertices of A but are not in A is called the *vertex boundary* of A in F and is denoted by $\delta_F A$.

Theorem 7.10. *Let s and n be such that $2n > s > 8D$. Let $\varphi(D,s) = \frac{s}{4D} - 2$. Then $G(n,s)$ contains an induced subgraph F with d_M-diameter $\geq n - \frac{s}{2}$, such that each subset $A \subset F$ of d_M-diameter $< n - \frac{s}{2}$ satisfies the condition: $\nu_n(\delta_F A) > \varphi(D,s)\nu_n(A)$.*

Proof of Theorem 7.10. Suppose that for some $n, s \in \mathbb{N}$ satisfying $2n > s > 8D$ there is no such subgraph in $G(n,s)$. Then for each induced subgraph F in $G(n,s)$ of d_M-diameter $\geq n - \frac{s}{2}$ we can find a subset $A \subset F$ of d_M-diameter $< n - \frac{s}{2}$ such that $\nu_n(\delta_F A) \leq \varphi(D,s)\nu_n(A)$. We start with $F_1 = G(n,s)$ (the definitions of M_n and μ_n imply that the d_M-diameter of M_n is $\geq n$), find a subset $A_1 \subset F_1$ of d_M-diameter $< n - \frac{s}{2}$ such that $\nu_n(\delta_{F_1} A_1) \leq \varphi(D,s)\nu_n(A_1)$, and delete $A_1 \cup \delta_{F_1} A_1$ from $G(n,s)$. If the obtained graph F_2 still has d_M-diameter $\geq n - \frac{s}{2}$, we find a subset A_2 in it such that $\nu_n(\delta_{F_2} A_2) \leq \varphi(D,s)\nu_n(A_2)$. We delete the subset $A_2 \cup \delta_{F_2} A_2$ from F_2. We continue in an obvious way until we get a set of d_M-diameter $< n - \frac{s}{2}$ (this should eventually happen since M_n is finite). We denote this set A_p, where p is the number of steps in the process.

Observe that each of the sets A_i has diameter $< n - \frac{s}{2}$, and that the d_M-distance between any A_i and A_j $(i \neq j)$ is at least s (see the observation above).

We introduce a family of 1-Lipschitz functions f_θ on M, where $\theta = \{\theta_i\}_{i=1}^p \in \Theta = \{-1, 1\}^p$ by the formula:

$$f_\theta(x) = \begin{cases} \theta_j\left(\dfrac{s}{2} - \operatorname{dist}(x, A_j)\right) & \text{if } \operatorname{dist}(x, A_j) < \frac{s}{2} \\ 0 & \text{if } \operatorname{dist}(x, \cup_{i=1}^p A_i) \geq \frac{s}{2}. \end{cases}$$

The function is well-defined since the inequality $\operatorname{dist}(x, A_j) < \frac{s}{2}$ cannot be satisfied for more than one value of j. Straightforward verification shows that this function is 1-Lipschitz.

We endow $\Theta = \{-1, 1\}^p$ with the natural probability measure $\mathcal{P}$ and introduce for each $x \in M$ a function $F_x \in L_1(\Theta, \mathcal{P})$ given by $F_x(\theta) = f_\theta(x)$. It is clear that the mapping $x \mapsto F_x$ is 1-Lipschitz.

Applying inequality (7.3) to this mapping we get

$$\begin{aligned} D &\geq \int_{M_n \times M_n} \|F_x(\theta) - F_y(\theta)\|_{L_1(\Theta, \mathcal{P})} d\mu_n(x, y) \\ &\geq \int_{M_n \times M_n} \int_\Theta |f_\theta(x) - f_\theta(y)| d\mathcal{P}(\theta) d\mu_n(x, y) \\ &\geq \int_{M_n \times M_n} \int_{\Psi(x,y)} |f_\theta(x)| d\mathcal{P}(\theta) d\mu_n(x, y), \end{aligned}$$

where $\Psi(x, y)$ is the subset of Θ for which $f_\theta(x)$ and $f_\theta(y)$ have different signs (we mean that signs have values in $\{-1, 0, 1\}$). Observe that the value of $|f_\theta(x)|$ does not depend on θ. We get

$$\begin{aligned} &\int_{M_n \times M_n} \int_{\Psi(x,y)} |f_\theta(x)| d\mathcal{P}(\theta) d\mu_n(x, y) \\ &\qquad \geq \int_{(\cup_{i=1}^p A_i) \times M_n} |f_\theta(x)| \int_{\Psi(x,y)} d\mathcal{P}(\theta) d\mu_n(x, y). \end{aligned}$$

Now we observe that for $x \in A_j$ and y satisfying $(x, y) \in \operatorname{supp}\mu_n$ we have $d_M(x, y) \geq n$ and therefore $d_M(y, A_j) \geq \frac{s}{2}$ (recall that the diameter of A_j is $< n - \frac{s}{2}$). Hence $\mathcal{P}(\Psi(x, y)) \geq \frac{1}{2}$ for each pair (x, y) from $\operatorname{supp}\mu_n$ satisfying $x \in \cup_{i=1}^p A_i$. We get

$$\begin{aligned} \int_{(\cup_{i=1}^p A_i) \times M_n} |f_\theta(x)| \int_{\Psi(x,y)} d\mathcal{P}(\theta) d\mu_n(x, y) &\geq \int_{(\cup_{i=1}^p A_i) \times M_n} \frac{s}{2} \cdot \frac{1}{2} d\mu_n(x, y) \\ &= \frac{s}{4} \nu_n\left(\cup_i A_i\right). \end{aligned}$$

Recalling the beginning of this chain of inequalities, we get

$$D \geq \frac{s}{4}\,\nu_n\left(\cup_i A_i\right). \tag{7.4}$$

Observe that

$$\nu_n(\cup_{i=1}^{p} A_i) + \nu_n(\cup_{i=1}^{p-1} \delta_{F_i} A_i) = 1 \quad \text{and} \quad \nu_n(\cup_i \delta_{F_i} A_i) \leq \varphi(D,s)\nu_n(\cup_i A_i).$$

Therefore

$$(1+\varphi(D,s))\nu_n\left(\cup_i A_i\right) \geq 1. \tag{7.5}$$

Combining (7.4) and (7.5) we get

$$D \geq \frac{s}{4(1+\varphi(D,s))},$$

or $\varphi(D,s) \geq \frac{s}{4D} - 1$, a contradiction. □

7.4 Exercises

Exercise 7.11. Construct a sequence $\{Y_n\}$ of graphs (endowed with their graph distances) and a metric space X with bounded geometry such that there exist 1-Lipschitz injective maps $f_n : Y_n \to X$ which are not uniformly coarse.

Exercise 7.12. Let X be a metric space containing isometrically a family $\{Y_n\}$ of expanders. Find sets M_n and a measure μ on $M_n \times M_n$ satisfying the conditions of Theorem 7.6.

There is a hint to this exercise.

Interested readers are encouraged to prove the same for a metric spaces with bounded geometry weakly containing a family of expanders.

Exercise 7.13. Let $\{Y_n\}$ be a family of expanders. Find a metric space X with bounded geometry which contains $\{Y_n\}$ weakly, but is such that no family $\{Z_n\}$ of expanders admits uniformly bilipschitz embeddings into X.

There is a hint to this exercise.

Exercise 7.14. Let $\{G_n\}_{n=1}^{\infty}$ be a family of k-regular graphs such that the girth $g(G_n)$ is at least $2n+1$. Show that for every set $A \subset V(G_n)$ of diameter $< n$ the inequality $|\delta_{G_n}(A)| \geq (k-2)|A|$ holds ($\delta_{G_n}(A)$ denotes the vertex boundary of A in G_n).

7.5 Notes and remarks

The notion of a bounded geometry space weakly containing a family of expanders, without this terminology, can be found in [202, pp. 350–351], where it is attributed to

Gromov and the reference is given to [178]. The same notion in a more general context (without the bounded geometry restriction) is introduced in [126, §3], where the name *coarse quasi-embedding* is used (for this more general notion), and the references are to [178] and [179]. We use the terminology introduced in [186, p. 260].

Theorem 7.6 was proved in [354, Theorem 2.4]. Another version of this theorem was independently proved by Tessera [425].

A proof of the von Neumann minimax theorem can be found, for example, in [418, p. 344] or in [422, §1.16].

Theorem 7.10 is from [355]. The exhaustion process used in the proof of Theorem 7.10 is similar to the one used in [293]. The idea of "random" signing of functions in a situation similar to the one in the proof of Theorem 7.10 was used in [389]. It should be mentioned that the conditions of the type of used in Theorem 7.10 are not sufficient for coarse non-embeddability into L_1. In fact, as we know (Theorem 5.57) there exist families of graphs with indefinitely growing girth which admit uniformly bilipschitz embeddings into L_1. On the other hand (Exercise 7.14), such families of graphs are *locally expanding* in the sense of Theorem 7.10. (An earlier example of a locally expanding family of graphs admitting uniformly coarse embeddings into a Hilbert space was given in [27].)

Theorem 7.10 was used in [84] to study the relationship between various large-scale geometric properties of metric spaces which have been important in recent work on Baum–Connes type conjectures. See [400], [401] for further related studies.

7.6 Hints to exercises

To Exercise 7.12. Use Theorem 4.7 and the argument preceding Corollary 4.10.

To Exercise 7.13. Use the observation that for each metric space (X, d_X) and each $0 < \varepsilon < 1$ the function $(x, y) \mapsto (d_X(x, y))^\varepsilon$ is also a metric on X.

Chapter 8

Applications of Markov chains to embeddability problems

8.1 Basic definitions and results on finite Markov chains

Let $(\Omega, \Sigma, \mathbb{P})$ be a probability space. We use $\mathbb{P}(B_1|B_2)$ to denote the conditional probability (that is $\mathbb{P}(B_1 \cap B_2)/\mathbb{P}(B_2)$).

Definition 8.1. Let $S = \{s_1, \dots, s_k\}$ be a finite set. A *Markov chain* with *state space* S is a sequence $\{X_0, X_1, \dots\}$ of S-valued random variables on a probability space $(\Omega, \Sigma, \mathbb{P})$ satisfying for all n, all s_i, s_j, and all D the condition

$$\mathbb{P}(X_{n+1} = s_j | X_n = s_i) = \mathbb{P}(X_{n+1} = s_j | (X_n = s_i) \cap D) = a_{i,j}, \tag{8.1}$$

where $a_{i,j}$ are numbers depending only on i and j (but not on n), and D is an event depending only on variables $X_{n-1}, \dots, X_0$ and such that $\mathbb{P}((X_n = s_i) \cap D) > 0$. The numbers $a_{i,j}$ (also denoted a_{ij}) are called *transition probabilities* and the $n \times n$ matrix $A := \{a_{i,j}\}_{i,j=1}^n$ is called the *transition matrix.*

Definition 8.2. The *initial distribution* is the set of numbers

$$\mathbb{P}(X_0 = s_1), \dots, \mathbb{P}(X_0 = s_k),$$

considered as a row vector $\mu^{(0)}$. The *distribution at time t* is the set of numbers $\mathbb{P}(X_t = s_1), \dots, \mathbb{P}(X_t = s_k)$, considered as a row vector $\mu^{(t)}$.

Observation 8.3. The following equality is a straightforward consequence of our definitions:

$$\mu^{(t)} = \mu^{(0)} A^t. \tag{8.2}$$

(In the right-hand side we mean a multiplication of a row vector by a matrix.)

Unless we would like to consider a state space with some additional structure, we assume that the state space is $\{1, \dots, k\}$.

Definition 8.4. A Markov chain $\{X_t\}_{t=0}^{\infty}$ on the state space $\{1, \dots, k\}$ with transition probabilities

$$a_{ij} := \mathbb{P}(X_{t+1} = j | X_t = i)$$

is called *symmetric* if $a_{ij} = a_{ji}$.

Definition 8.5. Let $\{X_t\}_{t=0}^{\infty}$ be a Markov chain on $S = \{1, \dots, k\}$ with a transition matrix $A = \{a_{ij}\}$. A probability distribution $\{\pi_i\}_{i=1}^{k}$ on S is called *reversible* for the chain $\{X_t\}$ if

$$\pi_i \, a_{ij} = \pi_j \, a_{ji} \tag{8.3}$$

for all $i, j \in \{1, \dots, k\}$. A Markov chain is called *reversible* if there exists a reversible distribution for it.

A probability distribution $\pi = \{\pi_i\}_{i=1}^{k}$ on S is called *stationary* for a Markov chain $\{X_t\}$, or for the corresponding transition matrix A, if

$$\pi A = \pi.$$

A Markov chain is called *stationary* if $\mu^{(t)}$ does not depend on t.

Example 8.6. A typical (and very useful) example of a symmetric Markov chain: consider a k-regular connected graph G with n vertices and let $S = \{s_1, \dots, s_n\}$ be the vertex set $V(G)$. We consider the following random variables: X_0 is distributed according to any distribution and $a_{i,j}$ are defined by

$$a_{i,j} = \begin{cases} \frac{1}{k} & \text{if } s_i \text{ and } s_j \text{ are adjacent} \\ 0 & \text{otherwise.} \end{cases}$$

This is clearly a symmetric Markov chain. One can easily check that this Markov chain is not stationary unless X_0 has uniform distribution.

Proposition 8.7. *A reversible distribution for a Markov chain is stationary.*

Proof. In fact, the only thing which we need to check is: $\sum_{i=1}^{k} \pi_i \, a_{ij} = \pi_j$. But this is immediate from (8.3), in fact (8.3) implies $\sum_{i=1}^{k} \pi_i \, a_{ij} = \sum_{i=1}^{k} \pi_j \, a_{ji} = \pi_j$ since $\sum_{i=1}^{k} a_{ji} = 1$, see Definition 8.1. □

Observation 8.8. Condition (8.3) is equivalent to the condition: the product

$$\begin{bmatrix} \pi_1 & \dots & 0 \\ \vdots & \ddots & \vdots \\ 0 & \dots & \pi_n \end{bmatrix} \begin{bmatrix} a_{11} & \dots & a_{1k} \\ \vdots & \ddots & \vdots \\ a_{n1} & \dots & a_{nn} \end{bmatrix} \tag{8.4}$$

is a symmetric matrix (where the first matrix is diagonal with $\pi_1, \dots, \pi_n$ on the diagonal).

Example 8.9. A very useful general example of a reversible Markov chain: consider a (not necessarily regular) finite connected graph G with k vertices and let $S = \{1, \dots, k\}$ be the vertex set $V(G)$. We consider the following random variables: X_0 is distributed according to the distribution π with

$$\pi_i = \frac{d_i}{\sum_{i=1}^{n} d_i},$$

where d_i is the degree of i in G, and $a_{i,j}$ are defined by

$$a_{i,j} = \begin{cases} \frac{1}{d_i} & \text{if } i \text{ and } j \text{ are adjacent} \\ 0 & \text{otherwise.} \end{cases}$$

Straightforward verification shows that π is a reversible distribution for this Markov chain.

8.2 Markov type

It turns out that Markov chains are very useful for problems of bilipschitz embeddability of metric spaces, because the behavior of Markov chains in different metric spaces is in a certain sense incompatible. One of the most useful tools for distinguishing the behavior of Markov chains in different metric spaces is introduced in the following definition.

Definition 8.10. Given a metric space (X,d) and $p \in [1,\infty)$, we say that X has *Markov type* p if there exists a constant $K > 0$ such that for every stationary reversible Markov chain $\{Z_t\}_{t=0}^{\infty}$ on $\{1,\dots,n\}$, every mapping $f : \{1,\dots,n\} \to X$ and every time $t \in \mathbb{N}$,

$$\mathbb{E}\, d(f(Z_t), f(Z_0))^p \le K^p\, t\, \mathbb{E}\, d(f(Z_1), f(Z_0))^p. \tag{8.5}$$

(Here and throughout, we omit some parentheses and write $\mathbb{E}\, x^p$ for $\mathbb{E}(x^p)$, etc.) The least such K is called the *Markov type* p *constant* of X, and is denoted $M_p(X)$.

Remark 8.11. Speaking of a stationary reversible Markov chain $\{Z_t\}_{t=0}^{\infty}$ we will always assume that the distribution

$$\mathbb{P}(Z_t = 1), \dots, \mathbb{P}(Z_t = n)$$

is reversible. (For some reversible Markov chains there exist stationary probability distributions which are not reversible, see Exercise 8.61. Some authors avoid such situations by adding irreducibility (see Exercise 8.59) to the definition of a reversible Markov chain.)

Theorem 8.12. *The Hilbert space has Markov type* 2 *with constant 1.*

Proof. It is clearly enough to prove the result for $L_2(0,1)$. Furthermore, it suffices to prove the result for the real line $\mathbb{R}$ (it passes to $L_2(0,1)$ by integration).

Let $\{Z_t\}_{t=0}^{\infty}$ be a stationary reversible Markov chain on $\{1,\dots,n\}$, with transition matrix $A = (a_{ij})$, and reversible distribution π. To prove that $\mathbb{R}$ has the Markov type 2 with constant 1 means to show that for every $x_1,\dots,x_n \in \mathbb{R}$ the inequality

$$\sum_{i,j} \pi_i\, (A^t)_{ij}\, (x_i - x_j)^2 \le t \sum_{i,j} \pi_i\, a_{ij}\, (x_i - x_j)^2 \tag{8.6}$$

holds, where A^t is the t-th power of the matrix A. In fact, the stronger inequality

$$\sum_{i,j} \pi_i \,(A^t)_{ij}\,(x_i - x_j)^2 \le \Lambda(t) \sum_{i,j} \pi_i \,a_{ij}\,(x_i - x_j)^2 \tag{8.7}$$

holds, where $\Lambda(t)$ is the value of the polynomial $1 + x + \cdots + x^{t-1}$ at the second largest eigenvalue of A in the case where such an eigenvalue is positive, and 1 if the second largest eigenvalue is negative. To make this statement meaningful we need to explain why the eigenvalues of A are real. This property follows from the fact that A is self-adjoint with respect to the inner product $\langle\cdot,\cdot\rangle_\pi$ introduced below.

To derive (8.6) from (8.7) it suffices to show that each eigenvalue of A is in the interval $[-1, 1]$. This statement follows from the fact that all entries of A are nonnegative and row sums are equal to 1. Therefore the absolute values of all entries of Az, where z is a column vector with coordinates $z_1, \ldots, z_n$, do not exceed $\max_{1\le i\le n} |z_i|$, the conclusion follows.

It is also clear that 1 is an eigenvalue of A and the all-ones vector is an eigenvector corresponding to the eigenvalue 1.

The assumption that the considered Markov chain is reversible and π is its reversible distribution implies $\pi_i\,(A^t)_{ij} = \pi_j\,(A^t)_{ji}$. This equality can be obtained using (8.4). In fact, denoting the diagonal matrix in (8.4) by D_π we have $D_\pi A = (D_\pi A)^*$, where $(D_\pi A)^*$ denotes the transposed matrix (the entries of the matrices are real numbers, so this notation is not confusing). Therefore we get

$$\begin{aligned} D_\pi A^t &= (D_\pi A)^*\, A^{t-1} \\ &= A^*\, D_\pi\, A^{t-1} \\ &= A^*\, A^*\, D_\pi\, A^{t-2} \\ &= \ldots \\ &= (A^*)^t\, D_\pi \\ &= (D_\pi A^t)^*. \end{aligned} \tag{8.8}$$

We introduce the inner product $\langle x, y\rangle_\pi$ on $\mathbb{R}^n$ by $\langle x, y\rangle_\pi = \sum_{i=1}^n \pi_i x_i y_i$.

By (8.8) matrices in both sides of (8.7) are symmetric, using this symmetry and expanding the squares, we get that (8.7) is equivalent to

$$\langle (I - A^t)x, x\rangle_\pi \le \Lambda(t)\,\langle (I - A)x, x\rangle_\pi\,. \tag{8.9}$$

If x is an eigenvector, then (8.9) holds for the following reasons: in the case where the eigenvalue is 1, inequality (8.9) is satisfied because both sides are 0, for all other eigenvalues it holds because the value of $\Lambda(t)$ cannot be less than the value of $1 + x + \cdots + x^{t-1}$ at any of the eigenvalues except 1. The general case follows by a spectral decomposition, since A is self-adjoint with respect to $\langle\cdot,\cdot\rangle_\pi$. □

The following immediate corollary of Theorem 8.12 is very important.

Corollary 8.13. *For every metric space (X, d), the Euclidean distortion of X satisfies:*

$$c_2(X, d) \geq M_2(X).$$

Proof. Let $f : X \to \ell_2$ be an embedding for which the Euclidean distortion is almost attained, that is

$$d(x, y) \leq \|f(x) - f(y)\| \leq d(x, y)(c_2(X, d) + \varepsilon)$$

for every $x, y \in X$. For every reversible stationary Markov chain $\{Z_t\}_{t=0}^{\infty}$ on X, the Markov type 2 property of ℓ_2 applied to the Markov chain $\{f(Z_t)\}_{t=0}^{\infty}$ gives

$$\begin{aligned}\mathbb{E}d^2(Z_t, Z_0) &\leq \mathbb{E}(\|f(Z_t) - f(Z_0)\|^2)\\ &\leq t\mathbb{E}(\|f(Z_1) - f(Z_0)\|^2)\\ &\leq t\,(c_2(X, d) + \varepsilon)^2\, \mathbb{E}d^2(Z_1, Z_0),\end{aligned}$$

therefore $M_2(X) \leq c_2(X, d) + \varepsilon$. Since $\varepsilon > 0$ is arbitrary the conclusion follows. □

8.3 First application of Markov type to embeddability problems: Euclidean distortion of graphs with large girth

Theorem 8.14. *Let G be a k-regular graph with girth g. Then*

$$c_2(G, d_G) \geq \frac{k-2}{k}\sqrt{\left\lfloor \frac{g}{2} \right\rfloor}.$$

Proof. By Corollary 8.13 it suffices to estimate $M_2(G)$ from below. To get such an estimate we consider the Markov chain introduced in Example 8.6, namely $\{Z_t\}_{t=0}^{\infty}$ is such that Z_0 is uniformly distributed on $V(G)$ and $\mathbb{P}(Z_{t+1} = v | Z_t = u)$ equals $1/k$ if u and v are adjacent, and 0 otherwise. Note that for every vertex $v \in V(G)$ the graph induced by the set of vertices $\{u : d_G(u, v) < \lfloor \frac{g}{2} \rfloor\}$ is a tree (otherwise there would be a cycle of length $< g$). Also, it is easy to check that as long as $t \leq \lfloor g/2 \rfloor - 1$, each step of the random walk $\{Z_t\}$ moves away from Z_0 with probability at least $\frac{k-1}{k}$ (we say "at least" because if $Z_t = Z_0$, then this probability is 1) and towards Z_0 with probability at most $\frac{1}{k}$. In other words, as long as $t \leq \lfloor g/2 \rfloor - 1$, the random walk, in expectation, moves away from Z_0. To get an estimate of $M_2(G)$ from below, we record the corresponding numbers: for every $1 \leq t \leq \lfloor g/2 \rfloor - 1$ we have

$$\begin{aligned}\mathbb{E}d_G(Z_{t+1}, Z_0) &\geq \frac{k-1}{k}(\mathbb{E}d_G(Z_t, Z_0) + 1) + \frac{1}{k}\,(\mathbb{E}d_G(Z_t, Z_0) - 1)\\ &= \mathbb{E}d_G(Z_t, Z_0) + \frac{k-2}{k}.\end{aligned}$$

Hence, for every $t \le \lfloor g/2 \rfloor - 1$ we have

$$\mathbb{E}d_G^2(Z_{t+1}, Z_0) \ge (\mathbb{E}d_G(Z_{t+1}, Z_0))^2 \ge \left(\frac{k-2}{k}\right)^2 (t+1)^2.$$

On the other hand, by the definition of Markov type 2,

$$\mathbb{E}d_G^2(Z_{t+1}, Z_0) \le M_2(G)^2(t+1)\mathbb{E}d_G^2(Z_1, Z_0) = M_2(G)^2(t+1).$$

Therefore $M_2(G) \ge \frac{k-2}{k}\sqrt{t+1}$. The result follows by taking $t = \lfloor g/2 \rfloor - 1$. □

8.4 Banach space theory: renormings of superreflexive spaces, q-convexity and p-smoothness

8.4.1 Definitions and duality

Now we develop some Banach space theory in order to prove generalizations of Theorem 8.14 for uniformly convex spaces. It is worth mentioning that the theory of uniformly convex and uniformly smooth spaces, and a related theory of superreflexivity are very developed and very extensive. We do not have the intention to present a detailed account of this theory. We prove only those results which we are going to use, we mention some other results to put our presentation in a suitable context. The Notes and Remarks section contains a more detailed survey of the theory and references to original sources and to monographs containing detailed presentations of other aspects of the theory.

Let $(X, \|\cdot\|)$ be a normed space. The *modulus of uniform convexity* of X is defined for $\varepsilon \in [0, 2]$ as

$$\delta_X(\varepsilon) = \inf\left\{1 - \frac{\|x+y\|}{2} : x, y \in X,\ \|x\| = \|y\| = 1,\ \|x-y\| = \varepsilon\right\}. \tag{8.10}$$

The normed space X is said to be *uniformly convex* if $\delta_X(\varepsilon) > 0$ for all $\varepsilon \in (0, 2]$. Furthermore, X is said to have *modulus of convexity of power type q* if there exists a constant c such that $\delta(\varepsilon) \ge c\,\varepsilon^q$ for all $\varepsilon \in [0, 2]$.

Remark 8.15. It is easy to see that the modulus of convexity of a Banach space is equal to the infimum of moduli of convexity of its two-dimensional subspaces. An immediate consequence of this is: If a Banach space X is finitely representable in a Banach space Y, then $\delta_X(\varepsilon) \ge \delta_Y(\varepsilon)$ for each $\varepsilon \in [0, 1]$. Combining this with Dvoretzky theorem (stated on page 42) we get that $\delta_{\ell_2}(\varepsilon) \ge \delta_Y(\varepsilon)$ for each infinite-dimensional Banach space Y. This statement is also known to be true for finite-dimensional spaces (see Notes and Remarks).

Example 8.16. The modulus of convexity of a Hilbert space is easily computable. In fact, using the parallelogram identity we get, for x and y satisfying the conditions of the definition of the modulus of convexity:

$$\|x + y\|^2 + \varepsilon^2 = 4.$$

Hence $\|x + y\| = \sqrt{4 - \varepsilon^2}$ and $\delta_{\ell_2} = 1 - \sqrt{1 - \frac{\varepsilon^2}{4}} \approx \frac{\varepsilon^2}{8}$.

Remark 8.17. Combining this example with Remark 8.15 we get that if a Banach space X has modulus of convexity of power type q, then $q \geq 2$.

Proposition 8.18. *A normed space X has modulus of convexity of power type q if and only if there exists a constant $K > 0$ such that for every $x, y \in X$*

$$2\,\|x\|^q + \frac{2}{K^q}\,\|y\|^q \leq \|x + y\|^q + \|x - y\|^q. \tag{8.11}$$

Here we proof only the "if" part, it is convenient to prove the "only if" part together with the dual statement which is presented below as Proposition 8.24.

Proof of the "if" part of Proposition 8.18. Replacing in (8.11) x by $\frac{x+y}{2}$ and y by $\frac{x-y}{2}$, we get the inequality

$$2\left\|\frac{x+y}{2}\right\|^q + \frac{2}{K^q}\left\|\frac{x-y}{2}\right\|^q \leq \|x\|^q + \|y\|^q. \tag{8.12}$$

Now, if $\|x\| = \|y\| = 1$ and $\|x - y\| = \varepsilon$, we get from here that

$$\left\|\frac{x+y}{2}\right\|^q \leq 1 - \left(\frac{\varepsilon}{2K}\right)^q.$$

and so

$$\delta_X(\varepsilon) \geq 1 - \left(1 - \left(\frac{\varepsilon}{2K}\right)^q\right)^{\frac{1}{q}} \geq \frac{1}{q}\left(\frac{\varepsilon}{2K}\right)^q,$$

where we use linear approximation and convexity of the function $1 - (1 - x)^{\frac{1}{q}}$. □

Remark 8.19. It is clear that we can "almost reverse" this argument. However this does not complete the proof of Proposition 8.18 because we get (8.11) only in the case where $\|x + y\| = \|x - y\|$.

Definition 8.20. A normed space X is called *q-uniformly convex* if there exists a constant K such that (8.11) holds. The least K for which (8.11) holds is called the *q-convexity constant* of X, and is denoted $K_q(X)$.

Remark 8.21. Sometimes it is more convenient to use q-uniform convexity in the equivalent form which is slightly different from the one obtained in (8.12):

$$\left\|\frac{x+y}{2}\right\|^q + \left\|\frac{x-y}{2K}\right\|^q \leq \frac{\|x\|^q + \|y\|^q}{2}. \tag{8.13}$$

The following definition was introduced with the purpose to get a notion which is in a certain sense dual to the notion of uniform convexity.

Definition 8.22. The *modulus of uniform smoothness* of a normed space X is defined for $\tau > 0$ as

$$\rho_X(\tau) = \sup\left\{\frac{\|x+\tau y\| + \|x-\tau y\|}{2} - 1 : \ x, y \in X, \ \|x\| = \|y\| = 1\right\}. \quad (8.14)$$

A normed space X is called *uniformly smooth* if $\lim_{\tau\to 0} \frac{\rho_X(\tau)}{\tau} = 0$. Furthermore, X is said to have *modulus of smoothness of power type p* if there exists a constant K such that $\rho_X(\tau) \le K\tau^p$ for all $\tau > 0$.

Remark 8.23. Combining Remark 8.17 with Propositions 8.18 and 8.24 and with Theorem 8.27 we get that in this case necessarily $p \le 2$.

Proposition 8.24. *A normed space X has modulus of smoothness of power type p if and only if there exists a constant $S > 0$ such that for every $x, y \in X$*

$$\|x+y\|^p + \|x-y\|^p \le 2\,\|x\|^p + 2\,S^p\,\|y\|^p. \quad (8.15)$$

Proof of the "if" part of Proposition 8.24. If (8.15) holds, then for all $x, y \in X$ with $\|x\| = \|y\| = 1$ and all $\tau > 0$ we have

$$\begin{aligned}\frac{\|x+\tau y\| + \|x-\tau y\|}{2} - 1 &\le \left(\frac{\|x+\tau y\|^p + \|x-\tau y\|^p}{2}\right)^{\frac{1}{p}} - 1\\ &\le (1 + S^p\|\tau y\|^p)^{\frac{1}{p}} - 1\\ &\le \frac{1}{p}(S\tau)^p,\end{aligned}$$

where we use linear approximation and concavity of the function $(1+x)^{\frac{1}{p}} - 1$. Therefore X has modulus of smoothness of power type p. □

The "only if" part of Proposition 8.24 is proved later, see page 238.

Definition 8.25. A normed space is called *p-uniformly smooth* if (8.15) holds for some $S > 0$. The least S for which (8.15) holds is called the *p-smoothness constant* of X, and is denoted $S_p(X)$.

Before completing the proofs of Propositions 8.18 and 8.24 we establish some very important and useful duality results. We start with a classical result:

Theorem 8.26 (D. Milman–Pettis). *Every uniformly convex Banach space is reflexive.*

Proof. Let X be a uniformly convex Banach space. It suffices to show that for each $h \in X^{**}$, $\|h\| = 1$ we have $h \in \pi(X)$, where $\pi : X \to X^{**}$ is the canonical embedding. By the Goldstine theorem there exists an ultrafilter $\mathcal{U}$ on S_X such that $w^*\!-\!\lim_{\mathcal{U}} x = h$, where $w^* - \lim$ denotes the limit in the weak* topology. To complete the proof it suffices to show that the ultrafilter converges in the strong topology of X. Since X is a complete space, it is enough to show that $\mathcal{U}$ is a Cauchy ultrafilter in the sense that for each $\varepsilon > 0$ there exists a subset $A \in \mathcal{U}$ such that $\|x - y\| < \varepsilon$ for each $x, y \in A$.

Let $\delta = \delta_X(\varepsilon) > 0$. Then $x, y \in X$, $\|x\|, \|y\| = 1$, and $\|x + y\| \geq 2 - \delta$ imply that $\|x - y\| < \varepsilon$. Let $f \in S_{X^*}$ be such that $|h(f) - 1| < \delta/2$. Let $V = \{g \in X^{**} : |g(f) - 1| < \delta/2\}$; V is a weak* neighborhood of h. Hence $V \cap S_X \in \mathcal{U}$. Let $x, y \in V \cap S_X$. Then $|f(x) + f(y)| > 2 - \delta$ and therefore $\|x + y\| \geq 2 - \delta$. Hence $\|x - y\| < \varepsilon$. Thus $\mathcal{U}$ is a Cauchy ultrafilter in the strong topology and its limit h is in $\pi(X)$. □

Theorem 8.27. *Let $\frac{1}{q} + \frac{1}{p} = 1$, $p, q \in (1, \infty)$.*

(a) A Banach space X is p-uniformly smooth if and only if its dual space X^* is q-uniformly convex. Furthermore, the p-uniform smoothness constant of X is equal to the q-uniform convexity constant of X^*.

(b) A Banach space X is q-uniformly convex if and only if its dual space X^* is p-uniformly smooth. Furthermore, the q-uniform convexity constant of X is equal to the p-uniform smoothness constant of X^*.

Proof. (a) Suppose that X^* is q-uniformly convex with q-uniform convexity constant K. Let $x, y \in X$. Let $f \in X^*$ be such that $\|f\| = 1$ and $f(x + y) = \|x + y\|$. Let $g \in X^*$ be such that $\|g\| = 1$ and $g(x - y) = \|x - y\|$. We are going to use the q-uniform convexity in the form (8.13) for suitably adjusted vectors f and g, namely for the vectors

$$\varphi = N^{-\frac{1}{q}} \|x + y\|^{p-1} f, \quad \zeta = N^{-\frac{1}{q}} \|x - y\|^{p-1} g,$$

where

$$N = \frac{\|x + y\|^p + \|x - y\|^p}{2}.$$

The adjustment is such that $\|\varphi\|^q + \|\zeta\|^q = 2$ and the first equality in the following formula holds

$$\begin{aligned}
\left(\frac{\|x + y\|^p + \|x - y\|^p}{2}\right)^{\frac{1}{p}} &= \frac{\varphi(x + y) + \zeta(x - y)}{2} \\
&= \left(\frac{\varphi + \zeta}{2}\right)(x) + \left(\frac{\varphi - \zeta}{2}\right)(y) \\
&\leq \left(\left\|\frac{\varphi + \zeta}{2}\right\|^q + \left\|\frac{\varphi - \zeta}{2K}\right\|^q\right)^{\frac{1}{q}} (\|x\|^p + \|Ky\|^p)^{\frac{1}{p}} \\
&\leq \left(\frac{\|\varphi\|^q + \|\zeta\|^q}{2}\right)^{\frac{1}{q}} (\|x\|^p + \|Ky\|^p)^{\frac{1}{p}} \\
&= (\|x\|^p + \|Ky\|^p)^{\frac{1}{p}}.
\end{aligned}$$

The first inequality follows from the Hölder inequality. To get the second inequality we use the q-uniform convexity of X^* in the form (8.13).

Now suppose that X is p-uniformly smooth with p-uniform smoothness constant K. Let $f, g \in X^*$ and $\omega > 0$. Let $x \in X$ be such that $\|x\| = 1$ and $\|f + g\| \leq (f + g)(x) + \omega$. Let $y \in X$ be such that $\|y\| = 1$ and $\|f - g\| \leq (f - g)(y) + \omega$. We are going to use the p-uniform smoothness for suitably adjusted vectors x and y, namely for the vectors

$$\xi = M^{-\frac{1}{p}} \left\| \frac{f+g}{2} \right\|^{q-1} x, \quad \psi = M^{-\frac{1}{p}} \left\| \frac{f-g}{2K} \right\|^{q-1} \cdot \frac{y}{K},$$

where

$$M = \left\| \frac{f+g}{2} \right\|^q + \left\| \frac{f-g}{2K} \right\|^q.$$

The adjustment is such that $2\|\xi\|^p + 2\|K\psi\|^p = 2$ and the first equality in the following formula holds

$$\begin{aligned}
&\left(\left\| \frac{f+g}{2} \right\|^q + \left\| \frac{f-g}{2K} \right\|^q \right)^{\frac{1}{q}} \\
&\leq M^{-\frac{1}{p}} \left(\frac{f+g}{2} \right)(x) \left\| \frac{f+g}{2} \right\|^{q-1} \\
&\quad + M^{-\frac{1}{p}} \left(\frac{f-g}{2} \right) \left(\frac{y}{K} \right) \left\| \frac{f-g}{2K} \right\|^{q-1} + R(\omega) \\
&= \left(\frac{f+g}{2} \right)(\xi) + \left(\frac{f-g}{2} \right)(\psi) + R(\omega) \\
&= f \left(\frac{\xi + \psi}{2} \right) + g \left(\frac{\xi - \psi}{2} \right) + R(\omega) \\
&\leq (\|f\|^q + \|g\|^q)^{\frac{1}{q}} \left(\left\| \frac{\xi + \psi}{2} \right\|^p + \left\| \frac{\xi - \psi}{2} \right\|^p \right)^{\frac{1}{p}} + R(\omega) \\
&\overset{(p\text{-smoothness})}{\leq} (\|f\|^q + \|g\|^q)^{\frac{1}{q}} \cdot \frac{1}{2} (2\|\xi\|^p + 2\|K\psi\|^p)^{\frac{1}{p}} + R(\omega) \\
&= \left(\frac{\|f\|^q + \|g\|^q}{2} \right)^{\frac{1}{q}} + R(\omega),
\end{aligned}$$

where $R(\omega) > 0$ is a function of ω satisfying $\lim_{\omega \downarrow 0} R(\omega)$. The second inequality follows from the Hölder inequality. To get the third inequality we use the p-uniform smoothness of X. Letting $\omega \downarrow 0$ we get that X^* is q-uniformly convex.

(b) Can be proved by combining part (a) with Theorem 8.26. If X is q-uniformly convex, then, by Theorem 8.26, $\pi_X(X) = X^{**}$, thus X^{**} is also q-uniformly convex. Applying part (a) we get that X^* is p-uniformly smooth.

On the other hand, if X^* is p-uniformly smooth, then, by part (a), X^{**} is q-uniformly convex and thus $X^{**} = \pi_X(X)$. □

There exists a very nice and simple formula for ρ_{X^*} in terms of δ_X.

Theorem 8.28. *For every Banach space X*

$$\rho_{X^*}(\tau) = \sup_{0\le\varepsilon\le 2}\Big(\frac{\tau\varepsilon}{2} - \delta_X(\varepsilon)\Big), (\tau > 0). \tag{8.16}$$

Proof. In fact

$$\begin{aligned} 2\rho_{X^*}(\tau) &= \sup_{f,g\in S_{X^*}} (\|f+\tau g\| + \|f-\tau g\| - 2) \\ &= \sup_{f,g\in S_{X^*}} \sup_{x,y\in S_X} ((f+\tau g)(x) + (f-\tau g)(y) - 2) \\ &= \sup_{x,y\in S_X} (\|x+y\| + \tau\|x-y\| - 2) \\ &= \sup_{0\le\varepsilon\le 2} \sup_{x,y\in S_X, \|x-y\|=\varepsilon} (\tau\varepsilon - (2 - \|x+y\|)) \\ &= \sup_{0\le\varepsilon\le 2} (\tau\varepsilon - 2\delta_X(\varepsilon)). \end{aligned}$$

□

Remark 8.29. It seems that there is no simple reverse relationship, that is no simple formula expressing $\delta_X(\varepsilon)$ in terms of $\rho_{X^*}(\tau)$.

Proof of the "only if" part of Proposition 8.24. Suppose that $\rho_X(\tau) \le (C\tau)^p$ for some constant C and $1 < p \le 2$. Then, modifying slightly the definition of ρ_X, we get that for all $\|x\| = 1$ and $\|y\| \le 1$ we have

$$\frac{\|x+y\| + \|x-y\|}{2} \le 1 + (C\|y\|)^p.$$

We need a similar estimate for

$$\left(\frac{\|x+y\|^p + \|x-y\|^p}{2}\right)^{\frac{1}{p}}.$$

In this connection we consider the difference:

$$\left(\frac{\|x+y\|^p + \|x-y\|^p}{2}\right)^{\frac{1}{p}} - \frac{\|x+y\| + \|x-y\|}{2}.$$

This difference can be written in the form

$$\begin{aligned} &\left(\frac{\|x+y\|^p + \|x-y\|^p}{2}\right)^{\frac{1}{p}} - \frac{\|x+y\| + \|x-y\|}{2} \\ &= b\left(\left(\frac{(1+\beta)^p + (1-\beta)^p}{2}\right)^{\frac{1}{p}} - 1\right), \end{aligned} \tag{8.17}$$

where the numbers b and β are given by

$$b := \frac{\|x+y\| + \|x-y\|}{2} \quad \text{and} \quad \beta := \frac{\|x+y\| - \|x-y\|}{\|x+y\| + \|x-y\|}.$$

The function

$$w(\beta) := \left(\frac{(1+\beta)^p + (1-\beta)^p}{2}\right)^{\frac{1}{p}} - 1$$

is such that $w(0) = 0$ and $w'(0) = 0$. Since this function is analytic at 0, we get that $w(\beta) \le C(p)\beta^2$ for $|\beta| \le 1$, where the constant $C(p)$ depends on p only. Since $|\beta| \le \frac{\|y\|}{b} \le \|y\|$ and $1 \le p \le 2$, we get

$$\left(\frac{\|x+y\|^p + \|x-y\|^p}{2}\right)^{\frac{1}{p}} \le 1 + (C\|y\|)^p + C(p)\|y\|^2 \le \left(1 + (S\|y\|)^p\right)^{\frac{1}{p}}$$

for all x and y with $\|y\| \le \|x\| = 1$, where the constant S depends only on C and p. Therefore

$$\frac{\|x+y\|^p + \|x-y\|^p}{2} \le \|x\|^p + (S\|y\|)^p \tag{8.18}$$

for all x and y with $\|y\| \le \|x\|$. Finally, since we may assume that $S \ge 1$, the inequality (8.18) holds for all x and y. □

Proof of the "only if" part of Proposition 8.18. Suppose that $\delta_X(\varepsilon) \ge (\varepsilon/C)^q$ for some constant C. Then, by (8.16),

$$\rho_{X^*}(\tau) = \sup_{0\le\varepsilon\le 2}\left(\frac{\tau\varepsilon}{2} - \delta_X(\varepsilon)\right) \le \sup_{0\le\varepsilon\le\infty}\left(\frac{\tau\varepsilon}{2} - \left(\frac{\varepsilon}{C}\right)^q\right) \le (C\tau)^p,$$

where $\frac{1}{p} + \frac{1}{q} = 1$. Here we use the well-known inequality $\alpha\beta \le \frac{\alpha^p}{p} + \frac{\beta^q}{q} \le \alpha^p + \beta^q$.

By Proposition 8.24 we get that X^* is p-uniformly smooth. Applying Theorem 8.27 we get that X is q-uniformly convex. □

Example 8.30 (Without proof). Well-known and very important examples of q-uniformly convex and p-uniformly smooth Banach spaces are the classical $L_p(0,1)$ spaces:

(a) $L_p(0,1)$ is p-uniformly smooth and 2-uniformly convex if $1 < p \le 2$.

(b) $L_p(0,1)$ is 2-uniformly smooth and p-uniformly convex if $2 \le p < \infty$.

8.4.2 Pisier theorem on renormings of uniformly convex spaces

The following class of Banach spaces plays a very important role in Banach space theory.

Definition 8.31. A Banach space X is called *superreflexive* if each Banach space Y which is finitely representable in X is reflexive.

Example 8.32. Each uniformly convex and each uniformly smooth Banach space is superreflexive. In fact, both uniform convexity and uniform smoothness are determined by two-dimensional subspaces, and hence any Banach space which is finitely represented in a uniformly convex (uniformly smooth) Banach space is uniformly convex (uniformly smooth). It remains to use the fact that uniformly convex and uniformly smooth Banach spaces are reflexive (Theorem 8.26 and Exercise 8.63).

The converse does not hold, it is not difficult to construct examples of superreflexive Banach spaces which are not uniformly convex or smooth. It becomes immediate if we observe that all finite-dimensional Banach spaces are superreflexive.

Another way of showing that the converse does not hold is based on the following. Banach spaces X and Y are called *isomorphic* if there exists a bijective bounded operator $T : X \to Y$ such that T^{-1} is also bounded (actually boundedness of T^{-1} is redundant, it follows from the open mapping theorem).

Proposition 8.33. *If X and Y are isomorphic and X is superreflexive, then Y is also superreflexive.*

Proof. Assume the contrary. Then there exists a nonreflexive space Z which is finitely representable in Y. By Proposition 2.33, there exists an ultrafilter $\mathcal{U}$ such that the ultrapower $Y^{\mathcal{U}}$ contains a subspace isometric to Z. Then (as is easy to check) the ultrapower $X^{\mathcal{U}}$ contains a subspace isomorphic to Z. It is well known (and follows, for example, from Theorem 6.11 (a)$\Leftrightarrow$(d)$\Leftrightarrow$(e)) that a Banach space isomorphic to a nonreflexive Banach space is nonreflexive. Therefore $X^{\mathcal{U}}$ contains a nonreflexive subspace. By Proposition 2.31, $X^{\mathcal{U}}$, and hence Z is finitely representable in X. Thus X is not superreflexive. We get a contradiction. □

Remark 8.34. Sometimes Proposition 8.33 is stated as: *superreflexivity is an isomorphic invariant.*

On the other hand, it is easy to see that a Banach space isomorphic to a uniformly convex space does not have to be uniformly convex.

Example 8.35. Consider the space ℓ_2 with the norm

$$|||\{x_i\}_{i=1}^{\infty}||| = \max\{\|\{x_i\}_{i=1}^{\infty}\|_{\ell_2}, |x_1| + |x_2|\}.$$

Denote this space by Y. It is easy to verify that Y is not uniformly convex. On the other hand, the identical embedding $T : \ell_2 \to Y$ is bounded and has bounded inverse, so Y is isomorphic to ℓ_2. Hence ℓ_2 with the norm $|||\cdot|||$ is superreflexive, but not uniformly convex.

A very important result on superreflexive spaces is the following theorem:

Theorem 8.36 (Enflo). *Each superreflexive Banach space is isomorphic to a uniformly convex Banach space and to a uniformly smooth Banach space.*

This theorem admits the following strengthening:

Theorem 8.37 (Pisier). *Each superreflexive Banach space is isomorphic to a q-uniformly convex Banach space for some $2 \le q < \infty$ and to a p-uniformly smooth Banach space for some $1 < p \le 2$.*

We are not going to give complete proofs of these theorems in this book, we shall prove only the following important for us part of Theorem 8.37.

Theorem 8.38. *Each uniformly convex Banach space is q-uniformly convex for some $2 \le q < \infty$.*

Using the duality (Theorems 8.26, 8.27, 8.28, and hints to Exercise 8.63) we get:

Corollary 8.39. *Each uniformly smooth Banach space is p-uniformly smooth for some $1 < p \le 2$.*

Proof of Theorem 8.38. First we introduce a notion which is very natural in the study of uniform convexity.

Let C be a closed bounded subset of a real Banach space X, $a \in \mathbb{R}$, and $f \in X^*$. Then the set $S = \{x \in C : f(x) > a\}$ is called a *slice* of C.

We define the following set-valued derivatives on bounded subsets of a Banach space.

Let $\varepsilon > 0$ and C be a bounded subset in a Banach space X. We define $C_\varepsilon^0 = C$ and

$$C_\varepsilon^1 = C'_\varepsilon = \{x \in C \text{ such that any slice of } C \text{ containing } x \text{ has diameter } > \varepsilon\}.$$

For a positive integer $n \ge 2$ the *dent-derivative* of C of order n is defined by

$$C_\varepsilon^n = (C_\varepsilon^{n-1})_\varepsilon^1.$$

Remark 8.40. Such set-valued derivatives are usually defined not only for positive integers n, but also for arbitrary ordinal numbers n, however we do not need this generalization here.

We define the *dentability index* of a bounded set in a Banach space by

$$\operatorname{Dent}(C, \varepsilon) = \begin{cases} \inf\{n : C_\varepsilon^n = \emptyset\} & \text{if such } n \in \mathbb{N} \text{ exists} \\ \infty & \text{otherwise.} \end{cases} \tag{8.19}$$

Lemma 8.41. *If X is a uniformly convex Banach space, then* $\mathrm{Dent}(C,\varepsilon) < \infty$ *for each bounded set $C \subset X$ and each $\varepsilon > 0$.*

Proof. It is clear that if $C \subset A$, then $C_\varepsilon^n \subset A_\varepsilon^n$ for each $n \in \mathbb{N}$ and each $\varepsilon > 0$ and therefore $\mathrm{Dent}(C,\varepsilon) \le \mathrm{Dent}(A,\varepsilon)$. In is also clear that $(rB_X)'_\varepsilon = \emptyset$ if $r \le \frac{\varepsilon}{2}$ (where rB_X is a multiple of a unit ball). Therefore it suffices to show that for each $r > \frac{\varepsilon}{2}$ we have

$$(rB_X)'_\varepsilon \subset (1-\omega(\varepsilon,r,X))rB_X \tag{8.20}$$

for some $0 < \omega(\varepsilon,r,X) < 1$ which depends only on ε, r, and X and is non-increasing in $r > 0$ for fixed ε and X.

The desired property of balls almost coincides with the definition of the uniform convexity. In fact, first let us show that a point $z \in B_X$ is contained in a slice of diameter $\le \varepsilon$ if $\|z\| > 1-\delta_X(\varepsilon)$. In fact, let z^* be such that $\|z^*\| = 1$ and $z^*(z) = \|z\|$. Let S be the slice $\{x \in X : z^*(x) > \|z\|\}$. Although $z \notin S$, the small perturbation argument implies that it suffices to show that $\mathrm{diam}(S) < \varepsilon$. Suppose not, then S contains points u and w such that $\|u-w\| \ge \varepsilon$. Using convexity of the norm we may assume that $\|u\| = \|w\| = 1$. In fact, we may get that one of the norms is equal to 1 just by extending the line segment joining u and w in the direction in which the functional z^* is non-decreasing, so we assume that $\|u\| = 1$ and $\|w\| \le 1$. Now we consider a line $w+tr$, where $t \in \mathbb{R}$ and $r \in \ker z^*$. This line intersects the unit sphere in two points. By convexity of the norm, the distance between u an at least one of the intersection points is $\ge \varepsilon$.

Also it is easy to see that we may assume $\|u-w\| = \varepsilon$ (just consider a continuous curve on the set $S_X \cap \{x \in X : z^*(x) > \|z\|\}$ joining u and w, it should contain a point satisfying this condition). By the definition of the modulus of convexity we get

$$\left\|\frac{u+w}{2}\right\| \le 1-\delta_X(\varepsilon).$$

On the other hand

$$z^*\left(\frac{u+w}{2}\right) > \|z\| > 1-\delta_X(\varepsilon).$$

We get a contradiction.

Thus we have proved that

$$(B_X)'_\varepsilon \subset (1-\delta_X(\varepsilon))B_X.$$

By homothety, this implies

$$(rB_X)'_{r\varepsilon} \subset (1-\delta_X(\varepsilon))rB_X.$$

Applying this result to $\varepsilon' = \frac{\varepsilon}{r}$ we get

$$(rB_X)'_\varepsilon \subset \left(1-\delta_X\left(\frac{\varepsilon}{r}\right)\right)rB_X$$

provided $\frac{\varepsilon}{r} \leq 2$. It remains to show that $\delta_X(t)$, $0 \leq t \leq 2$ is a non-decreasing function. By Remark 8.15, it suffices to do so for a two-dimensional space X. Let $0 < t_0 \leq 2$, let x and y be such that $\|x\| = \|y\| = 1$, $\|x - y\| = t_0$ and $\delta_X(t_0) = 1 - \left\|\frac{x+y}{2}\right\|$. It is enough to show that there exists $t_1 < t_0$ such that for each $s \in [t_1, t_0]$ there exist x' and y' such that $\|x'\| = \|y'\| = 1$, $\|x' - y'\| = s$ and

$$\left\|\frac{x' + y'}{2}\right\| \geq \left\|\frac{x + y}{2}\right\|.$$

To achieve this goal we introduce the functional $f \in X^*$ such $\|f\| = 1$ and

$$f\left(\frac{x + y}{2}\right) = \left\|\frac{x + y}{2}\right\|.$$

If

$$f(x) = f(y) = f\left(\frac{x + y}{2}\right), \tag{8.21}$$

the claim is obvious: We just move the line containing x and y in the direction of increase of f keeping it parallel to the original position until its intersection with B_X has length $t \in [0, t_0]$. Then the midpoint m of the obtained line segment satisfies:

$$\|m\| \geq f(m) \geq f\left(\frac{x + y}{2}\right) = \left\|\frac{x + y}{2}\right\|.$$

If (8.21) is not satisfied, we assume without loss of generality that $f(y) > f(x)$. Let $g \in X^*$ be such that $g(x) = g(y) = g\left(\frac{x+y}{2}\right) > 0$ and let $v = \left(\frac{x+y}{2}\right) / \left\|\frac{x+y}{2}\right\|$. Our choice of f implies that $f(v) \geq f(y)$. Therefore the line $n_1 = \{u \in X : f(u) = f(y)\}$ either has common points with the intersection of the interior of B_X with the half-plane $\{u : g(u) > g(x)\}$, or contains a line segment on the surface of B_X (the reader is advised to sketch a picture). The latter case cannot occur because we assumed that X is uniformly convex. In the former case the line $n_2 = \{u \in X : f(u) = f(x)\}$ cannot intersect the intersection of the interior of B_X with the half-plane $\{u : g(u) > g(x)\}$ and therefore the midpoints $m(c)$ of the line segments which are intersections of B_X and lines of the form $\{u : g(u) = c\}$ with $c > g(x)$ satisfy $f(m(c)) \geq f\left(\frac{x+y}{2}\right)$. This completes the proof of the lemma. □

Now we turn to the proof of Theorem 8.38. We let $N_k = \mathrm{Dent}(B_X, 2^{-k}) - 1$ and introduce the following function on X:

$$f(x) = \|x\| + \sum_{k=1}^{\infty} \sum_{n=1}^{N_k} \frac{2^{-k}}{N_k} d(x, B^n_{2^{-k}}),$$

where $B = B_X$, the sets $\{B^n_{2^{-k}}\}_{n,k}$ are its dent-derivatives, and $d(x, B^n_{2^{-k}})$ is the distance in the original norm of X from x to the dent-derivative.

Let $|||\cdot|||$ be the Minkowski functional of the convex symmetric set $C = \{x \in X : f(x) \le 1\}$, that is

$$|||x||| = \inf\{\lambda \in (0,\infty) : \frac{x}{\lambda} \in C\}.$$

Observe that $\|x\| > 1$ implies $f(x) > 1$ and that $\|x\| \le \frac{1}{2}$ implies $f(x) \le 1$. Therefore $\|x\| \le |||x||| \le 2\|x\|$ for all $x \in X$. As it is well known, the Minkowski functional of a bounded convex symmetric set is a norm, therefore $(X, |||\cdot|||)$ is a normed space, and the identical embedding of the Banach space $(X, \|\cdot\|)$ into $(X, |||\cdot|||)$ is an isomorphism. It remains to show that the Banach space $(X, |||\cdot|||)$ is q-uniformly convex for some $q \in [2,\infty)$.

Lemma 8.42. *If $\varepsilon > 0$ and $x, y \in X$ are such that $f(x) = f(y) = 1$ and $\|x - y\| \ge \varepsilon$, then*

$$f\left(\frac{x+y}{2}\right) \le 1 - \frac{\varepsilon^2}{32\left(\mathrm{Dent}\left(B_X, \frac{\varepsilon}{8}\right)\right)^2}. \tag{8.22}$$

Proof. Let $k \in \mathbb{N}$ be such that $\frac{\varepsilon}{8} \le 2^{-k} \le \frac{\varepsilon}{4}$ and $n \in \mathbb{N}$ be the largest number for which both $x \in B^n_{2^{-k}}$ and $y \in B^n_{2^{-k}}$. Assume that $x \in B^n_{2^{-k}} \setminus B^{n+1}_{2^{-k}}$. Observe that $\|x - y\| \ge \varepsilon$ implies that $n < N_k$ (because it implies $\frac{x+y}{2} \in B^{n+1}_{2^{-k}}$). Let $\gamma = \frac{\varepsilon}{4N_k}$.

The main idea of the proof is to show that the gap between the average of distances and the distance from the midpoint is sufficiently nontrivial for at least one of the terms on the suitable scale. We mean the following claim.

Claim 8.43. There exists $1 \le m \le N_k - n$ such that

$$\frac{1}{2}\left(d(x, B^{n+m}_{2^{-k}}) + d(y, B^{n+m}_{2^{-k}})\right) - d\left(\frac{x+y}{2}, B^{n+m}_{2^{-k}}\right) \ge \gamma. \tag{8.23}$$

Proof. First we show that the contrary to (8.23), that is,

$$\begin{aligned}\forall m \in \{1,\dots,N_k - n\} \quad &\frac{1}{2}\left(d(x, B^{n+m}_{2^{-k}}) + d(y, B^{n+m}_{2^{-k}})\right)\\ &- d\left(\frac{x+y}{2}, B^{n+m}_{2^{-k}}\right) < \gamma\end{aligned} \tag{8.24}$$

implies

$$\forall m \in \{1,\dots,N_k - n\} \quad \frac{1}{2}\left(d(x, B^{n+m}_{2^{-k}}) + d(y, B^{n+m}_{2^{-k}})\right) < m\gamma. \tag{8.25}$$

The idea of the argument is that the midpoint of two points in $B^{n+m}_{2^{-k}}$ which are at distance $> 2\cdot 2^{-k}$ from each other is in $B^{n+m+1}_{2^{-k}}$, and this leads to the desired estimate.

In fact, for $m = 1$ we have $x, y \in B^n_{2^{-k}}$ and $\|x - y\| \geq \varepsilon > 2 \cdot 2^{-k}$. Therefore $\frac{x+y}{2} \in B^{n+1}_{2^{-k}}$ and the assumption (8.24) for $m = 1$ implies the desired estimate (8.25) for $m = 1$.

To verify (8.25) for other values of m we use induction. So we assume that (8.25) holds for some value of m and verify the same statement for $m + 1$. Inequality (8.25) implies that there exist $\widetilde{x}, \widetilde{y} \in B^{n+m}_{2^{-k}}$ such that $\frac{1}{2}(\|x - \widetilde{x}\| + \|y - \widetilde{y}\|) < m\gamma$. This inequality implies that $\widetilde{x}$ and $\widetilde{y}$ are sufficiently far from each other:

$$\|\widetilde{x} - \widetilde{y}\| > \varepsilon - 2m\gamma \geq \frac{\varepsilon}{2} \geq 2 \cdot 2^{-k}.$$

Therefore $\frac{\widetilde{x}+\widetilde{y}}{2} \in B^{n+m+1}_{2^{-k}}$. Since

$$\left\| \frac{x+y}{2} - \frac{\widetilde{x}+\widetilde{y}}{2} \right\| < m\gamma,$$

we conclude that

$$d\left(\frac{x+y}{2}, B^{n+m+1}_{2^{-k}}\right) < m\gamma.$$

Together with the assumption that the inequality in (8.24) holds for m, this implies

$$\frac{1}{2}\left(d(x, B^{n+m+1}_{2^{-k}}) + d(y, B^{n+m+1}_{2^{-k}})\right) < (m+1)\gamma,$$

thus we have proved (8.25).

Now we apply the argument used in the inductive step to $m = N_k - n$. We find $\widetilde{x}, \widetilde{y} \in B^{N_k}_{2^{-k}}$ such that $\|\widetilde{x} - \widetilde{y}\| > \varepsilon - 2(N_k - n)\gamma \geq \frac{\varepsilon}{2}$. Therefore $\frac{\widetilde{x}+\widetilde{y}}{2} \in B^{N_k+1}_{2^{-k}}$. We get a contradiction because we assumed that $B^{N_k+1}_{2^{-k}}$ is empty. □

Using Claim 8.43 it is easy to complete the proof of Lemma 8.42. In fact, let k be such that $\frac{\varepsilon}{8} \leq 2^{-k} \leq \frac{\varepsilon}{4}$ and let m be such that (8.23) holds. Since all other terms having the same form as the left-hand side of (8.23) are nonnegative, and also $\frac{1}{2}(\|x\| + \|y\|) - \|\frac{x+y}{2}\| \geq 0$, inequality (8.23) implies

$$\frac{1}{2}(f(x) + f(y)) - f\left(\frac{x+y}{2}\right) \geq \frac{2^{-k}}{N_k}\gamma.$$

Recalling the choice of k and using the definition of γ and $f(x) = f(y) = 1$, we get

$$f\left(\frac{x+y}{2}\right) \leq 1 - \frac{\varepsilon^2}{32\left(\mathrm{Dent}\left(B_X, \frac{\varepsilon}{8}\right)\right)^2}.$$ □

Using Lemma 8.42 it is easy to derive that the Banach space X with the norm $|||\cdot|||$ is uniformly convex and to find an estimate for its modulus of convexity. To this end, let x and y be such that $|||x||| = |||y||| = 1$ and $|||x - y||| = \varepsilon$. Then $f(x) = f(y) = 1$ and $\|x - y\| \geq \frac{\varepsilon}{2}$ (the double sum in the definition of f cannot exceed the first term). By Lemma 8.42, we get

$$f\left(\frac{x+y}{2}\right) \leq 1 - \frac{\varepsilon^2}{128\left(\mathrm{Dent}\left(B_X, \frac{\varepsilon}{16}\right)\right)^2}.$$

The definition of f immediately implies that f, considered as a function on the space X endowed with the norm $\|\cdot\|$, is 2-Lipschitz. Therefore the $\|\cdot\|$-distance between $\frac{x+y}{2}$ and any point z satisfying $f(z) = 1$ is $\geq \frac{\varepsilon^2}{256(\mathrm{Dent}(B_X, \frac{\varepsilon}{16}))^2}$. Since $|||\cdot||| \geq \|\cdot\|$, the same applies to the $|||\cdot|||$-distance. Thus

$$|||\frac{x+y}{2}||| \leq 1 - \frac{\varepsilon^2}{256\left(\mathrm{Dent}\left(B_X, \frac{\varepsilon}{16}\right)\right)^2},$$

and the modulus of convexity of the Banach space X with the norm $|||\cdot|||$ is at least

$$\delta_{X,|||\cdot|||}(\varepsilon) \geq \frac{\varepsilon^2}{256\left(\mathrm{Dent}\left(B_X, \frac{\varepsilon}{16}\right)\right)^2}. \tag{8.26}$$

Therefore to complete the proof of Theorem 8.38 it suffices to show that for each uniformly convex Banach space X there are constants $0 < C, r < \infty$ such that

$$\mathrm{Dent}(B_X, \varepsilon) \leq \frac{C}{\varepsilon^r}. \tag{8.27}$$

Unfortunately, the proofs of (8.27) known at the moment are rather indirect. (More direct proofs would be very useful for our understanding of uniformly convex spaces.) What we present below is a roundabout approach to proving (8.27).

One of the reasons for which (8.27) resists a direct proof is that it seems difficult if at all possible to show the dentability index $\mathrm{Dent}(C, \varepsilon)$ satisfies the following submultiplicativity relation:

$$\mathrm{Dent}(B_X, \varepsilon \cdot \widetilde{\varepsilon}) \leq \mathrm{Dent}(B_X, \varepsilon) \cdot \mathrm{Dent}(B_X, \widetilde{\varepsilon}). \tag{8.28}$$

Such a relation, if true, would imply (8.27) because it is well known that submultiplicativity implies power-type estimates.

However there exists another index, called *weak Szlenk index*, which is similar to the dentability index, for which a similar submultiplicativity relation is quite easy to prove. On the other hand, there exists an inequality between the dentability index of X and the weak Szlenk index of the space $L_p(X)$ of vector valued functions. The proof of Theorem 8.38 is obtained by combining these facts and the result on uniform convexity of $L_p(X)$ for a uniformly convex X.

Let $\varepsilon > 0$ and C be a bounded subset in a Banach space X. We define $C_\varepsilon^{\langle 0\rangle} = C$ and

$$C_\varepsilon^{\langle 1\rangle} = \{x \in C \text{ such that any weak-open neighborhood of } C \text{ containing } x \text{ has diameter } > \varepsilon\}.$$

For a positive integer $n \geq 2$ the *weak Szlenk derivative* of C of order n is defined by

$$C_\varepsilon^{\langle n\rangle} = (C_\varepsilon^{\langle n-1\rangle})_\varepsilon^{\langle 1\rangle}.$$

We define the *weak Szlenk index* of a bounded set in a Banach space by

$$\operatorname{wSz}(C,\varepsilon) = \begin{cases} \inf\{n : C_\varepsilon^{\langle n\rangle} = \emptyset\} & \text{if such } n \in \mathbb{N} \text{ exists} \\ \infty & \text{otherwise.} \end{cases} \tag{8.29}$$

Observation 8.44. It is clear that $C_\varepsilon^{\langle 1\rangle} \subseteq C_\varepsilon^1$. Therefore $C_\varepsilon^{\langle n\rangle} \subseteq C_\varepsilon^n$ for each $n \in \mathbb{N}$. Hence $\operatorname{wSz}(C,\varepsilon) \leq \operatorname{Dent}(C,\varepsilon)$ and Lemma 8.41 implies $\operatorname{wSz}(C,\varepsilon)$ is finite for any bounded subset of a uniformly convex Banach space.

For weak Szlenk index the submultiplicativity relation is easy to get:

Lemma 8.45.

$$\operatorname{wSz}(B_X, \varepsilon \cdot \widetilde{\varepsilon}) \leq \operatorname{wSz}(B_X,\varepsilon) \cdot \operatorname{wSz}(B_X,\widetilde{\varepsilon}). \tag{8.30}$$

Proof. It suffices to prove

$$\forall n \in \mathbb{N} \quad (B_X)_{\varepsilon\widetilde{\varepsilon}}^{\langle n \operatorname{wSz}(X,\widetilde{\varepsilon})\rangle} \subset (B_X)_\varepsilon^{\langle n\rangle}.$$

We prove this by induction. The statement is obviously true for $n = 0$. We assume that it holds for n and prove it for $n + 1$.

Let $x \notin (B_X)_\varepsilon^{\langle n+1\rangle}$. We need to show that $x \notin (B_X)_{\varepsilon\widetilde{\varepsilon}}^{\langle (n+1)\operatorname{wSz}(X,\widetilde{\varepsilon})\rangle}$. Since the derivatives are decreasing, we may assume that $x \in (B_X)_\varepsilon^{\langle n\rangle}$. Therefore there is a weak-open neighborhood U of x such that $\operatorname{diam}\big(U \cap ((B_X)_\varepsilon^{\langle n\rangle})\big) \leq \varepsilon$. Therefore

$$U \cap \big((B_X)_\varepsilon^{\langle n\rangle}\big) \subset x + \varepsilon B_X.$$

On the other hand it is clear that

$$(\varepsilon B_X)_{\varepsilon\widetilde{\varepsilon}}^{\langle \operatorname{wSz}(B_X,\widetilde{\varepsilon})\rangle} = \emptyset.$$

Therefore

$$\big(U \cap ((B_X)_\varepsilon^{\langle n\rangle})\big)_{\varepsilon\widetilde{\varepsilon}}^{\langle \operatorname{wSz}(B_X,\widetilde{\varepsilon})\rangle} = \emptyset.$$

Since the intersection of two weak-open neighborhoods is a weak-open neighborhood, this implies that all points of the set $U \cap \left((B_X)_\varepsilon^{\langle n\rangle}\right)$ which are not in

$$\left(U \cap ((B_X)_\varepsilon^{\langle n\rangle})\right)_{\varepsilon\widetilde{\varepsilon}}^{\langle \mathrm{wSz}(B_X,\widetilde{\varepsilon})\rangle}, \quad \text{are also not in} \quad \left((B_X)_\varepsilon^{\langle n\rangle}\right)_{\varepsilon\widetilde{\varepsilon}}^{\langle \mathrm{wSz}(B_X,\widetilde{\varepsilon})\rangle}.$$

In particular, using the induction hypothesis we get

$$x \notin \left((B_X)_\varepsilon^{\langle n\rangle}\right)_{\varepsilon\widetilde{\varepsilon}}^{\langle \mathrm{wSz}(B_X,\widetilde{\varepsilon})\rangle} \supset (B_X)_{\varepsilon\widetilde{\varepsilon}}^{\langle (n+1)\mathrm{wSz}(X,\widetilde{\varepsilon})\rangle}. \qquad \square$$

Next, we establish the connection between values of the dentability index of X and the weak Szlenk index of $L_p(X)$.

Lemma 8.46. *Let X be a uniformly convex Banach space and let $1 < p < \infty$, then* $\mathrm{Dent}(B_X, \varepsilon) \le \mathrm{wSz}(L_p(X), \frac{\varepsilon}{2})$.

Remark 8.47. We restrict our attention to uniformly convex X and to $p \in (1, \infty)$ because only these cases will be needed for our proof.

Here $L_p(X)$ is the space of X-valued functions on $[0,1]$ endowed with its L_p norm. This space can be described as the completion of the linear space of simple functions with respect to the L_p-norm defined in the following way.

We define a *simple* X-valued function on $[0,1]$ as a function of the form $h(t) = \sum_{i=1}^n \mathbf{1}_{A_i}(t)x_i$ where measurable subsets $\{A_i\}_{i=1}^n$ of $[0,1]$ form a partition of $[0,1]$, $\mathbf{1}_{A_i}$ are the indicator functions, and x_i are vectors in X. The L_p*-norm* of h is defined by

$$\|h\| := \left(\sum_{i=1}^n \|x_i\|^p m(A_i)\right)^{\frac{1}{p}},$$

where $m(A_i)$ is the Lebesgue measure of the set A_i.

Proof of Lemma 8.46. It suffices to prove the claim that for each finite collection $x_1, \dots, x_m \in (B_X)_\varepsilon^n$, the function $f = \sum_{i=1}^m \mathbf{1}_{(\frac{i-1}{m}, \frac{i}{m}]} x_i$ is in $B_{\varepsilon/2}^{\langle n\rangle}$, where B is the unit ball of $L_p(X)$.

The claim is clearly true for $n = 0$. Now we assume that it is true for n and prove it for $n+1$. So let $x_1, \dots, x_m \in (B_X)_\varepsilon^{n+1}$. We need to prove that the function $f = \sum_{i=1}^m \mathbf{1}_{(\frac{i-1}{m}, \frac{i}{m}]} x_i$ is in $B_{\varepsilon/2}^{\langle n+1\rangle}$. The induction hypothesis implies that $f \in B_{\varepsilon/2}^{\langle n\rangle}$.

The fact that $x_i \in (B_X)_\varepsilon^{n+1}$ implies that the set of points of $(B_X)_\varepsilon^n$ with distance $> \frac{\varepsilon}{2}$ to x_i at cannot be separated from x_i by a linear functional (the separating functional could be used to get a slice of diameter $\le \varepsilon$ containing x_i). The well-known separation theorem for closed convex sets implies that x_i is in the closure in the strong topology of the convex hull of the set of points in $(B_X)_\varepsilon^n$ with distance $> \frac{\varepsilon}{2}$ to x_i. Thus for each $i \in \{1,\ldots,m\}$ and each $\gamma > 0$ there exist $\{x_{i,j}\}_{j=1}^{k(i)} \subset (B_X)_\varepsilon^n$ and positive real numbers $\{\alpha_{i,j}\}_{j=1}^{k(i)}$ such that

$$\left\| x_i - \sum_{j=1}^{k(i)} \alpha_{i,j} x_{i,j} \right\| < \gamma,$$

$\sum_{j=1}^{k(i)} \alpha_{i,j} = 1$, and $\|x_i - x_{i,j}\| > \frac{\varepsilon}{2}$ for all $j = 1,\ldots,k(i)$.

We use these vectors $x_{i,j}$ in order to construct a sequence $\{f_k\} \subset L_p(X)$ such that

- Each f_k is a simple function of the form $\sum_{p=1}^{s} \mathbf{1}_{(\frac{p-1}{s},\frac{p}{s}]} y_{p,k}$, where s and the vectors $y_{p,k} \in (B_X)_\varepsilon^n$ depend on k.
- The sequence $\{f_k\}$ converges weakly to a function $g \in L_p(X)$ satisfying $\|f - g\| < \gamma$.
- Each element of the sequence $\{f_k\}$ satisfies $\|f - f_k\| \geq \frac{\varepsilon}{2}$.

The existence of such a sequence leads to the desired conclusion. In fact, consider an arbitrary weak-open neighborhood V of f. Picking $\gamma > 0$ to be small enough we may assume that V contains the ball of radius γ centered at f. Since the sequence $\{f_k\}$ converges weakly to some element of this ball, the set V contains some members of the sequence $\{f_k\}$. By the induction hypothesis we have $f_k \in B_{\varepsilon/2}^{\langle n\rangle}$. Thus $\operatorname{diam}(V \cap B_{\varepsilon/2}^{\langle n\rangle}) \geq \frac{\varepsilon}{2}$. Therefore $f \in B_{\varepsilon/2}^{\langle n+1\rangle}$. So it remains to construct the sequence $\{f_k\}$.

First we do the construction in the case where $m = 1$ and $x = x_1$, and denote the obtained sequence by $\{f_k^{(0,1],x}\}_{k=1}^{\infty}$.

Observe that approximating $\alpha_{i,j}$ by rational numbers and repeating some of the vectors $x_{i,j}$ several times, we may assume that all of the coefficients $\alpha_{i,j}$ are equal to each other and that their common value is $\frac{1}{\ell}$, where $\ell \in \mathbb{N}$. So we assume that the inequalities $\|x_1 - x_{1,j}\| > \frac{\varepsilon}{2}$ and $\|x_1 - \frac{1}{\ell}\sum_{j=1}^{\ell} x_{1,j}\| < \gamma$ hold.

We introduce the function

$$\tau(t) = \sum_{j=1}^{\ell} \mathbf{1}_{(\frac{j-1}{\ell},\frac{j}{\ell}]}(t) x_{1,j}$$

and extend it as a periodic function with period 1 to the whole real line. We define the function $f_k^{(0,1],x} : (0,1] \to X$ by

$$f_k^{(0,1],x}(t) = \tau(kt).$$

All properties of this sequence except the statement about the weak limit are easy to check.

As for the weak limit, we claim that the weak limit of the sequence $\{f_k^{(0,1],x}\}_{k=1}^{\infty}$ is the function which takes the value $\frac{1}{\ell}\sum_{j=1}^{\ell} x_{i,j}$ at each point of the interval $(0,1]$. To verify this statement we need the description of the dual space of $L_p(X)$. The description is:

Theorem 8.48. *If the Banach space X is uniformly convex, the dual space to $L_p(X)$ is $L_q(X^*)$, where $\frac{1}{p}+\frac{1}{q}=1$ and X^* is the dual space of X. If $f \in L_p(X)$ and $F \in L_q(X^*)$, the action $F(f)$ is given by $F(f)=\int_0^1 (F(t))(f(t))dt$, where $F(t) \in X^*$, $f(t) \in X$, and $(F(t))(f(t))$ is the value of $F(t)$ at $f(t)$.*

The standard proof of this theorem is based on the theory of vector measures and is not included here, see Notes and Remarks for references.

Now let $x^* \in X^*$, $d \in \mathbb{N}$, and $j \in \{1,\dots,d\}$. We introduce $F \in L_q(X^*)$ by $F = \mathbf{1}_{(\frac{j-1}{d},\frac{j}{d}]}x^*$. A simple computation shows that we have

$$\lim_{k\to\infty} F(f_k^{(0,1],x}) = \frac{1}{d}\left(\frac{1}{\ell}\sum_{j=1}^{\ell} x^*(x_{i,j})\right) = F(g),$$

where g is the constant function on $(0,1]$ taking the value $\frac{1}{\ell}\sum_{j=1}^{\ell} x_{i,j}$.

The description of the dual space presented in Theorem 8.48 implies that the linear span of the functionals of the form $\mathbf{1}_{(\frac{j-1}{d},\frac{j}{d}]}x^*$ is dense in the space $L_q(X^*)$. The conclusion follows.

As for the general case, we observe that, as is easy to check, the functions

$$f_k = \sum_{i=1}^{m} f_k^{(\frac{i-1}{m},\frac{i}{m}],x_i}$$

form the desired sequence, where $f_k^{(a,b],x}$ is defined as a "compression" of $f_k^{(0,1],x}$ to the interval $(a,b]$, that is,

$$f_k^{(a,b],x}(t) = \begin{cases} f_k^{(0,1],x}\left(\frac{t-a}{b-a}\right) & \text{if } t \in (a,b] \\ 0 & \text{otherwise.} \end{cases}$$ □

The established relation between $\mathrm{Dent}(X,\varepsilon)$ and $\mathrm{wSz}(L_p(X),\varepsilon/2)$ can be useful only if we can prove that the indices $\mathrm{wSz}(L_p(X),\varepsilon/2)$ are finite. In this connection our next step is to show that $L_p(X)$ is uniformly convex.

Theorem 8.49. *If X is a uniformly convex space and $1 < p < \infty$, then the space $L_p(X)$ is also uniformly convex.*

Proof. First we show that for each $\varepsilon > 0$ there is $\delta(\varepsilon) > 0$ such that for $y, z \in X$ satisfying $1 = \|y\| \geq \|z\|$ and $\|y - z\| \geq \varepsilon$ we have

$$\left\| \frac{y+z}{2} \right\|^p \leq (1 - \delta(\varepsilon)) \frac{\|y\|^p + \|z\|^p}{2}.$$

Assume the contrary, then for some $\varepsilon > 0$ there are sequences $\{y_n\}_{n=1}^{\infty}, \{z_n\}_{n=1}^{\infty}$ in X such that $1 = \|y_n\| \geq \|z_n\|$, $\|y_n - z_n\| \geq \varepsilon$, and

$$\lim_{n\to\infty} \frac{\left\| \frac{y_n+z_n}{2} \right\|^p}{\frac{\|y_n\|^p + \|z_n\|^p}{2}} = 1. \tag{8.31}$$

Now we use the strict convexity of the function t^p, $t \geq 0$, $1 < p < \infty$ and get

$$\left(\frac{1+t}{2}\right)^p < \frac{1+t^p}{2}, \quad t \geq 0,\ t \neq 1. \tag{8.32}$$

By (8.31) and the triangle inequality we get

$$\lim_{n\to\infty} \frac{\left(\frac{1+\|z_n\|}{2}\right)^p}{\frac{1+\|z_n\|^p}{2}} = 1.$$

By (8.32), this implies $\lim_{n\to\infty} \|z_n\| = 1$. Let $u_n = z_n/\|z_n\|$. Then $\lim_{n\to\infty} \|u_n - z_n\| = 0$, therefore

$$\liminf_{n\to\infty} \|u_n - y_n\| = \liminf_{n\to\infty} \|z_n - y_n\| \geq \varepsilon.$$

On the other hand, applying (8.31) we get

$$\lim_{n\to\infty} \left\| \frac{y_n + u_n}{2} \right\| = 1.$$

This contradicts the uniform convexity of X.

If $y, z \in X$ are not both zero, we divide them by $\max\{\|y\|, \|z\|\}$, and immediately get

$$\left\| \frac{y+z}{2} \right\|^p \leq \left(1 - \delta\left(\frac{\|y-z\|}{\max\{\|y\|, \|z\|\}}\right)\right) \frac{\|y\|^p + \|z\|^p}{2}. \tag{8.33}$$

Now let $f, g \in L_p(X)$, $\varepsilon > 0$, $\|f\|_p = \|g\|_p = 1$ and $\|f - g\|_p \geq \varepsilon$. Let D be the subset of $[0,1]$ (recall that $L_p(X)$ is the set of functions on $[0,1]$) for which

$$\|f(x) - g(x)\|^p \geq \frac{\varepsilon^p}{4}(\|f(x)\|^p + \|g(x)\|^p) \geq \frac{\varepsilon^p}{4} \max\{\|f(x)\|^p, \|g(x)\|^p\}. \tag{8.34}$$

Applying (8.33), we get that for each $x \in D$

$$\left\| \frac{f(x)+g(x)}{2} \right\|^p \le \left(1 - \delta\left(\frac{\varepsilon}{4^{\frac{1}{p}}}\right)\right) \frac{\|f(x)\|^p + \|g(x)\|^p}{2}. \tag{8.35}$$

Inequality (8.34) implies

$$\int_{[0,1]\setminus D} \|f(x) - g(x)\|^p dx \le \frac{\varepsilon^p}{2}.$$

Therefore

$$\int_D \|f(x) - g(x)\|^p dx \ge \frac{\varepsilon^p}{2}. \tag{8.36}$$

Let $\tilde{f}, \tilde{g} \in L_p(X)$ be the functions which coincide with f and g, respectively, on D, and are equal to 0 on $[0,1]\setminus D$. By (8.36) we have

$$\|\tilde{f} - \tilde{g}\|_p \ge \frac{\varepsilon}{2^{\frac{1}{p}}},$$

and hence

$$\max\{\|\tilde{f}\|_p, \|\tilde{g}\|_p\} \ge \frac{\varepsilon}{2^{\frac{1}{p}+1}}. \tag{8.37}$$

We have

$$\begin{aligned}
&\int_{[0,1]} \frac{\|f(x)\|^p + \|g(x)\|^p}{2} dx - \int_{[0,1]} \left\| \frac{f(x)+g(x)}{2} \right\|^p dx \\
&\quad \ge \int_{[0,1]} \frac{\|\tilde{f}(x)\|^p + \|\tilde{g}(x)\|^p}{2} dx - \int_{[0,1]} \left\| \frac{\tilde{f}(x)+\tilde{g}(x)}{2} \right\|^p dx \\
&\quad \overset{(8.35)}{\ge} \int_{[0,1]} \delta\left(\frac{\varepsilon}{4^{\frac{1}{p}}}\right) \frac{\|\tilde{f}(x)\|^p + \|\tilde{g}(x)\|^p}{2} dx \\
&\quad \overset{(8.37)}{\ge} \delta\left(\frac{\varepsilon}{4^{\frac{1}{p}}}\right) \frac{\varepsilon^p}{2^{p+2}}.
\end{aligned}$$

Therefore

$$\left\| \frac{f+g}{2} \right\|_p \le \left(1 - \delta\left(\frac{\varepsilon}{4^{\frac{1}{p}}}\right) \frac{\varepsilon^p}{2^{p+2}}\right)^{\frac{1}{p}}. \tag{8.38}$$

Letting

$$\delta_1(\varepsilon) = 1 - \left(1 - \delta\left(\frac{\varepsilon}{4^{\frac{1}{p}}}\right) \frac{\varepsilon^p}{2^{p+2}}\right)^{\frac{1}{p}},$$

we get a lower estimate for the modulus of convexity of $L_p(X)$. It is clear that $\delta_1(\varepsilon) > 0$ if $\varepsilon > 0$. □

The argument used in the next lemma is well known, we already used a version of it in Lemma 2.58 (page 68).

Lemma 8.50. *For each uniformly convex space X there exist $C, r \in (0, \infty)$ such that* $\mathrm{wSz}(B_X, \varepsilon) \le \frac{C}{\varepsilon^r}$ *for all* $\varepsilon \in (0, 2]$.

Proof. It is clear that $\mathrm{wSz}(B_X, \frac{1}{2}) = 2^r$ for some $r \in (0, \infty)$. By the submultiplicativity Lemma 8.45, we get

$$\mathrm{wSz}\left(B_X, \left(\frac{1}{2}\right)^n\right) \le \frac{1}{((\frac{1}{2})^n)^r}. \tag{8.39}$$

Now we use the obvious fact that $\mathrm{wSz}(B_X, \varepsilon)$ is a non-increasing function. Letting $C = \max_{\varepsilon\in[\frac{1}{2},2]}(\mathrm{wSz}(B_X, \varepsilon)\cdot\varepsilon^r)$ we immediately get that the desired inequality holds for each $\varepsilon \in [\frac{1}{2}, 2]$. Now let $\varepsilon \in (0, \frac{1}{2})$. Then $\varepsilon = \left(\frac{1}{2}\right)^{\lfloor\log_2(1/\varepsilon)\rfloor} \cdot R$, where $R \in \left(\frac{1}{2}, 1\right]$. By Lemma 8.45 and (8.39),

$$\begin{aligned}
&\mathrm{wSz}(B_X, \varepsilon)\cdot\varepsilon^r \\
&\le \left(\mathrm{wSz}\left(B_X, \left(\frac{1}{2}\right)^{\lfloor\log_2(1/\varepsilon)\rfloor}\right)\cdot\left(\left(\frac{1}{2}\right)^{\lfloor\log_2(1/\varepsilon)\rfloor}\right)^r\right)\cdot(\mathrm{wSz}(B_X, R)\cdot R^r) \\
&\le 1\cdot C = C.
\end{aligned}$$

This completes the proof of Lemma 8.50. □

Now we can complete the proof of Theorem 8.38. Since X is uniformly convex, the space $L_p(X)$ $(1 < p < \infty)$ is also uniformly convex. Hence, by Lemmas 8.46 and 8.50, we get $\mathrm{Dent}(B_X, \varepsilon) \le \frac{C}{\varepsilon^r}$. Combining this estimate with the estimate for the modulus of convexity of $(X, |||\cdot|||)$ obtained in (8.26) we get that X is q-uniformly convex for $q = 2r + 2$. □

8.5 Markov type of uniformly smooth Banach spaces

Everywhere in this section X is a p-uniformly smooth Banach space with p-smoothness constant $S_p(X)$. The main result of this section is:

Theorem 8.51. *Let $1 < p \le 2$ and let X be a p-uniformly smooth Banach space with p-smoothness constant $S_p(X)$. Then the Markov type of X satisfies the following inequality*

$$M_p(X) \le \frac{8}{(2^{p+1} - 4)^{1/p}} S_p(X).$$

In the next results we use the notion of a Banach space valued random variable, martingale and a conditional expectation, the reader is referred to Section 1.3 for the definitions and basic facts.

Lemma 8.52. *Fix $1 < p \le 2$ and let $Z \in L_p(\Omega, \Sigma, \mathbb{P}, X)$. Then*

$$\mathbb{E}\|Z\|^p \le \|\mathbb{E}Z\|^p + \frac{S_p(X)^p}{2^{p-1}-1} \cdot \mathbb{E}\|Z - \mathbb{E}Z\|^p.$$

Proof. Let $\theta \ge 0$ be the largest constant such that for every $Z \in L_p(X)$,

$$\theta \left(\mathbb{E}\|Z\|^p - \|\mathbb{E}Z\|^p\right) \le \mathbb{E}\|Z - \mathbb{E}Z\|^p. \tag{8.40}$$

Our goal is to show that $\theta \ge (2^{p-1} - 1)S_p(X)^{-p}$. To this end, fix $\varepsilon > 0$ and $Z \in L_p(X)$ such that

$$(\theta + \varepsilon)\left(\mathbb{E}\|Z\|^p - \|\mathbb{E}Z\|^p\right) > \mathbb{E}\|Z - \mathbb{E}Z\|^p. \tag{8.41}$$

Applying Definition 8.25 of $S_p(X)$ to the vectors $x = (Z + \mathbb{E}Z)/2$ and $y = (Z - \mathbb{E}Z)/2$, we get the pointwise inequality:

$$\|Z\|^p + \|\mathbb{E}Z\|^p \le 2\left\|\frac{1}{2}Z + \frac{1}{2}\mathbb{E}Z\right\|^p + 2\, S_p(X)^p \left\|\frac{1}{2}Z - \frac{1}{2}\mathbb{E}Z\right\|^p. \tag{8.42}$$

We take expectations of both sides of (8.42) and get

$$\begin{aligned}
\frac{1}{\theta+\varepsilon}\mathbb{E}\|Z - \mathbb{E}Z\|^p &\overset{(8.41)}{<} \mathbb{E}\|Z\|^p - \|\mathbb{E}Z\|^p \\
&\overset{(8.42)}{\le} 2\left(\mathbb{E}\left\|\frac{1}{2}Z + \frac{1}{2}\mathbb{E}Z\right\|^p + S_p(X)^p\, \mathbb{E}\left\|\frac{1}{2}Z - \frac{1}{2}\mathbb{E}Z\right\|^p\right) - 2\|\mathbb{E}Z\|^p \\
&= 2\left(\mathbb{E}\left\|\frac{1}{2}Z + \frac{1}{2}\mathbb{E}Z\right\|^p - \left\|\mathbb{E}\left(\frac{1}{2}Z + \frac{1}{2}\mathbb{E}Z\right)\right\|^p\right) \\
&\quad + 2\, S_p(X)^p\, \mathbb{E}\left\|\frac{1}{2}Z - \frac{1}{2}\mathbb{E}Z\right\|^p \\
&\overset{(8.40)}{\le} \frac{2}{\theta}\mathbb{E}\left\|\frac{1}{2}Z - \frac{1}{2}\mathbb{E}Z\right\|^p + 2\, S_p(X)^p\, \mathbb{E}\left\|\frac{1}{2}Z - \frac{1}{2}\mathbb{E}Z\right\|^p.
\end{aligned}$$

It follows that $\frac{2^p}{\theta+\varepsilon} \le \frac{2}{\theta} + 2S_p(X)^p$. Letting ε tend to zero we get the desired inequality. □

We also need a conditional version of Lemma 8.52. For most of the applications it suffices to consider the case where the σ-subalgebras $\mathcal{F}_i$ are finite.

Lemma 8.53. *Fix $1 < p \le 2$, let $Z \in L_p(\Omega, \Sigma, \mathbb{P}, X)$ and let $\mathcal{F}$ be a σ-subalgebra of Σ. Then for almost every $\omega \in \Omega$*

$$\mathbb{E}(\|Z\|^p|\mathcal{F})(\omega) \le \|\mathbb{E}(Z|\mathcal{F})(\omega)\|^p + \frac{S_p(X)^p}{2^{p-1}-1} \cdot \mathbb{E}(\|Z - \mathbb{E}(Z|\mathcal{F})\|^p|\mathcal{F})(\omega).$$

Since in this book we are going to apply martingales only in the case where all σ-subalgebras are finite, we skip the proof of this lemma. In fact, in the case where the σ-algebra $\mathcal{F}$ is finite, $A \subset \Sigma$ is one of its atoms (with positive $\mathbb{P}(A)$), and $\omega \in A$, the inequality reduces to

$$\mathbb{E}_A \|Z\|^p \le \|\mathbb{E}_A Z\|^p + \frac{S_p(X)^p}{2^{p-1}-1} \cdot \mathbb{E}_A \|Z - \mathbb{E}_A Z\|^p,$$

where

$$\mathbb{E}_A W = \frac{\mathbb{E}(W \cdot \mathbf{1}_A)}{\mathbb{P}(A)}$$

for any random variable (scalar- or vector-valued). It is clear that the obtained inequality is the same as the inequality obtained in Lemma 8.52, but for a different probability space.

Theorem 8.54. *Fix $1 < p \le 2$ and let $\{M_k\}_{k=0}^n \subseteq L_p(\Omega, \Sigma, \mathbb{P}, X)$ be a martingale in X. Then*

$$\mathbb{E}\|M_n - M_0\|^p \le \frac{S_p(X)^p}{2^{p-1}-1} \cdot \sum_{k=0}^{n-1} \mathbb{E}\|M_{k+1} - M_k\|^p.$$

Proof. Assume that $\{M_k\}_{k=0}^n$ is a martingale with respect to the filtration $\mathcal{F}_0 \subseteq \mathcal{F}_1 \subseteq \cdots \subseteq \mathcal{F}_{n-1}$; that is, $M_i = \mathbb{E}(M_{i+1}|\mathcal{F}_i)$ for $i = 0, 1, \ldots, n-1$. By Lemma 8.53 applied to $Z = M_n - M_0$ and $\mathcal{F} = \mathcal{F}_{n-1}$, we get

$$\begin{aligned}
&\mathbb{E}\left(\|M_n - M_0\|^p | \mathcal{F}_{n-1}\right)(\omega) \\
&\quad \le \|\mathbb{E}(M_n - M_0|\mathcal{F}_{n-1})(\omega)\|^p \\
&\quad + \frac{S_p(X)^p}{2^{p-1}-1} \cdot \mathbb{E}(\|M_n - M_0 - \mathbb{E}(M_n - M_0|\mathcal{F}_{n-1})\|^p|\mathcal{F}_{n-1})(\omega).
\end{aligned}$$

Using linearity of conditional expectation, this inequality can be rewritten as a pointwise inequality between functions:

$$\mathbb{E}\left(\|M_n - M_0\|^p | \mathcal{F}_{n-1}\right) \le \|M_{n-1} - M_0\|^p + \frac{S_p(X)^p}{2^{p-1}-1} \cdot \mathbb{E}(\|M_n - M_{n-1}\|^p|\mathcal{F}_{n-1}).$$

Taking expectations of both sides we get

$$\mathbb{E}\|M_n - M_0\|^p \le \mathbb{E}\,\|M_{n-1} - M_0\|^p + \frac{S_p(X)^p}{2^{p-1}-1} \cdot \mathbb{E}\|M_n - M_{n-1}\|^p,$$

and the required inequality follows by induction. □

Lemma 8.55. *Let X be a normed space, $\{Z_t\}_{t=0}^\infty$ a stationary reversible Markov chain on $\{1, \ldots, n\}$ and $f : \{1, \ldots, n\} \to X$. Then for every $t \in \mathbb{N}$ there are two X-valued martingales $\{M_s\}_{s=0}^t$ and $\{N_s\}_{s=0}^t$ (with respect to two different filtrations) with the following properties:*

(a) *For every* $1 \le s \le t-1$ *the equality*

$$f(Z_{s+1}) - f(Z_{s-1}) = (M_{s+1} - M_s) - (N_{t-s+1} - N_{t-s}) \tag{8.43}$$

holds.

(b) *For every* $0 \le s \le t-1$ *and* $p \ge 1$,

$$\max\{\mathbb{E}\|M_{s+1} - M_s\|^p, \mathbb{E}\|N_{s+1} - N_s\|^p\} \le 2^p\, \mathbb{E}\|f(Z_1) - f(Z_0)\|^p. \tag{8.44}$$

Proof. Let $A = (a_{ij})$ be the transition matrix of the Markov chain $\{Z_t\}$, and let

$$\pi_i := \mathbb{P}(Z_0 = i).$$

As we have mentioned above (see Remark 8.11) we assume that the distribution $\{\pi_i\}_{i=1}^n$ is reversible.

Define the map $(Lf) : \{1, \dots, n\} \to X$ by

$$Lf(i) = \sum_{j=1}^{n} a_{ij}(f(j) - f(i)) = \sum_{j=1}^{n} a_{ij} f(j) - f(i).$$

We use $\mathbb{E}(f(Z_s)|Z_0, \dots, Z_{s-1})$ to denote the conditional expectation of the random variable $f(Z_s)$ with respect to the σ-algebra generated by random variables $Z_0, \dots, Z_{s-1}$, that is, with respect to the σ-algebra generated by all sets of the form $\{\omega : Z_0(\omega) = x_0, \dots, Z_{s-1}(\omega) = x_{s-1}\}$.

We have

$$\mathbb{E}(f(Z_s)|Z_0, \dots, Z_{s-1}) = \mathbb{E}(f(Z_s)|Z_{s-1}) = Lf(Z_{s-1}) + f(Z_{s-1}),$$

where the first equality follows from the definition of a Markov chain, and the second equality is obtained by making the computation for each possible value of Z_{s-1}.

We claim that since $\{Z_s\}_{s=0}^{\infty}$ is stationary, and its stationary distribution is reversible, we also have that for every $0 \le s < t$,

$$\mathbb{E}(f(Z_s)|Z_{s+1}, \dots, Z_t) = Lf(Z_{s+1}) + f(Z_{s+1}). \tag{8.45}$$

In fact, reversibility implies (by induction) that for each $k \ge 1$ and each collection $i_0, i_1, \dots, i_k \in \{1, \dots, n\}$ we have

$$\pi_{i_0} a_{i_0,i_1} a_{i_1,i_2}, \dots, a_{i_{k-1},i_k} = \pi_{i_k} a_{i_k,i_{k-1}} a_{i_{k-1},i_{k-2}}, \dots, a_{i_1,i_0}. \tag{8.46}$$

If $s \ge k$, the equality (8.46) can be written as

$$\begin{aligned} \mathbb{P}(Z_s = i_0, Z_{s+1} = i_1, \dots, Z_{s+k} = i_k) \\ = \mathbb{P}(Z_{s-k} = i_k, Z_{s-k+1} = i_{k-1}, \dots, Z_s = i_0). \end{aligned} \tag{8.47}$$

The equation (8.45) follows.

We define $M_0 = f(Z_0)$. For each $s \geq 1$ we define M_s by

$$M_s = f(Z_s) - \sum_{r=0}^{s-1} Lf(Z_r).$$

We have

$$\mathbb{E}(M_s | Z_0, \ldots, Z_{s-1}) = Lf(Z_{s-1}) + f(Z_{s-1}) - \sum_{r=0}^{s-1} Lf(Z_r) = M_{s-1},$$

where we used the observation that $\sum_{r=0}^{s-1} Lf(Z_r)$ is measurable with respect to the σ-algebra generated by $Z_0, \ldots, Z_{s-1}$. Therefore, $\{M_s\}_{s=0}^{\infty}$ is a martingale with respect to the natural filtration $\{\mathcal{F}_k\}_{k=1}^{\infty}$, where $\mathcal{F}_k$ induced by $\{Z_s\}_{s=0}^{k}$. Observe that each σ-algebra in this filtration is finite.

We also define $N_0 := f(Z_t)$ and

$$N_s := f(Z_{t-s}) - \sum_{r=t-s+1}^{t} Lf(Z_r).$$

for $1 \leq s \leq t$. As in the case of $\{M_s\}$, for $s \geq 1$ we have

$$\mathbb{E}(N_s | Z_{t-s+1}, \ldots, Z_t) = N_{s-1}.$$

Therefore $\{N_s\}_{s=0}^{t}$ is a martingale with respect to the natural filtration $\{\widetilde{\mathcal{F}}_k\}_{k=0}^{t}$, where $\widetilde{\mathcal{F}}_k$ is generated by $Z_t, Z_{t-1}, \ldots, Z_{t-k}$.

The equalities

$$\begin{aligned} M_{s+1} - M_s &= f(Z_{s+1}) - f(Z_s) - Lf(Z_s), \\ N_{s+1} - N_s &= f(Z_{t-s-1}) - f(Z_{t-s}) - Lf(Z_{t-s}). \end{aligned}$$

imply (8.43). To prove (8.44) observe that for every $s \geq 0$ and $q \geq 1$,

$$\begin{aligned} \mathbb{E}\|Lf(Z_s)\|^p &= \sum_{i=1}^{n} \pi_i \left\| \sum_{j=1}^{n} a_{ij} [f(j) - f(i)] \right\|^p \\ &\leq \sum_{i=1}^{n} \sum_{j=1}^{n} \pi_i a_{ij} \|f(j) - f(i)\|^p \\ &= \mathbb{E}\|f(Z_1) - f(Z_0)\|^p. \end{aligned}$$

Therefore,

$$\begin{aligned} &\mathbb{E}\|M_{s+1} - M_s\|^p = \mathbb{E}\|f(Z_{s+1}) - f(Z_s) - Lf(Z_s)\|^p \\ &\quad \leq 2^{p-1}\,\mathbb{E}\|f(Z_{s+1}) - f(Z_s)\|^p + 2^{p-1}\,\mathbb{E}\|Lf(Z_s)\|^p \leq 2^p\,\mathbb{E}\|f(Z_1) - f(Z_0)\|^p, \end{aligned}$$

and similarly, $\mathbb{E}\|N_{s+1} - N_s\|^p \leq 2^p\,\mathbb{E}\|f(Z_1) - f(Z_0)\|^p$. □

Proof of Theorem 8.51. Let $\{Z_s\}_{s=0}^t, f, \{M_s\}_{s=0}^t$ and $\{N_s\}_{s=0}^t$ be as in Lemma 8.55. Assume first that t is even, and write $t = 2m$. Summing the identity (8.43) over $s = 1, 3, 5, \ldots, 2m-1$ we get

$$f(Z_t) - f(Z_0) = \sum_{k=1}^{m}(M_{2k} - M_{2k-1}) - \sum_{k=1}^{m}(N_{2k} - N_{2k-1}).$$

Observe that the sequences

$$\left\{\sum_{k=1}^{s}(M_{2k} - M_{2k-1})\right\}_{s=1}^{m} \quad \text{and} \quad \left\{\sum_{k=1}^{s}(N_{2k} - N_{2k-1})\right\}_{s=1}^{m}$$

are martingales with respect to filtrations $\{\mathcal{F}_{2k}\}_{k=1}^m$ and $\{\widetilde{\mathcal{F}}_{2k}\}_{k=1}^m$, respectively. Applying Theorem 8.54 to these martingales (and the convexity of the function $X \mapsto \|X\|^p$ in the first step) we get

$$\begin{aligned}
&\mathbb{E}\|f(Z_t) - f(Z_0)\|^p \\
&\le 2^{p-1}\mathbb{E}\left\|\sum_{k=1}^{m}(M_{2k} - M_{2k-1})\right\|^p + 2^{p-1}\mathbb{E}\left\|\sum_{k=1}^{m}(N_{2k} - N_{2k-1})\right\|^p \\
&\le \frac{2^{p-1}S_p(X)^p}{2^{p-1}-1}\sum_{k=1}^{m}\left(\mathbb{E}\|M_{2k} - M_{2k-1}\|^p + \mathbb{E}\|N_{2k} - N_{2k-1}\|^p\right) \\
&\le \frac{2^{p-1}S_p(X)^p}{2^{p-1}-1}\cdot\frac{t}{2}\cdot 2^{p+1}\mathbb{E}\|f(Z_1) - f(Z_0)\|^p.
\end{aligned}$$

When t is odd, apply the above reasoning at time $t-1$, to get

$$\begin{aligned}
\mathbb{E}\|f(Z_t) - f(Z_0)\|^p &\le 2^{p-1}\mathbb{E}\|f(Z_{t-1}) - f(Z_0)\|^p + 2^{p-1}\mathbb{E}\|f(Z_t) - f(Z_{t-1})\|^p \\
&\le \left(\frac{2^{3p-2}S_p(X)^p}{2^{p-1}-1}(t-1) + 2^{p-1}\right)\mathbb{E}\|f(Z_1) - f(Z_0)\|^p \\
&\le \frac{2^{3p-2}S_p(X)^p}{2^{p-1}-1}\, t\,\mathbb{E}\|f(Z_1) - f(Z_0)\|^p.
\end{aligned}$$

This concludes the proof of the theorem. □

8.6 Applications of Markov type to lower estimates of distortions of embeddings into uniformly smooth Banach spaces

Our purpose in this section is to derive some consequences of the results proved in the previous section.

Corollary 8.56. *Let X be a p-uniformly smooth Banach space, $1 < p \le 2$, and G be a finite k-regular, $k \ge 3$, graph of girth $g = g(G)$. Then*

$$c_X(G) \ge C(p, X, k)(\lfloor g(G)/2 \rfloor)^{\frac{p-1}{p}}.$$

Proof. We consider the standard Markov chain $\{Z_t\}_{t=0}^{\infty}$ (see Example 8.6) on vertices of G. Assume that the initial distribution is uniform on the set of all vertices of G. As in the proof of Theorem 8.14, for every $1 \le t \le \lfloor g/2 \rfloor - 1$ we have

$$\mathbb{E} d_G(Z_{t+1}, Z_0) \ge \frac{k-2}{k}(t+1).$$

Since $p > 1$ and $\mathbb{E}(d_G(Z_1, Z_0))^p = 1$, this inequality implies

$$\begin{aligned} \mathbb{E}(d_G(Z_{t+1}, Z_0))^p &\ge (\mathbb{E} d_G(Z_{t+1}, Z_0))^p \\ &\ge \left(\frac{k-2}{k}\right)^p (t+1)^p\, \mathbb{E}(d_G(Z_1, Z_0))^p. \end{aligned} \tag{8.48}$$

Let $f : V(G) \to X$ be such that

$$d_G(u, v) \le \|f(u) - f(v)\| \le d_G(u, v)(c_X(G) + \varepsilon). \tag{8.49}$$

Using Theorem 8.51 we get

$$\begin{aligned} &\left(\frac{k-2}{k}\right)^p (t+1)^p \mathbb{E}(d_G(Z_1, Z_0))^p \\ &\quad \overset{(8.48)}{\le} \mathbb{E}(d_G(Z_{t+1}, Z_0))^p \\ &\quad \overset{(8.49)}{\le} \mathbb{E}\|f(Z_{t+1}) - f(Z_0)\|^p \\ &\quad \overset{\text{Theorem 8.51}}{\le} (M_p(X))^p (t+1) \mathbb{E}\|f(Z_1) - f(Z_0)\|^p \\ &\quad \overset{(8.49)}{\le} (M_p(X))^p (t+1)(c_X(G) + \varepsilon)^p \mathbb{E}(d_G(Z_1, Z_0))^p. \end{aligned}$$

We let $t = \lfloor g/2 \rfloor - 1$ and $\varepsilon \downarrow 0$. We get

$$c_X(G) \ge \left(\frac{k-2}{kM_p(X)}\right)(\lfloor g/2 \rfloor)^{\frac{p-1}{p}}.$$

□

Corollary 8.57. *Let X be a p-uniformly smooth Banach space, $1 < p \leq 2$, and G be a finite k-regular graph ($k \geq 3$) with n vertices and the absolute spectral gap of its normalized adjacency matrix equal to $1 - \alpha$. Then*

$$c_X(G) \geq C(p, X, k, \alpha)(\ln n)^{\frac{p-1}{p}}.$$

Proof. The proof follows the same lines as for Corollary 8.56. We consider the standard Markov chain $\{Z_t\}_{t=0}^{\infty}$ (see Example 8.6) on vertices of G. First we assume that the initial distribution μ_u is concentrated in some vertex u. Then the distribution of Z_t will be $\mu_u A_G^t$, where A_G is the normalized adjacency matrix of G. Our first observation is that the distributions $\mu_u A_G^t$ approach the uniform distribution σ very fast. In fact, since the matrix A is symmetric we can write everything for transposes (so that it looks more familiar):

$$\begin{aligned}
\|A^t \widetilde{\mu}_u - \widetilde{\sigma}\|_1 &\leq \sqrt{n}\, \|A^t \widetilde{\mu}_u - \widetilde{\sigma}\|_2 \\
&= \sqrt{n}\, \|A^t (\widetilde{\mu}_u - \widetilde{\sigma})\|_2 \\
&\leq \sqrt{n}\, \alpha^t \|\widetilde{\mu}_u - \widetilde{\sigma}\|_2 \\
&\leq 2\sqrt{n}\, \alpha^t,
\end{aligned}$$

where transposes are denoted using $\widetilde{\ }$. Let $t = \lceil \log_{1/\alpha}(2n^{3/2}) \rceil$. Then

$$\|A^t \widetilde{\mu}_u - \widetilde{\sigma}\|_1 \leq \frac{1}{n}.$$

This means that the probability that Z_t is in the set of $\frac{n}{2}$ closest to u vertices of G is $\leq \frac{1}{2} + \frac{1}{n}$. Now we recall the computation on page 110 which implies that there are at most $\frac{n}{2}$ vertices in G at distance $\leq \log_k \left(\frac{n}{2} - 1\right)$ to u. This immediately implies that

$$\mathbb{E} d_G(Z_t, Z_0) \geq \left(\frac{1}{2} - \frac{1}{n}\right) \log_k \left(\frac{n}{2} - 1\right) \geq ct\mathbb{E} d_G(Z_1, Z_0),$$

where the constant $c > 0$ depends only k and α provided $n > 4$. Since $p > 1$, we get

$$\mathbb{E}(d_G(Z_t, Z_0))^p \geq c^p t^p \mathbb{E}(d_G(Z_1, Z_0))^p, \tag{8.50}$$

where we use the fact that $d_G(Z_1, Z_0) = 1$ almost everywhere. So the standard Markov chain on an expander moves fast from its original position for the chosen value of t. It is clear from the definition of the Markov chain that this estimate remains true if we assume that the initial distribution is uniform.

We complete the proof as in Corollary 8.56. For the reader's convenience we provide the details. Let $f : V(G) \to X$ be such that

$$d_G(u, v) \leq \|f(u) - f(v)\| \leq d_G(u, v)(c_X(G) + \varepsilon). \tag{8.51}$$

Using Theorem 8.51 we get

$$\begin{aligned}
c^p t^p \mathbb{E}(d_G(Z_1, Z_0))^p &\overset{(8.50)}{\le} \mathbb{E}(d_G(Z_t, Z_0))^p \\
&\overset{(8.51)}{\le} \mathbb{E}\|f(Z_t) - f(Z_0)\|^p \\
&\overset{\text{Theorem 8.51}}{\le} (M_p(X))^p t\, \mathbb{E}\|f(Z_1) - f(Z_0)\|^p \\
&\overset{(8.51)}{\le} (M_p(X))^p t (c_X(G) + \varepsilon)^p \mathbb{E}(d_G(Z_1, Z_0))^p.
\end{aligned}$$

Recalling that our choice of t is $t = \lceil \log_{1/\alpha}(2n^{3/2}) \rceil$ and letting $\varepsilon \downarrow 0$, we get

$$c_X(G) \ge \left(\frac{c}{M_p(X)}\right) (\lceil \log_{1/\alpha}(2n^{3/2}) \rceil)^{\frac{p-1}{p}} \ge C(p, X, k, \alpha)(\ln n)^{\frac{p-1}{p}}. \qquad \square$$

Remark 8.58. Observation (b) on page 198 shows that the result of Corollary 8.57 can be extended to all expander families.

8.7 Exercises

Exercise 8.59. A Markov chain with transition matrix $A = \{a_{i,j}\}_{i,j=1}^n$ is called *irreducible* if for every $i, j \in \{1, \dots, n\}$ there exists $m \in \mathbb{N}$ such that the (i, j)-entry of A^m is strictly positive. Show that a reversible distribution of a reversible irreducible Markov chain is uniquely determined by the chain.

Exercise 8.60. Consider a Markov chain with the transition matrix

$$\begin{bmatrix} 0 & \frac{1}{3} & 0 & \frac{2}{3} \\ \frac{2}{3} & 0 & \frac{1}{3} & 0 \\ 0 & \frac{2}{3} & 0 & \frac{1}{3} \\ \frac{1}{3} & 0 & \frac{2}{3} & 0 \end{bmatrix}.$$

(1) Show that this Markov chain is not reversible. (2) Find a stationary distribution for it.

Exercise 8.61. Give an example of a reversible Markov chain which has a stationary distribution which is not a reversible distribution.

There is a hint to this exercise.

Exercise 8.62. Let X be a uniformly convex Banach space and Z be a closed subspace of X. Show that the quotient X/Z is also uniformly convex.

Exercise 8.63. Prove that a uniformly smooth Banach space is reflexive.

One of the ways is outlined in the Hints to exercises section.

8.8 Notes and remarks

Basic material on Markov chains can be found in Chapters 2–6 of [191]. A much more extensive introduction can be found in the book [281].

The idea of using the Markov chains for Lipschitz extension problems is due to Ball [36] who introduced the notion of Markov type and proved that Hilbert spaces have Markov type 2. Our proof of this fact follows [292] and [335].

Theorem 8.14 and its proof are due to Linial, Magen, and Naor [292]. The conjecture that graphs with large girth have large Euclidean distortion goes back to [290]. The authors of [292] present two proofs of this conjecture. In addition to the proof presented in Section 8.3, they give a proof based on the PSD-criterion (this criterion is described in Section 4.4 of this book). The proof uses the description of the adjacency matrix for k-distant vertices in a graph of girth $2k + 1$ (similar to the description of $F_r(A)$ in Section 5.12) and analysis of properties of relevant orthogonal polynomials. There are some related and still open problems, see Section 11.3 for their discussion.

The paper [42] contains some extensions of Theorem 8.14.

The theory of uniformly convex spaces is a classical direction of Banach space theory. It was initiated by Clarkson [101], who proved that the spaces L_p, $1 < p < \infty$ are uniformly convex. The fact that $\delta_{\ell_2}(\varepsilon) \geq \delta_Y(\varepsilon)$ for all $\varepsilon \in [0, 2]$ and for all Banach spaces Y of dimension at least 2 is due to Nordlander [342].

The notion of a uniformly smooth space was introduced by Day [115], who used the term *uniformly flattened space*. In the mentioned paper Day discovered the duality between uniform convexity and smoothness.

Theorem 8.26 is due to D. Milman [330] and Pettis [372]. The presented proof is a version of the proof in [119, p. 49], where it is attributed to J. Rainwater. To the best of my knowledge this John Rainwater is a non-existing person and the attribution means that the proof evolved in the coffee room of the math department at the University of Washington and too many took part for it to be published privately.

The Goldstine theorem is from [171]. Its statement: *$\pi(B_X)$ is weak* dense in $B_{X^{**}}$.* It can be found (with a proof different from the original) in [130, Theorem V.4.5].

Estimates for the modulus of uniform convexity of spaces L_p implying the statement of Example 8.30 can be found in Hanner [193] and Kadets [228]. Some of these estimates were known already to Clarkson [101].

Propositions 8.18 and 8.24 are from [39, Proposition 7], see also [148]. It was shown in [39] (see also [148]) that $K_2(L_p) \leq 1/\sqrt{p-1}$ for $1 < p \leq 2$, and $S_2(L_p) \leq \sqrt{p-1}$ for $2 \leq p < \infty$. The paper [39] also contains estimates for q-uniform convexity and p-uniform smoothness constants of Schatten classes.

There is another way of proving Remarks 8.17 and 8.23: In [149], [148] (see also [288], Theorem 1.e.16) it is shown that if a Banach space X has modulus of convexity of power type q then X also has cotype q. Similarly, if X has modulus of smoothness of power type p then X has type p. Combining this result with our knowledge of possible values of type-cotype (see Definition 2.37 and Remark 2.41) we get the results mentioned in the remarks. Observe that L_1 has cotype 2 (see Theorem 2.53), but it is clearly not uniformly convex (it is non-reflexive).

Results on duality between uniformly convex and uniformly smooth spaces were developed in [115], [283], [148], [39]. Theorem 8.28 was obtained in [283] (we use a slightly shorter presentation of [148, p. 122]), Theorem 8.27 was obtained in [39]. Remark 8.29 is discussed in [148, p. 122]. This remark is related to the fact that $\delta_X(\varepsilon)$ could be a non-convex function, as was shown by Liokumovich [294].

Theorem 8.36 was obtained in [135]. Theorem 8.37 was obtained in [377]. These result were built upon the theory of superreflexive spaces created by James [213, 214]. Another, more elementary approach to the proof of Theorem 8.37 was suggested by Maurey and published in [50, Part 4, Chapter IV], see also [117, Chapter IV]. Our proof of Theorem 8.38 follows the approach of Lancien [265], we also use the presentation of Lancien's proof in [166]. The weak Szlenk derivative is the analogue of the derivative introduced by Szlenk [420] for the weak topology. The uniform convexity of the space $L_p(X)$ for uniformly convex X and $1 < p < \infty$ was established in [114]. We follow the proof of McShane [322]. The dual space of $L_p(X)$ (Theorem 8.48) was determined by Bochner and Taylor [67]. The description holds under more general condition than the uniform convexity of X, see [121, Chapter IV, Section 1].

There are many other results on uniform convexity and smoothness, some of them are contained in the accounts of the theory of uniformly convex and uniformly smooth spaces which can be found in [50], [117], [119], [141], [288], and [383], to which we refer readers for further information.

Theorem 8.51 is due to Naor, Peres, Schramm, and Sheffield [335, Theorem 2.3]. The immediate consequence of this theorem is that $M_2(L_p) \leq 4\sqrt{p-1}$ for every $2 \leq p < \infty$.

The proof of Lemma 8.52 is a slight modification of the proof of Lemma 3.1 in [36]. Theorem 8.54 was first proved in [377] (without the explicit constant), it is a simple corollary of Lemma 8.52 (see also Proposition 3.3 in [36]).

Lemma 8.55 is motivated by the continuous martingale decompositions of stochastic integrals constructed in [305].

We would like to turn the attention of readers who are interested in embeddings into non-Banach metric spaces to the many classes of metric spaces that are known to have Markov type: trees, Gromov hyperbolic spaces, negatively curved Riemannian manifolds [335] (see this paper for precise statements of the results), planar graph metrics (that is, planar graphs with their graph distances) [123], Alexandrov spaces of nonnegative curvature [347].

8.9 Hints to exercises

To Exercise 8.61. One can show that such a Markov chain cannot be irreducible.

To Exercise 8.63. One can use the following plan:

- Prove the following version of Theorem 8.28:

$$\rho_X(\tau) = \sup_{0\le\varepsilon\le 2}\left(\frac{\tau\varepsilon}{2} - \delta_{X^*}(\varepsilon)\right), \ (\tau > 0).$$

- Derive that (X is uniformly smooth)$\Rightarrow$(X^* is uniformly convex).
- Use Theorem 8.26.
- Show that (X^* is reflexive)$\Rightarrow$(X is reflexive).

Chapter 9

Metric characterizations of classes of Banach spaces

9.1 Introduction

In connection with problems of embeddability of metric spaces into Banach spaces it would be interesting to find metric characterizations of well-known classes of Banach spaces. By a *metric characterization* in the most general sense we mean a characterization which refers only to the metric structure of a Banach space and does not involve the linear structure. Later we shall define more specific forms of metric characterization.

The most important impetus to the development of the theory of metric characterizations was given by the following result.

Theorem 9.1 (Ribe theorem). *Let Z and Y be Banach spaces. If Z and Y are uniformly homeomorphic, then Z and Y are crudely finitely representable in each other.*

Recall (see Definition 2.28) that X is *crudely finitely representable* in Y if $\exists C < \infty$ $\forall n \in \mathbb{N}$, for an arbitrary n-dimensional subspace $X_n \subset X$ there is an n-dimensional subspace $Y_n \subset Y$ and a linear map $T_n : X_n \to Y_n$ such that $\|T_n\| \cdot \|T_n^{-1}\| \leq C$. Saying X and Y are *uniformly homeomorphic* we mean that there exists a uniformly continuous bijection $F : X \to Y$ such that the map F^{-1} is also uniformly continuous.

At this point three different proofs of the Ribe theorem are available. We present one of them in Section 9.2.

Many important properties $\mathcal{P}$ of Banach spaces are so-called *hereditary local isomorphic invariants*. This means that if a Banach space X has property $\mathcal{P}$ and Y is crudely finitely representable in X, then Y also has property $\mathcal{P}$.

Remark 9.2. The complicated terminology is explained as follows: *hereditary* means that the property is inherited by closed subspaces, *local* means that it is determined by a finite-dimensional structure, and *isomorphic invariant* means that it is inherited by isomorphic spaces.

Some of the properties which we have studied in this book are hereditary local isomorphic invariants. We mean such properties as *having type p* and *having cotype q*. In this chapter we shall study one more important hereditary local isomorphic invariant. It is called *superreflexivity* and it can be defined as being isomorphic to a uniformly convex space. (Possibly it is not immediately clear why this property is inherited by crudely finitely representable spaces, we shall show this in Section 9.3.2.) There are several other important properties of Banach spaces which are hereditary local isomorphic invariants.

An immediate consequence of the Ribe theorem is that each such property is preserved under uniform homeomorphisms. Therefore it is feasible to describe the property in terms of the metric structure of Banach spaces having it. If we compare this statement, for example, with the definitions of type and cotype, we see that the original definitions involve not only the metric structure, but also inequalities between norms of **linear combinations** of vectors in the space. That is, the original definitions of these important hereditary local isomorphic invariants cannot be regarded as given in terms of metric structure. Bourgain found an equivalent definition of superreflexivity in terms of metric structure and formulated the program of searching for equivalent definitions of other hereditary local isomorphic invariants in terms of metric structure with the next step consisting in studying these metrical concepts in general metric spaces in an attempt to develop an analogue of the linear theory. This program came to be known as *Ribe program*. We describe some results belonging to the Ribe program below and give more references in the Notes and Remarks section.

9.2 Proof of the Ribe theorem through Bourgain's discretization theorem

We are going to present the proof of the Ribe theorem based on Bourgain's discretization theorem.

Recall that the infimum of distortions of bilipschitz embeddings of a metric space X into a metric space Y is denoted $c_Y(X)$. Let X be a finite-dimensional Banach space and Y be an infinite-dimensional Banach space.

Definition 9.3. For $\varepsilon \in (0,1)$ let $\delta_{X\hookrightarrow Y}(\varepsilon)$ be the supremum of those $\delta \in (0,1)$ for which every δ-net $\mathcal{N}_\delta$ in B_X satisfies $c_Y(X) \le \frac{c_Y(\mathcal{N}_\delta)}{1-\varepsilon}$. The function $\delta_{X\hookrightarrow Y}(\varepsilon)$ is called the *discretization modulus for embeddings of* X *into* Y.

At the moment it is not even clear that the discretization modulus is defined for any $\varepsilon \in (0,1)$.

Theorem 9.4 (Bourgain's discretization theorem). *There exists* $C \in (0,\infty)$ *such that for every two Banach spaces* X, Y *with* $\dim X = n < \infty$ *and* $\dim Y = \infty$, *and every* $\varepsilon \in (0,1)$, *we have*

$$\delta_{X\hookrightarrow Y}(\varepsilon) \ge e^{-(c_Y(X)/\varepsilon)^{Cn}}. \tag{9.1}$$

Remark 9.5. Bourgain's discretization theorem is often quoted with the conclusion that if δ is at most as large as the right-hand side of (9.1) and $\mathcal{N}_\delta$ is a δ-net of B_X, then X admits a *linear* embedding into Y whose distortion is at most $c_Y(\mathcal{N}_\delta)/(1-\varepsilon)$. We are going to prove the theorem in this linear form. It is worth mentioning that the well-known differentiation argument shows that the bilipschitz version implies the linear version, see Notes and Remarks, item (3) on page 306.

Observation 9.6. Bourgain's discretization theorem implies the Ribe theorem.

Lemma 9.7. *A uniform homeomorphism $f : Z \to Y$ between two Banach spaces is bilipschitz for large distances, that is, for every $d \in (0,\infty)$ there exist $L_1, L_2 \in (0,\infty)$ such that $L_1\|x-y\|_Z \le \|f(x)-f(y)\|_Y \le L_2\|x-y\|_Z$ whenever $x, y \in Z$ satisfy $\|x-y\|_Z \ge d$.*

Proof. First we observe that the modulus of continuity of maps between Banach spaces are *subadditive* in the sense that

$$\omega_f(t+s) \le \omega_f(t) + \omega_f(s). \tag{9.2}$$

In fact, if $u, v \in Z$ are such that $\|u-v\| \le t+s$, then there is $w \in Z$ such that $\|u-w\| \le t$ and $\|w-v\| \le s$. Therefore $\|f(u)-f(v)\| \le \|f(u)-f(w)\| + \|f(w)-f(v)\| \le \omega_f(t)+\omega_f(s)$. Inequality (9.2) follows.

Since f is uniformly continuous, the quantity $\omega_f(d)$ is finite for some $d > 0$. Using (9.2) we get that $\omega_f(d)$ is finite for all $d \in (0,\infty)$. Let $L_2 = 2\omega_f(d)/d$.

Let $x, y \in Z$ be such that $\|x-y\| \ge d$, so let $\|x-y\| = kd + r$, where $k \in \mathbb{N}$ and $r \in [0,d)$. Since x and y are in a Banach space, there are vectors $v_0 = x, v_1, \dots, v_k, v_{k+1} = y$ in Z such that $\|v_{i-1} - v_i\| = d$ for $i = 1,\dots,k$, and $\|v_k - v_{k+1}\| = r$. We have $\|f(v_{i-1}) - f(v_i)\| \le \omega_f(d)$ for $i = 1,\dots,k+1$, therefore

$$\|f(x)-f(y)\| \le (k+1)\omega_f(d) \le kL_2d \le L_2\|x-y\|.$$

Let $L_1 > 0$ be any number satisfying $\omega_{f^{-1}}(L_1d) \le \frac{d}{2}$. Then it is clear that $\|x-y\| \ge d$ implies $\|f(x)-f(y)\| \ge L_1d$. Therefore $\|f(x)-f(y)\| = k(L_1d) + r$ for some $k \in \mathbb{N}$ and $r \in [0, L_1d)$. Let $u_0 = f(x), u_1, \dots, u_k, u_{k+1} = f(y)$ be such that $\|u_{i-1} - u_i\| = L_1d$ for $i = 1,\dots,k$, and $\|u_k - u_{k+1}\| = r$. Then $\|f^{-1}(u_{i-1}) - f^{-1}(u_i)\| \le \frac{d}{2}$ for $i = 1,\dots,k+1$. Therefore $\|x-y\| \le \frac{(k+1)d}{2} \le \frac{1}{L_1}\|f(x)-f(y)\|$. □

Remark 9.8. The same argument works for all metric spaces having something like geodesics of a suitable kind.

Proof of Observation 9.6. Actually we are going to show that the fact that $\delta_{X\hookrightarrow Y}(\varepsilon)$ is positive for some $\varepsilon > 0$ implies the Ribe theorem (Theorem 9.1).

So let Z and Y be uniformly homeomorphic infinite-dimensional Banach spaces. By Lemma 9.7, if $X \subseteq Z$ is any finite-dimensional subspace and $d > 0$, then d-nets in rB_X embed into Y with distortion bounded by some constant C which does not depend on the choice of a finite-dimensional subspace $X \subseteq Z$ and r for every $r > d$. By rescaling, we get that the same assertion holds for δ-nets in B_X for every $\delta \in (0,1)$. By Bourgain's discretization theorem (see also Remark 9.5), X admits a linear embedding into Y with distortion bounded above by $\frac{C}{1-\varepsilon}$. Of course the same argument applies also to finite-dimensional subspaces of Y. □

9.2.1 Proving Bourgain's discretization theorem. Preliminary step: it suffices to consider spaces with differentiable norm

The linear map which we are going to construct in our proof will be obtained as a derivative of a non-linear map. In this connection, it will be convenient for us to assume that the norm of X is differentiable everywhere except 0. The purpose of this section is to show that we may assume this without loss of generality because any norm on $\mathbb{R}^n$ can be arbitrarily well approximated by an infinitely differentiable norm.

Proposition 9.9. *For each norm $\|\cdot\|$ on $\mathbb{R}^n$ and each $\alpha > 0$ there exists a norm $|||\cdot|||$ on $\mathbb{R}^n$ such that*

$$\forall x \in \mathbb{R}^n \quad (1-\alpha)\|x\| \le |||x||| \le (1+\alpha)\|x\|$$

and $|||x|||$ is an infinitely differentiable function on $\mathbb{R}^n$ everywhere except 0.

Proof. Denote by W the normed space $(\mathbb{R}^n, \|\cdot\|)$. Let $\{x_i\}_{i=1}^m$ be an α-net on $\{x \in \mathbb{R}^n : \|x\| = 1\}$. Let $\{x_i^*\}_{i=1}^m \subset W^*$ be such that $\|x_i^*\| = 1$ and $x_i^*(x_i) = 1$. Then

$$\|x\|_1 \stackrel{\text{def}}{=} \max_{1\le i\le m} |x_i^*(x)|$$

is a norm on $\mathbb{R}^n$. It is easy to check that it satisfies

$$\forall x \in \mathbb{R}^n \quad (1-\alpha)\|x\| \le \|x\|_1 \le \|x\|.$$

Let $p \in \mathbb{N}$ be an even integer satisfying $m^{1/p} \le (1+\alpha)$. Then for each collection $\{a_i\}_{i=1}^m$ of real numbers the inequality

$$\max_{1\le i\le m} |a_i| \le \left(\sum_{i=1}^m a_i^p\right)^{\frac{1}{p}} \le (1+\alpha) \max_{1\le i\le m} |a_i|$$

holds. Therefore

$$|||x||| \stackrel{\text{def}}{=} \left(\sum_{i=1}^m (x_i^*(x))^p\right)^{\frac{1}{p}}$$

is a desired norm. All verifications are immediate. □

9.2.2 First step: picking the system of coordinates

From now on $(X, \|\cdot\|_X)$ will be a fixed n-dimensional normed space ($n > 1$), with unit ball $B_X = \{x \in X : \|x\|_X \leq 1\}$ and unit sphere $S_X = \{x \in X : \|x\|_X = 1\}$. We need the following classical result.

Theorem 9.10 (John's theorem). *Let X be an n-dimensional normed space. Then $d(X, \ell_2^n) \leq \sqrt{n}$, where d denotes the Banach–Mazur distance.*

Proof. The proof goes as follows. Let $\mathcal{E}$ be an ellipsoid of maximal volume in B_X. (The volume in X depends on the choice of a basis, so it is defined up to a positive multiplicative constant. Therefore the notion of an ellipsoid of maximal volume inside a compact set in a finite-dimensional space is well-defined.) We pick the system of coordinates in X for which $\mathcal{E}$ is the Euclidean ball. It is clear that to prove Theorem 9.10 it suffices to show that the multiple $\sqrt{n}\,\mathcal{E}$ contains B_X.

Assume the contrary. Let e_1 be a vector for which the Euclidean norm introduced above is equal to 1 and there is $d > \sqrt{n}$ such that $de_1 \in B_X$. We pick an orthonormal basis $\{e_i\}_{i=1}^n$ for the described Euclidean structure in such a way that e_1 is the first vector of this basis. Let $\{t_i\}_{i=1}^n$ be the coordinates of a vector with respect to this basis. We are going to show that there exist numbers $a > 1$ and $b < 1$ such that the ellipsoid

$$\mathcal{C} = \left\{ \sum_{i=1}^n t_i e_i : \left(\frac{t_1}{a}\right)^2 + \sum_{i=2}^n \left(\frac{t_i}{b}\right)^2 \leq 1 \right\} \tag{9.3}$$

is contained in B_X and has a larger volume than $\mathcal{E}$. It is easy to verify that to achieve the second goal we need to have

$$ab^{n-1} > 1. \tag{9.4}$$

Lemma 9.11. *Let $a > 1$ and $b \in (0, 1)$ be such that*

$$\frac{a^2}{d^2} + b^2\left(1 - \frac{1}{d^2}\right) \leq 1. \tag{9.5}$$

Then the ellipsoid $\mathcal{C}$ given by (9.3) *is contained in B_X.*

Proof. Observe that the convex hull of de_1 and $\mathcal{E}$ is contained in B_X. We are going to show that $\mathcal{C}$ is contained in the convex hull of $\mathcal{E} \cup \{de_1\}$. Clearly it suffices to consider the two-dimensional case (just the plane containing the vectors e_1 and e_2). We circumscribe about $\mathcal{E}$ a rhombus R with one of the diagonals being the line segment joining de_1 and $-de_1$. Easy computation shows that the other diagonal is of length $2\sqrt{\frac{d^2}{d^2-1}}$. It is easy to check that since we are concerned only with the case where $b < 1$, it suffices to check that the ellipse $\mathcal{C}$ is contained in the rhombus.

Now we make one more linear transformation, mapping the ellipse $\left(\frac{t_1}{a}\right)^2+\left(\frac{t_2}{b}\right)^2 \le 1$ onto the unit disc in the plane. The rhombus considered above is mapped onto the rhombus with lengths of diagonals equal to $2\frac{d}{a}$ and $\frac{2}{b}\sqrt{\frac{d^2}{d^2-1}}$. The condition for such a rhombus to contain the unit disc is obtained by computing its area in two different ways, and is

$$\sqrt{\left(\frac{d}{a}\right)^2+\left(\frac{1}{b}\right)^2\cdot\frac{d^2}{d^2-1}} \le \frac{d}{a}\cdot\frac{1}{b}\sqrt{\frac{d^2}{d^2-1}}.$$

Squaring both sides and simplifying we get (9.5). □

By Lemma 9.11 it suffices to show that there are numbers $a \in (1,\infty)$ and $b \in (0,1)$ such that the conditions (9.4) and (9.5) are satisfied. For $a \ge 1$ consider the function

$$f(a) = a\left(\frac{d^2-a^2}{d^2-1}\right)^{\frac{n-1}{2}}.$$

This function represents the quantity in the left-hand side of (9.4) for a family of ellipsoids which, by Lemma 9.11, are contained in B_X. Observe that $f(1) = 1$. To complete the proof it suffices to show that $f'(1) > 0$. This is a straightforward computation:

$$f'(a) = \left(\frac{d^2-a^2}{d^2-1}\right)^{\frac{n-1}{2}-1}\cdot\frac{d^2-na^2}{d^2-1}.$$

We have $f'(1) > 0$ because $d^2 > n$. □

Now we return to the proof of Theorem 9.4. We identify X with $\mathbb{R}^n$. By John's theorem (Theorem 9.10) we assume without loss of generality that the standard Euclidean norm $\|\cdot\|_2$ on $\mathbb{R}^n$ satisfies

$$\forall\, x \in X, \quad \frac{1}{\sqrt{n}}\|x\|_2 \le \|x\|_X \le \|x\|_2. \tag{9.6}$$

In what follows, the volume of a Lebesgue measurable set $A \subseteq \mathbb{R}^n$ will be denoted $\mathrm{vol}(A)$.

Recall that the statement which we are proving is: the existence of a low-distortion embedding of a δ-net of B_X into Y implies the existence of a linear embedding of X into Y with a suitable bound. It is clearly enough to prove the statement for sufficiently small δ and ε. Fix $\varepsilon, \delta \in (0, 1/8)$ and let $\mathcal{N}_\delta$ be a fixed δ-net in B_X. We also fix $D \in (1,\infty)$, a Banach space $(Y, \|\cdot\|_Y)$, and a mapping $f : \mathcal{N}_\delta \to Y$ satisfying

$$\forall\, x, y \in \mathcal{N}_\delta, \quad \frac{1}{D}\|x-y\|_X \le \|f(x)-f(y)\|_Y \le \|x-y\|_X. \tag{9.7}$$

By translating f, we assume without loss of generality that $f(\mathcal{N}_\delta) \subseteq 2B_Y$. Our goal will be to show that provided δ is small enough, namely $\delta \le e^{-(D/\varepsilon)^{Cn}}$, there exists an injective linear operator $T : X \to Y$ satisfying $\|T\|\cdot\|T^{-1}\| \le (1+12\varepsilon)D$.

9.2.3 Second step: construction of a Lipschitz almost-extension

The next step is to construct a mapping $F : \mathbb{R}^n \to Y$ that is a Lipschitz almost-extension of f, that is, F is Lipschitz and on $\mathcal{N}_\delta$, it takes values that are close to the corresponding values of f. As a preliminary step we prove the following lemma.

Lemma 9.12. *Let $K \subseteq \mathbb{R}^n$ be a convex set and fix $\tau, \eta, L \in (0, \infty)$. Assume that we are given a mapping $h : K + \tau B_X \to Y$ satisfying*

$$\|h(x) - h(y)\|_Y \le L\,(\|x - y\|_X + \eta) \tag{9.8}$$

for all $x, y \in K + \tau B_X$. Define $H : K \to Y$ by

$$H(x) = \frac{1}{\tau^n \operatorname{vol}(B_X)} \int_{\tau B_X} h(x - y)dy.$$

Then $\|H(x) - H(y)\|_Y \le L\left(1 + \frac{n\eta}{2\tau}\right)\|x - y\|_X$ for all $x, y \in K$.

Proof. Consider two points in K. Translating K if needed, we may assume that one of the points is the origin and denote the other one by x. Put $B = \{z \in X : \|z\| \le \tau\}$ and $B_x = \{z \in X : \|z - x\| \le \tau\}$. Then $\tau^n \operatorname{vol}(B_X) = \operatorname{vol} B$ and

$$\begin{aligned} H(0) - H(x) &= \frac{1}{\operatorname{vol} B}\int_B h(y)dy - \frac{1}{\operatorname{vol} B}\int_{B_x} h(y)dy \\ &= \frac{1}{\operatorname{vol} B}\left(\int_{B\setminus B_x} h(y)dy - \int_{B_x\setminus B} h(y)dy\right). \end{aligned}$$

Consider the family $\{L_t = t + \mathbb{R}x\}_{t\in X}$ of straight lines parallel to x. Each L_t intersect the sets $B\setminus B_x$ and $B_x\setminus B$ in intervals of the same length (which does not exceed $\|x\|$, and is exactly $\|x\|$ in the case where L_t intersects $B \cap B_x$). Observe that here by the *length* of a line segment $[u, v]$ we mean $\|u - v\|_X$.

Let $c : (B\setminus B_x) \to (B_x\setminus B)$ be the transformation which translates each interval $L_t \cap (B\setminus B_x)$ onto the interval $L_t \cap (B_x\setminus B)$ along L_t. It is clear that it is a measure-preserving transformation.

Fix a hyperplane $Z \subset X$ such that $x \notin Z$ and let P be the projection from X to Z along x, and $w(t)$ $(t \in X)$ be the length of $L_t \cap B$. The Lebesgue measure on X is the product of any two appropriately normalized measures on Z and on the line $\mathbb{R} \cdot x$. We normalize the measure on $\mathbb{R} \cdot x$ by setting the measure of $[0, x]$ to be equal $\|x\|$. This and the choice of the standard Lebesgue measure on $\mathbb{R}^n$ determine the Lebesgue measure, which we denote ν, on Z, and we have

$$\int_{P(B)} w(z)d\nu(z) = \operatorname{vol} B.$$

Consider B as the disjoint union of two sets ("wide" and "narrow"):

$$\begin{aligned} W &:= \{t \in B : L_t \cap B \cap B_x \ne \emptyset\}, \\ N &:= \{t \in B : L_t \cap B \cap B_x = \emptyset\}. \end{aligned}$$

Clearly we have $N \subset (B \backslash B_x)$. It is easy to see that the length of the intersection $L_t \cap (B \backslash B_x)$ is equal to $\|x\|$ for $t \in W$ and is equal to $w(t)$ for $t \in N$. Also we have

$$\|t - c(t)\| = \begin{cases} w(t), & t \in W \cap (B \backslash B_x), \\ \|x\|, & t \in N. \end{cases}$$

We get

$$\begin{aligned}
\operatorname{vol} B \, \|H(0) - H(x)\| &= \left\| \int_{B \backslash B_x} h(y)dy - \int_{B_x \backslash B} h(y)dy \right\| \\
&= \left\| \int_{B \backslash B_x} (h(y) - h \circ c(y))dy \right\| \\
&\le \int_{B \backslash B_x} \|h(y) - h \circ c(y)\| dy \qquad (9.9)\\
&\overset{(9.8)}{\le} L \int_{B \backslash B_x} (\|y - c(y)\| + \eta)\, dy \\
&= L\left(\eta \operatorname{vol}(B \backslash B_x) + \int_{B \backslash B_x} \|y - c(y)\| dy \right),
\end{aligned}$$

where the second equality holds because c is a measure preserving map of $B \backslash B_x$ onto $B_x \backslash B$. The integral in the last line will be estimated separately on N and on $W \cap (B \backslash B_x)$:

$$\int_N \|y - c(y)\| dy = \int_N \|x\| dy = \|x\| \operatorname{vol} N,$$

$$\begin{aligned}
\int_{W \cap (B \backslash B_x)} \|y - c(y)\| dy &= \int_{W \cap (B \backslash B_x)} w(y)dy \\
&= \int_{P(W \cap (B \backslash B_x))} w(t)\|x\| d\nu(t) \\
&= \|x\| \int_{P(W \cap (B \backslash B_x))} w(t) d\nu(t) \\
&= \|x\| \operatorname{vol} W,
\end{aligned}$$

where the second equality follows from the normalization of measures and from the fact that each intersection of $W \cap (B \backslash B_x)$ and L_t is an interval of length $\|x\|$. Hence

$$\int_{B \backslash B_x} \|t - c(t)\| dy(t) = \|x\|(\operatorname{vol} W + \operatorname{vol} N) = \|x\| \operatorname{vol} B. \qquad (9.10)$$

Now we estimate $\operatorname{vol}(B \backslash B_x)$ in terms of $\|x\|$. Here we use the easily verified inclusion:

$$(B \backslash B_x) \cup (B_x \backslash B) \subset B\left(\frac{x}{2}, \tau + \frac{\|x\|}{2}\right) \Big\backslash B\left(\frac{x}{2}, \tau - \frac{\|x\|}{2}\right).$$

This inclusion in combination with $\mathrm{vol}(B\backslash B_x) = \mathrm{vol}(B_x\backslash B)$ implies

$$\begin{aligned}\mathrm{vol}(B\backslash B_x) &\le \frac{\mathrm{vol}B}{2\tau^n}\left(\left(\tau+\frac{\|x\|}{2}\right)^n-\left(\tau-\frac{\|x\|}{2}\right)^n\right)\\ &= \frac{n\mathrm{vol}B}{2\tau}\|x\|+o(\|x\|) \quad \text{as} \quad \|x\|\to 0.\end{aligned}$$

We combine this estimate with (9.9) and (9.10), and get

$$\|H(0)-H(x)\| \le L\left(1+\frac{n\eta}{2\tau}\right)\|x\|+o(\|x\|).$$

Since K is convex, we can derive the desired estimate from this estimate using the same argument as in Lemma 9.7: consider a sequence of points $v_0 = x, v_1, \dots, v_{m-1}$, $v_m = y$ in K such that $v_i - v_{i-1} = \frac{y-x}{m}$ for $i = 1, \dots, m$. Then

$$\begin{aligned}\|H(y)-H(x)\| &\le \limsup_{m\to\infty}\sum_{i=1}^{m}\|H(v_i)-H(v_{i-1})\|\\ &\le \limsup_{m\to\infty}\sum_{i=1}^{m}\left(L\left(1+\frac{n\eta}{2\tau}\right)\|v_i-v_{i-1}\|+o\left(\frac{1}{m}\right)\right)\\ &= L\left(1+\frac{n\eta}{2\tau}\right)\|y-x\|.\end{aligned}$$

□

Now we turn to the construction of a Lipschitz almost-extension of f mentioned before Lemma 9.12.

Lemma 9.13. *If $\delta < \frac{\varepsilon}{4n}$, then there exists a map $F : \mathbb{R}^n \to Y$ which is differentiable everywhere on $\mathbb{R}^n$ and has the following properties.*

(a) *F is supported on $3B_X$.*

(b) $\|F(x)-F(y)\|_Y \le 8\|x-y\|_X$ *for all $x, y \in \mathbb{R}^n$.*

(c) $\|F(x)-F(y)\|_Y \le (1+\varepsilon)\|x-y\|_X$ *for all $x, y \in \frac{1}{2}B_X$.*

(d) $\|F(x)-f(x)\|_Y \le \frac{13n\delta}{\varepsilon}$ *for all $x \in \mathcal{N}_\delta$.*

Proof. We deduce Lemma 9.13 from Lemma 9.12 using partition of unity argument. Let $\{\phi_p : \mathbb{R}^n \to [0,1]\}_{p\in\mathcal{N}_\delta}$ be a family of smooth functions satisfying $\sum_{p\in\mathcal{N}_\delta}\phi_p(x) = 1$ for all $x \in \left(1+\frac{\delta}{2}\right)B_X$ and $\phi_p(x) = 0$ for all $(p,x) \in \mathcal{N}_\delta \times \mathbb{R}^n$ with $\|x-p\|_X \ge 2\delta$. One of the standard constructions of such functions is obtained by taking a smooth function $\psi : \mathbb{R}^n \to [0,1]$ which equals 1 on $\frac{3}{2}B_X$ and vanishes outside $2B_X$, and defining $\psi_p(x) = \psi((x-p)/\delta)$ for $(p,x) \in \mathcal{N}_\delta \times \mathbb{R}^n$. Then we order elements of $\mathcal{N}_\delta = \{p_1, p_2, \ldots, p_N\}$ and define

$$\phi_{p_1} = \psi_{p_1},$$
$$\phi_{p_j} = \psi_{p_j} \prod_{i=1}^{j-1} (1 - \psi_{p_i}) \quad \text{for} \quad j \in \{2, \ldots, N\}.$$

It is easy to verify (for example, by induction) that the identity

$$\sum_{p \in \mathcal{N}_\delta} \phi_p = 1 - \prod_{p \in \mathcal{N}_\delta} (1 - \psi_p)$$

holds. Since every $x \in \left(1 + \frac{\delta}{2}\right) B_X$ satisfies $\|x - p\|_X \le \frac{3}{2}\delta$ for some $p \in \mathcal{N}_\delta$, this implies $\sum_{p \in \mathcal{N}_\delta} \phi_p(x) = 1$ for each $x \in \left(1 + \frac{\delta}{2}\right) B_X$.

We define $g : \left(1 + \frac{\delta}{2}\right) B_X \to Y$ by

$$g(x) = \sum_{p \in \mathcal{N}_\delta} \phi_p(x) f(p). \tag{9.11}$$

Let $\beta : [0, \infty) \to [0, 1]$ be a differentiable 2-Lipschitz non-increasing function which satisfies $\beta(t) = 1$ for $t \in [0, 1]$ and $\beta(t) = 0$ for $t \in [2, \infty)$. Let $r : X \to [1, \infty)$ be a differentiable function satisfying the conditions $r(x) = 1$ for $x \in B_X$ and

$$\frac{\|x\|_X}{\left(1 + \frac{\delta}{2}\right)} \le r(x) \le \|x\|_X$$

for $x \in \mathbb{R}^n \setminus B_X$.

Now we introduce the map $h : \mathbb{R}^n \to Y$ by

$$h(x) = \beta(\|x\|_X) g\left(\frac{x}{r(x)}\right). \tag{9.12}$$

It is clear that h is everywhere differentiable (recall that we assume that $\|x\|_X$ is differentiable everywhere except 0). Observe that our definitions imply that for $x \in \left(1 + \frac{\delta}{2}\right) B_X$ we have

$$\sum_{p \in \mathcal{N}_\delta \cap (x + 2\delta B_X)} \phi_p(x) = 1 \quad \text{and} \quad g(x) = \sum_{p \in \mathcal{N}_\delta \cap (x + 2\delta B_X)} \phi_p(x) f(p).$$

Therefore for $x, y \in \left(1 + \frac{\delta}{2}\right) B_X$ we have

$$\begin{aligned} g(x) - g(y) &= \sum_{p \in \mathcal{N}_\delta \cap (x + 2\delta B_X)} \phi_p(x) f(p) - \sum_{q \in \mathcal{N}_\delta \cap (y + 2\delta B_X)} \phi_q(y) f(q) \\ &= \sum_{\substack{p \in \mathcal{N}_\delta \cap (x + 2\delta B_X) \\ q \in \mathcal{N}_\delta \cap (y + 2\delta B_X)}} \phi_p(x) \phi_q(y) (f(p) - f(q)). \end{aligned} \tag{9.13}$$

Since for each nonzero term in the last sum we have $\|p - q\| \leq \|x - y\| + 4\delta$, combining (9.7) and (9.13) we get

$$\forall x, y \in \left(1 + \frac{\delta}{2}\right) B_X, \quad \|g(x) - g(y)\|_Y \leq \|x - y\|_X + 4\delta. \tag{9.14}$$

For $x, y \in B_X$ we have $h(x) = g(x)$ and $h(y) = g(y)$, therefore

$$\forall x, y \in B_X, \quad \|h(x) - h(y)\|_Y \leq \|x - y\|_X + 4\delta. \tag{9.15}$$

If $x \in B_X$ and $y \in \mathbb{R}^n \setminus B_X$, we use $f(\mathcal{N}_\delta) \subseteq 2B_Y$ and the fact that β is 2-Lipschitz, and get

$$\begin{aligned}
\|h(x) - h(y)\|_Y &\leq \left\| g(x) - g\left(\frac{y}{r(y)}\right) \right\|_Y + (1 - \beta(\|y\|_X)) \left\| g\left(\frac{y}{r(y)}\right) \right\|_Y \\
&\overset{(9.14)}{\leq} \left\| x - \frac{y}{r(y)} \right\|_X + 4\delta + 2\left(\|y\|_X - 1\right) \sup_{p \in \mathcal{N}_\delta} \|f(p)\|_Y \\
&\leq \|x - y\|_X + \left\| y - \frac{y}{r(y)} \right\|_X + 4\delta + 4\left(\|y\|_X - 1\right) \\
&\leq \|x - y\|_X + 5(\|y\|_X - 1) + 4\delta.
\end{aligned}$$

Since $\|y\|_X - 1 \leq \|x - y\|_X + \|x\|_X - 1 \leq \|x - y\|_X$, it follows that

$$\forall x \in B_X, \forall y \in \mathbb{R}^n \setminus B_X, \quad \|h(x) - h(y)\|_Y \leq 6\|x - y\|_X + 4\delta. \tag{9.16}$$

If $x, y \in \mathbb{R}^n \setminus B_X$ then

$$\begin{aligned}
\|h(x) - h(y)\|_Y &\leq \left\| g\left(\frac{x}{r(x)}\right) - g\left(\frac{y}{r(y)}\right) \right\|_Y \beta(\|x\|_X) \\
&\quad + \left\| g\left(\frac{y}{r(y)}\right) \right\|_Y |\beta(\|x\|_X) - \beta(\|y\|_X)| \\
&\overset{(9.14)}{\leq} \left\| \frac{x}{r(x)} - \frac{y}{r(y)} \right\|_X + 4\delta + 4\|x - y\|_X \\
&\leq \left\| \frac{x}{\|x\|_X} - \frac{y}{\|y\|_X} \right\|_X + \left\| \frac{x}{r(x)} - \frac{x}{\|x\|_X} \right\|_X \\
&\quad + \left\| \frac{y}{r(y)} - \frac{y}{\|y\|_X} \right\|_X + 4\delta + 4\|x - y\|_X \\
&\leq 2\|x - y\|_X + \frac{\delta}{2} + \frac{\delta}{2} + 4\delta + 4\|x - y\|_X \\
&= 6\|x - y\|_X + 5\delta.
\end{aligned} \tag{9.17}$$

(See Exercise 6.27 in connection with the last inequality.)

Set $\tau = 2n\delta/\varepsilon \in (0, 1/2)$ and define for $x \in \mathbb{R}^n$,

$$F(x) = \frac{1}{\tau^n \operatorname{vol}(B_X)} \int_{\tau B_X} h(x-y)dy. \tag{9.18}$$

Since h is differentiable everywhere and its derivative may be assumed to be continuous, it follows from (9.18) that F is differentiable everywhere. Since h is supported on $2B_X$, the function F is supported on $(2+\tau)B_X \subseteq 3B_X$, therefore the condition (a) of Lemma 9.13 is satisfied. Due to (9.15), (9.16), and (9.17), an application of Lemma 9.12 with $K = \mathbb{R}^n$, $L = 6$ and $\eta = \delta$ shows that F is $6\big(1+\frac{\varepsilon}{4}\big)$-Lipschitz on $\mathbb{R}^n$, proving the assertion (b) of Lemma 9.13. Due to (9.15), an application of Lemma 9.12 with $K = (1-\tau)B_X$ shows that F is $(1+\varepsilon)$-Lipschitz on $(1-\tau)B_X \supseteq \frac{1}{2}B_X$. This establishes the assertion (c) of Lemma 9.13. To prove assertion (d) of Lemma 9.13, fix $x \in \mathcal{N}_\delta$. Then,

$$\begin{aligned} \|F(x) - h(x)\|_Y &\le \frac{1}{\tau^n \operatorname{vol}(B_X)} \int_{\tau B_X} \|h(x-y) - h(x)\|_Y \, dy \\ &\overset{(9.15)\ \&\ (9.16)}{\le} 6\tau + 4\delta. \end{aligned} \tag{9.19}$$

Also,

$$\begin{aligned} \|h(x) - f(x)\|_Y &\le \sum_{p \in \mathcal{N}_\delta} \|f(p) - f(x)\|_Y \phi_p(x) \\ &\le \max_{p \in \mathcal{N}_\delta \cap (x + 2\delta B_X)} \|f(x) - f(p)\|_Y \\ &\overset{(9.7)}{\le} 2\delta. \end{aligned} \tag{9.20}$$

Recalling that $\tau = 2n\delta/\varepsilon$, assertion (d) of Lemma 9.13 follows from (9.19) and (9.20), if $n \ge 6\varepsilon$. □

9.2.4 Third step: further smoothing of the map using Poisson kernels

The brief outline of the rest of the proof is: we smoothen the map F further using the convolution with the Poisson kernel. Then we show that the obtained map has a point of differentiability such that the derivative at that point is the desired linear map.

Now we provide a very short summary of information on the Poisson kernel, in Notes and Remarks the reader can find references to sources containing more details. The Poisson kernel arises in the solution of the following problem. Let $f \in L_2(\mathbb{R}^n)$ (where $\mathbb{R}^n$ is endowed with the Lebesgue measure). The problem is to find a function in $\mathbb{R}^{n+1}_+ := \{(x,t) : x = \{x_i\}_{i=1}^n \in \mathbb{R}^n, t > 0\}$ satisfying the Laplace equation

$$\Delta u := \frac{\partial^2 u}{\partial t^2} + \sum_{i=1}^n \frac{\partial^2 u}{\partial x_i^2} = 0 \tag{9.21}$$

such that the functions $\{u(x,t)\}_{t>0}$ converge to $f(x)$ in $L_2(\mathbb{R}^n)$ as $t \downarrow 0$.

The solution of this problem can be written in the form

$$u(x,t) = \int_{\mathbb{R}^n} \widehat{f}(z) e^{-2\pi i \langle z,x\rangle} e^{-2\pi \|z\|_2 t} dz, \quad t > 0, \tag{9.22}$$

where $\widehat{f}$ is the Fourier transform of f. One can check that the integral in (9.22) converges absolutely. Straightforward differentiation shows that the function $v(x,t,z) := e^{-2\pi i \langle z,x\rangle} e^{-2\pi \|z\|_2 t}$ satisfies the equation $\frac{\partial^2 v}{\partial t^2} + \sum_{i=1}^n \frac{\partial^2 v}{\partial x_i^2} = 0$ for each fixed z. Therefore the function u given by (9.22) satisfies (9.21). The L^2 theory of Fourier transform implies that $\{u(x,t)\}_{t>0}$ converge to $f(x)$ in $L_2(\mathbb{R}^n)$ as $t \downarrow 0$.

It is easy to verify that the formula (9.22) can be rewritten as a convolution

$$u(x,t) = P_t * f := \int_{\mathbb{R}^n} P_t(s) f(x-s) ds, \tag{9.23}$$

where the *Poisson kernel* $P_t(s)$ is given by

$$P_t(s) = \int_{\mathbb{R}^n} e^{-2\pi i \langle z,s\rangle} e^{-2\pi \|z\|_2 t} dz, \quad t > 0. \tag{9.24}$$

Recall that a *convolution* of two functions on $\mathbb{R}^n$ (the same definition is used for any locally compact Abelian group with operations denoted by $+$ and $-$) is defined by

$$f * g(x) = \int_{\mathbb{R}^n} f(x-y) g(y) dy. \tag{9.25}$$

Using substitution one can easily see that

$$f * g(x) = g * f(x) = \int_{\mathbb{R}^n} f(y) g(x-y) dy. \tag{9.26}$$

Using quite nontrivial classical computation with integrals (see Notes and Remarks for references) one can find an explicit formula for the Poisson kernel:

$$P_t(x) = \frac{c_n t}{\left(t^2 + \|x\|_2^2\right)^{\frac{n+1}{2}}}, \quad \text{where} \quad c_n = \frac{\Gamma\left(\frac{n+1}{2}\right)}{\pi^{\frac{n+1}{2}}}. \tag{9.27}$$

Some properties of Poisson kernels:

(a) Formula (9.27) shows that $P_t(x) > 0$.

(b) Applying the Fourier inversion formula to (9.24) we get that for each fixed t the Fourier transform $\widehat{P_t}(x)$ is equal to $e^{-2\pi \|x\|_2 t}$. As a special case of this formula we get

$$\int_{\mathbb{R}^n} P_t(x) dx = \widehat{P_t}(0) = 1. \tag{9.28}$$

(c) The semigroup property $P_t * P_s = P_{t+s}$ for $t, s > 0$. This property can be derived from (b) and the well-known fact that Fourier transform maps convolutions onto products.

Assume from now on that $\delta < \frac{\varepsilon}{4n}$ and fix a mapping $F : \mathbb{R}^n \to Y$ satisfying the conclusion of Lemma 9.13. We consider the functions $P_t * F : \mathbb{R}^n \to Y$ given by $P_t * F(x) = \int_{\mathbb{R}^n} P_t(x-y)F(y)dy$ (it is a convolution of a scalar-valued and a vector-valued function). Observe that the convolution $P_t * F$ is everywhere differentiable (we can differentiate under the integral sign and P_t is an everywhere differentiable function). The derivative $T = (P_{t_0} * F)'(x)$ is a linear map of X into Y. Our goal is to show that there exists $t_0 \in (0, \infty)$ and $x \in \mathbb{R}^n$ such that the derivative $T = (P_{t_0} * F)'(x)$ is injective and satisfies $\|T\| \cdot \|T^{-1}\| \leq (1 + 12\varepsilon)D$. Since the functions $P_t * F$ converge to F in $L_2(\mathbb{R}^n, Y)$ (this property of Poisson kernels mentioned above remains true for vector-valued functions) and F itself is close to a bilipschitz map when restricted to the δ-net $\mathcal{N}_\delta$, one might expect this to happen for every small enough t. However, so direct a proof of the existence of such T is not known so far. The known proof of existence of t_0 is by contradiction, and it does not show how to get a point t_0 for which $(P_{t_0} * F)'(x)$ has the desired properties.

The estimate of the norm of $(P_{t_0} * F)'(x)$ from above is relatively easy, it is based on the following lemma.

Lemma 9.14. *Assume that* $0 < t < \frac{\varepsilon}{32\sqrt{n}}$. *Then for every* $x, y \in \frac{1}{4} B_X$ *we have*

$$\|P_t * F(x) - P_t * F(y)\|_Y \leq (1 + 2\varepsilon)\|x - y\|_X.$$

It is convenient to postpone the proof of this lemma to Section 9.2.5, where we present all of the required estimates for Poisson kernels.

The search for points x and values of t_0 for which the derivative $(P_{t_0} * F)'(x)$ satisfies the suitable lower estimates is much more complicated. We use the notation ∂_a $(a \in X)$ for the directional derivative of a map $\Phi : X \to Y$, that is,

$$\partial_a(\Phi)(x) = \lim_{t \to 0,\, t \in \mathbb{R}} \frac{\Phi(x + ta) - \Phi(x)}{t}, \quad a \in X.$$

As for the lower estimates, we prove the following estimate in average.

Lemma 9.15. *Under suitable conditions on* $t \in (0, 1/2]$, $R \in (0, \infty)$ *and* $\delta \in \left(0, \frac{\varepsilon}{4n}\right)$, *for every* $x \in \frac{1}{8} B_X$ *and* $a \in S_X$ *we have*

$$\left(\|\partial_a(P_t * F)\|_Y * P_{Rt}\right)(x) \geq \frac{1 - \varepsilon}{D}. \tag{9.29}$$

The "suitable conditions" mentioned in the previous sentence are

$$\delta \leq \frac{\varepsilon t \ln(3/t)}{2\sqrt{n}} \leq \frac{\varepsilon^4}{6n^{5/2}(80D)^2}, \tag{9.30}$$

and

$$\frac{1360n^{3/2}D^2\ln(3/t)}{\varepsilon^2} \le R \le \frac{\varepsilon}{48t\sqrt{n}}. \tag{9.31}$$

The proof of Lemma 9.15 is also postponed to Section 9.2.5.

Remark 9.16. The "suitable conditions" in Lemma 9.15 are the conditions under which the argument below works. The important point is that the considered parameters can satisfy all of them, we will see this below for the selected values of the parameters.

To get out of the "directional" lower estimates of Lemma 9.15 a lower estimate in "all directions" at some point x for some value of t we use the following lemma about comparison of averages over directions.

Lemma 9.17. *Let μ be a Borel probability measure on S_X. Fix $R, A \in (0,\infty)$ and $m \in \mathbb{N}$. Then there exists $t \in (0,\infty)$ satisfying*

$$A \cdot (R+1)^{-m-1} \le t \le A, \tag{9.32}$$

such that

$$\begin{aligned}\int_{S_X}\int_{\mathbb{R}^n} &\|\partial_a(P_t * F)(x)\|_Y\,dx\,d\mu(a)\\ &\le \int_{S_X}\int_{\mathbb{R}^n} \|\partial_a(P_{(R+1)t} * F)(x)\|_Y\,dx\,d\mu(a) + \frac{8\mathrm{vol}(3B_X)}{m}.\end{aligned} \tag{9.33}$$

Proof. If (9.33) fails for all t satisfying (9.32) then for every $k \in \{0,\dots,m+1\}$ we have

$$\begin{aligned}\int_{S_X}\int_{\mathbb{R}^n} &\left\|\partial_a\left(P_{A(R+1)^{k-m-1}} * F\right)(x)\right\|_Y dx\,d\mu(a)\\ &> \int_{S_X}\int_{\mathbb{R}^n} \left\|\partial_a\left(P_{A(R+1)^{k-m}} * F\right)(x)\right\|_Y dx\,d\mu(a) + \frac{8\mathrm{vol}(3B_X)}{m}.\end{aligned} \tag{9.34}$$

By iterating (9.34) we get the estimate

$$\begin{aligned}\int_{S_X}\int_{\mathbb{R}^n} &\left\|\partial_a\left(P_{A(R+1)^{-m-1}} * F\right)(x)\right\|_Y dx\,d\mu(a)\\ &> \int_{S_X}\int_{\mathbb{R}^n} \left\|\partial_a\left(P_{A(R+1)} * F\right)(x)\right\|_Y dx\,d\mu(a) + \frac{8(m+1)\mathrm{vol}(3B_X)}{m}.\end{aligned} \tag{9.35}$$

At the same time, F is differentiable everywhere and is 8-Lipschitz. Therefore for every $a \in S_X$ we have $\|\partial_a F\|_Y \le 8$ everywhere. Now we use the commutativity of convolution (9.26) and get

$$\begin{aligned}\partial_a\left(P_{A(R+1)^{-m-1}} * F\right)(x) &= \partial_a\left(\int_{\mathbb{R}^n} P_{A(R+1)^{-m-1}}(y)F(x-y)dy\right)\\ &= \int_{\mathbb{R}^n} P_{A(R+1)^{-m-1}}(y)\partial_a F(x-y)dy\\ &= (P_{A(R+1)^{-m-1}} * \partial_a F)(x).\end{aligned}$$

Since F is supported on $3B_X$, it follows that

$$\begin{aligned}&\int_{\mathbb{R}^n} \left\|\partial_a\left(P_{A(R+1)^{-m-1}} * F\right)(x)\right\|_Y dx\\ &\quad= \int_{\mathbb{R}^n} \left\|\left(P_{A(R+1)^{-m-1}} * \partial_a F\right)(x)\right\|_Y dx\\ &\quad\le \int_{\mathbb{R}^n}\int_{\mathbb{R}^n} P_{A(R+1)^{-m-1}}(x-y)\|\partial_a F(y)\|_Y dxdy\\ &\quad\overset{(9.28)}{=} \int_{3B_X} \|\partial_a F(y)\|_Y dy \le 8\mathrm{vol}(3B_X).\end{aligned} \tag{9.36}$$

We integrate (9.36) with respect to μ. Since μ is a probability measure we get a contradiction with (9.35) □

We also need the following well-known lemma on cardinalities of nets in balls of finite-dimensional normed spaces.

Lemma 9.18. *Let O be an arbitrary subset of the unit ball of an n-dimensional normed linear space X. Let $\alpha > 0$. Then there is an α-net for O with cardinality $\le \left(1+\frac{2}{\alpha}\right)^n$.*

Proof. Let $\{x_i\}_{i=1}^M$ be a maximal subset in O satisfying $\|x_i - x_j\| \ge \alpha$ for $i \ne j$. It is clear that $\{x_i\}_{i=1}^M$ is an α-net. It is easy to see that the open balls $U_i = \left\{x \in X : \|x - x_i\| < \frac{\alpha}{2}\right\}$ are disjoint and are contained in

$$U = \left\{x \in X : \|x\| < 1 + \frac{\alpha}{2}\right\}.$$

Therefore

$$\sum_{i=1}^M \mathrm{vol}U_i \le \mathrm{vol}U$$

or

$$M\left(\frac{\alpha}{2}\right)^n \mathrm{vol}B_X \le \left(1+\frac{\alpha}{2}\right)^n \mathrm{vol}B_X.$$

The desired inequality follows. □

Now we show how to conclude the proof of Theorem 9.4 using all lemmas stated above. (It will then remain to prove Lemmas 9.15 and 9.14, see Section 9.2.5.)

Proof of Theorem 9.4. Assume that $\delta \in (0,1)$ satisfies

$$\delta \le \left(\frac{\varepsilon}{cD}\right)^{(cD/\varepsilon)^{2n}}, \tag{9.37}$$

where $c = 300$ (this is an overestimate for the ensuing calculation). It is easy to see that since D can be chosen to be $\le 2c_Y(X)$, estimate (9.37) implies (9.1). We fix an (ε/D)-net $\mathcal{F}$ in S_X with $|\mathcal{F}| \le \left(1+\frac{2D}{\varepsilon}\right)^n < \left(\frac{3D}{\varepsilon}\right)^n$. Let μ be the uniform probability measure on $\mathcal{F}$. Let

$$A = \left(\frac{\varepsilon}{cD}\right)^{5n}, \quad R = \left(\frac{cD}{\varepsilon}\right)^{4n} - 1, \quad m = \left\lfloor \left(\frac{cD}{\varepsilon}\right)^{n+1} \right\rfloor - 1. \tag{9.38}$$

Applying Lemma 9.17 with these parameters, we get the existence of $t \in (0,\infty)$ satisfying

$$\left(\frac{\varepsilon}{cD}\right)^{(cD/\varepsilon)^{2n}} \le t \le \left(\frac{\varepsilon}{cD}\right)^{5n}, \tag{9.39}$$

such that

$$\begin{aligned}\sum_{a\in\mathcal{F}} \int_{\mathbb{R}^n} &\|\partial_a(P_t * F)(x)\|_Y dx \\ &\le \sum_{a\in\mathcal{F}} \int_{\mathbb{R}^n} \|\partial_a(P_{(R+1)t} * F)(x)\|_Y dx + \frac{8|\mathcal{F}|\mathrm{vol}(3B_X)}{m}.\end{aligned} \tag{9.40}$$

Straightforward verification shows that for δ satisfying (9.37), R as in (9.38), and any t satisfying (9.39), inequalities (9.30) and (9.31) are satisfied for $n \ge 2$ (for $n = 1$ the statement of Bourgain's discretization theorem is obvious). Thus the conclusion (9.29) of Lemma 9.15 holds true for all $a \in S_X$ and $x \in \frac{1}{8}B_X$.

Note that by convexity of the norm and properties of the convolution (commutativity: $f*g = g*f$ and associativity: $f*(g*h) = (f*g)*h$) and of the Poisson kernel (positivity and semigroup property) we have for every $a \in S_X$ and every $x \in \mathbb{R}^n$:

$$\begin{aligned}\|\partial_a(P_{(R+1)t} * F)(x)\|_Y &= \|\partial_a(P_{Rt} * (P_t * F))(x)\|_Y \\ &= \|\left(P_{Rt} * (\partial_a(P_t * F))\right)(x)\|_Y \\ &\le \left(\|\partial_a(P_t * F)\|_Y * P_{Rt}\right)(x).\end{aligned}$$

Thus $\|\partial_a(P_t * F)\|_Y * P_{Rt} - \|\partial_a(P_{(R+1)t} * F)\|_Y \ge 0$ and we have

$$\begin{aligned}&\frac{\varepsilon}{D}\mathrm{vol}\left\{x \in \frac{1}{8}B_X : \left(\|\partial_a(P_t * F)\|_Y * P_{Rt}\right)(x) - \|\partial_a(P_{(R+1)t} * F)(x)\|_Y \ge \frac{\varepsilon}{D}\right\} \\ &\quad\le \left(\int_{\mathbb{R}^n} \left(\left(\|\partial_a(P_t * F)\|_Y * P_{Rt}\right)(x) - \|\partial_a(P_{(R+1)t} * F)(x)\|_Y\right) dx\right).\end{aligned}$$

Therefore for each $a \in S_X$

$$\begin{aligned}
&\operatorname{vol}\Big\{x \in \frac{1}{8}B_X : \\
&\qquad (\|\partial_a(P_t * F)\|_Y * P_{Rt})(x) - \|\partial_a(P_{(R+1)t} * F)(x)\|_Y \geq \frac{\varepsilon}{D}\Big\} \\
&\leq \frac{D}{\varepsilon}\left(\int_{\mathbb{R}^n} \big((\|\partial_a(P_t * F)\|_Y * P_{Rt})(x) - \|\partial_a(P_{(R+1)t} * F)(x)\|_Y\big)\, dx\right) \\
&= \frac{D}{\varepsilon}\left(\int_{\mathbb{R}^n} \|\partial_a(P_t * F)(x)\|_Y dx - \int_{\mathbb{R}^n} \|\partial_a(P_{(R+1)t} * F)(x)\|_Y dx\right).
\end{aligned} \tag{9.41}$$

Hence,

$$\begin{aligned}
&\operatorname{vol}\Big\{x \in \frac{1}{8}B_X : \\
&\qquad \exists a \in \mathcal{F},\ (\|\partial_a(P_t * F)\|_Y * P_{Rt})(x) - \|\partial_a(P_{(R+1)t} * F)(x)\|_Y \geq \frac{\varepsilon}{D}\Big\} \\
&\overset{(9.41)}{\leq} \frac{D}{\varepsilon}\left(\sum_{a\in\mathcal{F}}\int_{\mathbb{R}^n} \|\partial_a(P_t * F)(x)\|_Y dx - \sum_{a\in\mathcal{F}}\int_{\mathbb{R}^n} \|\partial_a(P_{(R+1)t} * F)(x)\|_Y dx\right) \\
&\overset{(9.40)}{\leq} \frac{D}{\varepsilon}\cdot\frac{8|\mathcal{F}|\operatorname{vol}(3B_X)}{m} \\
&\overset{(9.38)}{\leq} \frac{16D}{\varepsilon}\left(\frac{3D}{\varepsilon}\right)^n\left(\frac{\varepsilon}{cD}\right)^{n+1}(24)^n\operatorname{vol}\left(\frac{1}{8}B_X\right) \\
&= \frac{16}{300}\cdot\left(\frac{6}{25}\right)^n\operatorname{vol}\left(\frac{1}{8}B_X\right) \\
&< \operatorname{vol}\left(\frac{1}{8}B_X\right).
\end{aligned}$$

(In the inequality where we use (9.38), we also use our estimate for the cardinality of $\mathcal{F}$ and the observation that for $\varepsilon \in (0,1)$, $D \in (1,\infty)$ and $c = 300$, the inequality $\frac{1}{m} \leq 2(\frac{\varepsilon}{cD})^{n+1}$ holds.) Consequently, there exists $x \in \frac{1}{8}B_X$ satisfying

$$\forall a \in \mathcal{F},\quad (\|\partial_a(P_t * F)\|_Y * P_{Rt})(x) - \|\partial_a(P_{(R+1)t} * F)(x)\|_Y < \frac{\varepsilon}{D}. \tag{9.42}$$

But we already argued that (9.29) holds as well, so (9.42) implies that

$$\forall a \in \mathcal{F},\quad \|\partial_a(P_{(R+1)t} * F)(x)\|_Y \geq \frac{1-2\varepsilon}{D}. \tag{9.43}$$

Note that by (9.38) and (9.39) we have $(R+1)t \leq (\varepsilon/(cD))^n < \varepsilon/\big(32\sqrt{n}\big)$. Now we let $T = (P_{(R+1)t} * F)'(x)$. By Lemma 9.14 we have $\|T\| \leq 1 + 2\varepsilon$, just because the Lipschitz condition implies the corresponding bound on the derivative. On the other hand, by (9.43), $\|Ta\|_Y \geq (1-2\varepsilon)/D$ for all $a \in \mathcal{F}$. For $z \in S_X$ take $a \in \mathcal{F}$ such that $\|z-a\|_X \le \varepsilon/D$. Then,

$$\|Tz\| \ge \|Ta\| - \|T\| \cdot \|z-a\|_X \ge \frac{1-2\varepsilon}{D} - (1+2\varepsilon)\frac{\varepsilon}{D} \ge \frac{1-4\varepsilon}{D}$$

(if $\varepsilon < 1/2$). Hence T is invertible and $\|T^{-1}\| \le D/(1-4\varepsilon)$. Thus $\|T\| \cdot \|T^{-1}\| \le \frac{1+2\varepsilon}{1-4\varepsilon} D \le (1+12\varepsilon)D$, since one of our assumptions was $\varepsilon \in \left(0, \frac{1}{8}\right)$. □

9.2.5 Poisson kernel estimates and proofs of Lemmas 9.14 and 9.15

We need the following standard estimate, which holds for all $r, t \in (0, \infty)$.

$$\int_{\mathbb{R}^n \setminus (rB_X)} P_t(x)dx \le \frac{t\sqrt{n}}{r}. \tag{9.44}$$

To get (9.44) we need the homogeneity of Poisson kernels of degree $-n$, that is:

$$P_t(x) = t^{-n} P_1(x/t). \tag{9.45}$$

This property is an immediate consequence of the explicit formula (9.27). Let s_{n-1} denote the surface area of the unit Euclidean sphere S^{n-1} (in $\mathbb{R}^n$).

We have

$$\begin{aligned}\int_{\|x\|_X \ge r} P_t(x)dx &\overset{(9.6)}{\le} \int_{\|x\|_2 \ge r} P_t(x)dx \overset{(9.45)}{=} \int_{\|x\|_2 \ge r/t} P_1(x)dx \\ &= c_n s_{n-1} \int_{r/t}^{\infty} \frac{s^{n-1}}{(1+s^2)^{\frac{n+1}{2}}} ds \le c_n s_{n-1} \int_{r/t}^{\infty} \frac{ds}{s^2} = \frac{c_n s_{n-1} t}{r}.\end{aligned}$$

Now we recall (see (9.27)) that $c_n = \Gamma\left(\frac{n+1}{2}\right)/\pi^{\frac{n+1}{2}}$ and the well-known fact that $s_{n-1} = n\pi^{\frac{n}{2}}/\Gamma\left(\frac{n}{2}+1\right)$. We also recall the Stirling formula for the gamma function:

$$\Gamma(m) = \frac{\sqrt{2\pi} m^{m-\frac{1}{2}}}{e^m} \cdot e^{\frac{\theta}{12m}}, \tag{9.46}$$

where $0 < \theta < 1$. This formula is a generalization of the Stirling formula which we used in Chapter 4 (recall the well-known identity $\Gamma(n+1) = n!$ for $n \in \mathbb{N}$). We get

$$\begin{aligned} c_n s_{n-1} &= \frac{\Gamma\left(\frac{n+1}{2}\right) n\pi^{\frac{n}{2}}}{\pi^{\frac{n+1}{2}} \Gamma\left(\frac{n}{2}+1\right)} \\ &= \frac{\left(\frac{n+1}{2}\right)^{\frac{n}{2}} e^{\frac{n}{2}+1} n}{\sqrt{\pi} e^{\frac{n+1}{2}} \left(\frac{n+2}{2}\right)^{\frac{n}{2}+\frac{1}{2}}} \cdot e^{\frac{\theta}{12(n+1)/2} - \frac{\theta}{12(n+2)/2}} \\ &= \sqrt{\frac{2e}{\pi}} \cdot \frac{n}{\sqrt{n+2}} \cdot \frac{1}{\left(1+\frac{1}{n+1}\right)^{\frac{n}{2}}} \cdot e^{\frac{\theta}{12(n+1)/2} - \frac{\theta}{12(n+2)/2}} \\ &\le \sqrt{n}. \end{aligned} \tag{9.47}$$

The estimate (9.44) follows.

Another standard estimate that we will use is the following Lipschitz-type condition: for every $y \in \mathbb{R}^n$ we have

$$\int_{\mathbb{R}^n} |P_t(x) - P_t(x+y)|\,dx \le \frac{\sqrt{n}\|y\|_2}{t}. \tag{9.48}$$

In is easy to verify that by the homogeneity of degree $-n$ of the Poisson kernels (see (9.45)), it suffices to prove (9.48) in the case where $t = 1$. We have

$$\begin{aligned}
\int_{\mathbb{R}^n} |P_1(x) - P_1(x+y)|\,dx &= \int_{\mathbb{R}^n} \left| \int_0^1 \langle \nabla P_1(x+sy), y\rangle ds \right| dx \\
&\le \|y\|_2 \int_{\mathbb{R}^n} \|\nabla P_1(x)\|_2\,dx \\
&\overset{(9.27)}{=} \|y\|_2 (n+1) c_n \int_{\mathbb{R}^n} \frac{\|x\|_2}{\left(1+\|x\|_2^2\right)^{\frac{n+3}{2}}} dx \\
&= \|y\|_2 (n+1) c_n s_{n-1} \int_0^\infty \frac{r^n}{(1+r^2)^{\frac{n+3}{2}}} dr \\
&= \|y\|_2 c_n s_{n-1},
\end{aligned}$$

where we used the fact that the derivative of $r^{n+1}/(1+r^2)^{\frac{n+1}{2}}$ equals $(n+1)r^n/(1+r^2)^{\frac{n+3}{2}}$. The required estimate (9.48) now follows from (9.47).

Proof of Lemma 9.14. We have,

$$\begin{aligned}
\|P_t * F(x) - P_t * F(y)\|_Y &\le \int_{\mathbb{R}^n} P_t(z)\|F(x-z) - F(y-z)\|_Y dz \\
&\overset{(*)}{\le} \left((1+\varepsilon) \int_{\frac{1}{4}B_X} P_t(z) dz + 8 \int_{\mathbb{R}^n \setminus (\frac{1}{4}B_X)} P_t(z) dz \right) \|x-y\|_X \\
&\overset{(9.44)}{\le} \left(1+\varepsilon+32t\sqrt{n}\right) \|x-y\|_X \\
&\le (1+2\varepsilon)\|x-y\|_X
\end{aligned}$$

for t in the range described in the statement of Lemma 9.14. In the inequality marked with $(*)$ we used the fact that F is $(1+\varepsilon)$-Lipschitz on $\frac{1}{2}B_X$ and 8-Lipschitz on $\mathbb{R}^n$. □

Lemma 9.19. *For every $t \in (0, 1/2]$ and every $x \in B_X$ we have*

$$\|(P_t * F)(x) - F(x)\|_Y \le 20\sqrt{n}\, t \ln\left(\frac{3}{t}\right).$$

Proof. Since F is supported on $3B_X$,

$$\begin{aligned}\|(P_t * F)(x) - F(x)\|_Y \\ \le \int_{x+3B_X} \|F(x-y) - F(x)\|_Y P_t(y)dy \\ + \|F(x)\|_Y \int_{y\in\mathbb{R}^n\setminus(x+3B_X)} P_t(y)dy.\end{aligned} \tag{9.49}$$

Since F is 8-Lipschitz and it vanishes outside $3B_X$, we have $\|F(x)\|_Y \le 24$. Moreover, if $\|y-x\|_X \ge 3$ then $\|y\|_X \ge \|x-y\|_X - \|x\|_X \ge 2$, and therefore

$$\|F(x)\|_Y \int_{y\in\mathbb{R}^n\setminus(x+3B_X)} P_t(y)dy \le 24 \int_{\mathbb{R}^n\setminus(2B_X)} P_t(y)dy \overset{(9.44)}{\le} 12t\sqrt{n}. \tag{9.50}$$

To bound the first term in the right-hand side of (9.49) note that $\|y-x\|_X \le 3$ implies $\|y\|_2 \le \sqrt{n}\|y\|_X \le 4\sqrt{n}$. Also we have $\|F(y-x) - F(x)\|_X \le 8\|y\|_X \le 8\|y\|_2$. Hence,

$$\begin{aligned}\int_{x+3B_X} \|F(y-x) - F(x)\|_Y P_t(y)dy \le 8 \int_{\|y\|_2\le 4\sqrt{n}} \|y\|_2 P_t(y)dy \\ \overset{(9.45)}{=} 8t \int_{\|y\|_2\le\frac{4\sqrt{n}}{t}} \|y\|_2 P_1(y)dy = 8tc_n s_{n-1} \int_0^{\frac{4\sqrt{n}}{t}} \frac{s^n}{(1+s^2)^{\frac{n+1}{2}}} ds.\end{aligned} \tag{9.51}$$

Straightforward computation shows that the maximum of $s^n/(1+s^2)^{\frac{n+1}{2}}$ is attained at $s = \sqrt{n}$, and therefore

$$\frac{s^n}{(1+s^2)^{\frac{n+1}{2}}} \le \frac{1}{\sqrt{\left(1+\frac{1}{n}\right)^n (1+n)}} \le \frac{1}{\sqrt{en}} \quad \forall s \in (0,\infty),$$

where in the last step we use the well-known inequality $\left(1+\frac{1}{n}\right)^{n+1} \ge e$. On the other hand, dropping 1 in the denominator we get

$$\frac{s^n}{(1+s^2)^{\frac{n+1}{2}}} \le \frac{1}{s} \quad \forall s \in (0,\infty).$$

Hence,

$$\int_0^{\frac{4\sqrt{n}}{t}} \frac{s^n}{(1+s^2)^{\frac{n+1}{2}}} ds \le 1 + \int_{\sqrt{en}}^{\frac{4\sqrt{n}}{t}} \frac{ds}{s} = 1 + \ln\left(\frac{4}{t\sqrt{e}}\right). \tag{9.52}$$

The required result now follows from substituting (9.50), (9.51), (9.52) into (9.49), and using the fact that $t \le 1/2$ and $c_n s_{n-1} \le \sqrt{n}$ (see (9.47)). □

Proof of Lemma 9.15. Define

$$\Theta = \frac{170D\sqrt{n}\,t\ln(3/t)}{\varepsilon}. \tag{9.53}$$

For $w, y \in \frac{1}{2}B_X$ let $p, q \in \mathcal{N}_\delta$ satisfy $\|p - w\|_X, \|q - y\|_Y \le \delta$. Assume that $\|w - y\|_X \ge \Theta$. Using the properties (c) and (d) of the almost extension F (in Lemma 9.13), together with Lemma 9.19, we have

$$\begin{aligned}
\|(P_t * F)(w) - (P_t * F)(y)\|_Y &\ge \|f(p) - f(q)\|_Y - \|F(p) - f(p)\|_Y \\
&\quad - \|F(q) - f(q)\|_Y - \|F(w) - F(p)\|_Y - \|F(y) - F(q)\|_Y \\
&\quad - \|(P_t * F)(w) - F(w)\|_Y - \|(P_t * F)(y) - F(y)\|_Y \\
&\overset{(9.7)}{\ge} \frac{\|p-q\|_Y}{D} - \frac{26n\delta}{\varepsilon} - 2(1+\varepsilon)\delta - 40\sqrt{n}\,t\ln\left(\frac{3}{t}\right) \\
&\ge \frac{\|w-y\|_X - 2\delta}{D} - \frac{26n\delta}{\varepsilon} - 2(1+\varepsilon)\delta - 40\sqrt{n}\,t\ln\left(\frac{3}{t}\right) \\
&\ge \frac{1-\varepsilon/3}{D}\|w-y\|_X,
\end{aligned} \tag{9.54}$$

where the last inequality in (9.54) uses the assumptions $\|w - y\|_X \ge \Theta$ and (9.30).

Note that the second inequality in (9.30) implies that $\Theta \le 1/4$. Therefore, since $\|a\|_X = 1$ it follows from (9.54) that for every $z \in \frac{1}{4}B_X$,

$$\begin{aligned}
\frac{1-\varepsilon/3}{D}\Theta &\le \|P_t * F(z + \Theta a) - P_t * F(z)\|_Y \\
&= \left\| \int_0^\Theta \partial_a (P_t * F)(z + sa)ds \right\|_Y \\
&\le \int_0^\Theta \|\partial_a (P_t * F)(z + sa)\|_Y \, ds.
\end{aligned} \tag{9.55}$$

Since in Lemma 9.15 we assume that $\|x\|_X \le 1/8$, we have

$$\begin{aligned}
&\frac{1}{\Theta}\int_0^\Theta \int_{\mathbb{R}^n} \|\partial_a (P_t * F)(x + sa - y)\|_Y P_{Rt}(y)dyds \\
&\qquad \overset{(9.55)}{\ge} \frac{1-\varepsilon/3}{D}\int_{\frac{1}{8}B_X} P_{Rt}(y)dy \\
&\qquad \overset{(9.44)}{\ge} \frac{1-\varepsilon/3}{D}(1 - 8Rt\sqrt{n}) \\
&\qquad \overset{(9.31)}{\ge} \frac{(1-\varepsilon/3)(1-\varepsilon/6)}{D} \ge \frac{1-\varepsilon/2}{D}.
\end{aligned} \tag{9.56}$$

Since F is 8-Lipschitz and everywhere differentiable, the inequality $\|\partial_a F\|_Y \le 8$, and therefore $\|\partial_a(P_t * F)\|_Y \le 8$ holds everywhere. Hence,

$$\begin{aligned}\int_0^\Theta \int_{\mathbb{R}^n} &\|\partial_a(P_t * F)(x-y)\|_Y \,(P_{Rt}(y+sa) - P_{Rt}(y))\,dyds \\ &\le 8\int_0^\Theta \int_{\mathbb{R}^n} |P_{Rt}(y+sa) - P_{Rt}(y)|\,dyds \\ &\overset{(9.48)}{\le} \frac{8\sqrt{n}\|a\|_2}{Rt}\cdot\frac{\Theta^2}{2} \\ &\overset{(9.6)}{\le} \frac{4n\Theta^2}{Rt} \\ &\overset{(9.31)\&(9.53)}{\le} \frac{\varepsilon\Theta}{2D}.\end{aligned} \tag{9.57}$$

We can now conclude the proof of Lemma 9.15 as follows.

$$\begin{aligned}(\|\partial_a(P_t * F)\|_Y * P_{Rt})\,(x) &= \frac{1}{\Theta}\int_0^\Theta \int_{\mathbb{R}^n} \|\partial_a(P_t * F)(x-y)\|_Y\, P_{Rt}(y)dyds \\ &= \frac{1}{\Theta}\int_0^\Theta \int_{\mathbb{R}^n} \|\partial_a(P_t * F)(x+sa-y)\|_Y\, P_{Rt}(y)dyds \\ &\quad -\frac{1}{\Theta}\int_0^\Theta \int_{\mathbb{R}^n} \|\partial_a(P_t * F)(x-y)\|_Y\,(P_{Rt}(y+sa) - P_{Rt}(y))\,dyds \\ &\overset{(9.56)\&(9.57)}{\ge} \frac{1-\varepsilon}{D}.\end{aligned}$$

□

9.3 Test-space characterizations

One of the directions in which the Ribe program develops is based on the notion of test-spaces.

Definition 9.20. Let $\mathcal{P}$ be a class of Banach spaces and let $T = \{T_\alpha\}_{\alpha\in A}$ be a set of metric spaces. We say that T is a set of *test-spaces* for $\mathcal{P}$ if the following two conditions are equivalent:

(a) $X \notin \mathcal{P}$.

(b) The spaces $\{T_\alpha\}_{\alpha\in A}$ admit bilipschitz embeddings into X with uniformly bounded distortion.

Remark 9.21. We write $X \notin \mathcal{P}$ rather than $X \in \mathcal{P}$ for terminological reasons: we would like to use terms "test-spaces for reflexivity, superreflexivity, etc." rather than "test-spaces for *non*reflexivity, *non*-superreflexivity, etc."

Remark 9.22. Of course, each collection $\{T_\alpha\}_{\alpha \in A}$ of metric spaces produces the corresponding class of Banach spaces, defined by the condition (b) of Definition 9.20. It seems that at least in some cases it could be interesting to start with a class of metric spaces and to try to understand what is the corresponding class of Banach spaces. In the spirit of the Ribe program, one has to focus on the problem of test-space characterizations of hereditary local isomorphic invariants of Banach spaces which have already shown their importance for Banach space theory. However this study can be extended beyond the limits of the Ribe program and we can look for test-space characterizations of classes which are not stable under uniform homeomorphisms.

There is an obvious restriction on a class $\mathcal{P}$ having a set of test-spaces:

(H) If $X \in \mathcal{P}$ and a Banach space Y admits a bilipschitz embedding into X, then $Y \in \mathcal{P}$ (in particular, if a Banach space Y admits an isomorphic embedding into X, then $Y \in \mathcal{P}$).

So we suggest the following general problem:

Problem 9.23. Given a class $\mathcal{P}$ of Banach space satisfying the condition **(H)** find a set of test-spaces for $\mathcal{P}$.

In the cases where some sets of test-spaces are known, one can ask for more. Some of the most natural questions which can be asked in this connection are:

(a) Can one characterize $\mathcal{P}$ using one metric space (that is, using a one-element set of test-spaces)?

(b) Can one characterize $\mathcal{P}$ using one infinite unweighted graph with its graph distance?

(c) Can one choose a set of test-spaces to be unweighted finite graphs with their graph distances?

(d) Can one additionally require that the graphs have uniformly bounded degrees of vertices (both in finite and infinite cases)? (This is related to the notion of a metric space with bounded geometry.)

Remark 9.24. One can introduce similar definitions for uniform and coarse embeddings. We do not do this because of the small amount of information available on such characterizations.

Some of the very important classes of Banach spaces can be characterized in terms of test-spaces. One of the most important of these classes is the class of superreflexive Banach spaces. Before proceeding with its test-space characterization we need to recall some important fact about superreflexive spaces. The test-space characterization of superreflexivity is the only one which we present in full detail. Descriptions of other known test-space characterizations can be found in the Notes and Remarks section.

9.3.1 More Banach space theory: superreflexivity

We recall the definition which we already introduced in Section 8.4.2:

Definition 9.25. A Banach space X is called *superreflexive* if each Banach space which is finitely representable in X is reflexive.

The class of superreflexive Banach spaces turns out to be extremely important. One of the reasons for this is that the defining property of superreflexivity turns out to be equivalent to many other geometric and analytic properties of a Banach space. Some of these properties were mentioned in Section 8.4.2. In this section we study those equivalent characterizations of superreflexivity that lead to its test-space characterization. We need the following definition.

Definition 9.26. A Banach space X has the *finite tree property* if there exist $\delta > 0$ such that for each $k \in \mathbb{N}$ the unit ball of X contains a finite sequence $\{x_i^k : i = 1, \dots, 2^k - 1\}$ such that $x_i^k = \frac{1}{2}(x_{2i}^k + x_{2i+1}^k)$ and $\|x_{2i}^k - x_i^k\| = \|x_{2i+1}^k - x_i^k\| \geq \delta$ for each $i = 1, \dots, 2^{k-1} - 1$.

In our test-space characterization of superreflexivity we use the following result on (linear) characterization of superreflexivity.

Theorem 9.27. *The following properties of a Banach space X are equivalent:*

(a) *X is not superreflexive.*

(b) *X has the finite tree property.*

(c) *There exists $\alpha \in (0, 1]$ such that for each $m \in \mathbb{N}$ the unit ball of the space X contains a finite sequence $x_1, x_2, \dots, x_m$ of vectors satisfying, for any subinterval A of $\{1, \dots, m\}$ and any real coefficients $a_1, \dots, a_m$, the condition*

$$\left\| \sum_{i=1}^{m} a_i x_i \right\| \geq \alpha \left| \sum_{i \in A} a_i \right|, \tag{9.58}$$

where by a subinterval we mean a subset consisting of consecutive integers.

Proof. (a) $\Rightarrow$ (b) Suppose that X is not superreflexive. Then there is a nonreflexive space Z which is finitely representable in X. We apply to Z the result on characterization of reflexivity which we proved in Section 6.3 (Theorem 6.11). So we pick $0 < \theta < 1$ and find in the unit ball of Z a sequence $\{z_i\}_{i=1}^{\infty}$ satisfying, for each $n \in \mathbb{N}$, the condition

$$d(\operatorname{conv}(\{z_i\}_{i=1}^{n}), \operatorname{conv}(\{z_i\}_{i=n+1}^{\infty})) \geq \theta. \tag{9.59}$$

Using finite pieces of this sequence it is easy to construct finite trees in Z. In our construction we label the trees using $\{t_i\}_{i=0}^{2^{j+1}-1}$ and introduce them as follows:

$$t_1 = \frac{1}{2^j}\sum_{i=1}^{2^j} z_i$$

$$t_2 = \frac{1}{2^{j-1}}\sum_{i=1}^{2^{j-1}} z_i, \quad t_3 = \frac{1}{2^{j-1}}\sum_{i=2^{j-1}+1}^{2^j} z_i.$$

$$t_4 = \frac{1}{2^{j-2}}\sum_{i=1}^{2^{j-2}} z_i, \quad t_5 = \frac{1}{2^{j-2}}\sum_{i=2^{j-2}+1}^{2^{j-1}} z_i,$$

$$t_6 = \frac{1}{2^{j-2}}\sum_{i=2^{j-1}+1}^{2^{j-1}+2^{j-2}} z_i, \quad t_7 = \frac{1}{2^{j-2}}\sum_{i=2^{j-1}+2^{j-2}+1}^{2^j} z_i.$$

We proceed in the obvious way. The property $t_i = \frac{1}{2}(t_{2i} + t_{2i+1})$ is immediate from the definition. The property $\|t_{2i} - t_i\| = \|t_{2i+1} - t_i\| \geq \theta/2$ follows immediately from (9.59). Since Z is finitely represented in X, it follows that X also has the finite tree property.

(b) $\Rightarrow$ (a). Let X be a Banach space having the finite tree property. Let $\{x_i^j\}_{i=1}^{2^j-1}$, $j \in \mathbb{N}$, be the corresponding finite trees. Let $\mathcal{U}$ be a free ultrafilter on $\mathbb{N}$ and $X^{\mathcal{U}}$ be the corresponding ultraproduct. As we know from Proposition 2.31 the space $X^{\mathcal{U}}$ is finitely representable in X. So it remains to show that $X^{\mathcal{U}}$ is nonreflexive.

First we observe that the unit ball of $X^{\mathcal{U}}$ contains a δ-tree in the sense of the following definition.

Definition 9.28. Let $\delta > 0$. A sequence $\{x_i\}_{i=1}^{\infty}$ is called a *δ-tree* if $x_i = \frac{1}{2}(x_{2i} + x_{2i+1})$ and $\|x_{2i} - x_i\| = \|x_{2i+1} - x_i\| \geq \delta$.

This is immediate from the definition of an ultraproduct, just consider the vectors $\{u_i\}_{i=1}^{\infty} \subset X^{\mathcal{U}}$ represented by sequences $\{v_j(i)\}_{j=1}^{\infty}$ given by

$$v_j(i) = \begin{cases} x_i^j & \text{if } i \leq 2^j - 1 \\ 0 & \text{otherwise.} \end{cases}$$

On the other hand, we show that a Banach space containing a bounded δ-tree is nonreflexive. Although this fact is well known and follows from many other results of the Banach space theory, it does not seem that any simple proof of it is available. Our proof is by contradiction, so we assume the contrary, that is: there is a reflexive Banach space Z containing a bounded δ-tree. Consider the closed convex hull C of the tree. By a well-known corollary of the Hahn–Banach theorem C is weakly closed. Since the space Z is reflexive, C is weakly compact. Now we use the following notion and theorem.

Definition 9.29. Let C be a convex set in a Banach space Z. A point $w \in C$ is called a *strongly exposed point* of C if there is $f \in Z^*$ such that

(a) $f(u) < f(w)$ for every $u \in C, u \neq w$.

(b) $f(w_n) \to f(w)$ and $\{w_n\}_{n=1}^{\infty} \subset C$ imply $\|w_n - w\| \to 0$.

Theorem 9.30. *Each weakly compact convex set in a separable Banach space has a strongly exposed point.*

Remark 9.31. The statement of Theorem 9.30 is known to be true in nonseparable spaces, too.

First we show how to complete our proof of (b) $\Rightarrow$ (a) using Theorem 9.30. Let w be a strongly exposed point of C (where C is the closed convex hull of a bounded δ-tree) and f be the corresponding functional. It is clear from the definition of a strongly exposed point that there exists $\alpha > 0$ such that the set $S = \{u \in C : f(u) \geq f(w) - \alpha\}$ has diameter $< \delta$. It is clear that this set should contain at least one point of the tree (it would not be in its closed convex hull otherwise). Let $u_i \in S$. We have $u_i = \frac{1}{2}(u_{2i} + u_{2i+1})$. It is clear that at least one of the points u_{2i} and u_{2i+1} is contained in S. We get a contradiction: $\|u_{2i} - u_i\| = \|u_{2i+1} - u_i\| \geq \delta$ contradicts $\operatorname{diam} S < \delta$. This completes our proof of (b) $\Rightarrow$ (a).

Proof of Theorem 9.30. Our proof uses the following obvious observation: the notion of a strongly exposed point depends only on the topology of the space and does not change if another norm on the space is consider that is equivalent to the original norm in the following sense.

Definition 9.32. Two norms $\|\cdot\|_1$ and $\|\cdot\|_2$ on a linear space X are called *equivalent* if there are constants $0 < c \leq C < \infty$ such that

$$c\|x\|_1 \leq \|x\|_2 \leq C\|x\|_1$$

for each $x \in X$.

We shall be looking for norms having the following property.

Definition 9.33. A norm of a Banach space X is called *locally uniformly convex* if for every $\varepsilon > 0$ and $x \in X$, $\|x\| = 1$, there is $\delta(x, \varepsilon) > 0$ such that for any vector $\|y\| \leq 1$, the inequality

$$\|x + y\| > 2 - \delta(x, \varepsilon)$$

implies

$$\|x - y\| < \varepsilon.$$

Observation 9.34. The definition of a locally uniformly convex norm can be restated as: For any $x \in S_X$ (the unit sphere of X) and for any sequence $\{x_n\}$ satisfying $\|x_n\| \le 1$ and $\lim_{n\to\infty} \|x + x_n\| = 2$, the condition $\lim_{n\to\infty} \|x - x_n\| = 0$ holds.

Let C be a convex weakly compact set in a separable Banach space X. We are going to replace the original norm $\|\cdot\|$ of the space X by an equivalent norm $|||\cdot|||$ satisfying the following two conditions:

- The norm is locally uniformly convex in the sense of Definition 9.33 (see below).
- The new norm $|||\cdot|||$ attains its maximum on the set C.

The existence of such a norm is sufficient for our purposes because of the following lemma.

Lemma 9.35. *The point w at which a locally uniformly convex norm $|||\cdot|||$ attains its maximum on C is a strongly exposed point of C.*

Proof. Let $f \in X^*$ be a functional satisfying $|||f||| = 1$ and $f(w) = |||w|||$. We are going to show that f satisfies the conditions of Definition 9.29. Assume the contrary. Then there is a sequence $\{w_n\}$ in C such that $f(w_n) \to f(w)$, but

$$|||w_n - w||| \nrightarrow 0. \tag{9.60}$$

The condition $f(w_n) \to f(w)$ implies $\limsup_{n\to\infty} |||w_n + w||| \ge 2|||w|||$. Since

$$|||w_n||| \le |||w|||, \tag{9.61}$$

this implies

$$|||w_n + w||| \to 2|||w|||. \tag{9.62}$$

The combination of (9.60), (9.61), and (9.62) contradicts the local uniform convexity of the norm $|||\cdot|||$. □

Now we turn to the construction of $|||\cdot|||$. We start by constructing an equivalent strictly convex norm $\|\cdot\|_1$ on X. Let $\{x_i\}_{i=1}^\infty$ be a dense sequence in the unit sphere of X. Let $\{x_i^*\}_{i=1}^\infty \subset X^*$ be functionals satisfying $\|x_i^*\| = 1$ and $x_i^*(x_i) = 1$. We introduce the norm $\|\cdot\|_1$ by

$$\|x\|_1 \stackrel{\text{def}}{=} \left(\|x\|^2 + \sum_{i=1}^{\infty} 2^{-i}(x_i^*(x))^2\right)^{\frac{1}{2}}.$$

It is clear that $\|x\| \le \|x\|_1 \le \sqrt{2}\|x\|$. Since ℓ_2 is uniformly convex it cannot happen that $\|x\|_1 = \|y\|_1 = \left\|\frac{x+y}{2}\right\|_1$. Thus $\|\cdot\|_1$ is a strictly convex norm.

In the next step we construct an equivalent locally uniformly convex norm on X. Let $\{X_i\}_{i=0}^{\infty}$ be an increasing sequence of finite-dimensional subspaces of X such that $X_0 = \{0\}$ and the union $\cup_{i=0}^{\infty} X_i$ is dense in X. We introduce a new norm on X by

$$\|x\|_2 = \left(\sum_{i=0}^{\infty} 2^{-i}(d(x, X_i))^2\right)^{\frac{1}{2}},$$

where $d(x, X_i)$ is the distance from a vector x to the subspace X_i in the distance induced by the norm $\|\cdot\|_1$. Let $\|x\|_2 = 1$ and $\{x_i\}_{i=1}^{\infty}$ be such a sequence that $\|x_i\|_2 \leq 1$ and

$$\lim_{i\to\infty} \|x + x_i\|_2 = 2. \tag{9.63}$$

We are going to show that $\{x_i\}_{i=1}^{\infty}$ contains a subsequence which converges to x. Since the same argument can be used for each subsequence of $\{x_i\}$, this would imply that $\{x_i\}$ itself converges to x.

By the uniform convexity of ℓ_2, condition (9.63) implies that $\lim_{i\to\infty} d(x_i, X_j) = d(x, X_j)$. Now we pick $\varepsilon > 0$. Since $\cup_{i=0}^{\infty} X_i$ is dense in X, there is $m \in \mathbb{N}$ such that $d(x, X_m) < \varepsilon$. Therefore $d(x_i, X_m) < \varepsilon$ for sufficiently large $i \in \mathbb{N}$. Also the sequence $\{x_i\}$ is bounded. Let $z_i \in X_m$ be such that $\|z_i - x_i\|_1 < \varepsilon$, $i \in \mathbb{N}$. The sequence $\{z_i\}_{i=1}^{\infty}$ is bounded. Therefore it contains a convergence subsequence z_{i_n}. Therefore the sequence $\{x_{i_n}\}$ satisfies $\|x_{i_n} - x_{i_t}\| \leq 3\varepsilon$ for sufficiently large i_n and i_t. We repeat the procedure for smaller ε for this subsequence, by the diagonal procedure we get a Cauchy sequence. Let y be its limit. By (9.63) we get $\|x + y\|_1 = 2$, we also have $\|y\|_1 \leq 1$. Since the norm $\|\cdot\|_1$ is strictly convex we get $x = y$. Therefore the norm $\|\cdot\|_2$ is locally uniformly convex.

Now we modify the norm $\|\cdot\|_2$ in order to get an equivalent norm $|||\cdot|||$ on X which is locally uniformly convex and attains its maximum on C. We do this as follows. We let $|||\cdot|||_0 = \|\cdot\|_2$ and construct simultaneously sequences of:

- Norms $\{|||\cdot|||_i\}_{i=0}^{\infty}$ satisfying $|||x|||_0 \leq |||x|||_1 \leq \cdots \leq |||x|||_n \leq \ldots$ for all x, and such that

$$|||x|||_n \leq 2|||x|||_0 \tag{9.64}$$

  for all $x \in X$ and $n \in \mathbb{N}$. Therefore

$$|||x||| = \lim_{i\to\infty} |||x|||_i$$

  exists (and is finite) for every $x \in X$.

- Vectors $\{x_i\}_{i=0}^{\infty} \subset C$.

- Functionals $\{f_i\}_{i=0}^{\infty} \subset X^*$.

We let $\{M_i\}_{i=0}^{\infty}$ be given by

$$M_i = \sup_{x \in C} |||x|||_i .$$

The purpose of our construction of sequences is to achieve the following: For each weak-limit point w of the sequence $\{x_i\}$ the condition

$$|||w||| = \lim_{i \to \infty} |||w|||_i = \lim_{i \to \infty} M_i \tag{9.65}$$

holds and therefore $||| \cdot |||$ attains its maximum on C at w. (We have $w \in C$ since C is weakly compact.)

To achieve the goal (9.65) we do the following:

- We pick sequences $\{\alpha_i\}_{i=0}^{\infty}$ and $\{\beta_i\}_{i=0}^{\infty}$ of positive numbers such that $\alpha_i \downarrow 0$ and $\beta_i \downarrow 0$ as $i \to \infty$. Later we shall describe some additional conditions on these sequences.
- We pick sequences $\{x_i\}_{i=0}^{\infty}$, $\{f_i\}_{i=0}^{\infty}$, and $\{||| \cdot |||_i\}_{i=0}^{\infty}$ as follows:
  - Let $x_0 \in C$ be such that $|||x_0|||_0 \geq M_0 - \alpha_0$.
  - Let $f_0 \in X^*$ be such that $|||f_0|||_0 = 1$ and $f_0(x_0) = |||x_0|||_0$.
  - Let $|||x|||_1 = |||x|||_0 + \beta_0 |||f_0(x)x_0|||_0$
  - Let $x_1 \in C$ be such that $|||x_1|||_1 \geq M_1 - \alpha_1$.
  - Let $f_1 \in X^*$ be such that $|||f_1|||_1 = 1$ and $f_1(x_1) = |||x_1|||_1$.
  - Let $|||x|||_2 = |||x|||_1 + \beta_1 |||f_1(x)x_1|||_1$.
  - We proceed in an obvious way.

The main point of this construction is that we can choose the numbers α_i and β_i in such a way that

$$|f_i(x_j)| \geq M_i - \delta_i, \quad j \geq i, \tag{9.66}$$

where $\{\delta_i\}_{i=0}^{\infty}$ is some sequence of positive numbers tending to 0. It is easy to show that (9.66) implies the fact that $||| \cdot |||$ attains its maximum on C at w. In fact (9.66) immediately implies $|||w|||_i \geq M_i - \delta_i$. Since obviously $\sup_{x \in C} |||x||| = \lim_{i \to \infty} M_i$, the desired conclusion is immediate.

So we pick an arbitrary sequence $\{\delta_i\}_{i=0}^{\infty}$ satisfying $\delta_i > 0$, $\delta_i \geq \delta_{i+1}$ and $\delta_i \downarrow 0$ as $i \to \infty$, and show that we can achieve (9.66) by picking $\{\alpha_i\}_{i=0}^{\infty}$ and $\{\beta_i\}_{i=0}^{\infty}$ in a suitable way. We have

$$|||x|||_j = |||x|||_i + \sum_{k=i}^{j-1} \beta_k |||f_k(x)x_k|||_k . \tag{9.67}$$

First one can easily see that by picking $\{\beta_i\}_{i=0}^{\infty}$ we can achieve the inequality (9.64). Also, (9.67) implies (for the second inequality we drop all but one terms from the sum)

$$\begin{aligned} M_j &\geq |||x_i|||_i + \sum_{k=i}^{j-1} \beta_k |||f_k(x_i)x_k|||_k \\ &\geq |||x_i|||_i + \beta_i |||x_i|||_i^2 \\ &\geq (M_i - \alpha_i) + \beta_i (M_i - \alpha_i)^2. \end{aligned}$$

Hence

$$\begin{aligned} |||x_j|||_j &= |||x_j|||_i + \sum_{k=i}^{j-1} \beta_k |||f_k(x_j)x_k|||_k \\ &\geq M_j - \alpha_j \\ &\geq (M_i - \alpha_i) + \beta_i (M_i - \alpha_i)^2 - \alpha_j, \end{aligned}$$

from which, since $|||x_j|||_i \leq M_i$ and $|||x_i|||_i \leq M_i$, we get

$$\beta_i |f_i(x_j)| \cdot |||x_i|||_i \geq -\alpha_i + \beta_i (M_i - \alpha_i)^2 - \alpha_j - \sum_{k=i+1}^{j-1} \beta_k |||f_k(x_j)x_k|||_k$$

and

$$|f_i(x_j)| \geq \frac{(M_i - \alpha_i)^2}{M_i} - \frac{\alpha_i + \alpha_j}{\beta_i M_i} - \sum_{k=i+1}^{j} \frac{\beta_k |||f_k(x_j)x_k|||_k}{\beta_i M_i}.$$

Now we are ready to describe suitable choices of $\{\alpha_i\}_{i=0}^{\infty}$ and $\{\beta_i\}_{i=0}^{\infty}$. First we pick $\{\beta_i\}_{i=0}^{\infty}$ in such a way that, in addition to the condition (9.64) (which we have already mentioned), the conditions

$$\sum_{k=i+1}^{\infty} \frac{\beta_k |||f_k(x_j)x_k|||_k}{\beta_i M_i} \leq \frac{\delta_i}{3}$$

are satisfied for all $i = 0, 1, \ldots$.

Next we pick $\{\alpha_i\}_{i=0}^{\infty}$ in such a way that the conditions:

$$\frac{(M_i - \alpha_i)^2}{M_i} \geq M_i - \frac{\delta_i}{3}$$

and

$$\frac{\alpha_i + \alpha_j}{\beta_i M_i} \leq \frac{\delta_i}{3}, \quad j \geq i,$$

are satisfied. It is clear that such choices are possible. It is also clear that these three $\frac{\delta_i}{3}$-conditions imply the condition (9.66).

Now we need to verify that the norm $|||\cdot|||$ is still locally uniformly convex. The reason for this is quite general: if we consider a sum of a locally uniformly convex norm and a seminorm which is estimated from above by a multiple of this locally uniformly convex norm, we get a locally uniformly convex norm. We do the estimates in the case which we consider. We have

$$|||x||| = |||x|||_0 + p(x),$$

where $p(x)$ is a seminorm satisfying $p(x) \le |||x|||_0$, see (9.64).

So we assume that $x \in X$ and $\{x_n\}_{n=1}^{\infty}$ are such that $|||x||| = 1$, $|||x_n||| \le 1$ and $\lim_{n\to\infty} |||x + x_n||| = 2$. We need to show that $\lim_{n\to\infty} |||x - x_n||| = 0$. It suffices to find a subsequence of $\{x_n\}$ converging to x.

Passing to subsequences if necessary we may assume that the limits

$$\lim_{n\to\infty} |||x_n|||_0 = \alpha \quad \text{and} \quad \lim_{n\to\infty} p(x_n) = \beta$$

exist. It is clear that $\alpha \ge \frac{1}{2}$ and $\alpha + \beta = 1$, and that we may assume $|||x_n|||_0 \le \alpha$. We split $x = \tilde{x} + \hat{x}$ in such a way that $|||\tilde{x}|||_0 = \alpha$ and $|||x||| = |||\tilde{x}||| + |||\hat{x}|||$ (this is possible if $|||x|||_0 \ge \alpha$, we split x_n otherwise).

We have

$$\begin{aligned}
2 &= \lim |||x + x_n||| \\
&\le \lim(|||\tilde{x} + x_n|||_0 + |||\hat{x}|||_0 + p(\tilde{x}) + p(\hat{x}) + p(x_n)) \\
&\le \lim(|||\tilde{x}|||_0 + |||x_n|||_0 + |||\hat{x}|||_0 + p(\tilde{x}) + p(\hat{x}) + p(x_n)) \\
&= \alpha + \alpha + |||x|||_0 - \alpha + p(x) + (1 - \alpha) \\
&= \lim(|||x||| + |||x_n|||) \\
&= 2.
\end{aligned}$$

Hence $\lim |||\tilde{x} + x_n|||_0 = 2\alpha$. Since the norm $|||\cdot|||_0$ is locally uniformly convex, we get

$$\lim_{n\to\infty} |||\tilde{x} - x_n|||_0 = 0.$$

Therefore

$$\lim_{n\to\infty} |||\tilde{x} - x_n||| = 0.$$

Therefore $|||\tilde{x}||| = 1$ and $\tilde{x} = x$. Thus $|||\cdot|||$ is locally uniformly convex. By Lemma 9.35, this completes the proof of (b) $\Rightarrow$ (a). □

(a)$\Rightarrow$(c). Let Y be a nonreflexive Banach space which is finitely representable in X. By Theorem 6.11, (a)$\Leftrightarrow$(c) for some $\theta \in (0,1)$ there are sequences $\{y_i\}_{i=1}^{\infty} \subset B_Y$ and $\{f_i\}_{i=1}^{\infty} \subset B_{Y^*}$ such that

$$f_n(y_k) = \begin{cases} \theta & \text{if } n \le k \\ 0 & \text{if } n > k. \end{cases}$$

This condition implies that for each finitely nonzero sequence $\{a_i\}$ of scalars we have

$$\left\|\sum_{i=1}^{\infty} a_i y_i\right\| \geq \theta \left|\sum_{i=n}^{\infty} a_i\right|. \tag{9.68}$$

Therefore for each finite subinterval $A \subset \mathbb{N}$ we have

$$2\left\|\sum_{i=1}^{\infty} a_i y_i\right\| \geq \theta \left|\sum_{i\in A} a_i\right|. \tag{9.69}$$

Since Y is finitely representable in X, for each $m \in \mathbb{N}$ and each $\varepsilon > 0$ there is a sequence $\{x_i\}_{i=1}^m$ in X satisfying (9.58) with $\alpha = \frac{\theta}{2(1+\varepsilon)}$.

(c)$\Rightarrow$(a). Let $\mathcal{U}$ be a free ultrafilter on $\mathbb{N}$. It is easy to see that the condition (c) implies that the unit ball of $X^{\mathcal{U}}$ contains an infinite sequence $\{u_i\}_{i=1}^{\infty}$ satisfying

$$\left\|\sum_{i} a_i u_i\right\| \geq \alpha \left|\sum_{i\in A} a_i\right| \tag{9.70}$$

for each finite finitely-nonzero sequence $\{a_i\}_{i=1}^{\infty}$ of real numbers and each subinterval A of $\mathbb{N}$. Therefore the functionals f_n given by

$$f_k\left(\sum_i a_i u_i\right) = \alpha\left(\sum_{i=k}^{\infty} a_i\right)$$

are contained in the unit ball. They also satisfy the condition

$$f_k(x_n) = \begin{cases} \alpha & \text{if } k \leq n \\ 0 & \text{if } k > n. \end{cases}$$

Using the characterization of reflexivity Theorem 6.11 and equivalence (a)$\Leftrightarrow$(c) we get that $X^{\mathcal{U}}$ is nonreflexive. Since, as we know (Proposition 2.31), $X^{\mathcal{U}}$ is finitely representable in X, we get that X is non-superreflexive. □

9.3.2 Characterization of superreflexivity in terms of diamond graphs

Several different test-space characterizations of the class of superreflexive spaces are known. We present a test-space characterization of superreflexivity which is in a certain sense the closest to the Banach space theoretical characterization of superreflexivity presented in the previous section.

The *diamond graph* of level 0 is denoted D_0. It is defined as a graph having two vertices joined by an edge of weight (also called *length*) 1. D_i is obtained from D_{i-1} as follows. Given an edge $uv \in E(D_{i-1})$, it is replaced by a quadrilateral u, a, v, b with edge lengths 2^{-i}. We endow D_n with their shortest path metrics. We consider the vertex set of D_n as a subset of the vertex set of D_{n+1}, it is easy to check that this defines an isometric embedding of D_n into D_{n+1}.

Theorem 9.36. *Let X be a Banach space. Then X is not superreflexive if and only if there is a constant $C < \infty$ such that $c_X(D_n) \le C$ for all $n \in \mathbb{N}$.*

Proof. Proof of the "if" part of Theorem 9.36 is an immediate consequence of the following lemma and the characterization of superreflexivity in terms of the finite tree property (Theorem 9.27).

Lemma 9.37. *If there exist uniformly bilipschitz embeddings of $\{D_n\}_{n=1}^{\infty}$ into a Banach space X, then X has the finite tree property.*

Proof. Let $f_n : D_n \to X$ be uniformly bilipschitz embeddings. Without loss of generality we assume that

$$\delta d_{D_n}(x,y) \le \|f_n(x) - f_n(y)\| \le d_{D_n}(x,y) \tag{9.71}$$

for some $\delta > 0$, all $n \in \mathbb{N}$ and all $x, y \in D_n$.

Let us show that this implies that the unit ball of X contains a sequence $\{x_i : i = 1, \dots, 2^n - 1\}$ required in the definition of a finite tree property. To get such a sequence we use the map f_n. Let $x_1 = f_n(u_0) - f_n(v_0)$, where $\{u_0, v_0\} = V(D_0)$.

Now we consider the quadrilateral u_0, a, v_0, b. Inequality (9.71) implies

$$\|f_n(a) - f_n(b)\| \ge \delta.$$

Consider two pairs of vectors (corresponding to two different paths joining u_0 and v_0 in D_1):

Pair 1. $f_n(v_0) - f_n(a)$, $f_n(a) - f_n(u_0)$. **Pair 2.** $f_n(v_0) - f_n(b)$, $f_n(b) - f_n(u_0)$.

The inequality $\|f_n(a) - f_n(b)\| \ge \delta$ implies that at least one of the following is true

$$\|(f_n(v_0) - f_n(a)) - (f_n(a) - f_n(u_0))\| \ge \delta$$

or

$$\|(f_n(v_0) - f_n(b)) - (f_n(b) - f_n(u_0))\| \ge \delta.$$

Suppose that the first inequality holds. We let

$$x_2 = 2(f_n(v_0) - f_n(a)) \quad \text{and} \quad x_3 = 2(f_n(a) - f_n(u_0)).$$

It is clear that $x_1 = \frac{1}{2}(x_2 + x_3)$ and $\|x_2 - x_3\| \ge 2\delta$, hence the conditions of Definition 9.26 are satisfied. Also, the condition (9.71) implies that $\|x_2\|, \|x_3\| \le 1$.

We continue our construction of a δ-tree in the unit ball of X in a similar manner. For example, to construct x_4 and x_5 we consider the corresponding quadrilateral a, a_1, v_0, b_1 in D_2. The inequality $\|f_n(a_1) - f_n(b_1)\| \ge \delta/2$ implies that at least one of the following is true

$$\|(f_n(v_0) - f_n(a_1)) - (f_n(a_1) - f_n(a))\| \ge \delta/2$$

or

$$\|(f_n(v_0) - f_n(b_1)) - (f_n(b_1) - f_n(a))\| \ge \delta/2.$$

Suppose that the second inequality holds. We let

$$x_4 = 4(f_n(v_0) - f_n(b_1)) \quad \text{and} \quad x_5 = 4(f_n(b_1) - f_n(a)).$$

It is clear that $x_2 = \frac{1}{2}(x_4 + x_5)$ and $\|x_4 - x_5\| \geq 2\delta$, hence the conditions of Definition 9.26 are satisfied. Also (9.71) implies that $\|x_4\|, \|x_5\| \leq 1$. Proceeding in an obvious way we get the set $\{x_i : i = 1, \dots, 2^n - 1\}$ required in the definition of the finite tree property. Clearly, this construction can be done for each $n \in \mathbb{N}$. □

To prove the "only if" part of Theorem 9.36 it is convenient to assume that all edges of the diamond D_n have length one (that is, we multiply all distances in the definition of D_n given above by 2^n). We start by describing the inductive construction of a 2-bilipschitz embedding of the diamond graph D_n into ℓ_1, the image of D_n will be a subset of $\ell_1^{2^n}$ contained in $\{0, 1\}^{2^n}$. The map which we construct is 1-Lipschitz and its inverse is 2-Lipschitz. We map vertices of D_0 onto 0 and 1 considered as elements of ℓ_1^1. It is clearly an isometric embedding.

Now we assume that we have already constructed a suitable embedding of D_{n-1} into $\ell_1^{2^{n-1}}$. We construct an embedding of D_n into $\ell_1^{2^n}$ in two steps. First we map those vertices of D_n which come from vertices of D_{n-1} in the following way. If the image of a vertex $v \in V(D_{n-1})$ is $(v_1, v_2, v_3, \dots, v_{2^{n-1}}) \in \ell_1^{2^{n-1}}$, we map the corresponding vertex in $V(D_n)$ onto $(v_1, v_1, v_2, v_2, v_3, v_3, \dots, v_{2^{n-1}}, v_{2^{n-1}}) \in \ell_1^{2^n}$. Recall that to get D_n an edge $uv \in E(D_{n-1})$ was replaced by a quadrilateral u, a, v, b. It is clear that the induction hypothesis implies that the images of u and v in $\ell_1^{2^n}$ differ in two consecutive coordinates. We map a and b onto different sequences each of which differs from images of u and v in one coordinate only. It is clear that this is a well-defined map.

Now we show that the constructed map is 2-bilipschitz. It is clear that the map $D_n \mapsto \ell_1^{2^n}$ is 1-Lipschitz (any two vertices which form an edge are mapped onto vectors at distance 1 in $\ell_1^{2^n}$). So it remains to show that the inverse image is 2-Lipschitz. The analysis of the diamond graphs needed to establish this will also be used for estimates of the Lipschitz constants for embeddings into general non-superreflexive spaces.

Let u, v be two adjacent vertices in D_k, u^ℓ, v^ℓ be their images in $\ell_1^{2^k}$. It is clear that u^ℓ, v^ℓ have only one different coordinate. We assume that the value of the coordinate is 0 for u^ℓ and 1 for v^ℓ. Let u_n, v_n be the corresponding vertices in D_n, $n > k$. Then the images u_n^ℓ, v_n^ℓ of u_n, v_n in $\ell_1^{2^n}$ differ on a set F of 2^{n-k} consecutive coordinates and coincide on the set R of the remaining coordinates. We say that the set of vertices in D_n, whose images in $\ell_1^{2^n}$ coincide with u_n^ℓ and v_n^ℓ on this set R of the remaining coordinates, forms a *subdiamond*. The vertex u_n is called the *bottom of the subdiamond* and the vertex v_n is called the *top of the subdiamond*.

It is easy to check that for any two vertices w and z of D_n there is a well-defined *smallest subdiamond of* D_n containing both w and z. Let $F = F_{w,z} \subset \{1, \dots, 2^n\}$ be a subset on which the coordinates of the images of vertices of the smallest subdiamond containing w and z may differ.

There are two possible cases:

(a) At least one of the vertices w, z is either top or bottom of this subdiamond.

(b) Neither w nor z is the top or bottom of the subdiamond.

It is easy to see by induction that in case (a) the distance between w and z in D_n is the same as the distance between their images in $\ell_1^{2^n}$. So it remains to consider case (b).

The vertex of the subdiamond whose image has ones in the left half of F and zeros in the right half of F is called the *leftmost* vertex of the subdiamond. Similarly we define the *rightmost* vertex of the subdiamond.

It is clear from the construction that each vertex in the subdiamond, except the top and the bottom, is either closer (with respect to the graph distance of D_n) to the leftmost vertex or to the rightmost vertex. In the first case we say that the vertex is in the *left side*, in the second case we say that it is in the *right side*. We need to consider two subcases of (b):

- w and z are on the same side.
- w and z are on different sides.

Consider the same-side case, assume that the considered side is the left side. We claim that in this case the $\ell_1^{2^n}$-image of one of the vertices, w or z, should have all of the coordinates greater than that of the image leftmost vertex; and the image of the other should have all of the coordinates less than that of the image leftmost vertex. In fact, if this is not the case, the vertices are contained in the smaller subdiamond formed by the leftmost vertex and either the top or the bottom of the subdiamond which we consider, and we get a contradiction.

Now, if the condition of the previous paragraph is satisfied, it is easy to see that the D_n-distance between w and z is the same as the $\ell_1^{2^n}$-distance between their images. (The shortest path goes through the leftmost vertex, and its length is exactly the amount of different $\ell_1^{2^n}$-coordinates.)

Now we consider the different-sides case. In this case either

(1) One of the vertices is closer to the top of the subdiamond, and the other is closer to the bottom, or

(2) Both vertices are at least as close to the top (bottom) as to the bottom (top).

In case (1) the image of the vertex which is closer to the top has either all of the coordinates in the left half of F equal to 1, or all of the coordinates in the right half of F equal to 1. We consider the left-half case only. In this case the image of the other vertex has all the coordinates in the left half of F equal to 0. Therefore the $\ell_1^{2^n}$-distance between the $\ell_1^{2^n}$-images is at least $\frac{1}{2}|F|$. Since $\frac{1}{2}|F|$ is half of the diameter of the subdiamond, this implies that the map inverse to the embedding of D_n into $\ell_1^{2^n}$ is 2-Lipschitz.

In case (2) we consider only the subcase where both vertices are at least as close to the top as to the bottom (the other case is similar). In this case the shortest path between the vertices w and z goes through the top t and the vectors $f(t) - f(w)$ and $f(t) - f(z)$ are disjoint, where f is our notation for the map $f : V(D_n) \to \ell_1^{2^n}$ which we constructed, therefore the ℓ_1-distance between $f(w)$ and $f(z)$ is the same as the graph distance between w and z.

To show that diamond graphs admit uniformly bilipschitz embeddings into an arbitrary non-superreflexive space, we use the characterization proved in Theorem 9.27. Namely, we use the fact that for each non-superreflexive space X there is a sequence $\{x_i\}_{i=1}^{2^n}$ in B_X satisfying (9.58). Now we map the vertex w of D_n corresponding to the sequence $\{w_i\}_{i=1}^{2^n} \in \ell_1^{2^n}$ onto the vector $\sum_{i=1}^{2^n} w_i x_i$. We denote $\sum_{i=1}^{2^n} w_i x_i$ by $\varphi(w)$. It remains to show that for $w, z \in V(D_n)$ the norms of $\varphi(w) - \varphi(z)$ in X can be estimated from below by norms of $f(w) - f(z)$ in $\ell_1^{2^n}$.

In the case where either w or z is the top or the bottom of the smallest subdiamond containing the images of w and z, the differences $\varphi(w) - \varphi(z)$ and $f(w) - f(z)$ have all coordinates of the same sign and therefore, by (9.58),

$$\|\varphi(w) - \varphi(z)\|_X \geq \alpha \|f(w) - f(z)\|_{\ell_1^{2^n}} = \alpha d_{D_n}(w, z).$$

In the case where w and z are on the same side of the subdiamond, the estimate is the same as is the reason for its validity.

In the different-sides case we may assume that the shortest path between w and z is through the bottom of the subdiamond, which we denote y, and we may assume that

$$d_{D_n}(w, y) \geq d_{D_n}(y, z). \tag{9.72}$$

We may also assume that w is on the left side. In this case we use (9.58) in the case where A is the left half of F. We write $\varphi(w) = \sum_{i=1}^{2^n} w_i x_i$, $\varphi(z) = \sum_{i=1}^{2^n} z_i x_i$, and $\varphi(y) = \sum_{i=1}^{2^n} y_i x_i$. Inequality (9.72) implies that z is closer to y than the rightmost vertex. For this reason all coordinates of z on A are the same as of y and, by (9.58), we get

$$\begin{aligned} \|\varphi(w) - \varphi(z)\|_X &\geq \alpha \left| \sum_{i \in A} (w_i - z_i) \right| \\ &= \alpha \left| \sum_{i \in A} (w_i - y_i) \right| \\ &\geq \frac{1}{2} \alpha d_{D_n}(w, y) \\ &\geq \frac{\alpha}{4} d_{D_n}(w, z). \end{aligned}$$

□

9.4 Exercises

Exercise 9.38. Let $0 < \alpha < 1$ and X be an n-dimensional normed linear space. Then the cardinality of an α-net for B_X is at least α^{-n}.

Exercise 9.39. Prove the following estimate for the discretization modulus:

$$\delta_{\ell_1^n \hookrightarrow L_2}(\varepsilon) \leq 2/\left((1-\varepsilon)^2 n\right).$$

There is a hint to Exercise 9.39.

Exercise 9.40. Construct an equivalent norm $|||\cdot|||$ on ℓ_2 such that $|||\cdot|||$ does not attain its maximum on the unit ball of ℓ_2 in its original norm.

Exercise 9.41. We introduce the infinite diamond as a kind of inductive limit of diamond graphs. Recall that the *diamond graph* of level 0 is denoted D_0. It has two vertices joined by an edge of length 1. D_i is obtained from D_{i-1} as follows. Given an edge $uv \in E(D_{i-1})$, it is replaced by a quadrilateral u, a, v, b with edge lengths 2^{-i}. We endow D_n with their shortest path metrics. We consider the vertex of D_n as a subset of the vertex set of D_{n+1}, it is easy to check that this defines an isometric embedding. We introduce the *infinite diamond* D_ω as the union of the vertex sets of $\{D_n\}_{n=0}^{\infty}$. For $u, v \in D_\omega$ we introduce $d_{D_\omega}(u, v)$ as $d_{D_n}(u, v)$ where $n \in \mathbb{N}$ is any integer for which $u, v \in V(D_n)$. Since the natural embeddings $D_n \to D_{n+1}$ are isometric, it is easy to see that $d_{D_n}(u, v)$ does not depend on the choice of n for which $u, v \in V(D_n)$.

Show that a Banach space admitting a bilipschitz embedding of the infinite diamond is nonreflexive.

There is a hint to Exercise 9.41.

Exercise 9.42. The *Laakso graph* of level 0 is denoted L_0. It consists of two vertices joined by an edge of length 1. The *Laakso graph* L_i is obtained from L_{i-1} as follows. Each edge $uv \in E(L_{i-1})$ of length 4^{-i+1} is replaced by a graph with 6 vertices u, t_1, t_2, o_1, o_2, v where o_1, t_1, o_2, t_2 form a quadrilateral, and there are only two more edges ut_1 and vt_2, with all edge lengths 4^{-i}. We endow L_n with their shortest path metrics.

Prove the following analog of the "only if" part of Theorem 9.36: Let X be a nonsuperreflexive Banach space. Then there is a constant $C < \infty$ such that $c_X(L_n) \leq C$ for all $n \in \mathbb{N}$.

9.5 Notes and remarks

The Ribe theorem was obtained in [393]. Some versions of Ribe's proof were presented in [136] and [55, pp. 222–224]. Another proof of the Ribe theorem was obtained by Heinrich and Mankiewicz [200], it is also presented in [55, p. 222]. The third proof of the Ribe theorem is due to Bourgain [77]. The paper [77] is very difficult reading, for this reason several authors worked on a simplification and clarification of Bourgain's argument. One of the steps in Bourgain's proof was simplified in [52]. The rest of Bourgain's proof was simplified and explained by Giladi, Naor, and Schechtman [164]. In Section 9.2 we present Bourgain's proof of the Ribe theorem following [164] and using [52]. Lemma 9.13 is a refinement of a result of Bourgain [77]. The proof of Bourgain's almost extension result has been significantly simplified by Begun [52], and our proof of Lemma 9.13 follows Begun's argument. Our proof of Lemma 9.12 is due to Begun [52].

Bourgain's proof of the Ribe theorem is, in my opinion, more complicated than the other proofs. I have chosen this proof for presentation in this chapter because the quantitative estimates obtained in Bourgain's proof (and further results along the same lines) are important for applications (see [164], [336]). Also, Bourgain's proof is the only one which is not presented in the well-known monograph [55]. There is an alternative proof of Bourgain's discretization theorem in the case where the target space is superreflexive, see [282]. There are interesting related results and open problems for specific target spaces in the paper [164].

Estimate (9.1) was proved in [164], Bourgain [77] proved a slightly weaker estimate

$$\delta_{X \hookrightarrow Y}(\varepsilon) \geq e^{-(n/\varepsilon)^{Cn}}. \tag{9.73}$$

The estimate (9.1) implies (9.73) since due to the Dvoretzky theorem (Theorem 2.30) $c_Y(\ell_2^n) = 1$, and therefore $c_Y(X) \leq \sqrt{n}$ by John's theorem (Theorem 9.10).

Lemma 9.7 is due to Corson and Klee [103]. Proposition 9.9 is a part of a well-developed direction devoted to approximation of arbitrary convex bodies by smooth ones, see [406] and references therein. Theorem 9.10 is due to John [218]. We give a proof which uses only basic calculus and n-dimensional analytic geometry. There are many other proofs which are more elegant and provide additional information, but use more advanced tools, see [37, Lecture 3], [380, Theorem 1.12], [381, Chapter 3].

Lemma 9.18 is a classical lemma, most possibly it was rediscovered many times by different authors. See, for example, the proof of Schur's theorem in [105, Theorem 36.14].

More details on the Poisson kernel can be found in [416, Chapter III, Section 2] and [428, Section X.3]. The computation of the surface area of the unit sphere can be found, for example, in [37, Lecture 1]. The Stirling formula for the gamma function is derived, for example, in [150, Section 540]. See [150, Section 531] for relations between the gamma function and factorial.

The program which came to be known as "the Ribe program" was suggested by Bourgain in [76]. After suggesting it Bourgain wrote: "A detailed exposition of this program will appear in J. Lindenstrauss's forthcoming survey paper [J. Lindenstrauss, Topics in the geometry of metric spaces, to appear]." Unfortunately, the mentioned paper of Lindenstrauss was never published and I have never seen even an unpublished version of it. Today the Ribe program is a very active research direction. See [38] for a short and [334] for an extensive survey on the Ribe program.

The theory of superreflexive spaces was initiated and developed by James in a very interesting series of papers, see [213] and [214]. In particular he proved Theorem 9.27. Further important contributions are due to Enflo [135] (see Theorem 8.36) and Pisier [377] (see Theorem 8.37).

Theorem 9.30 is due to Lindenstrauss [284]. The notion of a locally uniformly convex norm was introduced in [296]. Kadets [229] proved that each separable Banach space has an equivalent locally uniformly convex norm. We presented a simplified proof which is due to Davis and Johnson [111]. The nonseparable version of Theorem 9.30 is due to Troyanski [429].

There is an alternative proof of the fact that a space containing a δ-tree cannot be reflexive. It is based on the argument going back to Asplund and Namioka [29] and can be found in [50, pp. 232–235]. See [74] for further development of this method which also generalizes the above mentioned result of Troyanski.

Our proof of the existence of the equivalent norm $|||\cdot|||$ which attains its maximum on a weakly compact set uses the idea of Lindenstrauss [284], which was used by him to prove that operators attaining maximum of the norm on a weakly compact separable set are dense in the space of all operators. There is an alternative approach to the final part of our proof of Theorem 9.30. It is based on the notion of a farthest point, see [28], [269], [270].

The test-space characterization of superreflexivity which we present in Section 9.3.2 is due to Johnson and Schechtman [226], our proof of the "if" part follows [356]. The original proof in [226] uses Theorem 8.36. The paper [226] uses some ideas of the papers [83] and [276], in which diamond graphs were used to prove negative results about dimension reduction in ℓ_1. It seems that the first metric geometry paper dealing with diamond graphs is the paper of Gupta, Newman, Rabinovich, and Sinclair [190] (the conference version of this paper appeared in 1999). A closely related construction is due to Laakso [261].

Exercise 9.39 is from [164].

Exercise 9.41 is from [356]. Since spaces with the Radon–Nikodým property also do not contain bounded δ-trees (see e.g. [55]) we actually get a stronger result: if a Banach space admits a bilipschitz embedding of an infinite diamond, it does not have the Radon–Nikodým property. In [356] a similar result was proved for the Laakso space defined as an inductive limit of the graphs L_n (defined in Exercise 9.42): We consider the vertex of L_n as a subset of the vertex set of L_{n+1}, it is easy to check that this defines an isometric embedding. We introduce the *Laakso space* L_ω as the union of the vertex sets of $\{L_n\}_{n=0}^{\infty}$. For $u, v \in L_\omega$ we introduce $d_{L_\omega}(u, v)$ as $d_{L_n}(u, v)$ where $n \in \mathbb{N}$ is any integer for which $u, v \in V(L_n)$. Since the natural embeddings $L_n \to L_{n+1}$ are isometric, it is easy to see that $d_{L_n}(u, v)$ does not depend on the choice of n for which $u, v \in V(L_n)$. Laakso space was originally constructed in [261], our description of it is similar to the version presented in [267, p. 290]. The fact that Laakso space does not admit bilipschitz embeddings into a Banach space with the Radon–Nikodým property was known before, see [96].

Exercise 9.42 is from [226] who sketched a proof of the fact that a Banach space X is non-superreflexive if and only if there is a constant $C < \infty$ such that $c_X(L_n) \le C$ for all $n \in \mathbb{N}$. Their proof of the "if" part is not similar to our proof of the "if" part of Theorem 9.36. In an interesting related study Cheeger and Kleiner [97] proved a general result on embeddability into L_1 of a wide class of Laakso-space-type metric spaces.

9.5.1 Another test-space characterization of superreflexivity: binary trees

The following test-space characterization of superreflexivity predates the characterization which we presented in Section 9.3.2. It was proved by Bourgain in the same paper [76] in which he suggested the Ribe program. For $n \in \mathbb{N}$ we define the *binary tree of depth n* as a graph T_n whose vertex set is the set of sequences of 0 and 1 containing at most n elements, including an empty sequence. Two sequences are joined by an edge if and only if one of them is obtained from the other by adding one element at the right end.

Theorem 9.43. *A Banach space is non-superreflexive if and only if it admits uniformly bilipschitz embeddings of the graphs* $\{T_n\}_{n=1}^{\infty}$.

A proof of Theorem 9.43 can be found in [383]. See [313] for a result in a closely related direction.

Baudier [45] proved the following strengthening of the "only if" part of Theorem 9.43. We define the *infinite binary tree* T_∞ as an infinite graph whose vertex set is the set of all finite sequences of 0 and 1, including an empty sequence. Two sequences are joined by an edge if and only if one of them is obtained from the other by adding one element at the right end.

Theorem 9.44. *A Banach space X is non-superreflexive if and only if it admits a bilipschitz embedding of* T_∞.

Theorem 9.44 can be obtained by combining Theorem 9.43 with Theorem 2.6. It should be mentioned that 9.44 was proved before Theorem 2.6.

9.5.2 Further results on test-spaces

Test-spaces are known for several important classes of Banach spaces.

First we observe that the Maurey–Pisier theorem ([319], we stated it as Theorem 2.55), combined with

(1) The estimates of type and cotype of spaces L_p (Theorem 2.53);

(2) The result of Bretagnolle–Dacuhna-Castelle–Krivine [80] implying that for $1 \leq p \leq q \leq 2$ the space $L_q(0,1)$ is isometric to a subspace of $L_p(0,1)$;

(3) Combination of the result on differentiability of Lipschitz maps proved by Heinrich and Mankiewicz [200] (see also [55, Theorem 7.9 and Corollary 7.10]) with the *principle of local reflexivity* of Banach spaces ([286], [225]) (we state and prove the principle of local reflexivity in Theorem 10.18) implies that a finite-dimensional Banach space F admits a bilipschitz embedding into a Banach space X with distortion C if and only if F admits an isomorphic embedding $f : F \to X$ such that $\|f\| \cdot \|f^{-1}|_{f(F)}\| = C$;

implies the following results:

(a) The spaces $\{\ell_\infty^n\}_{n=1}^\infty$ form a collection of test-spaces for the class of spaces with nontrivial cotype.

(b) The spaces $\{\ell_1^n\}_{n=1}^\infty$ form a collection of test-spaces for the class of Banach spaces with nontrivial type.

(c) Item (b) is a part of a more general result: For any fixed $p \in [1,2)$ the spaces $\{\ell_p^n\}_{n=1}^\infty$ form a collection of test-spaces for the class of Banach spaces which have type q for some $q > p$.

The mentioned classes of Banach spaces and their test-spaces were studied in the spirit of Problem 9.23(a)–(d). We mention the results obtained on these lines:

(a) Bourgain, Milman, and Wolfson [79] proved that the Hamming cubes $\{\mathbb{F}_2^n\}_{n=1}^\infty$ are a collection of test-spaces for the class of Banach spaces with nontrivial type. See [379] for a simpler proof.

(b) Mendel and Naor [327] proved that *lattice graphs* $L_{m,n}$ defined as graphs whose vertex sets are $\{0, 1, \ldots, m\}^n$, two vertices are joined by an edge if and only if their ℓ_∞-distance is equal to 1, form a collection of test-spaces for the class of Banach spaces without cotype.

(c) Ostrovskii [357] proved that for each sequence $\{X_m\}_{m=1}^\infty$ of finite-dimensional Banach spaces there exists a sequence $\{H_n\}_{n=1}^\infty$ of finite connected unweighted graphs with maximum degree 3 such that the following conditions on a Banach space Y are equivalent:

- Y admits uniformly isomorphic embeddings of $\{X_m\}_{m=1}^{\infty}$.
- Y admits uniformly bilipschitz embeddings of $\{H_n\}_{n=1}^{\infty}$.

This result can be combined with the consequences of the Maurey–Pisier theorem mentioned above.

(d) Baudier [45] and Ostrovskii [361] showed that the classes mentioned above can be characterized using one test-space.

9.5.3 Further results on the Ribe program

We would like to mention classes of Banach spaces for which the Ribe program was completed (in some sense):

(a) Banach spaces with type p $(1 < p \leq 2)$ [326]; earlier the program was almost completed in [79], see [379] for a simplification of some proofs in [79]. Many interesting questions remain open, see [137], [163], [334].

(b) Banach spaces with cotype q $(2 \leq q < \infty)$ [327], some improvements in [162].

(c) q-uniformly convex Banach spaces [278] and [328]. The corresponding problem for p-uniformly smooth Banach spaces remains open, see [334].

9.5.4 Non-local properties

The Ribe program concerns only hereditary local isomorphic invariants. But the problem of a metric characterization and a test-space characterization can be posed for an arbitrary class of Banach spaces satisfying the condition **(H)** defined on page 288. Johnson [426, Problem 1.1] suggested this problem for reflexivity and the Radon–Nikodým property. Some results in this direction were obtained in [356] and [362]. We have already mentioned some of the results of [356] in Exercise 9.41 and our comment to it.

9.6 Hints to exercises

To Exercise 9.39. Use the fact that the metric space $(\ell_1^n, \sqrt{\|x-y\|_1})$ embeds isometrically into L_2 (see Corollary 1.43). Derive from here that any δ-net in $B_{\ell_1^n}$ embeds into L_2 with distortion at most $\sqrt{2/\delta}$. Derive from Theorem 4.23 the equality $c_{L_2}(\ell_1^n) = \sqrt{n}$. Combine these estimates.

To Exercise 9.41. Use the fact that a Banach space containing a bounded δ-tree is nonreflexive (page 290) and the argument used to prove the "if" part of Theorem 9.36.

Chapter 10

Lipschitz free spaces

10.1 Introductory remarks

A slight variation of the argument of Proposition 1.7 shows that each metric space X can be embedded into $\ell_\infty(X)$ and that this embedding is in a certain sense canonical, up to a choice of a base point. If we consider an n-element metric space, we can find its canonical isometric embedding into ℓ_∞^n. However, for applications it is important to embed into spaces which do not contain large-dimensional subspaces which are close to ℓ_∞^n.

In this connection it is interesting to analyze the second known canonical embedding of a metric space X into a Banach space. It was invented and reinvented by many different authors. For this reason the construction has several different names: Arens–Eells space, Kantorovich–Rubinshtein space, Lipschitz free space. I have decided to use the term *Lipschitz free space*, because I did not come to a definite conclusion on the priority problem, and prefer to use the term "Lipschitz free space", rather than to use different names for different versions and special cases.

This chapter has a modest purpose: to introduce some basic facts about the Lipschitz free spaces and to mention (in the Notes and Remarks section) some other known facts about these spaces. In the presentation of the basic facts we treat separately the case where the metric space is a graph with its graph distance. In this case the Lipschitz free space admits a natural description of terms of the cut space and the cycle space.

10.2 Lipschitz free spaces: definition and properties

Definition 10.1. Let X be a metric space. A *molecule* of X is a function $m : X \to \mathbb{R}$ which is supported on a finite set and which satisfies $\sum_{p\in X} m(p) = 0$. For $p, q \in X$ define the molecule m_{pq} by $m_{pq} = \mathbf{1}_p - \mathbf{1}_q$, where $\mathbf{1}_p$ and $\mathbf{1}_q$ are indicator functions of singleton sets $\{p\}$ and $\{q\}$. We endow the space of molecules with the seminorm

$$\|m\|_{\mathrm{LF}} = \inf\left\{\sum_{i=1}^{n} |a_i| d_X(p_i, q_i) : m = \sum_{i=1}^{n} a_i m_{p_i q_i}\right\}.$$

It is not difficult to see that this is actually a norm (this statement also follows from Theorem 10.2). The *Lipschitz free space* over X is defined as the completion of the space of all molecules with respect to the norm $\|\cdot\|_{\mathrm{LF}}$. We denote the Lipschitz free space over X by $\mathrm{LF}(X)$.

By a *pointed metric space* we mean a metric space with a distinguished point, denoted O. By $\mathrm{Lip}_0(X)$ we denote the space of all Lipschitz functions $f : X \to \mathbb{R}$ satisfying $f(O) = 0$, where O is the distinguished point of a pointed metric space X. It is not difficult to check that $\mathrm{Lip}_0(X)$ is a Banach space with respect to the norm $\|f\| = \mathrm{Lip}(f)$.

Theorem 10.2. *Let X be a pointed metric space. The dual space of the Lipschitz free space* $\mathrm{LF}(X)$ *can be described as*

$$\mathrm{LF}(X)^* = \mathrm{Lip}_0(X).$$

The duality for a given function $G \in \mathrm{Lip}_0(X)$ and a molecule $m = \sum_{i=1}^n a_i m_{p_i q_i}$ is given by

$$G(m) = \sum_{i=1}^n a_i (G(p_i) - G(q_i)). \tag{10.1}$$

Proof. Let $g \in \mathrm{LF}(X)^*$, we introduce a function on X by $G(p) = g(m_{pO})$. It is clear that $G(O) = 0$. Now we show that

$$\mathrm{Lip}(G) \leq \|g\|^*,$$

where $\|g\|^*$ is the norm of g in $\mathrm{LF}(X)^*$. In fact, we have $|G(p) - G(q)| = |g(m_{pO} - m_{qO})| \leq \|g\|^* \|m_{pO} - m_{qO}\|_{\mathrm{LF}} \leq \|g\|^* d_X(p,q)$.

On the other hand, for each $G \in \mathrm{Lip}_0(X)$, we can define the functional g on molecules of the form m_{pO} by $g(m_{pO}) = G(p)$, and extend it by linearity to the space of all molecules, so (10.1) holds (straightforward verification shows that this extension is well defined). To show that this functional extends to a bounded linear functional on $\mathrm{LF}(X)$ and to complete the proof it suffices to show that $\|g\|^* \leq \mathrm{Lip}(G)$.

Let $\varepsilon > 0$, let m be a molecule, and let $m = \sum_{i=1}^n a_i m_{p_i q_i}$ be its almost optimal representation, that is, $\|m\|_{\mathrm{LF}} \geq (1-\varepsilon) \sum_{i=1}^n |a_i| d_X(p_i, q_i)$. Since $m_{p_i q_i} = m_{p_i O} - m_{q_i O}$, we have

$$g(m) = \sum_{i=1}^n a_i (G(p_i) - G(q_i)) \leq \sum_{i=1}^n |a_i| \mathrm{Lip}(G) d_X(p_i, q_i) \leq \frac{\mathrm{Lip}(G)}{1-\varepsilon} \|m\|_{\mathrm{LF}}.$$

Letting $\varepsilon \downarrow 0$ we get $\|g\|^* \leq \mathrm{Lip}(G)$. □

Remark 10.3. The definition of $\|\cdot\|_{\mathrm{LF}}$ immediately implies that it is the largest seminorm on the space of molecules among all seminorms satisfying $\|m_{pq}\| \leq d_X(p,q)$.

Now we can prove the property which makes the Lipschitz free spaces important for metric geometry.

Theorem 10.4. *Let X be a pointed metric space. Then*

(a) *The map δ given by $\delta(p) = m_{pO}$ isometrically embeds X in* $\mathrm{LF}(X)$. *(We denote this map by δ because $\delta(p)$, considered as an element of the space* $(\mathrm{Lip}_0(X))^*$, *corresponds to the point evaluation at p.)*

(b) *Let E be a Banach space and let $T : X \to E$ be a Lipschitz map satisfying $T(O) = 0$. Then there is a unique bounded linear map $\tilde{T} : \mathrm{LF}(X) \to E$ such that $\tilde{T} \circ \delta = T$. This map satisfies $\|\tilde{T}\| = \mathrm{Lip}(T)$.*

Proof. (a) We need to prove $\|m_{pO} - m_{qO}\|_{\mathrm{LF}} = d(p,q)$. Since $m_{pO} - m_{qO} = m_{pq}$, the inequality $\|m_{pO} - m_{qO}\|_{\mathrm{LF}} \le d(p,q)$ follows immediately from the definition of $\|\cdot\|_{\mathrm{LF}}$. As for the inverse inequality, we can get it from the duality established in Theorem 10.2. In more detail: there exists a 1-Lipschitz function G in $\mathrm{Lip}_0(X)$ such that $G(q) - G(p) = d(p,q)$. The function $d(p,x) - d(p,O)$ is one such function. Therefore $\|m_{pO} - m_{qO}\|_{\mathrm{LF}} \ge |G(m_{pO} - m_{qO})| = |G(p) - G(q)| = d(p,q)$.

(b) Every molecule is uniquely expressible in the form $m = \sum_i a_i m_{p_i O}$, where points $\{p_i\}$ are distinct and none of them coincides with O. We define $\tilde{T}$ by

$$\tilde{T}(m) = \sum a_i T(p_i).$$

This definition immediately implies $\tilde{T} \circ \delta = T$. Since δ is an isometric embedding and $\mathrm{Lip}(\tilde{T}) = \|\tilde{T}\|$, this equality immediately implies $\|\tilde{T}\| \ge \mathrm{Lip}(T)$.

Is it easy to see that for $m = \sum_{i=1}^n a_i m_{p_i q_i}$ we have

$$\begin{aligned}\|\tilde{T}(m)\| &= \left\| \sum_{i=1}^n a_i \tilde{T}(m_{p_i q_i}) \right\| \\ &= \left\| \sum_{i=1}^n a_i (T(p_i) - T(q_i)) \right\| \\ &\le \sum_{i=1}^n |a_i| \mathrm{Lip}(T) d(p_i, q_i).\end{aligned}$$

Taking infimum over all representations of m in the form $\sum_{i=1}^n a_i m_{p_i q_i}$, we get $\|\tilde{T}(m)\| \le \mathrm{Lip}(T) \cdot \|m\|_{\mathrm{LF}}$. Thus $\|\tilde{T}\| \le \mathrm{Lip}(T)$. □

Remark 10.5. Theorem 10.4 shows that $\mathrm{LF}(X)$ is in a sense the “smallest” Banach space containing an isometric image of X. This “smallness” is however different from the natural Banach space theoretical “smallness”. It does not imply that any Banach space containing an isometric image of X contains an isometric copy of $\mathrm{LF}(X)$. One of the important examples of this type is presented in Theorem 10.16, see Remark 10.17.

Proposition 10.6. *Let X be a metric space and X_0 be a subset of X. The subset X_0 is regarded as a metric space with the metric inherited from X. The equality*

$$\|m\|_{\mathrm{LF}(X_0)} = \|m\|_{\mathrm{LF}(X)}$$

holds for each molecule m of X_0 and therefore there is a natural isometric embedding of $\mathrm{LF}(X_0)$ *into* $\mathrm{LF}(X)$.

Remark 10.7. At this point it becomes important that molecules are defined as real-valued functions (or real-valued finitely-supported measures). If we consider a complex-valued analogue of the spaces $\mathrm{LF}(X)$, the embedding of $\mathrm{LF}(X_0)$ into $\mathrm{LF}(X)$ does not have to be isometric.

Proof of Proposition 10.6. Let m be a molecule of X_0. It is immediate from the definition that $\|m\|_{\mathrm{LF}(X_0)} \geq \|m\|_{\mathrm{LF}(X)}$ (the infimum corresponding to the latter norm is over a larger set of representations).

To prove the inverse inequality we use the description of the dual space (Theorem 10.2) and the extension property of Lipschitz real-valued functions.

It is convenient to pick a distinguished point O in the smaller space X_0. By Theorem 10.2 there exists $f \in \mathrm{Lip}_0(X_0)$ such that $\mathrm{Lip}(f) = 1$ and $f(m) = \|m\|_{\mathrm{LF}(X_0)}$. By the extension property of Lipschitz real-valued functions (see Theorem 10.8 below), there exists $\tilde{f} \in \mathrm{Lip}_0(X)$ with $\mathrm{Lip}(\tilde{f}) = 1$ whose restriction to X_0 coincides with f. Therefore we have $\|m\|_{\mathrm{LF}(X)} \geq \tilde{f}(m) = f(m) = \|m\|_{\mathrm{LF}(X_0)}$. □

Theorem 10.8. *Let X_0 be a subset of a metric space (X, d), X_0 is also considered as a metric space with metric d. Let f be a real-valued Lipschitz function on X_0. Then there exists a Lipschitz function $\tilde{f} : X \to \mathbb{R}$ such that* $\mathrm{Lip}(\tilde{f}) = \mathrm{Lip}(f)$ *and the restriction $\tilde{f}|_{X_0}$ coincides with f. Furthermore, if f is a bounded function and $a \leq f(x) \leq b$ for some $a, b \in \mathbb{R}$ and all $x \in X_0$, then the function $\tilde{f}$ can be chosen to satisfy $a \leq \tilde{f}(x) \leq b$ for all $x \in X$.*

Proof. First we prove the first part of the theorem (without the boundedness condition). We can view this proof as follows. We assign values to $\tilde{f}(x)$ for each $x \in X \setminus X_0$, one at a time (formally the construction can actually be done like this only if $X \setminus X_0$ is at most countable, in the general case we use Zorn's lemma to complete the proof). We need to show that in each step there is at least one real number available as a suitable value for $\tilde{f}(x)$.

Straightforward verification shows that the only restriction for the choice of $\tilde{f}(x)$ is

$$\sup_{z \in X_0} (f(z) - \mathrm{Lip}(f) d(z, x)) \leq \tilde{f}(x) \leq \inf_{y \in X_0} (f(y) + \mathrm{Lip}(f) d(x, y)).$$

Therefore, to show that the desired choice is possible we need to prove that the inequality

$$f(z) - \mathrm{Lip}(f) d(z, x) \leq f(y) + \mathrm{Lip}(f) d(x, y)$$

holds for each $z, y \in X_0$. This is obvious, since the inequality can be rewritten as

$$f(z) - f(y) \le \mathrm{Lip}(f)(d(z,x) + d(x,y)),$$

and therefore it follows from the triangle inequality and

$$f(z) - f(y) \le \mathrm{Lip}(f)d(z,y),$$

which is just the Lipschitz condition.

This completes the first part of the proof in the case where $X \setminus X_0$ is finite or countable. In the general case we use Zorn's lemma (Lemma 2.5, page 34) as follows.

We consider the set E of all extensions of f to subsets of X containing X_0 and having the same Lipschitz constant as f. We regard E as a partially ordered set with the partial order $\le$ defined by: $g \le h$ if and only if the domain of h contains the domain of g and the restriction of h to the domain of g coincides with g. It is clear that each totally ordered subset H of this ordered set has an upper bound u defined by: the domain of u is the union U of all domains of elements of H, the value of u at $x \in U$ is defined as the value at x of any of the elements of H (the definition of the partial order is such that these values are the same). By Zorn's lemma, the set E has a maximal element $\widetilde{f}$. The definition of E implies that $\mathrm{Lip}(\widetilde{f}) = \mathrm{Lip}(f)$. The domain of $\widetilde{f}$ should coincide with X because otherwise, by the first part of the proof, we can extend the domain by one point.

In the case where $a \le f(x) \le b$ we construct $\widetilde{f}$ as before and then observe that if we consider the function

$$\widehat{f} = \min\big\{\max\{\widetilde{f}(x), a\},\ b\big\},$$

we get the desired function. □

10.3 The case where d_X is a graph distance

In the case where X is a graph with its graph distance, the construction of a Lipschitz free space can be obtained somewhat differently.

Let $G = (V(G), E(G)) = (V, E)$ be a graph. Denote by $C^1(G)$ and $C^0(G)$ the space of real-valued functions on the edge set and the vertex set, respectively. We consider some orientation of edges of G (see the corresponding definitions on page 136, recall that we denote by e^+ and e^- the head and the tail of an edge e, respectively). The corresponding *incidence matrix* D is defined as a matrix whose rows are labeled using vertices of G, whose columns are labeled using edges of G and the ve-entry is given by

$$d_{ve} = \begin{cases} 1, & \text{if } v = e^+, \\ -1, & \text{if } v = e^-, \\ 0, & \text{if } v \text{ is not incident to } e. \end{cases}$$

Remark 10.9. Almost all of the important results of this section do not depend on the choice of orientation, although many of the notions which we use do (for example, the incidence matrix).

Interpreting elements of $C^1(G)$ and $C^0(G)$ as column vectors, we may regard D as a matrix of a linear operator $D : C^1(G) \to C^0(G)$. We also consider the transpose matrix D^T and the corresponding operator $D^T : C^0(G) \to C^1(G)$.

It is easy to describe $\ker D^T$. In fact, for $f \in C^0(G)$ the value of $D^T(f) \in C^1(G)$ at an edge e is $f(e^+) - f(e^-)$, therefore $f \in \ker D^T$ if and only if it has the same value at the ends of each edge. It is clear that this happens if and only if f is constant on each of the connected components of G. Therefore the ranks of the operators D^T and D are equal to $|V| - c$, where c is the number of connected components of G.

We let $Z = \ker D$. This subspace of $C^1(G)$ is called the *cycle space* or *cycle subspace*. The name is chosen because, as we explain below, this subspace is spanned by the set of signed indicator functions of all cycles.

First we observe that for $f \in C^1(G)$ the function $Df \in C^0(G)$ is given by

$$(Df)(v) = \sum_{e,\, e^+=v} f(e) - \sum_{e,\, e^-=v} f(e). \tag{10.2}$$

It is convenient to view this formula in the following way. We regard f as a record of transportation of a certain commodity between storage places (vertices) and $f(e)$ is the amount of commodity transported directly from e^- to e^+. Then $(Df)(v)$ reflects the change in the amount of the commodity at the location labeled v. The function f is in $\ker D$ if and only if the amount of the commodity at each v remains unchanged.

Now we consider a cycle C in G. We consider one of the two possible orientations of C satisfying the following condition: each vertex of C is a head of exactly one edge and a tail of exactly one edge. Now we introduce the *signed indicator function* $\chi_C \in C^1(G)$ of the cycle C by

$$\chi_C(e) = \begin{cases} 1 & \text{if } e \in C \text{ and its orientations in } C \text{ and } G \text{ are the same} \\ -1 & \text{if } e \in C \text{ but its orientations in } C \text{ and } G \text{ are different} \\ 0 & \text{if } e \notin C. \end{cases} \tag{10.3}$$

It is immediate that $\chi_C \in \ker D$. Let us show that such functions span $\ker D$. To see this we observe that $\dim(\ker D) = |E(G)| - \operatorname{rank}(D) = |E(G)| - |V(G)| + c$. So it remains to show that there are $|E(G)| - |V(G)| + c$ cycles $\{C(i)\}_i$ such that the functions $\{\chi_{C(i)}\}_i$ are linearly independent.

To achieve this goal we pick a maximal subset of edges of G which does not form any cycles. Such a collection of edges of G induces a subgraph F of G which is called a *spanning forest of* G; in the case where G is connected, it is called a *spanning tree of* G. (These definitions could be slightly different from other definitions in the literature because one can answer differently the question: What to do with the isolated vertices of G?) It is a well-known (easy) fact that the number of edges in a spanning forest is $|V(G)| - c$. Therefore the number of remaining edges is $|E(G)| - |V(G)| + c$. Now we construct the cycles $C(i)$, $i \in E(G)\backslash E(F)$, with linearly independent $\{\chi_{C(i)}\}$ as follows. Each $i \in E(G)\backslash E(F)$ forms a cycle together with some edges of F. One can show that such a cycle is uniquely determined, but at this point it does not matter. We pick one such cycle and denote it $C(i)$. The fact that $\{\chi_{C(i)}\}$ are linearly independent is immediate, because the coefficient of an edge i in the linear combination $\sum_i a_i \chi_{C(i)}$ is equal to $\pm a_i$.

If we endow $C^1(G)$ with the ℓ_1 norm, we get a space which is natural to denote $\ell_1(E)$ (see page 160 for a more detailed definition).

Proposition 10.10. *If the graph G is connected, then the quotient space $\ell_1(E)/Z$, where Z is the cycle space, is isometric to the space* $\mathrm{LF}(V)$*, where V is endowed with its graph distance.*

Proof. Since the spaces $\ell_1(E)/Z$ and $\mathrm{LF}(V)$ are finite-dimensional, it suffices to show that the dual space $(\ell_1(E)/Z)^*$ can be identified in a natural way with the space $\mathrm{Lip}_0(V)$.

Similarly to $\ell_1(E)$ we introduce the spaces $\ell_\infty(E)$ and $\ell_2(E)$. The ℓ_∞-*norm* of a function $H : E(G) \to \mathbb{R}$ is defined by

$$\|H\|_\infty = \max_{e \in E(G)} |H(e)|.$$

The ℓ_2-*norm* of a function $H : E(G) \to \mathbb{R}$ is defined by

$$\|H\|_2 = \Big(\sum_{e \in E(G)} |H(e)|^2 \Big)^{\frac{1}{2}}.$$

It is clear that the space $\ell_2(E)$ is an inner product space, therefore we can associate with this space the notion of orthogonality. The subspace of $C^1(G)$ orthogonal to the cycle space Z is denoted by B and is called the *cut space* or *cut subspace*. So we have a direct, orthogonal in $\ell_2(E)$, decomposition

$$C^1(G) = Z \oplus B. \tag{10.4}$$

Now we are going to explain why B is called the cut space. The definition of a cut which we use here agrees with the definition of a cut given in Example 1.11, applied to the vertex set of G. By a *cut* we mean a partition R of V into two disjoint sets $(U, \bar{U})$, where $\bar{U} = V\backslash U$. We identify R with the set of edges of G with one end in U and the other end in $\bar{U}$. We associate with the cut R a *signed indicator function* given by picking of one of the sides of C (we pick U) and letting

$$\chi_R(e) = \begin{cases} 1 & \text{if } e \in R \text{ and its head is in } U \\ -1 & \text{if } e \in R \text{ and its tail is in } U \\ 0 & \text{if } e \notin R. \end{cases} \tag{10.5}$$

It is easy to verify that $\langle \chi_R, \chi_C \rangle = 0$ for each cycle C in G and each cut R in G. Therefore $\chi_R \in B$ for each cut R. Our next purpose is to show that B is spanned by $\{\chi_R : R \text{ is a cut in } G\}$. One of the ways of doing this is by counting the dimensions. We do this in the case where the graph G is connected. In this case $\dim Z = |E(G)| - |V(G)| + 1$ and therefore $\dim B = |V(G)| - 1$, so we need to find $|V(G)| - 1$ cuts whose signed indicator functions are linearly independent. This can be done as follows. Let T be any spanning tree in G. As we have already mentioned it has $|V(G)| - 1$ edges. Each of the edges $e \in E(T)$ separates T into two connected components, the corresponding partitions of $V(T) = V(G)$ determine cuts of G. The signed indicator functions of these cuts are linearly independent because the coefficient corresponding to e is nonzero for exactly one of the cuts.

Now we use the standard duality result (in general terms stating that the dual of the quotient space $\mathcal{X}/\mathcal{Y}$ is isometric to the subspace $\mathcal{Y}^{\perp} := \{f \in \mathcal{X}^* : \forall y \in \mathcal{Y},\ f(y) = 0\}$), the standard fact that $(\ell_1(E))^*$ can be identified with $\ell_\infty(E)$, and get that the dual space of the quotient space $\ell_1(E)/Z$ is naturally isomorphic to the space B_∞, where by B_∞ we mean the space B endowed with its ℓ_∞-norm.

To complete our argument it is convenient to use another description of B, namely to use the fact that B is the image of $C^0(G)$ under the action of D^T. We already know that $\dim B = \operatorname{rank} D^T$, so it suffices to show that each element of the image $D^T(C^0(G))$ is in B. Since B is the orthogonal complement of Z it suffices to show that $\langle z, x \rangle = 0$ for each $z \in Z$ and $x \in D^T(C^0(G))$, so let $x = D^T f$ for some $f \in C^0(G)$. We have

$$\langle z, x \rangle_{\ell_2(E)} = \langle z, D^T f \rangle_{\ell_2(E)} = \langle Dz, f \rangle_{\ell_2(V)} = \langle 0, f \rangle_{\ell_2(V)} = 0.$$

Therefore each $b \in B$ can be represented as $b = D^T f$, and so $b(e) = f(e^+) - f(e^-)$ for some $f \in C^0(G)$ and all $e \in E(G)$. It is clear that addition of a constant to the function f does not change $D^T f$, so we may assume $f(O) = 0$, so $f \in \mathrm{Lip}_0(V)$. It is clear that the Lipschitz constant of f is equal to

$$\mathrm{Lip}(f) = \max_{e \in E(G)} |f(e^+) - f(e^-)| = \|b\|_\infty.$$

Thus we have established a natural isometry between B_∞ and $\mathrm{Lip}_0(V)$. □

It seems worthwhile to describe explicitly the obtained isometric embedding K of (V, d_G) into $\ell_1(E)/Z$. This can be done in the following way. Denote by Q the quotient mapping $\ell_1(E) \to \ell_1(E)/Z$. We choose a vertex $O \in V$ and a spanning tree T in G. We let $K(O) = 0$ and $K(v) = Q\chi_{P(v)}$, where $P(v)$ is the uniquely determined path in T from O to v and $\chi_{P(v)}$ is its signed indicator function. The signed indicator function is defined similarly to (10.3): We start by orienting the path from O to v, this means that the head of each edge of the path is the vertex which is visited later if we travel on the path from O to v. After that, we set

$$\chi_{P(v)}(e) = \begin{cases} 1 & \text{if } e \in P(v) \text{ and its orientations in } P(v) \text{ and } G \text{ coincide} \\ -1 & \text{if } e \in P(v) \text{ but its orientations in } P(v) \text{ and } G \text{ differ} \\ 0 & \text{if } e \notin P(v). \end{cases} \tag{10.6}$$

Now we observe that the images of signed indicator functions of two oriented uv-paths in $\ell_1(E)/Z$ coincide: we factor over the cycle space Z and the difference of two signed indicator functions of uv-paths is obviously in $\ker D = Z$. Hence $\|K(v)\| \le d_G(O, v)$. In a similar way we prove

$$\|K(u) - K(v)\| \le d_G(u, v). \tag{10.7}$$

In fact, $\|K(u) - K(v)\| = \|Q(\chi_{P(v)} - \chi_{P(u)})\|$, and we can replace $P(v)$ by any other Ov-path and $P(u)$ by any other Ou-path. It is easy to verify that we can pick these paths, let us denote them $S(v)$ and $S(u)$, in such a way that $\chi_{S(v)} - \chi_{S(u)}$ is a signed indicator function of a shortest uv-path. Inequality (10.7) follows.

To finish our proof of the fact that K is an isometric embedding we need to estimate $\|K(u) - K(v)\|$ from below. Let $f \in \mathrm{Lip}_0(V)$ be such that $\mathrm{Lip}(f) = 1$ and $f(v) - f(u) = d_G(u, v)$. (It is easy to construct such a function: $f(w) = d_G(w, u)$ is one of the examples.) Let $b \in B$ be the corresponding element of the cut space. We have

$$\begin{aligned} \|K(v) - K(u)\|_{\ell_1(E)/Z} &\ge \langle b, \chi_{P(v)} - \chi_{P(u)} \rangle_{\ell_2(E)} \\ &= \langle D^T f, \chi_{P(v)} - \chi_{P(u)} \rangle_{\ell_2(E)} \\ &= \langle f, D\chi_{P(v)} - D\chi_{P(u)} \rangle_{\ell_2(V)} \\ &= \langle f, \mathbf{1}_v - \mathbf{1}_u \rangle_{\ell_2(V)} \\ &= f(v) - f(u) \\ &= d_G(u, v). \end{aligned}$$

In one of the steps we used an easy computation showing that $D\chi_{P(v)} - D\chi_{P(u)} = \mathbf{1}_v - \mathbf{1}_u$. This computation also shows a natural way of identification of the images of u and v in $\ell_1(E)/Z$ with molecules $\mathbf{1}_v - \mathbf{1}_u$. As we have shown above this identification is an isometry of $\ell_1(E)/Z$ and $\mathrm{LF}(V)$ (with V endowed with the metric d_G).

Remark 10.11. Combining Proposition 10.10 with the decomposition (10.4) we see that $\mathrm{LF}(X)$, in the case where X is a graph with its graph distance, may be regarded as the cut space endowed with the quotient ℓ_1 norm.

10.4 Lipschitz free spaces of some finite metric spaces

The purpose of this section is to analyze Lipschitz free spaces for some standard metric spaces. In particular, we show that $\mathrm{LF}(X)$ can be larger and more complicated (in a certain sense) than some other Banach spaces admitting bilipschitz (even isometric) embeddings of X.

Example 10.12. Let T be a graph which does not contain any cycles. Such a graph is called a *tree*. The metric space which we consider is the vertex set $V(T)$ of T endowed with the graph distance d_T of T. Since in this case the cycle space is $\{0\}$, Proposition 10.10 implies that $\mathrm{LF}(T)$ is isometric to $\ell_1(E(T))$.

Example 10.13. Let X be the complete graph K_n on n vertices with its graph distance. Then the $\mathrm{LF}(K_n)$ can be described in the following way. Consider the complete bipartite graph $K_{1,n}$. It is clear that if we multiply all distances between vertices of K_n by 2 we get a metric space isometric to the subset of $K_{1,n}$ consisting of all vertices in the part containing n points. Since $K_{1,n}$ is a tree, by Proposition 10.6 and Example 10.12, we get that $\mathrm{LF}(K_n)$ is isometric to a subspace of codimension 1 in $\mathrm{LF}(K_{1,n})$, and so to a subspace of codimension 1 in ℓ_1^n.

The two examples above can be generalized using the following terminology.

Definition 10.14. An $\mathbb{R}$-*tree* is a metric space (X, d) satisfying the following two conditions.

(a) For any points a and b in X, there exist a unique isometry φ of the closed interval $[0, d(a,b)]$ into X such that $\varphi(0) = a$ and $\varphi(d(a,b)) = b$.

(b) Any one-to-one continuous mapping $\psi : [0,1] \to T$ has the same range as the isometry φ associated to the points $a = \psi(0)$ and $b = \psi(1)$.

Theorem 10.15 (without proof). *Let X be a metric space. The following assertions are equivalent.*

- $\mathrm{LF}(X)$ *is isometric to a subspace of an L_1-space.*
- *X admits an isometric embedding into an $\mathbb{R}$-tree.*

The following example indicates that Lipschitz free spaces have a very complicated structure even for relatively simple metric spaces.

Theorem 10.16. *For some $C \geq 1$ and each $m \in \mathbb{N}$ there exists $N \in \mathbb{N}$ such that for $n \geq N$ the spaces $\mathrm{LF}(\mathbb{F}_2^n)$ contain subspaces F_m which are C-isomorphic to ℓ_∞^m.*

Remark 10.17. Theorem 10.16 illustrates the statement made in Remark 10.5. In fact, it is clear that the space $\mathbb{F}_2^n$ is isometric to a subset of the space ℓ_1^n: if we map each sequence of 0 and 1, as an element of $\mathbb{F}_2^n$ onto the same sequence, but now regarded as an element of ℓ_1^n, we get an isometry. On the other hand, since L_1 has cotype 2 (Theorem 2.53), it is easy to see that ℓ_1^n does not contain subspaces C-isomorphic to ℓ_∞^m for any fixed C and sufficiently large m.

Proof of Theorem 10.16. Let $\{(X_i, d_i)\}_i$ be a finite sequence of metric spaces. The ℓ_1-*sum* of $\{(X_i, d_i)\}_i$ is defined as a metric space (X, d), where X is the Cartesian product of $\{X_i\}_i$ and

$$d(\{u_i\}_i, \{v_i\}_i) = \sum_i d_i(u_i, v_i).$$

We use the notation $(\oplus_i X_i)_1$ for the ℓ_1-sum of $\{(X_i, d_i)\}_i$.

First we show that for each m and k there is an isometric embedding of the ℓ_1-sum of m cycles C_{2k} of lengths $2k$ into the metric space $\mathbb{F}_2^{km}$.

We label the vertices of the cycle C_{2k} by $1, 2, \ldots, 2k$. For $i \le k$ we map the vertex i onto the sequence $(\underbrace{1, \ldots, 1}_{i}, \underbrace{0, \ldots, 0}_{k-i}) \in \mathbb{F}_2^k$. For $i > k$ we map the vertex i onto the sequence $(\underbrace{0, \ldots, 0}_{i-k}, \underbrace{1, \ldots, 1}_{2k-i})$. It is easy to check that this map is an isometric embedding.

It is easy to check that $(\oplus_{i=1}^m \mathbb{F}_2^k)_1$ is isometric to $\mathbb{F}_2^{km}$. Therefore $(\oplus_{i=1}^m C_{2k})_1$ (we mean that each summand is equal to C_{2k}) is isometric to a subset of $\mathbb{F}_2^{km}$. Hence, by Proposition 10.6, it suffices to find C-isomorphic embeddings of ℓ_∞^n into $\mathrm{LF}\left((\oplus_{i=1}^{2^n} C_{2k})_1\right)$ for sufficiently large k, where $C \ge 1$ is some constant which does not depend on n.

Let $\mathbb{T}$ be the circle of length 1 with its natural geodesic metric. It is clear that after multiplication of all distances in C_{2k} by $\frac{1}{2k}$ we get a metric space isometric to a subspace of $\mathbb{T}$. It is also clear that if we multiply all distances in C_{2k} by $\frac{1}{2k}$, the only change in the Banach space $\mathrm{LF}\left((\oplus_{i=1}^{2^n} C_{2k})_1\right)$ will be that all of the norms of its vectors are multiplied by $\frac{1}{2k}$. Therefore, by Proposition 10.6, the spaces $\mathrm{LF}\left((\oplus_{i=1}^{2^n} C_{2k})_1\right)$ are isometric to subspaces of $\mathrm{LF}\left((\oplus_{i=1}^{2^n} \mathbb{T})_1\right)$. It is easy to see that, since the union of the images of the metric spaces $\left\{(\oplus_{i=1}^{2^n} C_{2k})_1\right\}_{k=2}^\infty$ is dense in the space $(\oplus_{i=1}^{2^n} \mathbb{T})_1$, the union of the natural images of the spaces $\left\{\mathrm{LF}\left((\oplus_{i=1}^{2^n} C_{2k})_1\right)\right\}_{k=2}^\infty$ in $\mathrm{LF}\left((\oplus_{i=1}^{2^n} \mathbb{T})_1\right)$ is dense in $\mathrm{LF}\left((\oplus_{i=1}^{2^n} \mathbb{T})_1\right)$. Therefore it suffices to find a C-isomorphic embedding of ℓ_∞^n into $\mathrm{LF}\left((\oplus_{i=1}^{2^n} \mathbb{T})_1\right)$ (with C independent of n).

We denote the space $\mathrm{Lip}_0\left((\oplus_{i=1}^{2^n} \mathbb{T})_1\right)$ by Lip, for short. Now we show that in order to find a C-isomorphic embedding of ℓ_∞^n into $\mathrm{LF}\left((\oplus_{i=1}^{2^n} \mathbb{T})_1\right)$ it is enough to find uniformly bounded operators

$$L_n : \mathrm{Lip}_0\left((\oplus_{i=1}^{2^n} \mathbb{T})_1\right) \to \ell_1^n$$

satisfying for some constant $D \in (0, \infty)$ the condition

$$DL_n(B_{\text{Lip}}) \supset B_{\ell_1^n}. \tag{10.8}$$

To do this we need the following result called the *principle of local reflexivity*. In this theorem we denote by X^* the dual space of X, by X^{**} the second dual of X, that is the dual of X^*. We identify X with the corresponding subspace of the second dual X^{**}.

Theorem 10.18. *Let X be a Banach space and E and F be finite-dimensional subspaces of X^{**} and X^*, respectively, and let $\varepsilon > 0$. Then there exists an injective linear operator $T : E \to X$ such that $\|T\|\|T^{-1}\| < 1 + \varepsilon$, where $T^{-1} : T(E) \to E$; the equality $x^*(Tx^{**}) = x^{**}(x^*)$ holds for all $x^{**} \in E$ and all $x^* \in F$, and $Tx = x$ for all $x \in E \cap X$.*

Lemma 10.19. *Let X, Y be Banach spaces and $T : X \to Y$ be a continuous linear operator with closed range. If $x^{**} \in X^{**}$ and $y \in Y$ are such that $T^{**}x^{**} = y$, where $T^{**} : X^{**} \to Y^{**}$ is the second conjugate operator, then for any $\delta > 0$ there exists $x \in X$ such that $\|x\| < (1 + \delta)\|x^{**}\|$ and $Tx = y$.*

Proof. This lemma can be derived from the Goldstine theorem which we have already mentioned (page 262). In fact, we may assume that $\|x^{**}\| \le 1$, so $y \in T^{**}(B_{X^{**}})$. By the Goldstine theorem $B_{X^{**}}$ is the weak* closure of B_X. Since T^{**} is weak* continuous, we get that y is in the weak* closure of $T(B_X)$. Both y and $T(B_X)$ are in Y, and the restriction of weak* topology to Y is just the weak topology of Y. Thus y is in the weak closure of $T(B_X)$. On the other hand $T(B_X)$ is a convex set, hence its weak and strong closures coincide. Therefore y is in the strong closure of $T(B_X)$. Since T has closed range, by the Baire category theorem, the strong closure of $T(B_X)$ has interior points in TX. This easily implies that for any $\alpha \in (0, 1)$ there exists a representation $y = \sum_{i=0}^{\infty} y_i$ such that $y_i \in \alpha^i T(B_X)$. Let $z_i \in B_X$ be such that $y_i = \alpha^i T(z_i)$. Then $y = T\left(\sum_{i=0}^{\infty} \alpha^i z_i\right)$, $\sum_{i=0}^{\infty} \alpha^i z_i \in X$, and $\left\|\sum_{i=0}^{\infty} \alpha^i z_i\right\| \le \frac{1}{1-\alpha}$. Picking α satisfying $\frac{1}{1-\alpha} = 1 + \delta$, we get the desired result. □

Proof of Theorem 10.18. Choose $\delta > 0$ so that if we let $\omega = \delta$ in Lemma 2.32, we get $\|T\| \cdot \|T^{-1}\| < 1 + \varepsilon$.

Choose norm one elements $a_1, a_2, \dots, a_m \in X^*$ containing a basis of F and such that

$$\|x^{**}\| < (1 + \delta) \sup_{j \in \{1,\dots,m\}} |x^{**}(a_j)|$$

for all $x^{**} \in E$ (this can be done using the argument employed in the proof of Proposition 9.9, see page 268). Choose $b_1, b_2, \dots, b_n$ to be a δ-net in the unit sphere of E such that $b_1, \dots, b_k$ is a basis for $X \cap E$ and $b_1, \dots, b_r$, $r \ge k$, is a basis for E. Then, for $1 \le p \le q \stackrel{\text{def}}{=} n - r$, we have the unique scalars $\{t_{p,i}\}$, $1 \le i \le r$, such that

$$b_{r+p} = \sum_{i \in \{1,\dots,r\}} t_{p,i} b_i.$$

Define for $1 \le p \le q$ numbers

$$s_{p,i} = \begin{cases} t_{p,i}, & i \le r, \\ -1, & i = r + p, \\ 0, & r < i \le n \text{ and } i \ne r + p. \end{cases}$$

Denote by X^n the ℓ_∞ direct sum of n copies of Banach space X. Define $A_0 : X^n \to X^{k+q}$ by

$$A_0(x_1, \dots, x_n) = \left(x_1, \dots, x_k, \sum_{i \in \{1,\dots n\}} s_{1,i} x_i, \dots, \sum_{i \in \{1,\dots n\}} s_{q,i} x_i \right).$$

It is easy to check that the operator A_0 is onto. Let Z be the ℓ_∞ direct sum of X^{k+q} and ℓ_∞^{nm}. Define $A : X^n \to Z$ by

$$A(x_1, \dots, x_n) = (A_0(x_1, \dots, x_n); (a_j(x_i))), 1 \le j \le m, 1 \le i \le n.$$

It is clear that A is a bounded linear operator with closed range. Let $A^{**} : (X^n)^{**} \to Z^{**}$ denote its second conjugate operator. It is easy to see that A^{**} is given by the same formula as A (but the entries are now in X^{**}) and that $(X^n)^{**}$ can be identified with $(X^{**})^n$. Straightforward verification shows that $A_0^{**}(b_1, \dots, b_n) = (b_1, \dots, b_k, 0, \dots, 0)$ and hence $A^{**}(b_1, \dots, b_n) \in Z$. Therefore, by Lemma 10.19, there exists $(d_1, \dots, d_n) \in X^n$ satisfying

$$\sup_i \|d_i\| < (1 + \delta) \sup_i \|b_i\| = 1 + \delta,$$

and such that

$$A(d_1, \dots, d_n) = A^{**}(b_1, \dots, b_n). \tag{10.9}$$

Define the operator $T : E \to X$ by $Tb_i = d_i$ for $1 \le i \le r$. For $1 \le p \le q$ the equalities $\sum_{1 \le i \le n} s_{p,i} b_i = 0$ and $\sum_{1 \le i \le n} s_{p,i} d_i = 0$ hold. This implies $Tb_i = d_i$ also for $r < i \le n$. Comparing the definition of A and (10.9) we get $d_i = b_i$ for $i = 1, \dots, k$. Therefore $Tx = x$ for $x \in E \cap X$.

Also the definition of A implies that $a_j(Tb_i) = a_j(d_i) = b_i(a_j)$. Therefore for each i we have

$$\|Tb_i\| \ge \sup_j |a_j(Tb_i)| = \sup_j |b_i(a_j)| \ge (1 + \delta)^{-1} \|b_i\| = (1 + \delta)^{-1} \ge 1 - \delta.$$

Now we recall the choice of δ (see the very beginning of the proof) and apply Lemma 2.32 (page 42). We get $\|T\| \|T^{-1}\| < 1 + \varepsilon$, this completes the proof. □

Now we return to the proof of Theorem 10.16. Straightforward verification shows that condition (10.8) together with the uniform boundedness of $\{L_n\}_{n=1}^\infty$

imply that the conjugate operators $L_n^* : \ell_\infty^n \to (\mathrm{Lip})^*$ are C-isomorphic embeddings for some $C \in (0,\infty)$. Now we apply the principle of local reflexivity with $X = \mathrm{LF}\left((\oplus_{i=1}^{2^n}\mathbb{T})_1\right)$, $E = L_n^*(\ell_\infty^n)$ and $F = \{0\}$. Denote the obtained operator by $T_n : E \to \mathrm{LF}\left((\oplus_{i=1}^{2^n}\mathbb{T})_1\right)$. It is clear that the compositions $T_n L_n^*$ are $C(1+\varepsilon)$-isomorphic embeddings of ℓ_∞^n into $\mathrm{LF}\left((\oplus_{i=1}^{2^n}\mathbb{T})_1\right)$.

Now we observe that for differentiable functions on the Cartesian product of 2^n copies of $\mathbb{T}$ the norm of the space Lip coincides with $\max_{s\in\{1,\dots,2^n\}}\max_\Theta\left|\frac{\partial f}{\partial\theta_s}(\Theta)\right|$, where by $\theta_1,\dots,\theta_{2^n}$ we denote the natural coordinates on the Cartesian product of $\mathbb{T}$ and $\Theta = (\theta_1,\dots,\theta_{2^n})$.

In fact, the inequality

$$\|f\|_{\mathrm{Lip}} \ge \max_{s\in\{1,\dots,2^n\}}\max_\Theta\left|\frac{\partial f}{\partial\theta_s}(\Theta)\right|$$

follows immediately from the definition of the partial derivative. The inequality

$$\|f\|_{\mathrm{Lip}} \le \max_{s\in\{1,\dots,2^n\}}\max_\Theta\left|\frac{\partial f}{\partial\theta_s}(\Theta)\right|$$

can be obtained in the following way. We have

$$\begin{aligned}
&|f(\theta_1,\dots,\theta_{2^n}) - f(\tilde\theta_1,\dots,\tilde\theta_{2^n})|\\
&\le \sum_{i=1}^{2^n}|f(\theta_1,\dots,\theta_{2^n-i},\theta_{2^n-i+1},\tilde\theta_{2^n-i+2}\dots,\tilde\theta_{2^n})\\
&\quad - f(\theta_1,\dots,\theta_{2^n-i},\tilde\theta_{2^n-i+1},\tilde\theta_{2^n-i+2}\dots,\tilde\theta_{2^n})|\\
&\le \sum_{i=1}^{2^n}\max_\Theta\left|\frac{\partial f}{\partial\theta_i}(\Theta)\right| d_{\mathbb{T}}(\theta_i,\tilde\theta_i)\\
&\le \max_{s\in\{1,\dots,2^n\}}\max_\Theta\left|\frac{\partial f}{\partial\theta_s}(\Theta)\right| d(\Theta,\tilde\Theta),
\end{aligned}$$

where the distance d in the last line is in $(\oplus_{i=1}^{2^n}\mathbb{T})_1$.

Now we introduce uniformly bounded operators $L_n : \mathrm{Lip} \to \ell_1^n$ satisfying (10.8). We start with the complex case. In this case we define the map $L_n : \mathrm{Lip} \to \ell_1^n$ by $f \mapsto \{k_j\hat f(k_j\sigma_j^{(1)},\dots,k_j\sigma_j^{(2^n)})\}_{j=1}^n$, where $\hat f(k_j\sigma_j^{(1)},\dots,k_j\sigma_j^{(2^n)})$ is the Fourier coefficient of f on $\mathbb{T}^{2^n}$ corresponding to $\exp(2\pi i(k_j\sigma_j^{(1)}\theta_1 + \cdots + k_j\sigma_j^{(2^n)}\theta_{2^n}))$, $k_j \in \mathbb{N}$, $\sigma_j^{(s)} = \pm 1$. It remains to specify $\sigma_j^{(s)}$ and k_j for which the map has the desired properties.

The numbers $\{k_j\}_{j=1}^n$ form a rapidly increasing sequence of integers. The exact meaning of "rapidly increasing" in this construction will be described later. As for

$\sigma_j^{(s)}$ where $j = 1, \ldots, n$ and $s = 1, \ldots, 2^n$ we pick them to be such that the collections $\{\sigma_j^{(s)}\}_{j=1}^n$ are all possible n-tuples consisting of ± 1.

First, let us show that (10.8) is satisfied. For this we consider the functions

$$\frac{1}{k_j}\exp(2\pi i(k_j\sigma_j^{(1)}\theta_1 + \cdots + k_j\sigma_j^{(2^n)}\theta_{2^n})) \in \text{Lip}, \quad j = 1, \ldots, n. \tag{10.10}$$

The partial derivative of such a function with respect to θ_s is equal to

$$2\pi i\sigma_j^{(s)}\exp(2\pi i(k_j\sigma_j^{(1)}\theta_1 + \cdots + k_j\sigma_j^{(2^n)}\theta_{2^n})), \quad j = 1, \ldots, n,$$

and therefore the norm of this function in the space Lip is 2π. On the other hand, it is clear that the image of this function under the operator L_n is the j-th unit vector of ℓ_1^n. This implies (10.8).

It remains to estimate the norm of L_n from above by a universal constant. Here is an outline of our argument: We consider a finite version of what is called the Riesz product. We get a nonnegative continuously differentiable function $A \in \text{Lip}$. To get the estimate of $\|L_n f\|_{\ell_1^n}$ for general $f \in \text{Lip}$ we consider the convolution $f * A$ (we use the natural group structure on $\mathbb{T}^{2^n}$). The norm $\|L_n f\|$ can be easily estimated in terms of $\|L_n(f * A)\|$: the Fourier coefficients of the convolution are the products of Fourier coefficients of the functions, and Fourier coefficients of A are immediately computable. On the other hand, the partial derivatives of the $f * A$ with respect to different θ_s contain different combinations of signs, therefore we can find a coordinate for which the signs agree with the signs of the terms by which they are multiplied. This leads to the estimate of $\|L_n f\|_{\ell_1^n}$ from above in terms of $\|f * A\|_{\text{Lip}}$.

Now the details. We let

$$A(\theta_1, \ldots, \theta_{2^n}) = \prod_{j=1}^{n}\left(1 + \cos\left(2\pi k_j \sum_{s=1}^{2^n}\sigma_j^{(s)}\theta_s\right)\right).$$

It is obvious that $A(\theta_1, \ldots, \theta_{2^n}) \geq 0$.

If $\{k_j\}$ are sufficiently rapidly increasing, the condition $\int A(\theta_1, \ldots, \theta_{2^n})d\Theta = 1$ holds. In fact, writing the trigonometric series of A using the formula

$$\cos x \cos y = \frac{\cos(x+y) + \cos(x-y)}{2}, \tag{10.11}$$

we get only terms of the forms

$$\cos\left(2\pi\sum_{s=1}^{2^n}\left(\sum_{j=1}^{n}k_j\Delta_j\sigma_j^{(s)}\right)\theta_s\right),$$

where each Δ_j is equal to $0, 1$, or -1. It is clear that if k_j are sufficiently rapidly increasing, the sums $\sum_{j=1}^n k_j\Delta_j\sigma_j^{(s)}$ are nonzero unless all Δ_j are equal to 0. Therefore the integrals $\int_0^1$ are equal to zero for all of the terms except the constant term.

Our next observation is that for f in Lip we have

$$\|f * A\|_{\mathrm{Lip}} \le \|f\|_{\mathrm{Lip}}. \tag{10.12}$$

In fact,

$$\begin{aligned} |f * A(\Theta_1) - f * A(\Theta_2)| &\le \int |f(\Theta_1 + \Gamma)A(-\Gamma) - f(\Theta_2 + \Gamma)A(-\Gamma)|d\Gamma \\ &\le \int |f(\Theta_1 + \Gamma) - f(\Theta_2 + \Gamma)| \cdot |A(-\Gamma)|d\Gamma \\ &\le \|f\|_{\mathrm{Lip}} \cdot d(\Theta_1, \Theta_2) \int |A(-\Gamma)|d\Gamma \\ &= \|f\|_{\mathrm{Lip}} \cdot d(\Theta_1, \Theta_2), \end{aligned}$$

where d is the distance in $\left(\oplus_{i=1}^{2^n}\mathbb{T}\right)_1$.

Now we write the complex Fourier series for A. We can do this by first replacing cosine using the formula

$$\cos z = \frac{1}{2}\exp(iz) + \frac{1}{2}\exp(-iz) \tag{10.13}$$

everywhere, and then by multiplying the results.

As is well known, to get Fourier coefficients of $f * A$ we multiply the respective Fourier coefficients of f and A. It is clear that $\widehat{A}(k_j\sigma_j^{(1)}, \dots, k_j\sigma_j^{(2^n)}) = \frac{1}{2}$, therefore

$$f \,\widehat{*}\, A(k_j\sigma_j^{(1)}, \dots, k_j\sigma_j^{(2^n)}) = \frac{1}{2}\widehat{f}(k_j\sigma_j^{(1)}, \dots, k_j\sigma_j^{(2^n)})$$

and

$$\|L_n(f * A)\| = \frac{1}{2}\|L_n f\|.$$

Recalling (10.12) we see that it remains to estimate $\|L_n(f * A)\|$ from above in terms of $\|f * A\|_{\mathrm{Lip}}$. It is easy to see that we have

$$\begin{aligned} f * A = \widehat{f}(0) + \sum_{j=1}^{n}\Bigg(D_{j,+}(\Theta)\exp\Bigg(2\pi i k_j\Bigg(\sum_{s=1}^{2^n}\sigma_j^{(s)}\theta_s\Bigg)\Bigg) \\ + D_{j,-}(\Theta)\exp\Bigg(-2\pi i k_j\Bigg(\sum_{s=1}^{2^n}\sigma_j^{(s)}\theta_s\Bigg)\Bigg)\Bigg) \end{aligned} \tag{10.14}$$

where $\widehat{f}(0) = \int f(\Theta)d\Theta$ and $D_{j,+}$, $D_{j,-}$ are functions whose complex Fourier series have zero coefficients outside the set

$$\left\{\left(\sum_{i<j}\varepsilon_i\sigma_i^{(1)}k_i, \dots, \sum_{i<j}\varepsilon_i\sigma_i^{(2^n)}k_i\right) \,\middle|\, \varepsilon_i = 0, 1, -1 \text{ for } 1 \le i < j\right\}.$$

Our second restriction on the growth of the sequence $\{k_j\}_{j=1}^n$ comes from the following considerations. We need the maximum of the partial derivative $\frac{\partial(f*A)}{\partial\theta_t}$ over $\mathbb{T}^n$ to be close to the maximum over $\mathbb{T}^n$ of the function

$$2\pi i\sum_{j=1}^{n}k_j\sigma_j^{(t)}\Big(D_{j,+}(\Theta)\exp\Big(2\pi i k_j\Big(\sum_{s=1}^{2^n}\sigma_j^{(s)}\theta_s\Big)\Big) - D_{j,-}(\Theta)\exp\Big(-2\pi i k_j\Big(\sum_{s=1}^{2^n}\sigma_j^{(s)}\theta_s\Big)\Big)\Big).$$

The possibility of choosing $\{k_j\}_{j=1}^n$ satisfying this condition can be shown in the following way. We write

$$D_{j,+}(\Theta) = \sum_{\substack{\{\varepsilon_1,\dots,\varepsilon_{j-1}\}\\ \in\{0,1,-1\}^{j-1}}}\beta(\varepsilon_1,\dots,\varepsilon_{j-1})\exp\Big(2\pi i\Big(\sum_{s=1}^{2^n}\sum_{i<j}\varepsilon_i\sigma_i^{(s)}k_i\theta_s\Big)\Big).$$

The partial derivative with respect to θ_t of the product

$$D_{j,+}(\Theta)\exp\Big(2\pi i k_j\Big(\sum_{s=1}^{2^n}\sigma_j^{(s)}\theta_s\Big)\Big)$$

is equal to the sum of

$$2\pi i k_j\sigma_j^{(t)}D_{j,+}(\Theta)\exp\Big(2\pi i k_j\Big(\sum_{s=1}^{2^n}\sigma_j^{(s)}\theta_s\Big)\Big) \tag{10.15}$$

and

$$\sum_{\substack{\{\varepsilon_1,\dots,\varepsilon_{j-1}\}\\ \in\{0,1,-1\}^{j-1}}}\Big(\beta(\varepsilon_1,\dots,\varepsilon_{j-1})\exp\Big(2\pi i\Big(\sum_{s=1}^{2^n}\sum_{i<j}\varepsilon_i\sigma_i^{(s)}k_i\theta_s\Big)\Big) \times\Big(2\pi i\sum_{i<j}\varepsilon_i\sigma_i^{(t)}k_i\Big)\Big) \times\exp\Big(2\pi i k_j\Big(\sum_{s=1}^{2^n}\sigma_j^{(s)}\theta_s\Big)\Big). \tag{10.16}$$

Observe that the sums in (10.15) and (10.16) contain the same exponentials and for a rapidly increasing sequence $\{k_j\}$ the coefficients in (10.16) are negligible in comparison with the coefficients in (10.15). Combining this with the well-known fact that the functionals on $C(\mathbb{T}^n)$ which map a function onto its Fourier coefficients are uniformly bounded (even uniform boundedness is not needed) we get that by picking $\{k_j\}$ sufficiently rapidly increasing we can make the maximum over $\Theta \in \mathbb{T}^n$ of the function (10.16) negligible in comparison with the similar maximum for (10.15). A similar argument can be used for $D_{j,-}$

Now we recall that $\{\{\sigma_j^{(s)}\}_{j=1}^n\}_{s=1}^{2^n}$ contains all possible n-tuples consisting of ± 1 and use the easy fact that

$$\max_{\sigma_j=\pm 1}\left|\sum_{j=1}^n \sigma_j\alpha_j\right| \geq \frac{1}{2}\sum_{j=1}^n |\alpha_j|$$

for any collection $\{\alpha_j\}_{j=1}^n$ of complex numbers. We get that for each fixed Θ we have

$$\begin{aligned}\max_t \Bigg| 2\pi i \sum_{j=1}^n k_j\sigma_j^{(t)}\Bigg(&D_{j,+}(\Theta)\exp\Bigg(2\pi i k_j\Bigg(\sum_{s=1}^{2^n}\sigma_j^{(s)}\theta_s\Bigg)\Bigg)\\ &- D_{j,-}(\Theta)\exp\Bigg(-2\pi i k_j\Bigg(\sum_{s=1}^{2^n}\sigma_j^{(s)}\theta_s\Bigg)\Bigg)\Bigg)\Bigg|\\ \geq \pi\sum_{j=1}^n k_j\Bigg|&D_{j,+}(\Theta)\exp\Bigg(2\pi i k_j\Bigg(\sum_{s=1}^{2^n}\sigma_j^{(s)}\theta_s\Bigg)\Bigg)\\ &- D_{j,-}(\Theta)\exp\Bigg(-2\pi i k_j\Bigg(\sum_{s=1}^{2^n}\sigma_j^{(s)}\theta_s\Bigg)\Bigg)\Bigg|.\end{aligned}$$

The maximum over $\Theta \in \mathbb{T}^n$ of the absolute value in the right-hand side is not less than

$$\left|\int D_{j,+}(\Theta)d\Theta\right|.$$

To see this we multiply the formula inside the absolute value by

$$\exp\Bigg(-2\pi i k_j\Bigg(\sum_{s=1}^{2^n}\sigma_j^s\theta_s\Bigg)\Bigg),$$

integrate, and use the fact that the integrals of the remaining terms are equal to 0.

Using (10.12) we get from here

$$\|f\|_{\mathrm{Lip}} \geq c\sum_{j=1}^n k_j\left|\int D_{j,+}(\Theta)d\Theta\right|. \tag{10.17}$$

Now we observe that $\int D_{j,+}(\Theta)d\Theta$ is just the Fourier coefficient of $f * A$ at the 2^n-tuple

$$(k_j\sigma_j^{(1)},\ldots,k_j\sigma_j^{(2^n)}).$$

So we have $\|f\|_{\mathrm{Lip}} \geq c\|L_n(f * A)\|$. This completes the proof in the complex case.

In the case of real scalars we define the map $L_n : \mathrm{Lip} \to \ell_1^n$ by

$$f \mapsto \{k_j(\hat{f}(k_j\sigma_j^{(1)},\dots,k_j\sigma_j^{(2^n)}) + \hat{f}(-k_j\sigma_j^{(1)},\dots,-k_j\sigma_j^{(2^n)}))\}_{j=1}^n$$

and use the same reasoning. □

10.5 Exercises

Exercise 10.20. Recall that *lattice graphs* $L_{m,n}$ are defined as graphs whose vertex sets are $\{0,1,\dots,m\}^n$, two vertices are joined by an edge if and only if their ℓ_∞-distance is equal to 1.

Use Bourgain's discretization theorem (Theorem 9.4) to show that if a family of metric spaces $\{X_k\}_{k=1}^\infty$ contains uniformly bilipschitz images of $\{L_{n,n}\}_{n=1}^\infty$, then there exists $C < \infty$ such that for each $m \in \mathbb{N}$ there is $k \in \mathbb{N}$ such that the space $\mathrm{LF}(X_k)$ contains a subspace C-isomorphic to ℓ_∞^n.

Exercise 10.21. Construct (unweighted) graphs $\{U_n\}_{n=1}$ with maximum degree 3 such that there is an absolute constant $0 < C < \infty$ such that $L_{n,n}$ admits a C-bilipschitz embedding into U_n.

There is a hint to Exercise 10.21.

Exercise 10.22. Construct a family of expanders $\{G_n\}_{n=1}$ (allowing loops) such that there is a constant $C < \infty$ such that the spaces $\{\mathrm{LF}(G_n)\}_{n=1}^\infty$ contain subspaces C-isomorphic to $\{\ell_\infty^m\}_{m=1}^\infty$. More precisely: for each $m \in \mathbb{N}$ there is $n \in \mathbb{N}$ such that ℓ_∞^m is C-isomorphic to a subspace of $\mathrm{LF}(G_n)$.

There is a hint to Exercise 10.22.

10.6 Notes and remarks

Material of Section 10.2 is mostly due to Arens and Eells [24]. Some of the results were proved later, in [285], [231], and Weaver [435, Section 2.2]. See [435, Section 1.5 and Proposition 2.2.5] in connection with our Remark 10.7 on complex-valued Lipschitz free spaces. Theorem 10.8 is due to McShane [321].

We refer to [61] and [62] for results and discussions related to the notions of the cycle subspace and cut subspace. The reader can find the basic results on spanning trees in, for example [122, Chapter 1], the generalization to spanning forests is immediate because a spanning forest is the union of spanning trees of connected components.

Examples 10.12, 10.13, and Theorem 10.15 are from [165], Example 10.13 was analyzed earlier in [231]. The main result of Section 10.4 is due to Bourgain [76]. The principle of local reflexivity was proved in [286] (initial version) and [225] (the version stated as Theorem 10.18). Our proof of the principle of local reflexivity follows [415]. See [447] concerning the theory of Riesz products.

Further results indicating that Lipschitz free spaces can be very complicated even for relatively simple metric spaces can be found in [249] and [336].

The space which we call the *Lipschitz free space* is sometimes called the *Arens–Eells space*. There is also a closely related work of Kantorovich [239],[240] and Kantorovich–Rubinshtein [241],[242]. It is clear that molecules can be regarded as finite measures with 0 value on the whole space X. The papers [239],[240],[241],[242] treat the case of Borel measures on compact sets, for which a norm similar to $\|\cdot\|_{\mathrm{LF}}$ is introduced and a duality result similar to Theorem 10.2 is proved. For this reason the norm $\|\cdot\|_{\mathrm{LF}}$ is sometimes called the *Kantorovich–Rubinshtein norm*. See [153], [167], [192], [237, Sections 2.2 and 3.2], [324], [369], [432], and [444] for related results and discussions. We turn the reader's attention to the paper of Godefroy and Kalton [167] in which Lipschitz free spaces were used to prove important results on non-linear classification of Banach spaces. This paper initiated the study of Banach space theoretical properties of Lipschitz free spaces, see [131], [169], [168], and [266] for further results in this direction.

Exercises 10.20–10.22 are based on [356, Theorem 2.1 and Corollary 2.3].

10.7 Hints to exercises

To Exercise 10.21. One of the approaches: Replace each vertex of $L_{n,n}$ by a binary tree of depth corresponding to the degree of the vertex. Replace each edge by a lengthy path.

To Exercise 10.22. Paste the graphs constructed in Exercise 10.21 and some family of expanders (as in Proposition 5.77). Add loops to vertices having smaller degrees (if needed). Use Exercise 10.20.

Chapter 11

Open problems

11.1 Embeddability of expanders into Banach spaces

Combining the Frechét proposition (Proposition 1.17) on isometric embeddability of any n-element metric space into ℓ_∞^n and the characterization of spaces with no nontrivial cotype (Theorem 2.56) we get that each space with no nontrivial cotype contains $(1+\varepsilon)$-bilipschitz images of any sequence of expanders. Besides this result and its immediate consequences, no results on existence of embeddings of families of expanders into Banach spaces are known. Existence of such embeddings is important in several contexts.

Apparently the first problem suggesting to investigate existence of other ways to embed expanders into Banach spaces was posed by Linial in the 2002 edition of the collection [317]:

Problem 11.1 ([317, Problem 3.1] Expanders and ℓ_∞ (N. Linial)). In simple terms I am asking whether there is a nontrivial way in which expanders can be embedded in a normed space. A daring version of the question is: Let X be an infinite-dimensional Banach space. Suppose there is a family of constant-degree expanders G_k on n_k vertices, $n_k \to \infty$ as $k \to \infty$, that embed in X with distortion $o(\log n_k)$. Does this imply that X contains copies of ℓ_∞^d with distortion $1+\varepsilon$ for every $\varepsilon > 0$? (By a theorem of Maurey this entails that X has no cotype.)

Problem 11.1 is still open (the collection [317] was last updated in August 2011). It follows from results of Naor–Peres–Schramm–Sheffield [335] that no family of expanders admits uniformly bilipschitz embeddings into a uniformly convex Banach space (this result is presented in Chapter 8 of this book). It is not known whether one can extend this result to any Banach space with nontrivial type.

In recent years the notion of uniformly coarse embeddability attracted a lot of attention. It turns out that even uniformly coarse embeddings of expanders are difficult to construct, and the following problem is also open:

Problem 11.2. Does there exist a family $\{G_n\}$ of expanders and a Banach space with nontrivial cotype such that $\{G_n\}$ embeds uniformly coarsely into X?

A closely related problem was suggested by Ozawa [363, p. 563].

Several "restricted" versions of Problems 11.1 and 11.2 also attracted much attention. There is significant progress on these lines, but many of the problems remain open. The purpose of the remainder of this section is to mention some solved and some still open "restricted" versions of Problems 11.1 and 11.2.

In connection with the results of Kasparov and Yu [243] (see Theorem 1.68 in this book) the following problem became important: *Does there exist a space with bounded geometry which is not coarsely embeddable into a uniformly convex space?* This problem was solved in the positive by Lafforgue [263] who proved that there exists a family of expanders which is not uniformly coarsely embeddable into any uniformly convex space. An alternative proof of this fact was given by Mendel and Naor [329]. In this connection the following problems were suggested:

Problem 11.3. Does there exist a family of expanders which are uniformly coarsely embeddable into a uniformly convex space?

Problem 11.3 was posed in [329, Section 1.1] and [345, Open question 5.6.6]. It is worth mentioning that the papers [264] and [329] contain constructions of families of expanders which are not uniformly coarsely embeddable into any Banach space with nontrivial type. The constructions of [263],[264],[329] left the following (suggested in [329, Section 1.1]) problem open:

Problem 11.4. It is unknown whether there exist families of finite graphs of bounded degree and girth $\geq$ log of the size, which do not admit uniformly coarse embeddings into any uniformly convex Banach space.

Mendel and Naor [329, Section 1.1] mention that this question is of particular interest in the context of the potential application to the Novikov conjecture that was suggested by Kasparov and Yu in [243]. One of the reasons for this interest is that Gromov's construction [179] of a finitely generated group whose Cayley graph does not embed coarsely into a Hilbert space is based on expanders with large girth.

Another problem closely related to Linial's Problem 11.1 was suggested by the author in [426, Section 2.4]:

Problem 11.5. Let $\{G_n\}$ be a family of expanders and $\{A_n\}$ be the sequence of the corresponding Lipschitz free spaces. Is it true that $\{A_n\}$ contain ℓ_∞^k of growing dimensions? (Of course, the answer is known to be "Yes" for families of expanders containing uniformly bilipschitz images of nets in $\{\ell_\infty^n\}_{n=1}^\infty$, see Exercises in Chapter 10 for examples.)

Comparing the results of Ozawa [363] (Theorem 6.1 in this book) and Raynaud [390] (Theorem 6.20 in this book) we see that for each nonreflexive Banach space with nontrivial type it is interesting to understand whether it admits uniformly coarse embeddings of some family of expanders. Furthermore, as we see from results of [264] and [329] mentioned above, only the problem of existence of embeddable families is unclear, the existence of non-embeddable families has been proved in the works just mentioned. Here it is worth mentioning that nonreflexive Banach spaces with nontrivial type are difficult to construct (see Notes and Remarks to Chapter 6 for references). It seems that the simplest and, in a certain sense, the most explicit known construction of such space is due to Pisier and Xu [384] (see also [383] and Section 6.5 in Chapter 6 of this book). The Pisier–Xu construction has an easier description of the norm than other known constructions, so it is more natural to expect that one should be able to answer the embeddability problem for this space rather than for other constructions of nonreflexive spaces with nontrivial type. Therefore we suggest:

Problem 11.6. Does there exist a family of expanders which are uniformly coarsely embeddable into the nonreflexive space with nontrivial type constructed in [384]? (The construction is presented in Section 6.5 of this book.)

11.2 Obstacles for coarse embeddability of spaces with bounded geometry into a Hilbert space

11.2.1 The main problem

In this section we recall the problem which was the subject of Chapter 7, but is still open. Let us recall the related definitions and the statement of the problem.

Definition 11.7. A uniformly discrete metric space A is said to have a *bounded geometry* if for each $r > 0$ there exists a positive integer $M(r)$ such that each ball in A of radius r contains at most $M(r)$ elements.

Definition 11.8. Let X be a space with bounded geometry and $\{Y_n\}_{n=1}^{\infty}$ be a family of expanders. We say that X *weakly contains* $\{Y_n\}$ if there are maps $f_n : Y_n \to X$ satisfying (with some abuse of notation we use Y_n to denote the vertex set of Y_n):

- Lipschitz constants $\mathrm{Lip}(f_n)$ are uniformly bounded
- $\displaystyle \lim_{n\to\infty} \max_{y\in Y_n} \frac{|f_n^{-1}(f_n(y))|}{|Y_n|} = 0.$

The images of Y_n in X are called *weak expanders*.

Problem 11.9. Suppose that a metric space M with bounded geometry is not coarsely embeddable into ℓ_2. Does it follow that M weakly contains a family of expanders?

11.2.2 Comments

This problem was stated for finitely generated groups in [186, p. 261], some versions of Problem 11.9 were posed in [345, Open question 5.7.5], [354, p. 166], and [425, p. 87]. All these versions of the problem are also still open.

A more vague problem: *under the same condition: find some expander-like structures in M*, is also of interest. This problem was studied in [354],[355],[425]. In particular the following result was proved (see Theorem 7.6 in this book):

Theorem 11.10. *Let (M, d_M) be a locally finite metric space which is not coarsely embeddable into a Hilbert space. Then there exists a constant D, depending on M only, such that for each $n \in \mathbb{N}$ there exists a finite set $M_n \subset M$ and a probability measure μ_n on $M_n \times M_n$ such that*

- $d_M(u, v) \geq n$ *for each* $(u, v) \in \operatorname{supp}\mu_n$.
- *For each Lipschitz function $f : M \to L_1$ we have*

$$\int_{M_n \times M_n} \|f(u) - f(v)\|_{L_1} d\mu_n(u, v) \leq D\operatorname{Lip}(f). \tag{11.1}$$

The structures of the type described in Theorem 11.10 were called *generalized expanders* in [425]. In the opinion of the present author this notion of generalized expanders is not completely satisfactory, because the presence of such expanders in M does not imply coarse non-embeddability of M into ℓ_p for $p > 2$. (The last statement can be derived from the fact [224],[327] that ℓ_p $(p > 2)$ does not embed coarsely into ℓ_2.)

A minimal requirement which I would impose on the notion of an expander-like structures: their presence implies coarse non-embeddability into ℓ_p for $1 \leq p < \infty$. The main motivation here is that under this requirement a proof of existence of such structures would prove nonexistence of the construction required in the following open problem suggested by Yu [443, Problem 3.5]:

Problem 11.11. Let $p > q \geq 2$. Construct a discrete metric space with bounded geometry which admits a (coarse) uniform embedding into ℓ_p but not into ℓ_q.

It is worth mentioning that in [355] it was shown that presence of the generalized expanders (see the paragraph following Theorem 11.10) implies a certain "limited" expansion property (the corresponding results are presented in Section 7.3). It is natural to say that this is a "limited" expansion property because expansion is guaranteed for subsets of "relatively small" diameter only. However, by now it is known (follows from results of [27] as well as from results of [359]) that the "limited" expansion property considered in [355] does not imply coarse nonembeddability into ℓ_2 (reason: graphs of growing girth have the "limited" expansion property and could be uniformly coarsely embeddable into ℓ_2). In this connection it is natural to look for stronger "limited" expansion properties, as it was suggested in the following problem, posed by Johnson, see [426, Section 2.2].

Problem 11.12. We know that the usual expansion property of a family of graphs with bounded degrees implies that the family is not uniformly coarsely embeddable into a Hilbert space. Characterize "limited" expansion properties for which such non-embeddability still holds.

11.3 Embeddability of graphs with large girth

As with expanders, talking about graphs with large girth we usually mean families $\{G_n\}_{n=1}^{\infty}$ of k-regular connected graphs, $k \geq 3$ with $g(G_n) \to \infty$ ($g(G_n)$ denotes the girth of G_n). Observe that each graph of girth ≥ 3 is simple (does not have loops or parallel edges). The case $k = 2$ is excluded because each family of 2-regular connected graphs with growing girths is just a family of cycles with growing length. Some papers deal with a more general situation considering graphs with all degrees ≥ 3 or graphs with average degrees $\geq 2 + \varepsilon$ for some $\varepsilon > 0$.

Systematic study of metric properties of graphs with large girth was initiated in [387]. Earlier an example of their application in the theory of metric embeddings was found in [41, Proof of Theorem 15]. Linial, London, and Rabinovich [290, Problem 8.7] suggested the following problem:

Problem 11.13. Let all vertices in G have degree ≥ 3. Does every embedding of G in a Euclidean space (of any dimension) have distortion $\geq \Omega(girth(G))$?

This problem is still open. Substantial progress on this problem was achieved in [292],[42]. In [335] it was proved that families of graphs with large girth do not admit uniformly bilipschitz embeddings into uniformly convex spaces.

In connection with the progress achieved in [292], Linial restated Problem 11.13 in the following way [317, Problem 2.3]:

Problem 11.14 (Linial [317, Problem 2.3]). Linial–London–Rabinovich [290] conjectured that $c_2(G) \geq \Omega(g(G))$. Linial–Magen–Naor [292] proved

$$c_2(G) \geq \Omega(\sqrt{g(G)})$$

for regular graphs. What is the true bound?

Our next problem on embeddings of k-regular families of graphs with growing girths arises from the following observation: Graphs with large girth and uniformly bounded ℓ_1 distortions constructed in [359] are such that

$$\operatorname{diam}(G_n) \approx \exp(g(G_n)).$$

Problem 11.15. How important is such growth of diameters for the example constructed in [359]? In particular, is it possible to find families $\{G_n\}$ of k-regular graphs with uniformly bounded ℓ_1 distortions and girths tending to infinity, such that the diameter of G_n is bounded from above by a polynomial in $g(G_n)$?

The next problem deals with uniformly coarse embeddability of one family of graphs into another. Saying that a family $\mathcal{F}_1$ of metric spaces is *uniformly coarsely embeddable* into a family $\mathcal{F}_2$ of metric spaces we mean that there exist ρ_1 and ρ_2 such that each metric space in $\mathcal{F}_1$ is coarsely embeddable into some metric space of the family $\mathcal{F}_2$, and the corresponding inequality (see Definition 1.45) holds with the same ρ_1 and ρ_2.

For problems of uniformly coarse embeddability of expander families it is very useful to study uniformly coarse embeddability of one expander family into another. In Section 5.15 we present a version of the example of Mendel and Naor [329, Section 9.1] of two families of expanders such that one of them is not uniformly coarsely embeddable into another. In the Mendel–Naor example the families $\mathcal{F}_1$, $\mathcal{F}_2$ are such that elements of $\mathcal{F}_1$ have many cycles of different lengths and elements of $\mathcal{F}_2$ have growing girth.

It would be very interesting to find obstacles for such embeddings in cases where different lengths of cycles cannot create an obstacle. In this connection we suggest the following problem.

Problem 11.16. Let $\mathcal{F}_1 = \{G_n\}$ be a family of expanders of large girth:

$$\mathrm{girth}(G_n) \geq c \log(|V(G_n)|),$$

and $\mathcal{F}_2$ be an arbitrary family of expanders. Is it always possible to find a uniformly coarse embedding of $\mathcal{F}_1$ into $\mathcal{F}_2$?

We would like to finish this section by mentioning the fact that expander families with large girth play a very important part in metric theory also because of their role in Gromov's [179] construction of random group, see Problem 11.4 in this connection.

11.4 Coarse embeddability of a Hilbert space into Banach spaces

Yu [442] and Kasparov–Yu [243] in their work on the Baum–Connes conjecture and the Novikov conjecture use conditions of coarse embeddability of metric spaces into "reasonably good" Banach spaces (see Section 1.7 for more information). In this connection it is natural to compare embeddability into different Banach spaces.

Because ℓ_2 is, in many respects, the 'best' space, and because of the Dvoretzky theorem [132] (Theorem 2.30 in this book) it is natural to expect that ℓ_2 is among the most difficult spaces to embed into. The strongest possible result in this direction would be a positive solution to the following problem.

Problem 11.17. Does ℓ_2 embed coarsely into an arbitrary infinite-dimensional Banach space?

I would like to mention that for applications considered in [442] and [243] the answer was given in Theorem 2.62 which immediately implies the following corollary:

Let A be a locally finite metric space which embeds coarsely into a Hilbert space, and let X be an infinite-dimensional Banach space. Then there exists a coarse embedding $f : A \to X$.

Here we should mention that the answer to Problem 11.17 is known to be positive in several special cases:

- The space ℓ_2 is coarsely embeddable into spaces containing isomorphic copies of c_0. It is a very special case of a result of Aharoni [5] who proved that every separable metric space admits a bilipschitz embedding into c_0. (See [30], [238], and [366] for simplifications of the proof and for improvements of the result of Aharoni.).

- Kalton [235] showed that the class of Banach spaces admitting coarse embeddings of any separable metric spaces is much wider than the class of spaces containing isomorphic copies of c_0, for example it contains some spaces with the Schur property.

- Mendel–Naor [325, Remark 5.10] found explicit formulas for coarse embeddings of L_2 into L_p $(p \geq 2)$, see Example 1.40 in this book.

- Nowak [344] and Ostrovskii [354, Theorem 5.1] used Mazur maps, the work of Odell–Schlumprecht [346] on sphere homeomorphisms, and the Dadarlat–Guentner [108] criterion for coarse embeddability into a Hilbert space (the criterion is applied to ℓ_2). The most general result here is for spaces containing subspaces with nontrivial cotype and unconditional bases.

As we see, the problem is open for many spaces which are, in a certain sense, between c_0 and "good" spaces. Examples of spaces for which the result is not known are: the original Tsirelson space [430] and nonreflexive spaces with notrivial type, such examples were found in [215],[217],[216],[384]. See the book [383] and Section 6.5 of the present book for more information on the space constructed in [384].

If the answer to Problem 11.17 is negative, the obstruction should be of somewhat different type from the known obstructions to coarse embeddability of Banach spaces, see Johnson–Randrianarivony [224], Kalton [236], and Mendel–Naor [327]: Obstructions found in these papers, in a certain sense, support the statement that the Hilbert space is the most difficult to embed into.

Bibliography

[1] M. Abért, Y. Glasner, B. Virág, Kesten's theorem for invariant random subgroups, (2012), arXiv:1201.3399.

[2] I. Abraham, Y. Bartal, O. Neiman, Nearly tight low stretch spanning trees, *49th Annual IEEE Symposium on Foundations of Computer Science*, 2008, arXiv:0808.2017.

[3] I. Abraham, Y. Bartal, O. Neiman, Advances in metric embedding theory, *Adv. Math.*, **228** (2011), no. 6, 3026–3126.

[4] D. Achlioptas, Database-friendly random projections: Johnson–Lindenstrauss with binary coins. Special issue on PODS 2001 (Santa Barbara, CA), *J. Comput. System Sci.*, **66** (2003), no. 4, 671–687.

[5] I. Aharoni, Every separable metric space is Lipschitz equivalent to a subset of c_0^+, *Israel J. Math.*, **19** (1974), 284–291.

[6] N. Ailon, B. Chazelle, The fast Johnson–Lindenstrauss transform and approximate nearest neighbors, *SIAM J. Comput.*, **39** (2009), no. 1, 302–322.

[7] M. Ajtai, Recursive construction for 3-regular expanders, *Combinatorica*, **14** (1994), no. 4, 379–416.

[8] F. Albiac, F. Baudier, Embeddability of snowflaked metrics with applications to the nonlinear geometry of the spaces L_p and ℓ_p for $0 < p < \infty$, *J. Geom. Anal.*, to appear, arXiv:1206.3774.

[9] F. Albiac, N. J. Kalton, *Topics in Banach Space Theory*, Graduate Texts in Mathematics, **233**, Springer, New York, 2006.

[10] N. Alon, Eigenvalues and expanders, *Combinatorica* **6** (1986), 83–96.

[11] N. Alon, Tools from higher algebra, in: *Handbook of Combinatorics*, Vol. **2**, pp. 1749–1783, Elsevier, Amsterdam, 1995.

[12] N. Alon, Problems and results in extremal combinatorics, I, *Discrete Math.* **273** (2003), 31–53.

[13] N. Alon, Perturbed identity matrices have high rank: proof and applications, *Combin. Probab. Comput.*, **18** (2009), no. 1–2, 3–15.

[14] N. Alon, A. Lubotzky, A. Wigderson, Semi-direct products in groups and zig-zag products in graphs: connections and applications, in: 42nd IEEE Symposium on Foundations of Computer Science (Las Vegas, NV, 2001), pp. 630–637, IEEE Comput. Soc. Press, Los Alamitos, CA, 2001.

[15] N. Alon, V. D. Milman, λ_1, isoperimetric inequalities for graphs and superconcentrators, *J. Combin. Theory*, Ser. B **38** (1985), 73–88.

[16] N. Alon, Y. Roichman, Random Cayley graphs and expanders, *Random Structures Algorithms*, **5** (1994), 271–284, `doi:10.1002/rsa.3240050203`.

[17] N. Alon, O. Schwartz, A. Shapira, An elementary construction of constant-degree expanders, *Combin. Probab. Comput.*, **17** (2008), 319–327.

[18] D. Amir, *Characterizations of Inner Product Spaces*, Operator Theory: Advances and Applications **20**, Birkhäuser Verlag, Basel, 1986.

[19] A. Amit, S. Hoory, N. Linial, A continuous analogue of the girth problem, *J. Combin. Theory* Ser. B, **84** (2002), no. 2, 340–363.

[20] A. Amit, N. Linial, Random graph coverings. I. General theory and graph connectivity, *Combinatorica*, **22** (2002), no. 1, 1–18.

[21] A. Amit, N. Linial, Random lifts of graphs: edge expansion, *Combin. Probab. Comput.*, **15** (2006), 317–332.

[22] A. Amit, N. Linial, J. Matoušek, Random lifts of graphs III: Independence and chromatic number, *Random Structures and Algorithms*, **20** (2002) 1–22.

[23] D. Angluin, A. Gardiner, Finite common coverings of pairs of regular graphs, *J. Combin. Theory*, Ser. B, **30** (1981), no. 2, 184–187.

[24] R. F. Arens, J. Eells, Jr., On embedding uniform and topological spaces, *Pacific J. Math.*, **6** (1956), 397–403.

[25] S. Arora, J. R. Lee, A. Naor, Euclidean distortion and the sparsest cut, *J. Amer. Math. Soc.*, **21** (2008), 1-21; Conference version: *STOC'05: Proceedings of the 37th Annual ACM Symposium on Theory of Computing*, pp. 553–562, ACM, New York, 2005.

[26] S. Arora, S. Rao, U. Vazirani, Expander flows, geometric embeddings and graph partitioning, *Journal of the ACM*, **56** (2009), No. 2, Article 5; Conference version: *Proceedings of the 36th Annual ACM Symposium on Theory of Computing*, pp. 222–231, ACM, New York, 2004.

[27] G. Arzhantseva, E. Guentner, J. Špakula, Coarse non-amenability and coarse embeddings, *Geom. Funct. Anal.*, **22** (2012), 22–36, `arXiv:1101.1993`.

[28] E. Asplund, Farthest points in reflexive locally uniformly rotund Banach spaces, *Israel J. Math.*, **4** (1966), 213–216.

[29] E. Asplund, I. Namioka, A geometric proof of Ryll–Nardzewski's fixed point theorem, *Bull. Amer. Math. Soc.*, **73** (1967), 443–445.

[30] P. Assouad, Remarques sur un article de Israel Aharoni sur les prolongements lipschitziens dans c_0 (Israel J. Math. 19 (1974), 284–291), *Israel J. Math.*, **31** (1978), no. 1, 97–100.

[31] P. Assouad, Plongements Lipschitziens dans $\mathbb{R}^n$, *Bull. Soc. Math. Fr.*, **111** (1983) 429–448.

[32] H. Auerbach, *On the area of convex curves with conjugate diameters* (in Polish), Ph. D. thesis, University of Lwów, 1930.

[33] Y. Aumann, Y. Rabani, An $O(\log k)$ approximate min-cut max-flow theorem and approximation algorithm, *SIAM J. Comput.*, **27** (1998), no. 1, 291–301.

[34] G. Averkov, N. Düvelmeyer, Embedding metric spaces into normed spaces and estimates of metric capacity, *Monatsh. Math.*, **152** (2007), 197–206.

[35] K. Ball, Isometric embedding in ℓ_p-spaces, *European J. Combin.*, **11** (1990), no. 4, 305–311.

[36] K. Ball, Markov chains, Riesz transforms and Lipschitz maps, *Geom. Funct. Anal.*, **2** (1992), 137–172.

[37] K. Ball, An elementary introduction to modern convex geometry, in: *Flavors of Geometry*, Math. Sci. Res. Inst. Publ., **31**, pp. 1–58, Cambridge University Press, Cambridge, 1997.

[38] K. Ball, The Ribe programme, *Séminaire Bourbaki*, (2012), exposé 1047, 12 pp., *Asterisque*, to appear

[39] K. Ball, E. A. Carlen, E. H. Lieb, Sharp uniform convexity and smoothness inequalities for trace norms, *Invent. Math.*, **115** (1994), no. 3, 463–482.

[40] S. Banach, *Théorie des opérations lińeaires*, Monografje Matematyczne, Warszawa, 1932; Ukrainian (1948) and English (1987) translations are available.

[41] Y. Bartal, Probabilistic approximation of metric spaces and its algorithmic applications, in: *37th Annual Symposium on Foundations of Computer Science* (Burlington, VT, 1996), pp. 184–193, IEEE Comput. Soc. Press, Los Alamitos, CA, 1996.

[42] Y. Bartal, N. Linial, M. Mendel, A. Naor, On metric Ramsey-type phenomena, *Ann. of Math.*, **162** (2005), 643–709.

[43] H. Bass, Covering theory for graphs of groups, *J. Pure Appl. Algebra*, **89** (1993), no. 1–2, 3–47.

[44] S. Bates, W. B. Johnson, J. Lindenstrauss, D. Preiss, G. Schechtman, Affine approximation of Lipschitz functions and nonlinear quotients, *Geom. Funct. Anal.*, **9** (1999), no. 6, 1092–1127.

[45] F. Baudier, Metrical characterization of super-reflexivity and linear type of Banach spaces, *Archiv Math.*, **89** (2007), no. 5, 419–429.

[46] F. Baudier, Embeddings of proper metric spaces into Banach spaces, *Houston J. Math.*, **38** (2012), no. 1, 209–223.

[47] F. Baudier, Quantitative nonlinear embeddings into Lebesgue sequence spaces, `arXiv:1210.0588`.

[48] F. Baudier, G. Lancien, Embeddings of locally finite metric spaces into Banach spaces, *Proc. Amer. Math. Soc.*, **136** (2008), 1029–1033.

[49] B. Baumslag, M. Tretkoff, Residually finite HNN extensions, *Comm. Algebra*, **6** (1978), 179–194.

[50] B. Beauzamy, *Introduction to Banach Spaces and their Geometry*, North-Holland Mathematics Studies **68** and Notas de Matematica **86**, North Holland Publishing Co., Amsterdam, 1982.

[51] B. Beauzamy, J.-T. Lapresté, *Modèles étalés des espaces de Banach*, Travaux en Cours, Hermann, Paris, 1984.

[52] B. Begun, A remark on almost extensions of Lipschitz functions, *Israel J. Math.*, **109** (1999), 151–155.

[53] B. Bekka, P. de la Harpe, A. Valette, *Kazhdan's Property* (T), New Mathematical Monographs **11**, Cambridge University Press, Cambridge, 2008.

[54] I. Benjamini, Y. Makarychev, Dimension reduction for hyperbolic space, *Proc. Amer. Math. Soc.*, **137** (2009), no. 2, 695–698.

[55] Y. Benyamini, J. Lindenstrauss, *Geometric Nonlinear Functional Analysis*, Vol. 1, Colloquium Publications **48**, American Mathematical Society, Providence, RI, 2000.

[56] J. Bergh, J. Löfström, *Interpolation Spaces. An Introduction*, Grundlehren der mathematischen Wissenschaften **223**, Springer-Verlag, Berlin, 1976.

[57] J. Bergh, J. Peetre, On the spaces V_p $(0 < p \leq \infty)$, *Bolletino della Unione Matematica Italiana*, **10** (1974), 632–648.

[58] N. L. Biggs, Girth, valency, and excess, *Linear Algebra Appl.*, **31** (1980), 55–59.

[59] N. L. Biggs, Girth and residual finiteness, *Combinatorica*, **8** (1988), no. 4, 307–312.

[60] N. L. Biggs, Cubic graphs with large girth, in: *Combinatorial Mathematics: Proceedings of the Third International Conference* (New York, 1985), Ann. New York Acad. Sci., **555**, pp. 56–62, New York Acad. Sci., New York, 1989.

[61] N. L. Biggs, *Algebraic Graph Theory*, Second edition, Cambridge Mathematical Library, Cambridge University Press, Cambridge, 1993.

[62] N. L. Biggs, Algebraic potential theory on graphs, *Bull. London Math. Soc.*, **29** (1997), no. 6, 641–682.

[63] N. L. Biggs, E. K. Lloyd, R. J. Wilson, *Graph Theory. 1736–1936*, Second edition, The Clarendon Press, Oxford University Press, New York, 1986.

[64] M. Blum, R. M. Karp, O. Vornberger, C. H. Papadimitriou, M. Yannakakis, The complexity of testing whether a graph is a superconcentrator, *Inform. Process. Lett.*, **13** (1981), no. 4–5, 164–167.

[65] L. M. Blumenthal, *Theory and Applications of Distance Geometry*, Oxford, Oxford University Press, 1953; Reprinted by Chelsea Publishing Company, New York, 1970.

[66] L. M. Blumenthal, K. Menger, *Studies in Geometry*, W. H. Freeman and Co., San Francisco, CA, 1970

[67] S. Bochner, A. E. Taylor, Linear functionals on certain spaces of abstractly-valued functions, *Ann. of Math.*, (2) **39** (1938), no. 4, 913–944.

[68] B. Bollobás, *Extremal Graph Theory*, London Mathematical Society Monographs **11**, Academic Press, London New York, 1978.

[69] B. Bollobás, *Extremal Graph Theory with Emphasis on Probabilistic Methods*, CBMS Regional Conference Series in Mathematics **62**, American Mathematical Society, Providence, R I, 1986.

[70] B. Bollobás, The isoperimetric number of random regular graphs, *European J. Combin.*, **9** (1988), no. 3, 241–244.

[71] J. A. Bondy, U. S. R. Murty, *Graph Theory*, Graduate Texts in Mathematics, **244**, Springer, New York, 2008.

[72] C. Borell, On the integrability of Banach space valued Walsh polynomials, in: *Séminaire de Probabilités*, XIII (Univ. Strasbourg, Strasbourg, 1977/78), Lecture Notes in Math., **721**, pp. 1–3, Springer, Berlin, 1979.

[73] N. Bourbaki, Éléments de mathématique. Les structures fondamentales de l'analyse. 3. Topologie générale. 1. Structures topologiques. 2. Structures uniformes (French), Paris: Hermann & Cie. VIII, 129 p. (1940). Newer editions as well as English and Russian translations are available.

[74] J. Bourgain, Strongly exposed points in weakly compact convex sets in Banach spaces, *Proc. Amer. Math. Soc.*, **58** (1976), 197–200.

[75] J. Bourgain, On Lipschitz embedding of finite metric spaces in Hilbert space, *Israel J. Math.*, **52** (1985), no. 1-2, 46–52.

[76] J. Bourgain, The metrical interpretation of superreflexivity in Banach spaces, *Israel J. Math.*, **56** (1986), no. 2, 222–230.

[77] J. Bourgain, Remarks on the extension of Lipschitz maps defined on discrete sets and uniform homeomorphisms, in: *Geometrical Aspects of Functional Analysis* (1985/86), Lecture Notes in Math., **1267**, pp. 157–167 Springer, Berlin, 1987.

[78] J. Bourgain, A. Gamburd, Uniform expansion bounds for Cayley graphs of $SL_2(\mathbb{F}_p)$, *Ann. of Math.* **167** (2008), no. 2, 625–642.

[79] J. Bourgain, V. Milman, H. Wolfson, On type of metric spaces, *Trans. Amer. Math. Soc.*, **294** (1986), no. 1, 295–317.

[80] J. Bretagnolle, D. Dacunha-Castelle, J.-L. Krivine, Lois stables et espaces L^p, *Ann. Inst. H. Poincaré*, Sect. B (N.S.), **2** (1965/1966) 231–259.

[81] E. Breuillard, B. Green, T. Tao, Suzuki groups as expanders, *Groups Geom. Dyn.*, **5** (2011), no. 2, 281–299.

[82] M. R. Bridson, A. Haefliger, *Metric spaces of non-positive curvature*, Grundlehren der Mathematischen Wissenschaften, **319**, Springer-Verlag, Berlin, 1999.

[83] B. Brinkman, M. Charikar, On the impossibility of dimension reduction in ℓ_1, *J. ACM*, **52** (2005), no. 5, 766–788; Conference version: *Proceedings of the 44th Annual IEEE Symposium on Foundations of Computer Science* (2003), pp. 514–523, IEEE Comput. Soc. Press, Los Alamitos, CA, 2003.

[84] J. Brodzki, G. Niblo, J. Špakula, R. Willett, N. Wright, Uniform local amenability, *J. Noncommutative Geom.*, to appear, `arXiv:1203.6169`.

[85] N. Brown, E. Guentner, Uniform embeddings of bounded geometry spaces into reflexive Banach space, *Proc. Amer. Math. Soc.*, **133** (2005), no. 7, 2045–2050.

[86] Y. Brudnyi, N. Kruglyak, *Interpolation Functors and Interpolation Spaces*. Vol. I. Translated from the Russian by Natalie Wadhwa. With a preface by Jaak Peetre. North-Holland Mathematical Library, 47. North-Holland Publishing Co., Amsterdam, 1991.

[87] D. Burago, Y. Burago, S. Ivanov, *A Course in Metric Geometry*, Graduate Studies in Mathematics, **33**, American Mathematical Society, Providence, RI, 2001.

[88] G. Calinescu, H. J. Karloff, Y. Rabani, Approximation algorithms for the 0-extension problem, *SIAM J. Comput.*, **34** (2004/05), no. 2, 358–372; Conference version: *Proceedings of the Twelfth Annual ACM-SIAM Symposium on Discrete Algorithms* (Washington, DC, 2001), pp. 8–16, SIAM, Philadelphia, PA, 2001.

[89] H. Cartan, Théorie des filtres, *C. R. Acad. Sci. Paris*, **205** (1937), 595–598.

[90] H. Cartan, Filtres et ultrafiltres, *C. R. Acad. Sci. Paris*, **205** (1937), 777–779.

[91] F. Chaatit, On uniform homeomorphisms of the unit spheres of certain Banach lattices, *Pacific J. Math.*, **168** (1995), 11–31.

[92] L. S. Chandran, A high girth graph construction, *SIAM J. Discrete Math.*, **16** (2003), no. 3, 366–370.

[93] M. Charikar, A. Sahai, Dimension reduction in the ℓ_1 norm, in: *Proceedings of the 43rd Annual IEEE Conference on Foundations of Computer Science*, pp. 251–260, IEEE Computer Society, Los Alamitos, CA, 2002.

[94] S. Chawla, A. Gupta, H. Räcke, Embeddings of negative-type metrics and an improved approximation to generalized sparsest cut, *ACM Trans. Algorithms*, **4** (2008), no. 2, Art. 22, 18 pp.; Conference version: *Proceedings of the Sixteenth Annual ACM-SIAM Symposium on Discrete Algorithms*, pp. 102–111, ACM, New York, 2005.

[95] J. Cheeger, A lower bound for the smallest eigenvalue of the Laplacian, in: *Problems in Analysis* (Papers dedicated to Salomon Bochner, 1969), pp. 195–199. Princeton University Press, Princeton, NJ, 1970.

[96] J. Cheeger, B. Kleiner, Differentiability of Lipschitz maps from metric measure spaces to Banach spaces with the Radon–Nikodým property, *Geom. Funct. Anal.*, **19** (2009), no. 4, 1017–1028, `arXiv:0808.3249`.

[97] J. Cheeger, B. Kleiner, Realization of metric spaces as inverse limits, and bilipschitz embedding in L_1, *Geom., Funct. Anal.*, **23** (2013), 96–133, `arXiv:1110.2406`.

[98] J. Cheeger, B. Kleiner, A. Naor, Compression bounds for Lipschitz maps from the Heisenberg group to L_1, *Acta Math.*, **207** (2011), 291–373; Conference version: A $(\log n)^{\Omega(1)}$ integrality gap for the sparsest cut SDP, in *50th Annual IEEE Symposium on Foundations of Computer Science* (FOCS 2009), pp. 555–564, IEEE Computer Soc., Los Alamitos, CA, 2009.

[99] X. Chen, Q. Wang, G. Yu, The maximal coarse Baum–Connes conjecture for spaces which admit a fibred coarse embedding into Hilbert space, (2012), `arXiv:1208.4543`.

[100] F. R. K. Chung, *Spectral Graph Theory*, CBMS Regional Conference Series in Mathematics **92**, American Mathematical Society, Providence, RI, 1997.

[101] J. A. Clarkson, Uniformly convex spaces, *Trans. Amer. Math. Soc.*, **40** (1936), no. 3, 396–414.

[102] R. R. Coifman, M. Cwikel, R. Rochberg, Y. Sagher, G. Weiss, A theory of complex interpolation for families of Banach spaces, *Adv. in Math.*, **43** (1982), no. 3, 203–229.

[103] H. Corson, V. Klee, *Topological Classification of Convex Sets*, Proc. Sympos. Pure Math., Vol. **VII** pp. 37–51, Amer. Math. Soc., Providence, R.I., 1963.

[104] H. S. M. Coxeter, W. O. J. Moser, *Generators and Relations for Discrete Groups*, Third edition. Ergebnisse der Mathematik und ihrer Grenzgebiete, **14**, Springer-Verlag, New York-Heidelberg, 1972. (Fourth edition and a Russian translation are available.)

[105] C. W. Curtis, I. Reiner, *Representation Theory of Finite Groups and Associative Algebras*. Reprint of the 1962 original. AMS Chelsea Publishing, Providence, RI, 2006.

[106] D. M. Cvetković, M. Doob, H. Sachs, *Spectra of Graphs. Theory and Application*, Pure and Applied Mathematics, **87**. Academic Press, Inc., New York-London, 1980. Newer editions are also available.

[107] D. Dacunha-Castelle, J.-L. Krivine, Applications des ultraproduits à l'étude des espaces et des algèbres de Banach, *Studia Math.*, **41** (1972), 315–334.

[108] M. Dadarlat, E. Guentner, Constructions preserving Hilbert space uniform embeddability of discrete groups, *Trans. Amer. Math. Soc.*, **355** (2003), no. 8, 3253–3275.

[109] S. Dasgupta, A. Gupta, An elementary proof of the Johnson–Lindenstrauss lemma, *Random Structures & Algorithms*, **22** (2002), 60–65.

[110] G. Davidoff, P. Sarnak, A. Valette, *Elementary Number Theory, Group Theory and Ramanujan Graphs*, London Mathematical Society Student Texts **55**, Cambridge University Press, Cambridge, 2003.

[111] W. J. Davis, W. B. Johnson, A renorming of nonreflexive Banach spaces, *Proc. Amer. Math. Soc.*, **37** (1973), 486–488.

[112] W. J. Davis, W. B. Johnson, J. Lindenstrauss, The ℓ_1^n problem and degrees of non-reflexivity, *Studia Math.*, **55** (1976), no. 2, 123–139.

[113] W. J. Davis, J. Lindenstrauss, The ℓ_1^n problem and degrees of non-reflexivity. II, *Studia Math.*, **58** (1976), no. 2, 179–196.

[114] M. M. Day, Some more uniformly convex spaces, *Bull. Amer. Math. Soc.*, **47** (1941), 504–507.

[115] M. M. Day, Uniform convexity in factor and conjugate spaces, *Ann. of Math.*, **45** (1944), no. 2, 375–385.

[116] M. M. Day, Polygons circumscribed about closed convex curves, *Trans. Amer. Math. Soc.*, **62** (1947), 315–319.

[117] R. Deville, G. Godefroy, V. Zizler, *Smoothness and Renormings in Banach Spaces*, Pitman Monographs and Surveys in Pure and Applied Mathematics, **64**, Longman Scientific & Technical, Harlow; copublished in the United States with John Wiley & Sons, Inc., New York, 1993.

[118] M. Deza, M. Laurent, *Geometry of Cuts and Metrics*, Algorithms and Combinatorics **15**, Springer-Verlag, New York, 1997.

[119] J. Diestel, *Geometry of Banach Spaces – Selected Topics*, Lecture Notes in Mathematics, Vol. **485**, Springer-Verlag, Berlin-New York, 1975.

[120] J. Diestel, H. Jarchow, A. Tonge, *Absolutely Summing Operators*, Cambridge Studies in Advanced Mathematics, **43**, Cambridge University Press, Cambridge, 1995.

[121] J. Diestel, J. J. Uhl, Jr. *Vector Measures*, With a foreword by B. J. Pettis. Mathematical Surveys, No. 15. American Mathematical Society, Providence, RI, 1977.

[122] R. Diestel, *Graph Theory*, Third edition, Graduate Texts in Mathematics, **173**, Springer-Verlag, Berlin, 2005.

[123] J. Ding, J. R. Lee, Y. Peres, Markov type and threshold embeddings, (2012), `arXiv:1208.6088`.

[124] A. Dmitriyuk, Y. Gordon, Generalizing the Johnson–Lindenstrauss lemma to k-dimensional affine subspaces, *Studia Math.*, **195** (2009), no. 3, 227–241.

[125] J. Dodziuk, Difference equations, isoperimetric inequality and transience of certain random walks, *Trans. Amer. Math. Soc.*, **284** (1984), no. 2, 787–794.

[126] A. N. Dranishnikov, Lipschitz cohomology, Novikov conjecture, and expanders, (Russian) *Tr. Mat. Inst. Steklova*, **247** (2004), Geom. Topol. i Teor. Mnozh., 59–73; English transl.: *Proc. Steklov Inst. Math.* **247**, no. 4 (2004), 50–63, `arXiv:math/0205172`.

[127] A. N. Dranishnikov, G. Gong, V. Lafforgue, G. Yu, Uniform embeddings into Hilbert space and a question of Gromov, *Canad. Math. Bull.*, **45** (2002), no. 1, 60–70.

[128] Y. Drier, N. Linial, Minors in lifts of graphs, *Random Structures Algorithms*, **29** (2006), 208–225

[129] E. Dubinsky, A. Pełczyński, H. P. Rosenthal, On Banach spaces X for which $\Pi_2(\mathcal{L}_\infty, X) = B(\mathcal{L}_\infty, X)$, *Studia Math.*, **44** (1972), 617–648.

[130] N. Dunford, J. T. Schwartz, *Linear Operators*. Part I: General Theory, Pure and Applied Mathematics **7**, Interscience Publishers, New York, 1958.

[131] Y. Dutrieux, V. Ferenczi, The Lipschitz free Banach spaces of $C(K)$-spaces, *Proc. Amer. Math. Soc.*, **134** (2006), no. 4, 1039–1044.

[132] A. Dvoretzky, Some results on convex bodies and Banach spaces, in: *Proc. Internat. Sympos. Linear Spaces* (Jerusalem, 1960), pp. 123–160, Jerusalem Academic Press, Jerusalem; Pergamon, Oxford, 1961; Announcement: A theorem on convex bodies and applications to Banach spaces, *Proc. Nat. Acad. Sci. U.S.A.*, **45** (1959) 223–226; erratum, 1554.

[133] P. Enflo, On the nonexistence of uniform homeomorphisms between L_p-spaces, *Arkiv Mat.*, **8** (1969), 103–105.

[134] P. Enflo, On a problem of Smirnov, *Arkiv Mat.*, **8** (1969), 107–109.

[135] P. Enflo, Banach spaces which can be given an equivalent uniformly convex norm, *Israel J. Math*, **13** (1972), 281–288.

[136] P. Enflo, Uniform homeomorphisms between Banach spaces, *Séminaire d'Analyse Fonctionnelle* (1975–1976), Exp. No. 18, 6 pp., École Polytech., Palaiseau, 1976.

[137] P. Enflo, On infinite-dimensional topological groups, *Séminaire sur la Géométrie des Espaces de Banach* (1977–1978), Exp. No. 10–11, 11 pp., École Polytech., Palaiseau, 1978.

[138] P. Erdős, Graph theory and probability, *Canad. J. Math.*, **11** (1959) 34–38.

[139] P. Erdős, H. Sachs, Reguläre Graphen gegebener Taillenweite mit minimaler Knotenzahl, *Wiss. Z. Martin-Luther-Univ. Halle-Wittenberg Math.-Natur. Reihe*, **12** (1963), 251–257.

[140] G. Exoo, R. Jajcay, On the girth of voltage graph lifts, *European J. Combin.* **32** (2011), no. 4, 554–562.

[141] M. Fabian, P. Habala, P. Hájek, V. Montesinos Santalucía, J. Pelant, V. Zizler, *Functional Analysis and Infinite-Dimensional Geometry*, CMS Books in Mathematics, **8**, Springer-Verlag, New York, 2001.

[142] J. Fakcharoenphol, S. Rao, K. Talwar, A tight bound on approximating arbitrary metrics by tree metrics, *J. Comput. System Sci.*, **69** (2004), no. 3, 485–497; Conference version: *Proceedings of the Thirty-Fifth Annual ACM Symposium on Theory of Computing*, pp. 448–455, ACM, New York, 2003.

[143] J. Fakcharoenphol, K. Talwar, An improved decomposition theorem for graphs excluding a fixed minor, in: *Approximation, Randomization, and Combinatorial Optimization*, Lecture Notes in Comput. Sci., **2764**, pp. 36–46, Springer, Berlin, 2003.

[144] J. Farahat, Exemples d'espaces B-convexes non réflexifs, d'après James et Lindenstrauss, Séminaire Maurey-Schwartz 1974–1975: Espaces L^p, applications radonifiantes et géométrie des espaces de Banach, Exp. No. XX, 12 pp. (erratum, p. 3) Centre Math., École Polytech., Paris, 1975. Séminaire d'Analyse Fonctionelle, 1974–1975.

[145] J. E. Farahat, On the problem of k structure, *Israel J. Math.*, **28** (1977), no. 1–2, 141–150.

[146] W. Feller, *An Introduction to Probability Theory and its Applications*, Vol. I, Third edition, John Wiley & Sons, Inc., New York-London-Sydney, 1968.

[147] W. Feller, *An Introduction to Probability Theory and its Applications*, Vol. II, Second edition, John Wiley & Sons, Inc., New York-London-Sydney, 1971.

[148] T. Figiel, On the moduli of convexity and smoothness, *Studia Math.*, **56** (1976), no. 2, 121–155.

[149] T. Figiel, G. Pisier, Séries aléatoires dans les espaces uniformément convexes ou uniformément lisses, *C. R. Acad. Sci. Paris*, Sér. A, **279** (1974), 611–614.

[150] G. M. Fikhtengolts (=Fikhtengol'ts), *A Course in Differential and Integral Calculus* (Russian), Vol. II, Moscow, Nauka, 1970; German translation: G. M. Fichtenholz, *Differential- und Integralrechnung*. II. Tenth edition. Translated from the Russian by Brigitte Mai and Walter Mai. Hochschulbücher für Mathematik, **62**. VEB Deutscher Verlag der Wissenschaften, Berlin, 1990.

[151] R. J. Fleming, J. E. Jamison, *Isometries on Banach Spaces: Function Spaces*, Chapman & Hall/CRC Monographs and Surveys in Pure and Applied Mathematics **129**, Chapman & Hall/CRC, Boca Raton, FL, 2003.

[152] R. J. Fleming, J. E. Jamison, *Isometries on Banach Spaces*, Vol. 2. *Vector-valued Function Spaces*, Chapman & Hall/CRC Monographs and Surveys in Pure and Applied Mathematics **138**, Chapman & Hall/CRC, Boca Raton, FL, 2008.

[153] J. Flood, *Free topological vector spaces*, Dissertationes Math. (Rozprawy Mat.) **221** (1984), 95 pp.

[154] P. Frankl, H. Maehara, The Johnson–Lindenstrauss lemma and the sphericity of some graphs, *J. Combin. Theory*, Ser. B, **44** (1988), no. 3, 355–362.

[155] M. Fréchet, Les dimensions d'un ensemble abstrait, *Math. Ann.*, **68** (1910), no. 2, 145–168.

[156] J. Friedman, A proof of Alon's second eigenvalue conjecture and related problems, *Mem. Amer. Math. Soc.*, **195** (2008), no. 910, viii+100 pp.

[157] A. Gamburd, S. Hoory, M. Shahshahani, A. Shalev, B. Virág, On the girth of random Cayley graphs, *Random Structures and Algorithms*, **35** (2009), no. 1, 100–117, arXiv:0707.1833v1

[158] A. Gardiner, Extremal regular expanders, *Ars Combin.*, **20A** (1985), 119–124.

[159] D. J. H. Garling, Stable Banach spaces, random measures and Orlicz function spaces, in: *Probability Measures on Groups* (Oberwolfach, 1981), Lecture Notes in Math., **928**, pp. 121–175, Springer, Berlin-New York, 1982.

[160] D. J. H. Garling, *Inequalities: A Journey into Linear Analysis*, Cambridge University Press, Cambridge, 2007.

[161] D. P. Giesy, On a convexity condition in normed linear spaces, *Trans. Amer. Math. Soc.* **125** (1966), 114–146; Additions and corrections to "On a convexity condition in normed linear spaces", *Trans. Amer. Math. Soc.* **140** (1969), 511–512.

[162] O. Giladi, M. Mendel, A. Naor, Improved bounds in the metric cotype inequality for Banach spaces, *J. Funct. Anal.* **260** (2011), 164–194.

[163] O. Giladi, A. Naor, Improved bounds in the scaled Enflo type inequality for Banach spaces, *Extracta Math.*, **25** (2010), no. 2, 151–164.

[164] O. Giladi, A. Naor, G. Schechtman, Bourgain's discretization theorem, *Annales Mathematiques de la faculte des sciences de Toulouse*, Vol. **XXI** (2012), no. 4, 817–837, arXiv:1110.5368.

[165] A. Godard, Tree metrics and their Lipschitz-free spaces, *Proc. Amer. Math. Soc.*, **138** (2010), no. 12, 4311–4320, arXiv:0904.3178.

[166] G. Godefroy, Renormings of Banach spaces, in: *Handbook of the Geometry of Banach spaces*, Vol. I, pp. 781–835, North-Holland, Amsterdam, 2001.

[167] G. Godefroy, N. J. Kalton, Lipschitz-free Banach spaces, *Studia Math.*, **159** (2003), no. 1, 121–141.

[168] G. Godefroy, G. Lancien, V. Zizler, The non-linear geometry of Banach spaces after Nigel Kalton, (2012), arXiv:1207.2958.

[169] G. Godefroy, N. Ozawa, Free Banach spaces and the approximation properties, *Proc. Amer. Math. Soc.*, to appear, arXiv:1201.0847.

[170] C. Godsil, G. Royle, *Algebraic Graph Theory*, Graduate Texts in Mathematics, **207**. Springer-Verlag, New York, 2001.

[171] H. H. Goldstine, Weakly complete Banach spaces, *Duke Math. J.*, **4** (1938), 125–131.

[172] L. Gottlieb, R. Krauthgamer, A nonlinear approach to dimension reduction, preprint (2009), `arXiv:0907.5477`; Conference version: Proceedings of the Twenty-Second Annual ACM-SIAM Symposium on Discrete Algorithms, pp. 888–899, SIAM, Philadelphia, PA, 2011.

[173] W. T. Gowers, B. Maurey, The unconditional basic sequence problem, *J. Amer. Math. Soc.*, **6** (1993), no. 4, 851–874.

[174] M. Gromov, Filling Riemannian manifolds, *J. Differential Geom.*, **18** (1983), no. 1, 1–147.

[175] M. Gromov, Asymptotic invariants of infinite groups, in: A. Niblo, M. Roller (eds.) *Geometric Group Theory*, London Math. Soc. Lecture Notes, **182**, pp. 1–295, Cambridge University Press, Cambridge, 1993.

[176] M. Gromov, Problems (4) and (5), in: *Novikov Conjectures, Index Theorems and Rigidity*. Vol. 1. Including papers from the conference held at the Mathematisches Forschungsinstitut Oberwolfach, Oberwolfach, September 6–10, 1993, S. C. Ferry, A. Ranicki, J. Rosenberg, (eds.), London Mathematical Society Lecture Note Series, **226**, p. 67, Cambridge University Press, Cambridge, 1995.

[177] M. Gromov, *Metric Structures for Riemannian and Non-Riemannian Spaces*, Based on the 1981 French original. With appendices by M. Katz, P. Pansu and S. Semmes. Translated from the French by S. M. Bates. Progress in Mathematics **152**, Birkhäuser, Boston, MA, 1999.

[178] M. Gromov, Spaces and questions, GAFA 2000 (Tel Aviv, 1999), *Geom. Funct. Anal.* 2000, Special Volume, Part I, 118–161.

[179] M. Gromov, Random walk in random groups, *Geom. Funct. Anal.*, **13** (2003), no. 1, 73–146.

[180] M. Gromov, L. Guth, Generalizations of the Kolmogorov–Barzdin embedding estimates, *Duke Math. J.*, **161** (2012), 2549–2603, `arXiv:1103.3423`.

[181] M. L. Gromov, V. A. Rokhlin, Imbeddings and immersions in Riemannian geometry (Russian), *Uspehi Mat. Nauk*, **25** (1970) no. 5 (155), 3–62; English transl.: *Russian Math. Surveys*, **25** (1970), no. 5, 1–57.

[182] J. L. Gross, Every connected regular graph of even degree is a Schreier coset graph *J. Combin. Theory*, Ser. B, **22** (1977), no. 3, 227–232.

[183] J. L. Gross, Voltage graphs, in: *Handbook of Graph Theory*, eds. J. L. Gross, J. Yellen, pp. 661–683, CRC Press, Boca Raton, FL, 2004.

[184] J. L. Gross, T. W. Tucker, Generating all graph coverings by permutation voltage assignments, *Discrete Math.*, **18** (1977), 273–283.

[185] J. L. Gross, T. W. Tucker, *Topological Graph Theory*, Wiley-Interscience Series in Discrete Mathematics and Optimization, John Wiley & Sons, Inc., New York, 1987.

[186] E. Guentner, J. Kaminker, Geometric and analytic properties of groups, in: *Noncommutative Geometry*, eds. S. Doplicher and R. Longo, Lecture Notes Math. **1831**, pp. 253–262, Springer-Verlag, Berlin, 2004.

[187] S. Guerre-Delabriére, *Classical Sequences in Banach Spaces*, With a foreword by H. P. Rosenthal, Monographs and Textbooks in Pure and Applied Mathematics, **166**, Marcel Dekker, Inc., New York, 1992.

[188] A. Gupta, Embedding trees into low dimensional Euclidean spaces, *Discrete Comput. Geom.*, **24** (2000), 106–116.

[189] A. Gupta, R. Krauthgamer, J. R. Lee, Bounded geometries, fractals, and low-distortion embeddings, *FOCS*, 2003, pp. 534–543.

[190] A. Gupta, I. Newman, Y. Rabinovich, A. Sinclair, Cuts, trees and ℓ_1-embeddings of graphs, *Combinatorica*, **24** (2004) 233–269; Conference version in: *40th Annual IEEE Symposium on Foundations of Computer Science*, 1999, pp. 399–408, IEEE Comput. Soc. Press, Los Alamitos, CA, 1999.

[191] O. Häggström, *Finite Markov Chains and Algorithmic Applications*, London Mathematical Society Student Texts **52**, Cambridge University Press, Cambridge, 2002.

[192] L. G. Hanin, Kantorovich–Rubinstein norm and its application in the theory of Lipschitz spaces, *Proc. Amer. Math. Soc.*, **115** (1992), no. 2, 345–352.

[193] O. Hanner, On the uniform convexity of L^p and ℓ^p, *Ark. Mat.*, **3** (1956), 239–244.

[194] G. H. Hardy, J. E. Littlewood, G. Pólya, *Inequalities*, Cambridge University Press, Cambridge, 1934; Russian translation: Moscow, 1948.

[195] P. de la Harpe, *Topics in Geometric Group Theory*, Chicago Lectures in Mathematics, University of Chicago Press, Chicago and London, 2000.

[196] J. M. Heinonen, *Lectures on Analysis on Metric Spaces*, Universitext, Springer-Verlag, New York, 2001.

[197] J. M. Heinonen, Geometric embeddings of metric spaces. Report. University of Jyväskylä Department of Mathematics and Statistics, **90**. University of Jyväskylä, Jyväskylä, 2003.

[198] J. M. Heinonen, Nonsmooth calculus, *Bull. Amer. Math. Soc.* (N.S.) **44** (2007), no. 2, 163–232.

[199] S. Heinrich, Ultraproducts in Banach space theory, *J. Reine angew. Math.*, **313** (1980), 72–104.

[200] S. Heinrich, P. Mankiewicz, Applications of ultrapowers to the uniform and Lipschitz classification of Banach spaces, *Studia Math.*, **73** (1982), no. 3, 225–251.

[201] E. Helly, Über Systeme linearer Gleichungen mit unendlich vielen Unbekannten, *Monatsh. Math. Phys.*, **31** (1921), no. 1, 60–91.

[202] N. Higson, V. Lafforgue, G. Skandalis, Counterexamples to the Baum–Connes conjecture, *Geom. Funct. Anal.*, **12** (2002) 330–354.

[203] M. J. Hoare, On the girth of trivalent Cayley graphs, in: *Graphs and Other Combinatorial Topics* (Prague, 1982), Teubner-Texte Math., **59**, pp. 109–114, Teubner, Leipzig, 1983.

[204] J. Hoffmann-Jørgensen, Sums of independent Banach space valued random variables, *Studia Math.*, **52** (1974), 159–186.

[205] S. Hoory, N. Linial, A. Wigderson, Expander graphs and their applications, *Bull. Amer. Math. Soc.*, **43** (2006), 439–561.

[206] W. Imrich, Explicit construction of regular graphs without small cycles, *Combinatorica*, **4** (1984), no. 1, 53–59.

[207] P. Indyk, Algorithmic applications of low-distortion geometric embeddings, in: *Proc 42nd IEEE Symposium on Foundations of Computer Science*, 2001, pp. 10–33, the paper can be downloaded from: `http://theory.lcs.mit.edu/~indyk/`

[208] P. Indyk, J. Matoušek, Low-distortion embeddings of finite metric spaces, in: *Handbook of Discrete and Computational Geometry*, eds. J. E. Goodman and J. O'Rourke, pp. 177–196, Boca Raton, Chapman and Hall/CRC, 2004.

[209] P. Indyk, R. Motwani, Approximate nearest neighbors: towards removing the curse of dimensionality, in: *STOC '98* (Dallas, TX), pp. 604–613, ACM, New York, 1999.

[210] A. Jaffe, J. R. Lee, M. Moharrami, On the optimality of gluing over scales, *Discrete Comput. Geom.*, **46** (2011), no. 2, 270–282.

[211] R. C. James, Uniformly non-square Banach spaces, *Ann. of Math.*, **80** (1964), 542–550.

[212] R. C. James, Weak compactness and reflexivity, *Israel J. Math.*, **2** (1964), 101–119.

[213] R. C. James, Some self-dual properties of normed linear spaces, in: *Symposium on Infinite-Dimensional Topology* (Louisiana State Univ., Baton Rouge, La., 1967), Ann. of Math. Studies, No. **69**, pp. 159–175, Princeton University Press, Princeton, N. J., 1972.

[214] R. C. James, Super-reflexive Banach spaces, *Canad. J. Math.*, **24** (1972), 896–904.

[215] R. C. James, A nonreflexive Banach space that is uniformly nonoctahedral, *Israel J. Math.*, **18** (1974), 145–155.

[216] R. C. James, Nonreflexive spaces of type 2, *Israel J. Math.*, **30** (1978), 1–13.

[217] R. C. James, J. Lindenstrauss, The octahedral problem for Banach spaces, in: *Proceedings of the Seminar on Random Series, Convex Geometry and Geometry of Banach Spaces*, Aarhus, October 14–October 20, 1974; pp. 100–120, Aarhus Universitet, 1975.

[218] F. John, Extremum problems with inequalities as subsidiary conditions, in: *Courant Anniversary Volume*, pp. 187–204, Interscience, New York, 1948, (reprinted in F. John, *Collected Papers*, Vol. 2, pp. 543–560, Birkhäuser, Boston, 1985).

[219] W. B. Johnson, J. Lindenstrauss, Extensions of Lipschitz mappings into a Hilbert space, in: *Conference in Modern Analysis and Probability* (New Haven, Conn., 1982), Contemp. Math., **26**, pp. 189–206, Amer. Math. Soc., Providence, RI, 1984.

[220] W. B. Johnson, J. Lindenstrauss, Basic concepts in the geometry of Banach spaces, in: *Handbook of the Geometry of Banach Spaces* (W. B. Johnson and J. Lindenstrauss, eds.) Vol. 1, pp. 1–84, Elsevier, Amsterdam, 2001.

[221] W. B. Johnson, J. Lindenstrauss, eds., *Handbook of the Geometry of Banach Spaces*, Vol. 1, North-Holland, Amsterdam, 2001.

[222] W. B. Johnson, J. Lindenstrauss, eds., *Handbook of the Geometry of Banach Spaces*, Vol. 2, North-Holland, Amsterdam, 2003.

[223] W. B. Johnson, A. Naor, The Johnson–Lindenstrauss lemma almost characterizes Hilbert space, but not quite, *Discrete Comput. Geom.*, **43** (2010), no. 3, 542–553, doi:10.1007/s00454-009-9193-z.

[224] W. B. Johnson, L. Randrianarivony, $\ell_p(p > 2)$ does not coarsely embed into a Hilbert space, *Proc. Amer. Math. Soc.*, **134** (2006), no. 4, 1045–1050.

[225] W. B. Johnson, H. P. Rosenthal, M. Zippin, On bases, finite dimensional decompositions and weaker structures in Banach spaces, *Israel J. Math.*, **9** (1971), 488–506.

[226] W. B. Johnson, G. Schechtman, Diamond graphs and super-reflexivity, *J. Topol. Anal.*, **1** (2009), no. 2, 177–189.

[227] P. Jordan, J. von Neumann, On inner products in linear, metric spaces, *Ann. Math.,* (2) **36** (1935), 719–723.

[228] M. I. Kadets, Unconditionally convergent series in a uniformly convex space (Russian), *Uspehi Mat. Nauk* (N.S.) **11** (1956), no. 5, 185–190.

[229] M. I. Kadets, Spaces isomorphic to a locally uniformly convex space (Russian), *Izv. Vyssh. Uchebn. Zaved. Matematika* (1959), no. 6, 51–57; Also: Letter to the editor (Russian), *Izv. Vyssh. Uchebn. Zaved. Matematika* (1961), no. 6, 186–187.

[230] M. I. Kadets, A. Pełczyński, Basic sequences, biorthogonal systems and norming sets in Banach and Fréchet spaces (Russian), *Studia Math.*, **25** (1965), 297–323.

[231] V. M. Kadets, Lipschitz mappings of metric spaces (Russian), *Izv. Vyssh. Uchebn. Zaved. Mat.* **1985**, no. 1, 30–34; English transl.: *Soviet Math.* (Iz. VUZ), **29** (1985), no. 1, 36–41.

[232] J.-P. Kahane, Sur les sommes vectorielles $\sum \pm u_n$, *C. R. Acad. Sci. Paris*, **259** (1964) 2577–2580.

[233] J.-P. Kahane, *Some Random Series of Functions*, D. C. Heath and Co., Raytheon Education Co., Lexington, Mass. 1968; Russian translation: Mir, Moscow, 1973; Second edition: Cambridge University Press, Cambridge, 1985.

[234] S. Kakutani, Concrete representation of abstract (L)-spaces and the mean ergodic theorem, *Ann. Math.*, **42** (1941), 523–537.

[235] N. J. Kalton, Spaces of Lipschitz and Hölder functions and their applications, *Collect. Math.*, **55** (2004), 171–217.

[236] N. J. Kalton, Coarse and uniform embeddings into reflexive spaces, *Quart. J. Math. Oxford*, **58** (2007), 393–414.

[237] N. J. Kalton, The nonlinear geometry of Banach spaces, *Rev. Mat. Comput.*, **21** (2008), no. 1, 7–60.

[238] N. J. Kalton, G. Lancien, Best constants for Lipschitz embeddings of metric spaces into c_0, *Fund. Math.*, **199** (2008), 249–272.

[239] L. V. Kantorovich, On mass transportation (Russian), *Doklady Acad. Naus SSSR*, (N.S.) **37**, (1942), 199–201; English transl.: *J. Math. Sci.* (N.Y.), **133** (2006), no. 4, 1381–1382.

[240] L. V. Kantorovich, On a problem of Monge (Russian), *Uspekhi Mat. Nauk,* **3** (1948), no. 2, 225–226; English transl.: *J. Math. Sci.* (N.Y.), **133** (2006), no. 4, 1383.

[241] L. V. Kantorovich, G. S. Rubinstein, On a functional space and certain extremum problems (Russian), *Dokl. Akad. Nauk SSSR* (N.S.), **115** (1957), 1058–1061.

[242] L. V. Kantorovich, G. S. Rubinstein, On a space of completely additive functions (Russian), *Vestnik Leningrad. Univ.*, **13** (1958), no. 7, 52–59.

[243] G. Kasparov, G. Yu, The coarse geometric Novikov conjecture and uniform convexity, *Adv. Math.*, **206** (2006), no. 1, 1–56.

[244] M. Kassabov, Symmetric groups and expander graphs, *Invent. Math.*, **170** (2007), no. 2, 327–354.

[245] M. Kassabov, A. Lubotzky, N. Nikolov, Finite simple groups as expanders, *Proc. Nat. Acad. Sci. USA*, **103** (2006), no. 16, 6116–6119.

[246] D. A. Kazhdan, On the connection of the dual space of a group with the structure of its closed subgroups (Russian), *Funktsional. Anal. i Prilozhen.*, **1** (1967), 71–74.

[247] C. Kelleher, D. Miller, T. Osborn, A. Weston, Strongly non-embeddable metric spaces, *Topology Appl.*, **159** (2012), no. 3, 749–755.

[248] A. Khinchin, Über dyadiche Bruche, *Math. Zeitschrift*, **18** (1923), 109–116.

[249] S. Khot, A. Naor, Nonembeddability theorems via Fourier analysis, *Math. Ann.*, **334** (2006), 821–852.

[250] B. Klartag, S. Mendelson, Empirical processes and random projections, *J. Funct. Anal.*, **225** (2005), no. 1, 229–245.

[251] P. Klein, S. Plotkin, S. Rao, Excluded minors, network decomposition, and multicommodity flow, in: *Proc. 25th Annual ACM Symposium on the Theory of Computing*, pp. 682–690, ACM, New York, 1993.

[252] A. Koldobsky, H. König, Aspects of the isometric theory of Banach spaces, in: *Handbook of the Geometry of Banach Spaces*, Vol. 1, eds., W. B. Johnson and J. Lindenstrauss, North-Holland Publishing Co., 2001, 899–939.

[253] A. N. Kolmogorov, Y. M. Barzdin, On the realization of nets in the 3-dimensional space (Russian), *Problemy Kibernetiki*, **8** (1967), 261–268; English translation in: *Selected Works of A. N. Kolmogorov*, Vol. 3, pp. 194–202 (and a remark on page 245), Kluwer Academic Publishers, Dordrecht, 1993.

[254] R. Krauthgamer, J. R. Lee, The intrinsic dimensionality of graphs, *Combinatorica*, **27** (2007), no. 5, 551–585; Conference version: *35th Symposium on Theory of Computing* (STOC'03), pp. 438–447, ACM, New York, June 2003.

[255] R. Krauthgamer, J. R. Lee, M. Mendel, A. Naor, Measured descent: A new embedding method for finite metrics, *Geom. Funct. Anal.*, **15** (2005), 839–858.

[256] J.-L. Krivine, B. Maurey, Espaces de Banach stables, *Israel J. Math.*, **39** (1981), 273–295.

[257] K. Kunugui, Applications des espaces a une infinité de dimensions a la théorie des ensembles, *Proc. Imp. Acad. Japan*, **11** (1936), 351–353.

[258] K. Kuratowski, Quelques problèmes concernant les espaces métriques nonseparables, *Fundam. Math.*, **25** (1935), 534–545.

[259] K. Kuratowski, *Topology*, Vol. I, New edition, revised and augmented, Translated from the French by J. Jaworowski, Academic Press, New York-London; Państwowe Wydawnictwo Naukowe, Warsaw 1966.

[260] S. Kwapień, Isomorphic characterizations of inner product spaces by orthogonal series with vector valued coefficients, *Studia Math.*, **44** (1972), 583–595.

[261] T. J. Laakso, Ahlfors Q-regular spaces with arbitrary $Q > 1$ admitting weak Poincare inequality, *Geom. Funct. Anal.*, **10** (2000), no. 1, 111–123.

[262] H. E. Lacey, *The Isometric Theory of Classical Banach Spaces*, Die Grundlehren der mathematischen Wissenschaften, Band **208** Springer-Verlag, New York-Heidelberg, 1974.

[263] V. Lafforgue, Un renforcement de la propriété (T), *Duke Math. J.*, **143** (2008), no. 3, 559–602.

[264] V. Lafforgue, Propriété (T) renforcée banachique et transformation de Fourier rapide, *J. Topol. Anal.*, **1** (2009), no. 3, 191–206.

[265] G. Lancien, On uniformly convex and uniformly Kadec–Klee renormings, *Serdica Math. J.*, **21** (1995), no. 1, 1–18.

[266] G. Lancien, E. Pernecká, Approximation properties and Schauder decompositions in Lipschitz-free spaces, (2012), `arXiv:1207.1583`.

[267] U. Lang, C. Plaut, Bilipschitz embeddings of metric spaces into space forms, *Geom. Dedicata*, **87** (2001), 285–307.

[268] R. Latała, K. Oleszkiewicz, On the best constant in the Khinchin–Kahane inequality, *Studia Math.*, **109** (1994), no. 1, 101–104.

[269] K.-S. Lau, Farthest points in weakly compact sets, *Israel J. Math.*, **22** (1975), no. 2, 168–174.

[270] K.-S. Lau, A remark on strongly exposing functionals, *Proc. Amer. Math. Soc.*, **59** (1976), no. 2, 242–244.

[271] M. Ledoux, *The Concentration of Measure Phenomenon*, Mathematical Surveys and Monographs **89**, American Mathematical Society, Providence, RI, 2001.

[272] M. Ledoux, M. Talagrand, *Probability in Banach Spaces. Isoperimetry and Processes*, Ergebnisse der Mathematik und ihrer Grenzgebiete (3), Vol. **23**, Springer-Verlag, Berlin, 1991.

[273] J. R. Lee, *Distance scales, embeddings, and metrics of negative type*, unpublished manuscript, 2006, available at: `http://www.cs.washington.edu/homes/jrl/`; Conference version: *Proceedings of the Sixteenth Annual ACM-SIAM Symposium on Discrete Algorithms*, pp. 92–101, SIAM, Philadelphia, PA, USA, 2005.

[274] J. R. Lee, M. Mendel, A. Naor, Metric structures in L_1: dimension, snowflakes, and average distortion, *European J. Combin.*, **26** (2005), no. 8, 1180–1190.

[275] J. R. Lee, A. Naor, *Metric decomposition, smooth measures, and clustering*, unpublished manuscript, 2003.

[276] J. R. Lee, A. Naor, Embedding the diamond graph in L_p and dimension reduction in L_1, *Geom. Funct. Anal.*, **14** (2004), 745–747.

[277] J. R. Lee, A. Naor, Extending Lipschitz functions via random metric partitions, *Invent. Math.*, **160** (2005), no. 1, 59–95.

[278] J. R. Lee, A. Naor, Y. Peres, Trees and Markov convexity, *Geom. Funct. Anal.*, **18** (2009) 1609–1659.

[279] T. Leighton, Finite common coverings of graphs, *J. Combin. Theory*, Ser. B, **33** (1982), no. 3, 231–238.

[280] T. Leighton, S. Rao, Multicommodity max-flow min-cut theorems and their use in designing approximation algorithms, *J. ACM*, **46** (1999), no. 6, 787–832; Conference version: *Proceedings of IEEE Symposium on Foundations of Computer Science*, 1988, pp. 422–431, IEEE Comput. Soc. Press, Los Alamitos, CA, 1988.

[281] D. A. Levin, Y. Peres, E. L. Wilmer, *Markov Chains and Mixing Times*, With a chapter by James G. Propp and David B. Wilson, American Mathematical Society, Providence, RI, 2009.

[282] S. Li, A. Naor, Discretization and affine approximation in high dimensions, *Israel J. Math.*, to appear, `arXiv:1202.2567`

[283] J. Lindenstrauss, On the modulus of smoothness and divergent series in Banach spaces, *Michigan Math. J.*, **10** (1963), 241–252.

[284] J. Lindenstrauss, On operators which attain their norm, *Israel J. Math.*, **1** (1963), 139–148.

[285] J. Lindenstrauss, On nonlinear projections in Banach spaces, *Michigan Math. J.*, **11** (1964), 263–287.

[286] J. Lindenstrauss, H. P. Rosenthal, The $\mathcal{L}_p$ spaces, *Israel J. Math.*, **7** (1969), 325–349.

[287] J. Lindenstrauss, L. Tzafriri, *Classical Banach Spaces*, Volume I, Sequence Spaces, Springer-Verlag, Berlin, 1977.

[288] J. Lindenstrauss, L. Tzafriri, *Classical Banach Spaces*, Volume II, Function Spaces, Springer-Verlag, Berlin, 1979.

[289] N. Linial, Finite metric spaces–combinatorics, geometry and algorithms, in: *Proceedings of the International Congress of Mathematicians*, Vol. III (Beijing, 2002), pp. 573–586, Higher Ed. Press, Beijing, 2002.

[290] N. Linial, E. London, Y. Rabinovich, The geometry of graphs and some of its algorithmic applications, *Combinatorica*, **15** (1995), no. 2, 215–245.

[291] N. Linial, A. Magen, Least-distortion Euclidean embeddings of graphs: products of cycles and expanders, *J. Combin. Theory*, Ser. B **79** (2000), no. 2, 157–171.

[292] N. Linial, A. Magen, A. Naor, Girth and Euclidean distortion, *Geom. Funct. Anal.*, **12** (2002), 380–394.

[293] N. Linial, M. Saks, Low diameter graph decompositions, *Combinatorica*, **13** (1993), 441–454.

[294] V.I. Liokumovich, Existence of B-spaces with a nonconvex modulus of convexity (Russian), *Izv. Vyssh. Ucebn. Zaved., Matematika* 1973, no. 12(139), 43–49.

[295] J.E. Littlewood, On bounded bilinear forms in an infinite number of variables, *Quarterly Journ.* (Oxford Series), **1** (1930), 164–174.

[296] A.R. Lovaglia, Locally uniformly convex Banach spaces, *Trans. Amer. Math. Soc.*, **78** (1955), 225–238.

[297] L. Lovász, *Combinatorial Problems and Exercises*, Second edition, AMS Chelsea Publishing, Providence, RI, 2007. (Original edition: North-Holland Publishing Co., Amsterdam-New York, 1979.)

[298] A. Lubotzky, Locally symmetric graphs of prescribed girth and Coxeter groups, *SIAM J. Discrete Math.*, **3** (1990), no. 2, 277–280.

[299] A. Lubotzky, *Discrete Groups, Expanding Graphs, and Invariant Measures*, Progress in Mathematics, Vol. **125**, appendix by J.D. Rogawski, Birkhäuser, Basel-Boston, 1994.

[300] A. Lubotzky, Cayley graphs: eigenvalues, expanders and random walks, in: *Surveys in Combinatorics*, 1995 (Stirling), London Math. Soc. Lecture Note Ser., **218**, pp. 155–189, Cambridge University Press, Cambridge, 1995.

[301] A. Lubotzky, Expander graphs in pure and applied mathematics, *Bull. Amer. Math. Soc.*, **49** (2012), 113–162.

[302] A. Lubotzky, R. Phillips, P. Sarnak, Ramanujan graphs, *Combinatorica*, **8** (1988), 261–277.

[303] A. Lubotzky, B. Weiss, Groups and expanders, in: *Expanding Graphs*, pp. 95–109, Princeton, NJ, 1992; DIMACS Ser. Discrete Math. Theoret. Comput. Sci., **10**, Amer. Math. Soc., Providence, RI, 1993.

[304] A. Lubotzky, A. Żuk, *On Property* (τ), book manuscript, 2003.

[305] T.J. Lyons, T.S. Zhang, Decomposition of Dirichlet processes and its applications, *Ann. Probab.*, **22** (1994), 494–524.

[306] C.C. MacDuffee, *The Theory of Matrices*, Corrected reprint of first edition, Chelsea Publishing Company, 1946 (Original edition: Ergebnisse der Mathematik und ihrer Grenzgebiete, **5**, Springer-Verlag, 1933).

[307] F.J. MacWilliams, N.J.A. Sloane, *The Theory of Error-correcting Codes* North-Holland Mathematical Library, Vol. **16**, North-Holland Publishing Co., Amsterdam-New York-Oxford, 1977.

[308] G.A. Margulis, Explicit constructions of expanders (Russian), *Problemy Peredachi Informatsii*, **9** (1973), no. 4, 71–80; English translation: *Problems of Inform. Transmission*, **9** (1973), no. 4, 325–332 (1975).

[309] G.A. Margulis, Explicit constructions of graphs without short cycles and low density codes, *Combinatorica*, **2** (1982), no. 1, 71–78.

[310] G. A. Margulis, Explicit group-theoretic constructions of combinatorial schemes and their applications in the construction of expanders and concentrators (Russian), *Problemy Peredachi Informatsii*, **24** (1988), no. 1, 51–60; English translation: *Problems Inform. Transmission*, **24** (1988), no. 1, 39–46.

[311] J. Matoušek, On the distortion required for embedding finite metric spaces into normed spaces, *Israel J. Math.*, **93** (1996), 333–344.

[312] J. Matoušek, On embedding expanders into ℓ_p spaces, *Israel J. Math.*, **102** (1997), 189–197.

[313] J. Matoušek, On embedding trees into uniformly convex Banach spaces, *Israel J. Math.*, **114** (1999), 221–237.

[314] J. Matoušek, *Lectures on Discrete Geometry*, Graduate Texts in Mathematics **212**, Springer-Verlag, New York, 2002.

[315] J. Matoušek, Embedding finite metric spaces into normed spaces, an updated version of Chapter 15 of the book [314], prepared for a Japanese translation published in 2005. English version is available at `http://kam.mff.cuni.cz/~matousek/`

[316] J. Matoušek, On variants of the Johnson–Lindenstrauss lemma, *Random Structures Algorithms*, **33** (2008), no. 2, 142–156.

[317] J. Matoušek (ed.), since June 2010 maintained jointly with A. Naor; Last revision: August 2011; *Open problems on embeddings of finite metric spaces*, can be downloaded from `http://kam.mff.cuni.cz/~matousek/`

[318] B. Maurey, Type, cotype and K-convexity, in: *Handbook of the Geometry of Banach Spaces*, Vol. 2, pp. 1299–1332, North-Holland, Amsterdam, 2003.

[319] B. Maurey, G. Pisier, Séries de variables aléatoires vectorielles indépendantes et propriétés géométriques des espaces de Banach, *Studia Math.*, **58** (1976), no. 1, 45–90.

[320] S. Mazur, Une remarque sur l'homéomorphie des champs fonctionels (French), *Studia Math.*, **1** (1929), 83–85.

[321] E. J. McShane, Extension of range of functions, *Bull. Amer. Math. Soc.*, **40** (1934), no. 12, 837–842.

[322] E. J. McShane, Linear functionals on certain Banach spaces, *Proc. Amer. Math. Soc.*, **1** (1950), 402–408.

[323] J. Meier, *Groups, Graphs and Trees. An Introduction to the Geometry of Infinite Groups*, London Mathematical Society Student Texts, **73**, Cambridge University Press, Cambridge, 2008.

[324] J. Melleray, F. V. Petrov, A. M. Vershik, Linearly rigid metric spaces and the embedding problem, *Fund. Math.*, **199** (2008), no. 2, 177–194.

[325] M. Mendel, A. Naor, Euclidean quotients of finite metric spaces, *Adv. Math.*, **189** (2004), no. 2, 451–494.

[326] M. Mendel, A. Naor, Scaled Enflo type is equivalent to Rademacher type, *Bull. London Math. Soc.*, **39** (2007), 493–498.

[327] M. Mendel, A. Naor, Metric cotype, *Ann. Math.*, **168** (2008), 247–298.

[328] M. Mendel, A. Naor, Markov convexity and local rigidity of distorted metrics, *J. Eur. Math. Soc.* (JEMS), **15** (2013), no. 1, 287–337; Conference version: *Computational Geometry* (SCG'08), pp. 49–58, ACM, New York, 2008.

[329] M. Mendel, A. Naor, Nonlinear spectral calculus and super-expanders, *Publ. Math. Inst. Hautes Études Sci.*, to appear: `arXiv:1207.4705`; Conference version: Towards a calculus for non-linear spectral gaps, in: *Proceedings of the Twenty-First Annual ACM-SIAM Symposium on Discrete Algorithms*, ed. M. Charikar, pp. 236–255, SIAM, Philidelphia, 2010, `arXiv:0910.2041`

[330] D. P. Milman, On some criteria for the regularity of spaces of the type (B), *Dokl. Acad. Nauk SSSR*, New Ser., **20** (1938), 243–246.

[331] D. P. Milman, V. D. Milman, The geometry of nested families with empty intersection. Structure of the unit sphere of a nonreflexive space (Russian), *Matem. Sbornik*, **66** (1965), no. 1, 109–118; English transl.: *Amer. Math. Soc. Transl.* (2) v. **85** (1969), 233–243.

[332] V.D. Milman, G. Schechtman, *Asymptotic Theory of Finite Dimensional Normed Spaces*, Lecture Notes in Mathematics **1200**, Springer-Verlag, Berlin, 1986.

[333] A. Naor, L_1 embeddings of the Heisenberg group and fast estimation of graph isoperimetry, in: *Proceedings of the International Congress of Mathematicians*, 2010, Hyderabad India, Vol. III, pp. 1549–1575, Hindustan Book Agency, New Delhi, 2011.

[334] A. Naor, An introduction to the Ribe program, *Jpn. J. Math.*, **7** (2012), no. 2, 167–233.

[335] A. Naor, Y. Peres, O. Schramm, S. Sheffield, Markov chains in smooth Banach spaces and Gromov hyperbolic metric spaces, *Duke Math. J.*, **134**, no. 1 (2006), 165–197.

[336] A. Naor, G. Schechtman, Planar Earthmover is not in L_1, *SIAM J. Computing*, **37** (2007), 804–826.

[337] W. D. Neumann, On Leighton's graph covering theorem, *Groups Geom. Dyn.*, **4** (2010), no. 4, 863–872.

[338] I. Newman, Y. Rabinovich, Lower bound on the distortion of embedding planar metrics into Euclidean space, *Discrete Comput. Geom.* **29** (2003), no. 1, 77–81.

[339] I. Newman, Y. Rabinovich, Hard metrics from Cayley graphs of Abelian groups, *Theory of Computing*, **5** (2009), 125–134; Conference version: *24th International Symposium on Theoretical Aspects of Computer Science* 2007, Lecture Notes in Comput. Sci., **4393**, pp. 157–162, Springer, Berlin, 2007.

[340] M. Newman, *Integral Matrices*, Pure and Applied Mathematics, Vol. **45**, Academic Press, New York-London, 1972.

[341] A. Nilli, On the second eigenvalue of a graph, *Discrete Math.*, **91** (1991), no. 2, 207–210.

[342] G. Nordlander, The modulus of convexity in normed linear spaces, *Ark. Mat.*, **4** (1960), 15–17.

[343] G. Nordlander, On sign-independent and almost sign-independent convergence in normed linear spaces, *Ark. Mat.*, **4** (1962), 287–296.

[344] P. W. Nowak, On coarse embeddability into ℓ_p-spaces and a conjecture of Dranishnikov, *Fund. Math.*, **189** (2006), no. 2, 111–116.

[345] P. W. Nowak, G. Yu, *Large Scale Geometry*, European Mathematical Society, Zürich, 2012.

[346] E. Odell, T. Schlumprecht, The distortion problem, *Acta Math.*, **173** (1994), no. 2, 259–281.

[347] S.-I. Ohta, Markov type of Alexandrov spaces of non-negative curvature, *Mathematika*, **55** (2009), no. 1–2, 177–189.

[348] W. Orlicz, Über unbedingte Konvergenz in Funktionenräumen, I, *Studia Math.*, **4** (1933), 33–37.

[349] W. Orlicz, Über unbedingte Konvergenz in Funktionenräumen, II, *Studia Math.*, **4** (1933), 41–47.

[350] M. I. Ostrovskii, Finite dimensional characteristics related to superreflexivity of Banach spaces, *Bull. Aust. Math. Soc.*, **69** (2004), 289–295.

[351] M. I. Ostrovskii, Sobolev spaces on graphs, *Quaestiones Mathematicae*, **28** (2005), 501–523.

[352] M. I. Ostrovskii, Coarse embeddings of locally finite metric spaces into Banach spaces without cotype, *C. R. Acad. Bulgare Sci.*, **59** (2006), no. 11, 1113–1116.

[353] M. I. Ostrovskii, *On comparison of the coarse embeddability into a Hilbert space and into other Banach spaces*, unpublished manuscript, 2006, available at `http://facpub.stjohns.edu/ostrovsm`

[354] M. I. Ostrovskii, Coarse embeddability into Banach spaces, *Topology Proc.*, **33** (2009), 163–183.

[355] M. I. Ostrovskii, Expansion properties of metric spaces not admitting a coarse embedding into a Hilbert space, *C. R. Acad. Bulgare Sci.*, **62** (2009), 415–420; Expanded version: `arXiv:0903.0607`

[356] M. I. Ostrovskii, On metric characterizations of some classes of Banach spaces, *C. R. Acad. Bulgare Sci.*, **64** (2011), no. 6, 775–784.

[357] M. I. Ostrovskii, Different forms of metric characterizations of classes of Banach spaces, *Houston. J. Math.*, to appear, `arXiv:1112.0801`.

[358] M. I. Ostrovskii, Auerbach bases and minimal volume sufficient enlargements, *Canadian Math. Bull.*, **54** (2011), 726–738.

[359] M. I. Ostrovskii, Low-distortion embeddings of graphs with large girth, *J. Funct. Anal.*, **262** (2012), 3548–3555.

[360] M. I. Ostrovskii, Embeddability of locally finite metric spaces into Banach spaces is finitely determined, *Proc. Amer. Math. Soc.*, **140** (2012), 2721–2730.

[361] M. I. Ostrovskii, Test-space characterizations of some classes of Banach spaces, Algebraic Methods in Functional Analysis, The Victor Shulman Anniversary Volume, I.G. Todorov, L. Turowska (Eds.), Operator Theory: Advances and Applications, Vol. 233, Birkhäuser, Basel, 2013, to appear, `arXiv:1112.3086`.

[362] M. I. Ostrovskii, On metric characterizations of the Radon–Nikodým and related properties of Banach spaces, preprint (2013), `arXiv:1302.5968`.

[363] N. Ozawa, A note on non-amenability of $\mathcal{B}(\ell_p)$ for $p = 1, 2$, *Internat. J. Math.* **15** (2004), no. 6, 557–565.

[364] J. Peetre, *A Theory of Interpolation of Normed Spaces*, Notas de Matemática, No. **39**, Instituto de Matemática Pura e Aplicada, Conselho Nacional de Pesquisas, Rio de Janeiro, 1968.

[365] J. Peetre, Sur la transformation de Fourier des fonctions à valeurs vectorielles, *Rend. Sem. Mat. Univ. Padova*, **42** (1969), 15–26.

[366] J. Pelant, Embeddings into c_0, *Topology Appl.*, **57** (1994), 259–269.

[367] A. Pełczyński, Projections in certain Banach spaces, *Studia Math.*, **19** (1960), 209–228.

[368] A. Pełczyński, A note on the paper of I. Singer "Basic sequences and reflexivity of Banach spaces", *Studia Math.*, **21** (1961/1962), 371–374.

[369] V. G. Pestov, Free Banach spaces and representations of topological groups (Russian), *Funktsional. Anal. i Prilozhen.*, **20** (1986), no. 1, 81–82.

[370] J. Petersen, Die Theorie der regulären Graphen, *Acta Math.*, **15** (1891), no. 1, 193–220.

[371] F. V. Petrov, D. M. Stolyarov, P. B. Zatitskiy, On embeddings of finite metric spaces in ℓ_∞^n, *Mathematika*, **56** (2010), no. 1, 135–139, `arXiv:0903.4355`

[372] B. J. Pettis, A proof that every uniformly convex space is reflexive, *Duke Math. J.*, **5** (1939), no. 2, 249–253.

[373] A. Pietsch, *History of Banach Spaces and Linear Operators*, Birkhäuser, Boston, MA, 2007.

[374] M. S. Pinsker, On the complexity of a concentrator, in: *7th International Teletraffic Conference*, pp. 318/1–318/4, 1973.

[375] G. Pisier, Sur les espaces de Banach qui ne contiennent pas uniformément de ℓ_n^1, *C. R. Acad. Sci. Paris*, Sér. A-B **277** (1973), A991–A994; Seminar version: *Séminaire Maurey-Schwartz* (1973–1974), *Espaces L^p, applications radonifiantes et géométrie des espaces de Banach*, Exp. No. 7, 19 pp. (errata, p. E.1) Centre de Math., École Polytech., Paris, 1974.

[376] G. Pisier, Type des espaces normés, *C. R. Acad. Sci.* Paris Sér. A-B **276** (1973), A1673–A1676; Seminar version: *Séminaire Maurey-Schwartz* (1973–1974), *Espaces L^p, applications radonifiantes et géométrie des espaces de Banach*, Exp. No. 3, 12 pp. (errata, p. E.1) Centre de Math., École Polytech., Paris, 1974.

[377] G. Pisier, Martingales with values in uniformly convex spaces, *Israel J. Math.*, **20** (1975), 326–350.

[378] G. Pisier, Les inégalités de Khintchine–Kahane, d'après C. Borell, in: *Séminaire sur la Géométrie des Espaces de Banach* (1977–1978), Exp. No. 7, 14 pp., École Polytech., Palaiseau, 1978.

[379] G. Pisier, Probabilistic methods in the geometry of Banach spaces, in: *Probability and Analysis (Varenna, 1985)*, Lecture Notes in Math., **1206**, pp. 167–241, Springer, Berlin, 1986.

[380] G. Pisier, *Factorization of Linear Operators and Geometry of Banach Spaces*, CBMS Regional Conference Series in Mathematics **60**, AMS, Providence, 1986.

[381] G. Pisier, *The Volume of Convex Bodies and Banach Space Geometry*, Cambridge Tracts in Mathematics, **94**, Cambridge University Press, Cambridge, 1989.

[382] G. Pisier, Complex interpolation between Hilbert, Banach and operator spaces, *Memoirs Amer. Math. Soc.* **208** (2010), no. 978.

[383] G. Pisier, Martingales in Banach spaces (in connection with type and cotype), Lecture notes of a course given at l'Institut Henri Poincaré, February 2–8, 2011, 242 pp; see the web site: `http://perso-math.univ-mlv.fr/users/banach/Winterschool2011/`

[384] G. Pisier, Q. Xu, Random series in the real interpolation spaces between the spaces v_p, in: *Geometrical Aspects of Functional Analysis. Israel Seminar*, 1985-86, J. Lindenstrauss and V.D. Milman (eds.), Lecture Notes Math., **1267**, pp. 185–209, Springer-Verlag, Berlin, 1987.

[385] A. Plichko, On the volume method in the study of Auerbach bases of finite-dimensional normed spaces, *Colloq. Math.*, **69** (1995), 267–270.

[386] V. Pták, Biorthogonal systems and reflexivity of Banach spaces, *Czechoslovak Math. J.*, **9** (1959), 319–326.

[387] Y. Rabinovich, R. Raz, Lower bounds on the distortion of embedding finite metric spaces in graphs, *Discrete Comput. Geom.*, **19** (1998), no. 1, 79–94.

[388] A. Ralston, P. Rabinowitz, *A First Course in Numerical Analysis*. Reprint of the 1978 second edition. Dover Publications, Inc., Mineola, NY, 2001.

[389] S. Rao, Small distortion and volume preserving embeddings for planar and Euclidean metrics, in: *Proceedings of the Fifteenth Annual Symposium on Computational Geometry (Miami Beach, FL, 1999)*, pp. 300–306, ACM, New York, 1999.

[390] Y. Raynaud, Espaces de Banach superstables, distances stables et homéomorphismes uniformes, *Israel J. Math.*, **44** (1983), 33–52.

[391] O. Reingold, L. Trevisan, S. P. Vadhan, Pseudorandom walks on regular digraphs and the **RL** vs. **L** problem, in: *STOC'06: Proceedings of the 38th Annual ACM Symposium on Theory of Computing*, J. M. Kleinberg (ed.), pp. 457–466, ACM, New York, 2006, `doi:10.1145/1132516.1132583`.

[392] O. Reingold, S. Vadhan, A. Wigderson, Entropy waves, the zig-zag graph product, and new constant-degree expanders, *Ann. of Math.* (2) **155** (2002), no. 1, 157–187.

[393] M. Ribe, On uniformly homeomorphic normed spaces, *Ark. Mat.*, **14** (1976), no. 2, 237–244.

[394] J. Roe, *Lectures on Coarse Geometry*, University Lecture Series, **31**, American Mathematical Society, Providence, RI, 2003.

[395] H. P. Rosenthal, On subspaces of L^p, *Ann. Math.* (2) **97** (1973), 344–373.

[396] H. L. Royden, *Real Analysis*, Second edition, Mcmillan, New York, 1968.

[397] E. Rozenman, A. Shalev, A. Wigderson, Iterative construction of Cayley expander graphs, *Theory of Computing*, **2** (2006), 91–120.

[398] H. Sachs, Regular graphs with given girth and restricted circuits, *J. London Math. Soc.*, **38** (1963), 423–429.

[399] H. Sachs, On regular graphs with given girth, in: *Theory of Graphs and its Applications*, ed. M. Fiedler, (Proc. Sympos. Smolenice, 1963), pp. 91–97, Publ. House Czechoslovak Acad. Sci., Prague, 1964.

[400] H. Sako, A generalization of expander graphs and local reflexivity of uniform Roe algebras, (2012), `arXiv:1208.5642v4`.

[401] H. Sako, Translation C^*-algebras and property A for uniformly locally finite spaces, (2012), `arXiv:1212.5900v1`.

[402] L. Saloff-Coste, *Aspects of Sobolev-type Inequalities*, London Mathematical Society Lecture Note Series **289**, Cambridge University Press, Cambridge, 2002.

[403] M. Sapir, Aspherical groups and manifolds with extreme properties, (2011), `arXiv:1103.3873`.

[404] P. Sarnak, *Some Applications of Modular Forms*, Cambridge Tracts in Mathematics **99**, Cambridge University Press, Cambridge-New York, 1990.

[405] N. Sauer, Extremaleigenschaften regulärer Graphen gegebener Taillenweite. I, II, *Österreich. Akad. Wiss. Math.-Natur. Kl. S.-B.* II, **176** (1967), 9–25; ibid. **176** (1967) 27–43.

[406] M. Schmuckenschlaeger, A simple proof of an approximation theorem of H. Minkowski, *Geom. Dedicata*, **48** (1993), no. 3, 319–324.

[407] I. J. Schoenberg, Metric spaces and positive definite functions, *Trans. Amer. Math. Soc.*, **44** (1938), 522–553.

[408] A. Schrijver, *Theory of Linear and Integer Programming*, Wiley-Interscience Series in Discrete Mathematics, John Wiley & Sons, Ltd., Chichester, 1986.

[409] F. Shahrokhi, D. W. Matula, The maximum concurrent flow problem, *J. Assoc. Comput. Mach.*, **37** (1990), no. 2, 318–334.

[410] Y. Shalom, Bounded generation and Kazhdan's property (T), *Inst. Hautes Études Sci. Publ. Math.* No. **90** (1999), 145–168 (2001).

[411] D. B. Shmoys, Cut problems and their application to divide-and-conquer, in: *Approximation Algorithms for NP-hard Problems*, (D. S. Hochbaum, ed.), pp. 192–235, PWS, Boston, USA, 1997.

[412] I. Singer, Basic sequences and reflexivity of Banach spaces, *Studia Math.*, **21** (1961/1962), 351–369.

[413] P. Solé, The second eigenvalue of regular graphs of given girth, *J. Combin. Theory* Ser. B, **56** (1992), 239–249.

[414] J. M. Steele, *The Cauchy–Schwarz Master Class. An Introduction to the Art of Mathematical Inequalities*, MAA Problem Books Series, Mathematical Association of America, Washington, DC; Cambridge University Press, Cambridge, 2004.

[415] C. Stegall, A proof of the principle of local reflexivity, *Proc. Amer. Math. Soc.*, **78** (1980), no. 1, 154–156.

[416] E. M. Stein, *Singular Integrals and Differentiability Properties of Functions*, Princeton Mathematical Series, No. **30** Princeton University Press, Princeton, NJ, 1970.

[417] J. C. Stillwell, *Classical Topology and Combinatorial Group Theory*, Graduate Texts in Mathematics, **72**, Springer-Verlag, New York-Berlin, 1980 (Second edition: 1993).

[418] G. Strang, *Linear Algebra and its Applications*, Second edition, Academic Press, New York, 1980.

[419] B. Sudakov, J. Verstraëte, Cycle lengths in sparse graphs, *Combinatorica*, **28** (3) (2008) 357–372.

[420] W. Szlenk, The non-existence of a separable reflexive Banach space universal for all separable reflexive Banach spaces, *Studia Math.*, **30** (1968), 53–61.

[421] R. M. Tanner, Explicit concentrators from generalized N-gons, *SIAM J. Algebraic Discrete Methods*, **5** (1984), no. 3, 287–293.

[422] T. Tao, *Structure and Randomness: Pages from Year One of a Mathematical Blog*, AMS, Providence, 2008.

[423] T. Tao, Notes for the course "Expansion in groups" (2012) at `http://terrytao.wordpress.com/`

[424] A. E. Taylor, A geometric theorem and its application to biorthogonal systems, *Bull. Amer. Math. Soc.*, **53** (1947), 614–616.

[425] R. Tessera, Coarse embeddings into a Hilbert space, Haagerup property and Poincaré inequalities, *J. Topol. Anal.*, **1** (2009), 87–100.

[426] *Texas 2009 Nonlinear Problems*, open problems suggested at the seminar "Nonlinear geometry of Banach spaces", Texas A & M University, August 2009. Available at: `http://facpub.stjohns.edu/ostrovsm/Texas2009nonlinearproblems.pdf`

[427] E. C. Titchmarsh, Reciprocal formulae involving series and integrals, *Math. Z.*, **25** (1926), 321–381.

[428] A. Torchinsky, *Real-variable Methods in Harmonic Analysis*, Pure and Applied Mathematics, **123**, Academic Press, Inc., Orlando, FL, 1986 (Reprinted by Dover in 2004).

[429] S. L. Troyanski, On locally uniformly convex and differentiable norms in certain non-separable Banach spaces, *Studia Math.*, **37** (1970/71), 173–180.

[430] B. S. Tsirelson, It is impossible to imbed ℓ_p of c_0 into an arbitrary Banach space (Russian), *Funkcional. Anal. i Prilozhen.*, **8** (1974), no. 2, 57–60; English translation: *Functional Anal. Appl.*, **8** (1974), 138–141.

[431] V. V. Vazirani, *Approximation Algorithms*, Springer, Berlin-New York, 2001.

[432] A. M. Vershik, The Kantorovich metric: the initial history and little-known applications (Russian), *Zap. Nauchn. Sem. S.-Peterburg. Otdel. Mat. Inst. Steklov.* (POMI), **312** (2004), Teor. Predst. Din. Sist. Komb. i Algoritm. Metody. 11, 69–85, 311; translation in *J. Math. Sci.* (N.Y.), **133** (2006), no. 4, 1410–1417, `arXiv:math/0503035`.

[433] H. Walther, Über reguläre Graphen gegebener Taillenweite mit minimaler Knotenpunktzahl, *Wiss. Z. Techn. Hochsch. Ilmenau*, **11** (1965), 93–96.

[434] H. Walther, Eigenschaften von regulären Graphen gegebener Taillenweite und minimaler Knotenpunktanzahl, *Wiss. Z. Techn. Hochsch. Ilmenau*, **11** (1965), 167–168.

[435] N. Weaver, *Lipschitz Algebras*, World Scientific Publishing Co., Inc., River Edge, NJ, 1999.

[436] J. H. Wells, L. R. Williams *Embeddings and Extensions in Analysis*, Ergebnisse der Mathematik und ihrer Grenzgebiete **84**, Springer-Verlag, Berlin, 1975.

[437] J. Wenzel, Superreflexivity and J-convexity of Banach spaces, *Acta Math. Univ. Comenian.* (N.S.), **66** (1997), 135–147.

[438] D. P. Williamson, D. B. Shmoys, *The Design of Approximation Algorithms*, Cambridge University Press, Cambridge, 2011.

[439] J. S. Wilson, On characteristically simple groups, *Math. Proc. Cambridge Philos. Soc.*, **80** (1976), no. 1, 19–35.

[440] H. S. Witsenhausen, Minimum dimension embedding of finite metric spaces, *J. Combin. Theory*, Ser. A **42** (1986), no. 2, 184–199.

[441] D. Wolfe, Imbedding a finite metric set in an N-dimensional Minkowski space, *Nederl. Akad. Wet., Proc.*, Ser. A **70**, (1967), 136–140.

[442] G. Yu, The coarse Baum–Connes conjecture for spaces which admit a uniform embedding into Hilbert space, *Inventiones Math.*, **139** (2000), 201–240.

[443] G. Yu, Higher index theory of elliptic operators and geometry of groups, in: *International Congress of Mathematicians*, Vol. II, pp. 1623–1639, Eur. Math. Soc., Zürich, 2006.

[444] P. B. Zatitskiy, On the coincidence of the canonical embeddings of a metric space into a Banach space, *J. Math. Sci.*, New York, **158** (2009), No. 6, 853–857; translation from *Zap. Nauchn. Semin. POMI*, **360** (2008), 153–161.

[445] M. Zorn, A remark on method in transfinite algebra, *Bull. Amer. Math. Soc.* **41** (1935), 667–670.

[446] A. Żuk, Property (T) and Kazhdan constants for discrete groups, *Geom. Funct. Anal.*, **13** (2003), no. 3, 643–670.

[447] A. Zygmund, *Trigonometric Series*, Cambridge University Press, Cambridge, 1959.

Author index

Subject index

Bei Fragen zur Produktsicherheit wenden Sie sich bitte an:
If you have any questions regarding product safety,
please contact:

Walter de Gruyter GmbH
Genthiner Straße 13
10785 Berlin
productsafety@degruyterbrill.com